AF622665

2nd IWA Leading-Edge Conference on Water and Wastewater Treatment Technologies

WATER AND ENVIRONMENTAL MANAGEMENT SERIES

2nd IWA Leading-Edge Conference on Water and Wastewater Treatment Technologies

Edited by Mark Van Loosdrecht and Jonathan Clement

Publishing

Published by IWA Publishing, Alliance House, 12 Caxton Street, London SW1H 0QS, UK
Telephone: +44 (0) 20 7654 5500; Fax: +44 (0) 20 7654 5555; Email: publications@iwap.co.uk
Web: www.iwapublishing.com

First published 2004
© 2004 IWA Publishing

Layout by The Clyvedon Press Ltd, Cardiff, UK (www.clyvedonpress.com)

Apart from any fair dealing for the purposes of research or private study, or criticism or review, as permitted under the UK Copyright, Designs and Patents Act (1998), no part of this publication may be reproduced, stored or transmitted in any form or by any means, without the prior permission in writing of the publisher, or, in the case of photographic reproduction, in accordance with the terms of licences issued by the Copyright Licensing Agency in the UK, or in accordance with the terms of licenses issued by the appropriate reproduction rights organization outside the UK. Enquiries concerning reproduction outside the terms stated here should be sent to IWA Publishing at the address printed above.

The publisher makes no representation, express or implied, with regard to the accuracy of the information contained in this book and cannot accept any legal responsibility or liability for errors or omissions that may be made.

British Library Cataloguing in Publication Data
A CIP catalogue record for this book is available from the British Library

Library of Congress Cataloging- in-Publication Data
A catalog record for this book is available from the Library of Congress

Disclaimer

The information provided and the opinions given in this publication are not necessarily those of IWA or of the editors, and should not be acted upon without independent consideration and professional advice. IWA and the editors will not accept responsibility for any loss or damage suffered by any person acting or refraining from acting upon any material contained in this publication.

ISBN: 1 84339 508 8
ISSN: 1476-1785

Contents

Part Two: Wastewater

Preface

Wastewater and drinking water treatment are an essential element of urban infrastructure. In the course of the last century there has been a large technical development. This has been so successful that for the general public in industrialized countries this infrastructure is hardly recognized. Nevertheless there is an ongoing activity to further improve the existing processes. The IWA Leading Edge Technology conference aims to stimulate this development and the dissemination of the results. A selection of the 60 presentation during the conference are included in this volume of WEMS. Clearly wastewater and drinking water treatment are two separate fields due to the very different boundary conditions put to the organisations. Nevertheless several issues as membrane processes, removal of micropollutants and water reuse are of crucial importance for both fields. This gives the added value of the LET conference where emerging issues in the water field can be rapidly discussed and solved.

Mark Van Loosdrecht and Jonathan Clement
December 2004

Part One: Drinking Water

Influence of NOM and Membrane Surface Charge on UF-membrane fouling

G. Galjaard[*], J. C.Kruithof[*], G. Raspati[**]

[*]NV PWN Water Supply Company North-Holland, PO Box 2113, 1990 AC VELSERBROEK, The Netherlands (E-mail: *gilbert.galjaard@pwn.nl, joop.kruithof@pwn.nl*)
[**]UNESCO-IHE, Westvest 7, 2601 DA DELFT, The Netherlands
(E-mail: *raspa1@unesco-ihe.org*)

Abstract In literature NOM-fouling of UF-membranes is described due to the deposition of high molecularweight organics like polysaccharides with a molecular weight exceeding the MWCO-value of the membrane. In a former PWN study it was hypothesized that not these high molecular weight organics but low molecular weight organics and metal complexes were responsible for the acceleration of film formation and therefore fouling. In this study experiments on bench-scale were carried out to reduce the fouling potential of IJssel Lake water by:

- removing high molecular weight (HMW) organics like polysaccharides, with enhanced coagulation;
- removing low molecular weight (LMW) organics like acids and humics, with ion exchange;
- reducing the negative surface charge of the membrane or affix a positive charge to promote electrostatic exclusion.

From this study it can be concluded that the removal of LMW organics (MW lower then MWCO-value of membrane) and HMW organics leads to a much lower fouling potential then the removal of HMW organics alone. The combination of a relatively high positively charged hydrophilic membrane and the removal of LMW fractions with an anion resin (MIEX) results to a high gross flux rate at almost no fouling. However results have been achieved during a relatively short period on a bench–scale size. Therefore extended pilot research is scheduled.

Keywords NOM-fouling, surface charge, surface water, ultrafiltration

Introduction

Since start up of the Heemskerk plant continuous research into the fouling mechanisms is carried out to reduce the necessary pre-treatment for the IJssel Lake water, not only for the Heemskerk plant but also for separate direct treatment of IJssel Lake water. In 2001 pre-coat (EPCE) experiments were carried out in parallel with in-line coagulation experiments, both feeding the raw water directly into UF-membranes. This study resulted in the hypothesis that irreversible fouling is not caused by pore blockage but by attachment of the deposited material including colloidal polymers. These polymers were hypothesized to be formed by natural organic matter (NOM) from the IJssel Lake and organic metal complexes. The interaction of these polymer films with the membrane material (hydrophilic negatively charged) made it difficult to reverse fouling [Galjaard, et al. 2002].

In literature NOM-fouling of UF-membranes is very often attributed to the deposition of high molecular weight (HMW) organics, like polysaccharides and polyhydroxyaromatics on the membrane surface larger then the MWCO-value of the UF membrane [Laine, et al,

© 2004 IWA Publishing. *2nd IWA Leading-Edge Conference on Water and Wastewater Treatment Technologies* edited by Mark Van Loosdrecht & Jonathan Clement. ISBN: 1 84339 508 8.

2002]. In some cases rejection of smaller organics like carboxylic acids and humics was found and attributed to the deposition of polysaccharides on the membrane surface making the "pores" of the membrane smaller, reaching the cut off value of the acids and humics [Amy, et al, 2001].

Our hypothesis is different and proposes that the larger organics interact at high concentrations (at the membrane surface) forming long polymers. The low molecular weight (LMW) organics like carboxylic acids and humics combine with the HMW organics by electrostatic forces and function as an accelerator to form a film the same way organic metal complexes do. The membrane may become irreversible fouled rapidly, if the formed film is oppositely charged as the membrane surface since then the film is adsorbed by the membrane According to our hypothesis two solutions exist to reduce the fouling potential of IJssel Lake water, being:

- remove LMW organics like the hydrophilic acids and humics and avoid organic metal complexes (by not using coagulants like ferric or alum) to prevent fast polymerisation and adsorption to the membrane surface (with ion exchange);
- reduce the negative surface charge of the membrane or affix an opposite charge to promote electrostatic exclusion of the formed film.

Background

Natural Organic Matter (NOM)

NOM describes the complex mixture of organic material, such as humic and hydrophilic acids present in all drinking water sources. NOM might be harmless on itself but it can cause major problems in the water treatment by reacting with chlorine to form disinfection by-products (DBP's), lowering the UV-transmission, etc. This leads to higher regeneration frequencies of activated carbon filters, to membrane fouling, etc.

With regards to its solubility in water, NOM can be categorized into three large groups, they are hydrophobic, transphilic, and hydrophilic. A hydrophobic species tends to be electrically neutral and nonpolar, and thus preferring neutral and nonpolar solvents or molecular environments. Hydrophobic is used interchangeably with "lipophilic". A hydrophilic molecule or portion of a molecule is one that typically is electrically polarized and capable of H-bonding, enabling it to dissolve more readily in water than in oil or other "non-polar" solvents. Transphilic represents a group of NOM with neutral charge [Amy, 2001].

NOM and membranes

As a heterogeneous mixture, NOM represents a complex solute that interacts with the membrane surface and pores. For UF membranes, there are three possible interactions with NOM: steric exclusion (rejection), electrostatic exclusion (rejection), and adsorption onto membrane surface/pores (fouling).

Amy and co-workers [Amy et al., 1999] found that electrostatic exclusion plays an important role in NOM rejection. It was concluded that the rejection increased with humic content, while a clear trend was not found for molecular weight (MW). It was also noteworthy that there was a preferential rejection of UVA_{254} over DOC, suggesting an effective removal of hydrophobic acids (humic and fulvic) as measured by SUVA.

The lack of a clear trend between NOM rejection and MW supported the assumption that *electrostatic exclusion* plays a more significant role than the *steric exclusion.* This conclusion was based on the fact that the MW range of the NOM is lower than the membrane's MWCO. This leads to the concept of *effective MWCO*, which describes the specific NOM MW corresponding to 90 % rejection. This means UF membranes with relatively high manufacture-reported MWCO values are able to achieve significant NOM removal by exhibiting an 'effective MWCO' significantly lower than manufacturer-reported MWCO.

Membrane properties

Since the development of the asymmetric membrane most ultrafiltration membranes for large scale water applications are hollow fibres made of polymers. These fibres need very low trans membrane pressure to produce relatively high amounts of water. The modules of these membranes contain a lot of membrane surface, are chemically resistant and relatively cheap to manufacture making them suitable for large scale applications in the potable water industry. All of these fibres are made by phase inversion processes. Phase inversion is a process whereby a polymer is transformed in a controlled manner from liquid to a solid state. The process of solidification is very often initiated by the transition from one liquid state into two liquids. At a certain stage during demixing, one of the liquid phases (the high polymer concentration phase) will solidify so that a solid matrix is formed. The used solvent and the polymer are electrically polarized creating a membrane with, in most cases, a negative charge (the solvent dilutes into water). Polymers used as membrane material are (listed in order from most hydrophilic to most hydrophobic), [Mulder 1991]:

- Cellulose esters;
- Polycarbonate (PC);
- Polyacrylonitrile (PAN);
- Polyether sulphone blended with 1 % polyvinylpropylene (PES/PVP);
- Polyether sulphone (PES);
- Polysulphone (PS);
- Teflon (PTFE);
- Polyvinylidene fluoride (PVDF);
- Polypropylene (PP).

For our study we used standard PES/PVP membranes with slight alterations were used. In addition PAN membranes were selected since they are more hydrophilic at a surface charge closer to zero.

Experimental set-up

The pilot consists of two independently treated feed streams, fed into the small bench scale UF-unit consisting of 4 parallel tested UF-membranes (see figure 1).

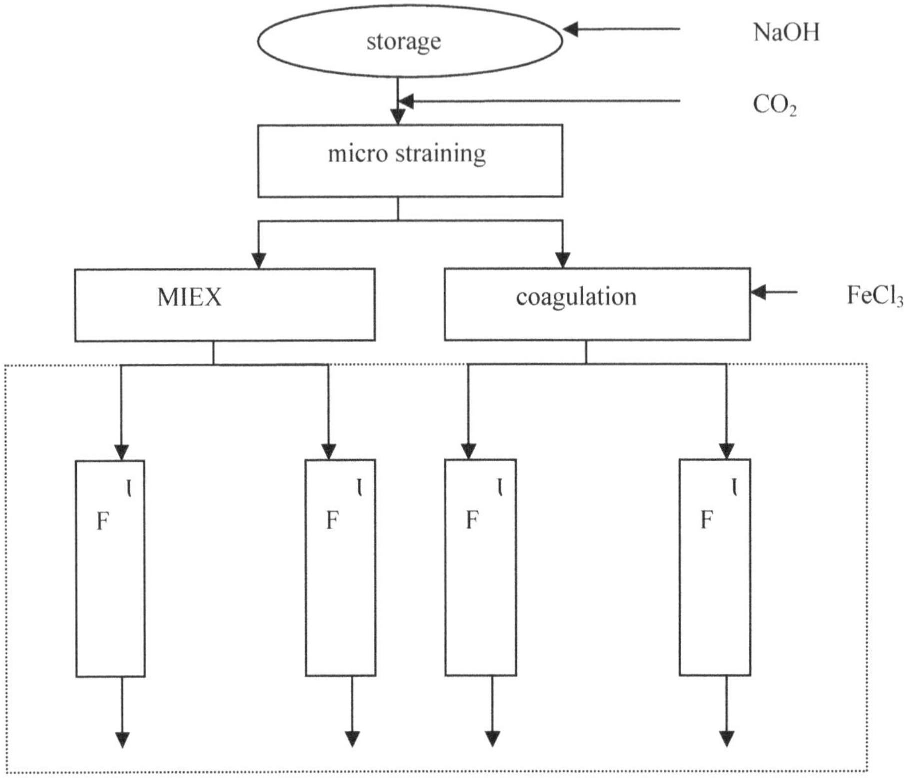

Figure 1 Treatment scheme pilot

IJssel Lake water is stored into a reservoir and chemically softened with NaOH. After storage of several days pH is adjusted from 9.3 till around 8 by a CO_2-dosage and the water is led over micro strainers with a pore size of 30 μm. Up till here water from the existing full-scale treatment plant is used. In the pilot facility the stream is split up into an ion exchange unit (MIEX® pilot) and a coagulation tank. The product of both treatment steps are then fed into a bench-scale UF-installation, the so called Quick-Scan. Each stream is treated by two membranes which differ in material or charge or cut off.

Quick-Scan and membranes

A semi-pilot scale UF Unit known as the Quick-Scan is used to carry out the UF-experiments. This installation can test up to four elements simultaneously. The membranes can be fed by one or by two feed pumps creating the possibility to experiment with two different feed water types. The controls and the data-acquisition are fully automated.

In total 6 different membranes have been tested from two different manufactures. In table 1 a summary is given of the membrane characteristics.

Table 1 Membrane characteristics

Mem.	Material	MWCO (kDa)	ID/OD (mm)	Permeability [L/(h.m^2.bar)]	Comments/expected or measured charge by manufacture
1	PES/VP	150	0.8/1.3	300	slightly hydrophilic negative
2	PES/VP	250	0.8/1.3	600	slightly hydrophilic negative
3	PES/VP	150	0.8/1.3	200	hydrophobic less PVP
4	PES/VP	150	0.8/1.3	400	hydrophobic more PVP
5	PAN	13	0.8/1.4	200	hydrophilic negative (– 1mV)
6	PAN	13	0.8/1.4	200	hydrophilic positive (+4 mV)

Results

Initial membrane charge by streaming potential

To verify the charges and the charge differences streaming potential (SP) measurements were carried out. The used streaming potential measurement is not a standardized measurement so protocol and equipment had to be developed. The technique is derived from colloidal chemistry books to measure porous media like glass beads in tubes.

The streaming potential measurements where carried out on NOM and suspended matter free water at a specific conductivity, pH and pressure.

In table 2 the measured initial charges are summarized.

The SP measurement was used to verify relative differences in charges. In general the PES/PVP membrane properties seem to be in agreement with the manufacturers information; the membranes with less PVP seem to have a lower negative charge and visa versa. The only exception is membrane 4, where the highest negative value was expected but the lowest was found. The charge of the PAN membranes follow the manufacturers information. The measured values of the manufacturers and the measured SP values differ but are in the same order.

Table 2 Results streaming potential measurements initial membrane charge

Mem.	Material	Comments/expected or measured charge by manufacture	Measured SP at standard conditions (mV)
1	PES/VP	slightly hydrophilic negative	–40 to –60
2	PES/VP	slightly hydrophilic negative open	–60 to –100
3	PES/VP	tends to hydrophobic less PVP	–20 to –25
4	PES/VP	More hydrophilic more PVP	–20 to –25
5	PAN	hydrophilic negative (–1 mV)	–50 to –100
6	PAN	hydrophilic positive (+4 mV)	+200 to +400

**Zeta potential measured with the flow potential method by Anton-Paar ElectroKiniticAnalyzer*

NOM analysis of the raw water and UF feed water types

The ILC feed stream was optimized to a maximum UVT (82 %) and an as low as possible zeta potential value to reduce possible adsorption onto the membranes. For this very difficult to coagulate raw water this meant a ferric dose of 15 mg Fe^{3+}/L at a pH-value of 7.0 with a contact time of 5 minutes. The ion exchange process (MIEX) was optimised to produce a UVT of approximately 92 %. For both pre-treatment the total DOC content was reduced from around 5 - 6 to 3 mg/L. To determine the removal of the different NOM fractions analysis was performed of the raw water, the ILC water and the MIEX treated water. Table 3 gives hydrophilic chromatographic fractionation.

Table 3 Chromatographic fractionation from HMW (PS) to LMW (acids)

	Polysaccharides (μg/L)	Humics (μg/L)	Building blocks (μg/L)	Neutrals + amphiphilics (μg/L)	Acids (μg/L)
Raw	839	1988	1152	630	62
After ILC	385	789	1143	530	66
After MIEX	848	509	1197	511	0

The results are very interesting, both pre-treatments produce more or less the same total amount of DOC but the fractionation is completely different. After ILC most of the larger compounds like the polysaccharides and a big part of the humics are removed. After ion-exchange only the negatively charged smaller DOC compounds like the acids and most of the humics are removed. The goal to produce water with a lower content of HMW compounds like polysaccharides and a water with the polysaccharides but without the LMW charged matter seemed to be succesfull.

Experiments with PES/PVP membranes

The first set of experiments was carried out with the standard PES/PVP membranes and the PES/PVP membranes with a lower negative charge. The flux was increased from 50 – 100 L/(h.m^2) if a stable operation occurred.

Both membranes foul quickly. The backwashes and the enhanced backwashes are not capable to restore the TMP sufficiently during the experiment. The overall fouling rate is more or less the same for both membranes. At the beginning of the experiment pressure increased rapidly with the standard PES/PVP module which did not happen with the lower charged one. The fouling rate within a filtration cycle was lower for the lower charged PES/PVP membrane. Hypothesis for the cause of the rapid increase and poor restoration after the washes, the attachment of colloidal polymers was confirmed. Visually long chains of material were released from the fibre which seemed to be very difficult to flush out or to dissolve with the used chemicals (see also figure 4). Although the fouling rate seems to be higher for the more negatively charged membrane rapid fouling of both membranes took place.

In figure 3 the TMP in time of both PES/PVP membranes with the lower negative charge are plotted to compare the influence of the two feed water types.

There is a significant difference in fouling behaviour between the two feed water types. TMP restoration after a backwash for the MIEX treated water is better then for the coagulated water. Both membranes foul but the fouling rate on the MIEX treated water is lower. Visually there was also a remarkable difference. No long chains of suspended matter were released from the membrane fibres into the MIEX stream. The fibres were as clean as at the start (see also figure 5). The MIEX treated matrix does not seem to form a thin film as the matrix in the coagulated stream. In the NOM analysis of the permeate of the membranes (table 4) we see that both membranes rejected a big part of the polysaccharides. The smaller polar compounds (building blocks and acids) also seem to be rejected while their size is to small to be size excluded by the membrane. The amount and the difference between the two streams is too small to conclude anything significant about the impact of small polar organics on the formation of this thin film but there is a clear difference in membrane performance and visual formation of long chains in the fibres.

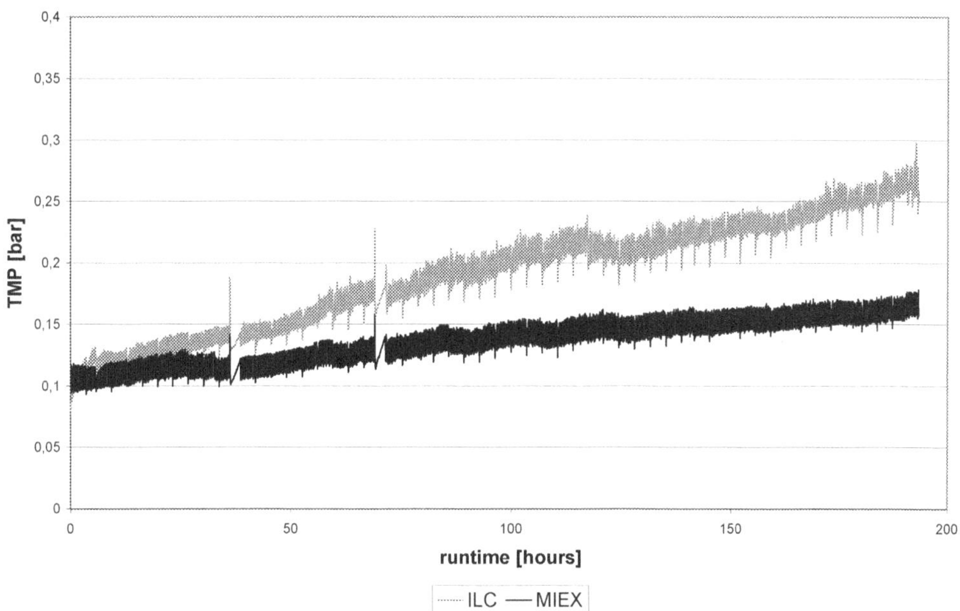

Figure 3 TMP over time at flux 50 L/m^2.h, PES/PVP membranes with less PVP (-20 mV) feed ILC and MIEX water

Figure 4 Feed stream ILC, backwash

Figure 5 Feed stream MIEX, backwash

Table 4 Chromatographic fractionation from HMW (PS) to LMW (acids) for membrane permeate

	Polysaccharides (μg/L)	Humics (μg/L)	Building blocks (μg/L)	Neutrals + amphiphilics (μg/L)	Acids (μg/L)
PermeateILC	85	765	1037	480	60
PermeateMIEX	181	401	1118	441	0

In figure 6 normalized clean water fluxes (CWF) are displayed of this experiment with the PES/PVP membranes (–40 and –20 mV) on both pre-treatment streams at the beginning of the experiments and after the experiments when they have been cleaned intensively.

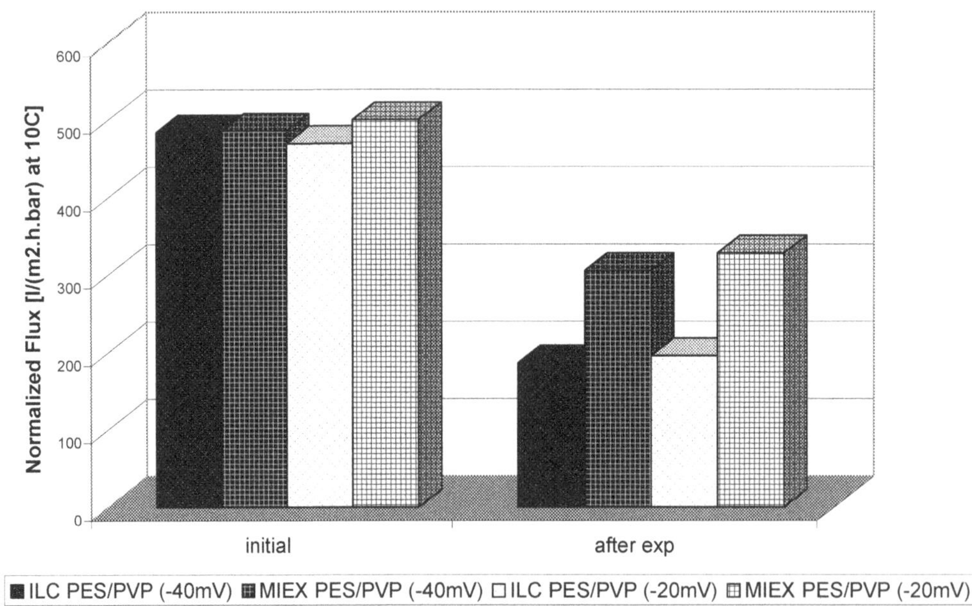

Figure 6 CWF initially and after experiments

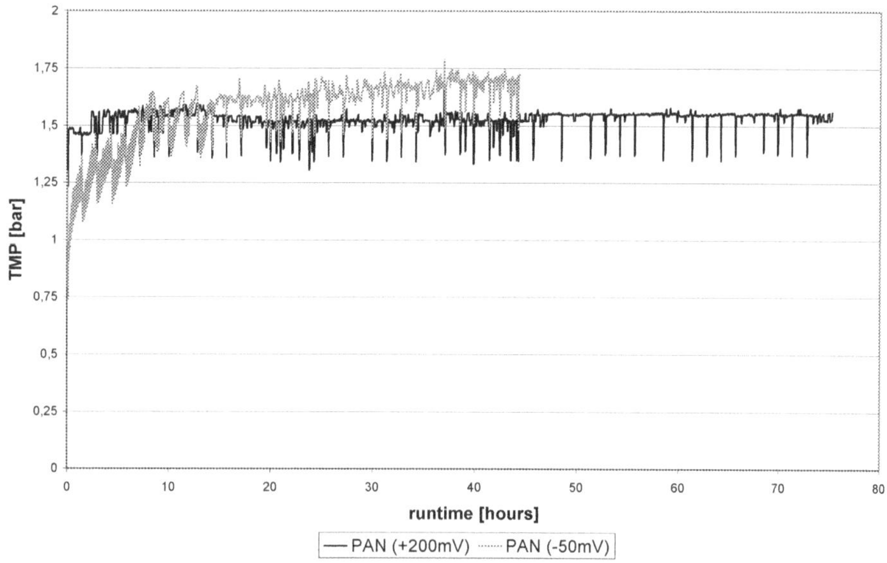

Figure 7 TMP in time of PAN modules on MIEX pre-treated water

The irreversible fouling after cleaning for both membranes and both pre-treatments is significant. In all cases initial CWF could not be restored. The irreversible fouling for the coagulated water is higher then for the ion exchanged water. Both standard membranes with the higher negative charge seem to be a bit lower in flux restoration.

Experiments with PAN membranes

The results of the standard PAN modules were more or less the same as of the PES/PVP modules on the two different pre-treatments. Also with the PAN modules the fouling rate of the coagulated water was much higher then the MIEX treated water. There was a remarkable difference; on MIEX treated water the positively charged module did not seem to foul. In figure 7 the TMP in time is given of two different PAN modules on the MIEX treated water. The gross flux during this experiment was 100 L/(h.m^2).

After two days the negatively charged membrane unit was switched off, while at constant flux the maximum system pressure was reached. The positively charged membrane did not seemed to be bothered by anything. The TMP is not even increasing during a filtration cycle. An explanation could be the positive charge of the membrane since the MIEX exchanged a lot of negatively charged dissolved and suspended matter for negatively charged ions, the charge of the remaining dissolved matter is neutral..

Conclusions

From this orientating study on the influence of the membrane charge and NOM fractions on the membrane fouling behaviour in treating Lake IJssel water the following can be concluded:

- enhanced coagulation is a good pre-treatment method to remove the larger part of the high molecular weight (HMW) fraction of the NOM, like polysaccharides (PS);
- ion-exchange with an anion resin like the MIEX removes all negatively charged low molecular weight (LMW) fraction of the NOM, like humics and acids;

- removal of HMW (PS) organic matter does lead to a higher fouling potential compared to the removal of LMW negatively charged fractions;
- the combination of a relatively high positively charged hydrophilic membrane and the removal of LMW fractions with MIEX lead to a high gross flux rate and almost no fouling.

References

Amy G, Cho J: (1999). Interactions between Natural Organic Matter (NOM) and Membranes: Rejection and Fouling. *Water Science and Technology* **40** (9), 131–139.

Amy G. et al. (2001). Natural Organic Matter (NOM) rejection by, and fouling of, NF and UF membranes. ISBN 1-58321-106-3.

Cho J., Amy G., Pellegrino J. (2000). Membrane Filtration of Natural Organic Matter: Comparison of Flux Decline, NOM Rejection, and Foulants during Filtration with Three UF Membranes. *J. Desalination* **127**, 283–298.

Cho J., Amy G., Pellegrino J. (2000). Membrane Filtration of Natural Organic Matter: Factors and Mechanisms Affecting Rejection and Flux Decline with Charged Ultra filtration (UF). *J. Membrane Science* **164**, 89–110.

Cho J., Amy G., Pellegrino J., Yoon Y. (1998). Characterization of Clean and Natural Organic Matter (NOM) Fouled NF and UF Membranes, and Foulants Characterization. *J. Desalination* **118**, 101–108.

Galjaard G., Kruithof J.C. (2002). Enhanced Pre-Coat Engineering (EPCE) for MF and UF: Steps to Full-Scale Application. Proceedings IWA, ISSN 0941-0961.

Kaiya Y., Itoh Y., Takizawa S., Fujita K., Tagawa T. (2000). Analysis of organic matter causing membrane fouling in drinking water treatment. *Water Science and Technology* **41** (10–11), 59–67.

Laine J.M., Campos C., Baudin .I, Janex M.L. (2002). Understanding Membrane Fouling: A Review of Over a Decade of Researc. Proceedings IWA, ISSN 0941-0961.

Lee H., Amy G., Cho J., Yoon Y., Moon S.H., Kim I.S. (2001). Cleaning Strategies for Flux Recovery of An Ultra filtration Membrane Fouled by Natural Organic Matter. *J. Water Resources* **35** (14), 3301–3308.

Lee S., Shim Y., Kim I.S., Sohn J., Cho J. (2002). Determination of Mass Transport Characteristics for Natural Organic Matter (NOM) in Ultrafiltration (UF) and Nanofiltration (NF) Membranes. J. *Water Science and Technology* **2** (2), 151–160.

Mulder M. (1991). Basis Principles of Membrane Technology. ISBN 0-7923-0978-2.

Shim Y., Lee H.J., Lee S., Moon S.H., Cho J. (2002). Effects of natural organic matter and ionic strength on membrane surface charge. *J. Environmental Science Technology* **36** (17), 3864–3871.

Stein H.N. (1995). The preparation of dispersions in liquids. ISBN 0-8247-9674-8.

The use of advanced analytical tools for the characterization of organic foulant of MF and UF membranes

H. Habarou[*,**], **J.P Croué**[**], **M. Pontié**[***], **H. Suty**[*]

*Anjou Recherche Maisons Laffitte - VÉOLIA WATER, France
**LCEE UMR CNRS 6008, Université de Poitiers, France
***E.N.S.C.P, LECA UMR CNRS n°7575, France

Abstract The goal of our work was to characterize and identify the organic foulants responsible for flux declines of MF membranes. Two different natural waters (one low DOC high alkalinity content and one high DOC low alkalinity content) have been filtered through hollow fiber PVDF membranes. Using a large panel of analyticals tools (AFM, FESEM, streaming potential, ESCA, TOF SIMS) membrane autopsy confirmed that the organic foulant that remains after backwash and chemical cleaning was a mixture of several biopolymers (proteins, polysaccharides, aminosugars and polyhydroxyaromatics). Aminosugars and polysaccharides which originate from microbial activity (microbial cell wall residues and/or exopolymers from DOM constituents and/or biofilm) appeared to be the major constituents of the fouling material

Key words: organic foulant, hollow fiber, membrane autopsy, microfiltration

1. Introduction

Successful utilization of membrane technology has been greatly limited by membrane fouling. Fouling increases operation and maintenance costs by deteriorating membrane performance (flux decline versus time) and ultimately shortening membrane life. Natural organic matter (NOM) plays an important role in membrane fouling. Humic substances as well as polysaccharides and proteins have been shown to exert fouling properties towards low and high pressure membranes.

The overall objective of our work is to better characterize the nature of the fouling material that is responsible for the reduction of permeability observed during low pressure membrane filtration (MF) of raw and chemically treated surface waters using hollow fiber modules. Our paper focused on the membrane autopsy conducted on hollow fibers operated with raw and clarified waters using a large panel of analytical tools including novel technologies such as ESCA (Electron Spectroscopy for Chemical Analysis), TOF SIMS (Time of flight – Secondary Ion Mass Spectrometry), pyrolysis gaz chromatography/mass spectrometry in addition to FESEM (Field Emission Gun Scanning Electronic Microscopy), AFM (Atomic Force Microscopy), contact angle and streaming potential measurements.

© 2004 IWA Publishing. *2nd IWA Leading-Edge Conference on Water and Wastewater Treatment Technologies* edited by Mark Van Loosdrecht & Jonathan Clement. ISBN: 1 84339 508 8.

2. Materials and methods

2.1. Hollow fiber module

A Polyvinyl Di Fluoride MF membrane with 0.20 µm Molecular Weight Cut Off was used for our study. The general characteristics of the home-made hollow fiber module are given in table 1.

Table 1 General characteristics of the hollow fiber modules

Material	PVDF
Inside diameter (µm)	390
Outside diameter (µm)	650
Fiber length (cm)	43
Number of fibers	21
Surface (m^2)	0.0184
Mean pore size (µm)	0.2 µm
Range of pH	2 à 12
Filtration	Outside to Inside

2.2. Feed water

The experiments were conducted with :

- the prefiltered (1 µm) Marne River raw water (France) sampled in March 2003 at the Annêt sur Marne drinking water treatment plant during a very shinny period which favored the increase of the biological activity,
- the clarified Urne (Sand Filtered Water) water collected at different periods of the year 2003 from the Magenta drinking water treatment plant.

The physical chemical characteristics of the studied waters are given in table 2. The Marne River raw water can be considered as a low DOC, low humic content, high alkalinity surface water as compared to the Urne clarified water that is more enriched in DOC with low hardness and alkalinity.

Table 2 Main characteristics of the feed water (raw surface water prefiltered on 1µm)

Parameters	DOC (mg C/L)	UV_{254nm} (m^{-1})	SUVA (L/mg C.m)	Ca^{2+} mg/L	Alkalinity mg/LCaCO_3	Si (mgSiO_2/L)
Marne River RW	1.6–1.9	3.2–3.7	1.7–2.3	85	200	1.5
Urne River SFW	3.2–8.0	5.2–12.9	1.3–2.3		35	15–16

RW: Raw Water prefiltered through 1µm; SFW: Sand Filtered Water

2.3. Filtration protocol

Prior to any filtration, module integrity was checked by soaking the module into ultrapure water and flowing compressed air into the permeate channel. Filtration tests were operated with lab-scale pilot test units equipped with hollow fiber modules in a dead-end filtration mode at constant pressure and constant temperature (20°C). MF membranes were performed at -350 mbar. Transmembrane pressure, permeate flow rate as well as feed water temperature were recorded with data logger every minute.

The membrane autopsy was conducted on membranes that showed a 40 % flux reduction obtained after several days of operation. Back wash (MilliQ water) was performed every day and a final chemical cleaning (NaOCl solution of 200 ppm of free chlorine at 20°C followed by a 20 g/L citric acid solution at 35°C) was operated at the end of the experiment. Hollow fibers were extracted before and after the chemical cleaning for membrane autopsy. Figure 1 shows a picture of the pilot unit used in this study.

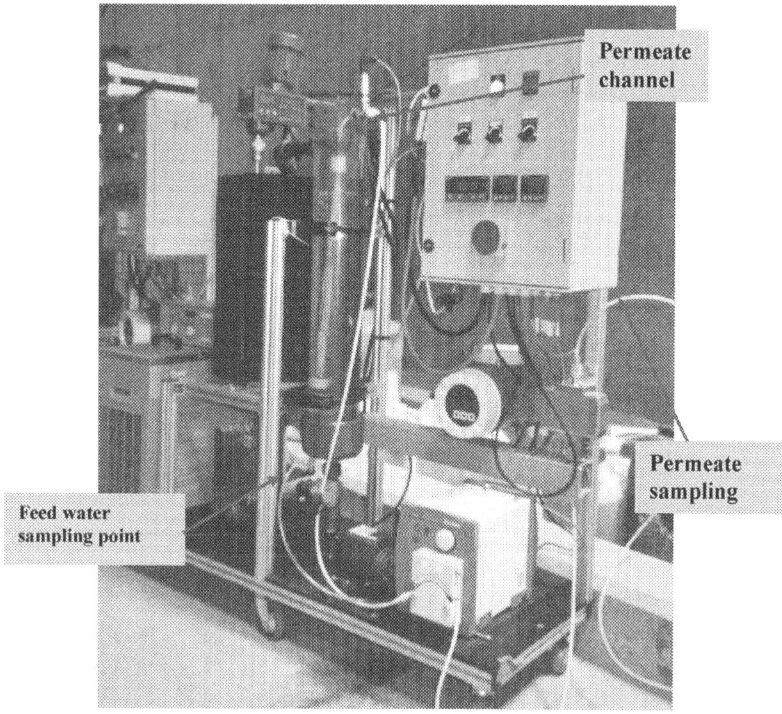

Figure 1 MF filtration unit

2.4. Membrane autopsy

Membrane autopsies were performed using two approaches on both backwashed and chemically cleaned hollow fibers:

Characterization by pyrolysis GC/MS of the fouling material isolated from the hollow fibers using sonication (hollow fibers that were cut into small pieces were subjected to sonication in the presence of 50 mL of MQ water. After filtration through a 1.2 μm porosity glass fiber membrane, the solution was lyophilized),

Characterization of the membrane surface using contact angle measurement, AFM, FESEM, ESCA and TOF SIMS analyses.

2.5. Analytical methods

Dissolved Organic Carbon (DOC) measurements were performed using a Shimadzu TOC 5000A. UV Absorbance measurements at 254 nm were conducted using a UV-VIS SAFAS Double Energy System 190(UP) spectrophotometer.

Foulant material (approximately 1 mg of dry sample deposited into 100 µL quartz tubes) were submitted to flash pyrolysis–GC/MS according to the method described by Bruchet et al. (1990) using a Pyroprobe2000 (Chemical Data Systems, Oxford, Pa.) filament pyrolyzer coupled with a Hewlett Packard G1800 A.

Zeta potential values are calculated from a homemade streaming potential apparatus, as previously described (Pontié *et al,* 1997). A pair of silver-chloride electrodes (Radiometer-analytical, France) are judiciously introduced, one in the permeate channel and the other one in the retentate compartment. The electrodes are connected to a high impedance voltmeter.

The contact angle measurements were carried out by the sessile drop technique (MQ water). The contact angle of the droplet with the surface is measured with KG1 apparatus purchased from KRUSS (Deutschland). The angles are measured using the goniometer of the apparatus. These measurements were carried out on dried fiber membrane samples.

Atomic Force Microscopy (A.F.M.) analyses were performed with membrane samples that were preliminary dried at room temperature. The samples are fixed by a self-adhesive tape and the external side of the membrane fiber is analysed with a Nanoscope IIIA from Digital Instruments (multimode). All samples are analysed in tapping mode and phase contract in order to characterize samples morphology and to measure the Mean Surface Roughness (RMS).

Field Emission Scanning Electron Microscopy (F.E.S.E.M) was conducted using a LEO1501 apparatus. Before analysis all samples are fixed with carbon self adhesive tape and then recovered of a ultra-thin layer of Pt/Pd (4 nm) using an evaporating technique. Beam energy was below 3 keV.

Electron Spectroscopy for Chemical Analysis (E.S.C.A) is also called XPS (X-Ray Photoelectron Spectroscopy) and provides information about the elemental composition of the membrane surface. ESCA analysis were conducted with a Scienta 200 (source: Al Kα monochromatic - 450 Watt), area : 0.5 x 4 mm^2, detection angle: 90°, penetration thickness of 10 nm). All samples were dried before being analysed. Results presented correspond to the mean average of the analyses of three different parts of the membrane: the top, the middle and the end of the fiber.

Time Of Flight Secondary Ion Mass Spectrometry (TOF SIMS) provides information about functional groups involved in the chemistry of membrane surface. TOF SIMS analyses were conducted on dried membranes with a ION TOF, TOF IV (two primary ion sources: Ga^+ (25keV) and Cs^+ (10 keV). For each element either positive or negative, results correspond to the mean area of three points analysed in the middle of the fiber.

3. Results and discussion

Whatever the tested water, a rapid reduction of the relative permeability as a function of time was observed as a consequence of the fouling properties (not shown). The flux was only partly recovered after daily backflushing. The objective of a 40 % flux decline was obtained after several days of filtration, followed by a final backwash and a chemical cleaning. Some fibers were sampled before and after chemical cleaning for autopsy. Notice that usually unsignificant removals of DOC and UV abs were observed during filtration in accordance with Wiesner *et al.*(1992).

Figure 2 shows as examples 2D FESEM images of clean and fouled (after backwash) MF-PVDF membranes obtained after filtration of the 1μm prefiltered Marne river raw water. Fouling materials were observed at the surface but also in the pores for all fouled membranes analyzed after backwash or after chemical cleaning.

Furthermore, evidence of the presence of foulants at the membrane surface and within the pores was confirmed by the modification of the membrane characteristics. For instance, table 3 compares the water permeability, contact angle, zétâ potential and roughness of virgin, fouled (after the final backwash) and chemically cleaned membranes.

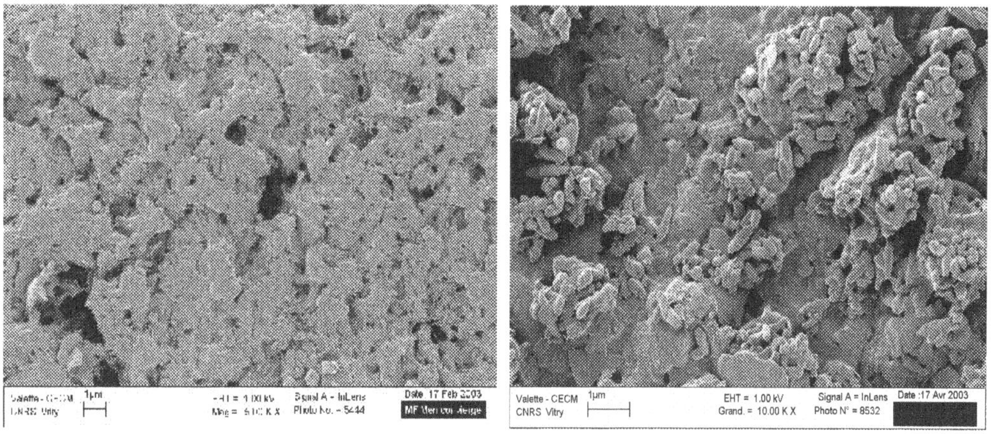

Figure 2 2D FESEM images of clean (left) and fouled (right) MF-PVDF membranes (Marne Raw Water, 1 μm)

Table 3 Evolution of the water permeability, contact angle, zétâ potential and surface mean roughness (Rms roughness) of virgin, fouled (after the final backwash) and chemically cleaned PVDF membrane (Marne Raw Water, 1 μm)

Parameters	Virgin	Fouled	After chemical cleaning
MQ water permeability *(L/h.m^2.bar)*	430	10	150
Contact angle *(°)*	79 ± 2	52	77
Zétâ Potential *(mV)* (±1)	-18	-2	-8
Roughness *(nm)* (20*20 μm^2)	190 ± 40	220 ± 38	240 ± 48

The MF PVDF membrane is characterized by a rough and hydrophobic character. It is important to notice that the high roughness observed for the MF-PVDF membrane may significantly influence the contact angle measurements, as reported by Neumann and Good (1972).

After several days of filtration and a final backwash, the MilliQ water permeability was found to be severely reduced for all studied waters, a slight decrease of its contact angle was generally observed (reduction of the hydrophobic character). For all studied waters (Figure 3) the increase in roughness (more important with the Urne SFW in winter as compared to the other waters) confirmed the accumulation of materials at the membrane surface. The strong attenuation of the charge (less negative) as deduced from the streaming potential measurements also indicates that non charged or positively charged foulants are also present in the membrane pores. After chemical cleaning the initial permeability of the MF modules was only partially recovered (only shown for the Marne raw water, table 3). Furthermore, this procedure has a variable efficiency depending on the parameter taken into account. For example, the treatment helped to approach the initial roughness while the initial Zêta potential could not be reached. These results seem to indicate that both alkaline and acid cleanings did not totally remove the fouling materials.

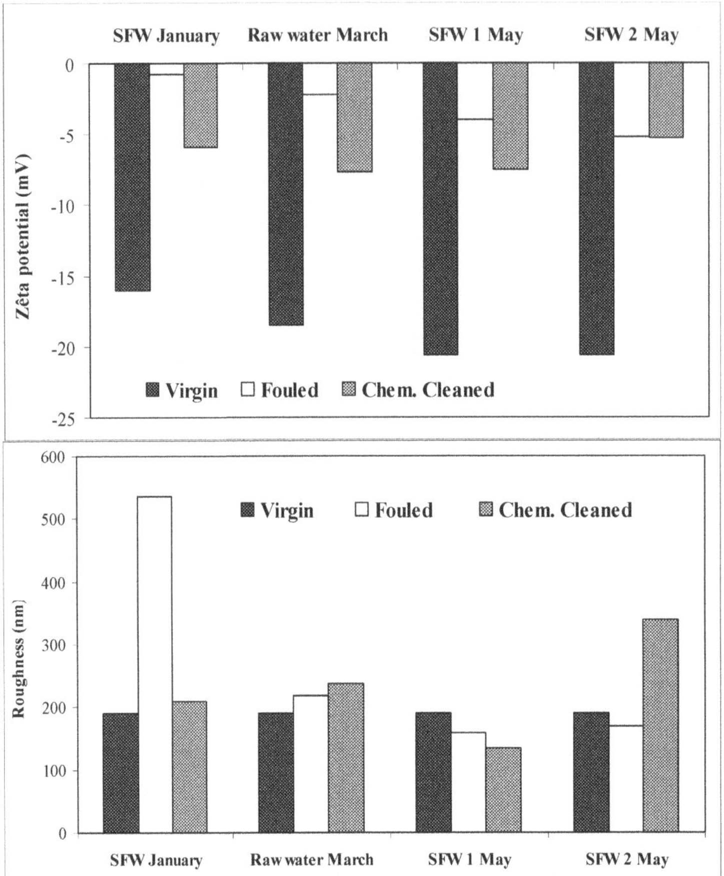

Figure 3 Zêta potential and roughness measurements of MF PVDF at different steps: virgin membrane, fouled (after backwash) membranes and chemically cleaned fibers.

ESCA provides the elemental composition of the "extreme" surface (10 nm) of the membrane allowing the identification of potential foulant. It also provides insight of the chemical structure of the organics with the possible identification of functional groups. Table 5 gives the elemental composition (expressed in per cent, sum of all the elements identified is equal to 100 %) of virgin hollow fibers and hollow fibers isolated from the module after backwash and chemical cleaning.

Table 4 Elemental composition (ESCA analysis) of clean and fouled membranes

MF PVDF		% C	% F	% O	% N	% Si	% Ca
SFW January	virgin	49,61	43,02	6,11	1,26	-	-
	fouled	52,74	30,39	**12,91**	**3,06**	0,55	0,35
	chem. cleaned	48,94	44,57	6,49	-	-	-
RW March	virgin	49,61	43,02	6,11	1,26	-	-
	fouled	49,69	21,48	**22,4**	**2,34**	2,83	1,26
	chem. cleaned	49,16	39,36	9,02	0,91	1,18	0,37

The elemental composition determined by ESCA analysis revealed the presence of nitrogen and oxygen for virgin MF PVDF, element that do not enter as a constituent of the membrane polymer. These elements may be part of chemicals commonly used by manufacturers to make membrane surface more hydrophilic.

Elemental compositions of the fouled membranes were found to be significantly different as compared to virgin membrane and enriched in oxygen and nitrogen, result that confirms the presence of organic deposits. In addition, the decrease of the relative abundance of fluoride is a good indicator of additional organic constituents (i.e. foulants). The average elemental composition of aquatic NOM (i.e. 45-50 % C; 35-40 % O; 3-5 % N; 4-5 % H; 1-2 % S) agrees with the observed changes of the elemental composition of the membrane that are an increase of N (85%) and O (366%) for all membranes, a constant C content for PVDF. ESCA analyses also clearly underlined that N was present at the surface of the membrane as amine and amide functional groups. After chemical cleaning, traces of organic foulants still remain on the membrane surface as shown by the deviation noticed for some elements (C, O and F) with the virgin membrane. However these results confirmed the beneficial impact of the cleaning procedure. ESCA results also underline the contribution of inorganic species such as calcium and silica in the composition of the fouling material. As already mentioned in the literature, these ions may act as bridges between the negatively charged membrane and organic matters.

TOF SIMS results (Figure 4) confirmed the enrichment of the membrane surface in N and O included in amine, amide and carboxylic functional groups especially during the spring (not shown) and the summer periods. The literature indicates that these structures are abundant in microbial exopolymer mixture.

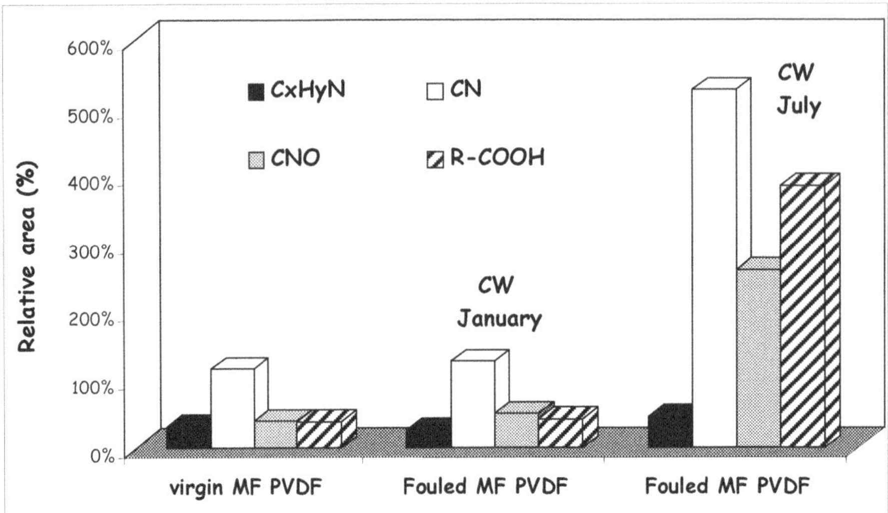

Figure 4 TOF SIMS analyses of fouled (after backwash) membranes

Part of the foulant was recovered from the hollow fibers using sonication in the presence of MilliQ water, lyophilized and subjected to flash pyrolysis GC/MS analysis. Figure 5 shows the GC/MS pyrochromatogram of the foulant isolated from the MF PVDF after backwash and chemical cleaning after filtration of the Marne raw water. As already shown by others (Mallevialle et al. 1989 ; Kaiya et al., 2000) the material isolated after backwash is a mixture of aminosugars (acetamide peak), polysaccharides (hydroxypropanone, furfural, and cyclopentenone peaks), proteins (acetonitrile, toluene, styrene, pyrrole and pyrridine peaks) and polyhydroxyaromatic structures i.e. humic-type material (phenol and cresol peaks). The large predominance of the acetamide peak in both chromatograms indicates that the fouling material is mainly of microbial origin (living and/or nonliving residues plus soluble microbial products). The identification of the furane methanol peak is a strong indicator (DNA pyrolysis derivative) of microbial activity. Similar fingerprints were obtained from the other studied waters with intense aminosugar signals during the summer period. These conclusions are in accordance with the information deduced from the ESCA and TOF SIMS results discussed above. Pyrochromatograms also showed specific peaks which derivate from the pyrolysis of membrane residues (contaminants) extracted during the sonication of the fibers (especially visible in the material isolated from chemical cleaning fibers).

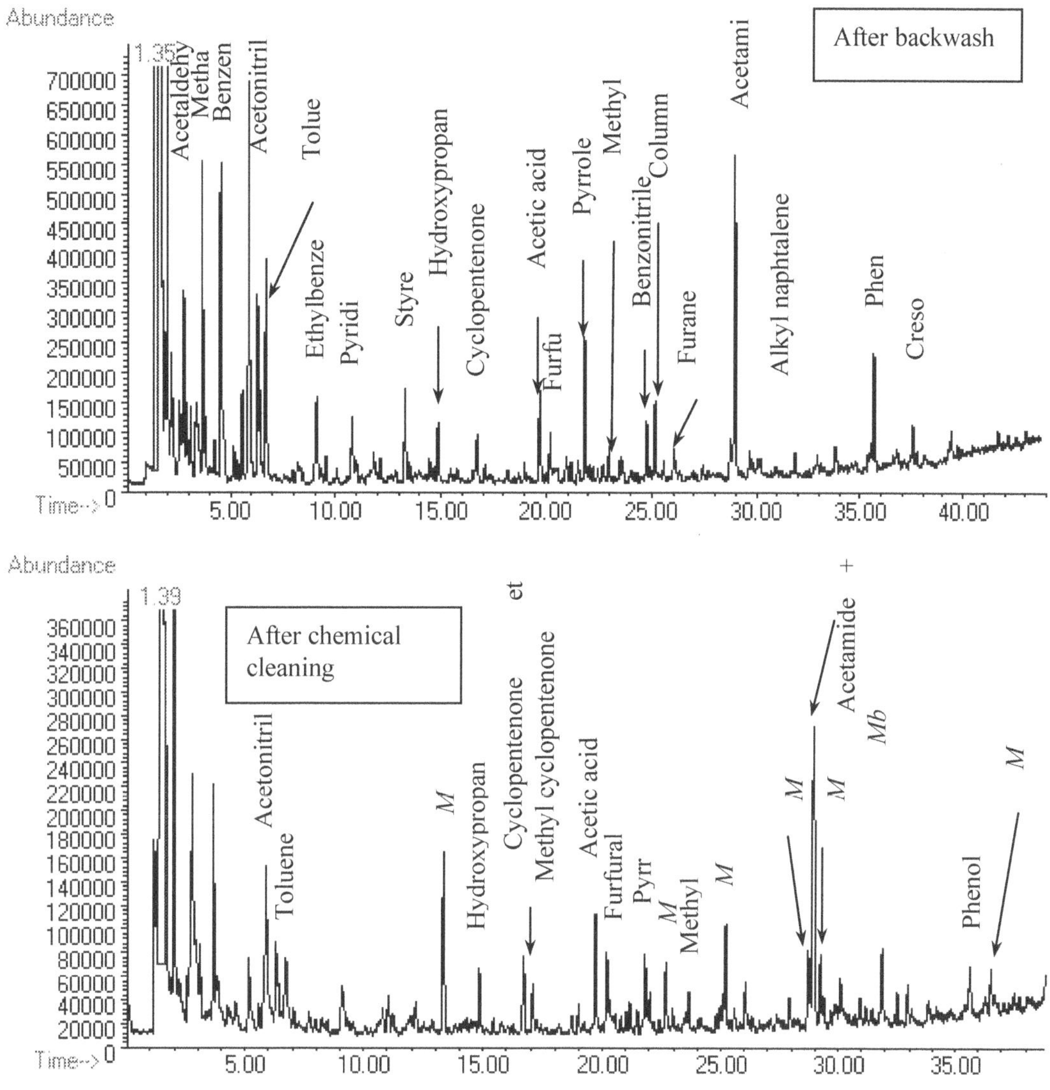

Figure 5 Pyrochromatogram of fouling material isolated from MF PVDF hollow fiber after backwash and after chemical cleaning (Marne Raw Water prefiltered through 1 μm).

As a support of our hypothesis, the presence of bacteria at the surface of the membrane was clearly observed on FESEM and AFM images recorded from MF PVDF after backwash (not shown).

4. Conclusion

As already observed by others, DOC and UV_{254} remained almost unchanged after microfiltration of natural waters which confirmed that foulants represent a minor part of the organic constituents of the feed water.

The presence of fouling material at the surface of the membrane or within the pores after backwash or chemical cleaning was confirmed based on contact angle, AFM and Zêta potential measurements. Pyrolysis GC/MS analysis operated on organics isolated (sonication) from the membrane indicated that the fouling material made of polyhydroxyaromatics, polysaccharides, aminosugars and proteinaceous structures is mainly of bacterial origin (living and/or nonliving residues plus soluble microbial products). TOF SIMS analyses support the hypothesis of the mainly microbial origin of the organic foulants (abundance of functional groups such as amine, amide and carboxylic acids known to compose microbial extracellular organic matter). ESCA analyses conducted at the membrane surface confirmed the presence organic materials with amine and amide functional groups and also underline the presence of inorganic species which could promote interactions (adsorption) between membrane and organic matters. Moreover part of the foulants remains after chemical cleaning.

Acknowledgements

J L Pastol and A. Valette for the FESEM analysis (CECM, UPR CNRS n°2801, Vitry, France), Françoise SOMMER (Biophy Research, Fuveau, France) for ESCA, AFM and fruitful discussions sions and Karima KECIL for experimental helps.

5. References

Bruchet, A., C. Rousseau, and J. Mallevialle. (1990). Pyrolysis–GC–MS for Investigating High-Molecular-Weight THM Precursors and Other Refractory Organics. *J. AWWA* **82**(9):66–74.

Kaiya Y., Itoh Y., Takizawa S., Fujita K. And Tagawa T. (2000). Analysis of organic matter causing membrane fouling in drinking water treatment. *Water Science and Technology* **41** (10–11), 59–67.

Mallevialle, J., Anselme, C. and Marsigny, O. (1989). Effects of Humic Substances on Membrane Processes, *In Aquatic Humic Substances: Influence on Fate and Treatment of Pollutants*, Ed. By Suffet, I. H. and MacCarthy, P., *ACS*, Washington, DC, pp. 749–767.

Neumann A.W., Good R.J. (1972). Thermodynamics of contact angles: I-Heteregeneous solid surfaces. *J. Coll. Int. Sci.* **38** (2), 341–358.

Wiesner, M.R., Clarck M.M, Jacangelo J.G., Lykins B.W., Marinas B.J, O'Melia C.R, Ritmann B.E and Semmens M.J. (1992). Committee report: membrane processes in potable water treatment. *J. AWWA*, 59–67.

Ceramic MF with Submicron-sized Activated Carbon Adsorption and Coagulation Pretreatments for Rapid and Effective NOM Removal

Yoshihiko Matsui*, Yoshitaka Fukuda*, Ryota Murase*, Nobuhiro Aoki, Satoru Mima**, Takanobu Inoue*, and Taku Matsushita***

* Department of Civil Engineering, Gifu University, Yanagido 1-1, Gifu 501-1193 Japan, Phone: +81-58-293-2429, Fax: +81-58-239-0163, E-mail: y-matsui@cc.gifu-u.ac.jp
** Research & Development Department, NGK Insulators, Ltd., Maegata-cho 1, Handa 475-0825, Japan

Abstract To enhance the removal of natural organic matter (NOM) by ceramic microfiltration (MF) of water samples pretreated with powdered activated carbon and a coagulant, manufacturer-supplied normal powdered activated carbon (PAC) was crushed to produce micron- and submicron-sized particles. The submicron-sized PAC removed NOM much better than the normal PAC did, owing to improvements in both adsorption capacity and kinetics. When submicron-sized PAC was used for pretreatment, the dosage was one-quarter that required for the normal PAC, and no membrane fouling or transmembrane pressure buildup due to the PAC itself was observed. The dosing sequence PAC, coagulant, and MF showed slightly better NOM removal than the sequence coagulant, PAC, MF.

Keywords: microfiltration, membrane, natural organic matter, humic substances.

Introduction

The addition of a metal coagulant, a powdered activated carbon (PAC), or both to a membrane filtration influent is a simple, cost-effective way to remove dissolved contaminants that cannot be removed by the membrane itself (Clark *et al.*, 1996; Wiesner and Laîné, 1996). Because the particle size cut-off of the membrane is on the order of micrometers or less, the formation of large floc particles during coagulation pretreatment is unnecessary. After the addition of the coagulant, microfloc particles form rapidly, so a long mixing time is not needed (Letterman *et al.*, 1999; Matsui *et al.*, 2003). In contrast, the removal of contaminants by adsorption onto PAC is slow and is dependent on the kinetics of slow adsorption. The residence times in drinking-water treatment plants are not long enough for efficient use of PAC, and the residence times for PAC/membrane systems are even shorter. To compensate for the short contact time, a high dosage is needed. Another strategy would be to use smaller PAC particles, since they would provide faster adsorption kinetics (Weber *et al.*, 1983; Najim *et al.*, 1990).

When both PAC and a coagulant are used, incorporation of the PAC particles into the flocs slows the adsorption kinetics even further. To prevent incorporation and ensure longer PAC contact time, PAC can be added to a contactor, and then the coagulant can be added directly to the membrane inlet. However, from the viewpoint of competitive adsorption,

© 2004 IWA Publishing. *2nd IWA Leading-Edge Conference on Water and Wastewater Treatment Technologies* edited by Mark Van Loosdrecht & Jonathan Clement. ISBN: 1 84339 508 8.

macromolecules that adversely affect the adsorption of smaller molecules by blocking pores should be removed by coagulation before PAC is added.

Accordingly, the objectives of this study were (1) to investigate the effect of PAC size on adsorptive removal of natural organic matter (NOM) during microfiltration (MF) and (2) to investigate the feasibility of MF of samples pretreated with both a coagulant and submicron-sized PAC.

Materials and Methods

Coagulant

The coagulant, polyaluminum chloride (PACl; 10% Al_2O_3, 62.5% basicity, Sumitomo Chemical Co. Ltd., Tokyo), was diluted before use so that a fixed quantity could be added to each experimental mixture.

Adsorbents

Submicron- and micrometer-sized PACs were obtained by grinding thermally activated wood-based PAC (Taikou-W, Futamura Chemical Industries Co., Ltd., Gifu, Japan) in a novel mill that is currently under development (NGK Insulators, Ltd.). To determine the effects of PAC size and the effectiveness of submicron-sized PAC in adsorption experiments, we used both the ground PACs and the as-received PAC. All PACs were dried in an oven at 105 °C and stored in a desiccator before use. The size distributions of the PAC particles (Fig. 1) were determined with laser light–scattering instruments (LMS-30, Seishin Enterprise Co., Ltd., and Microtrac HRA, Nikkiso Co., Ltd., Tokyo, Japan). The PACs are designated 0.8-μm PAC, 3.8-μm PAC, and 33-μm PAC on the basis of their median diameters.

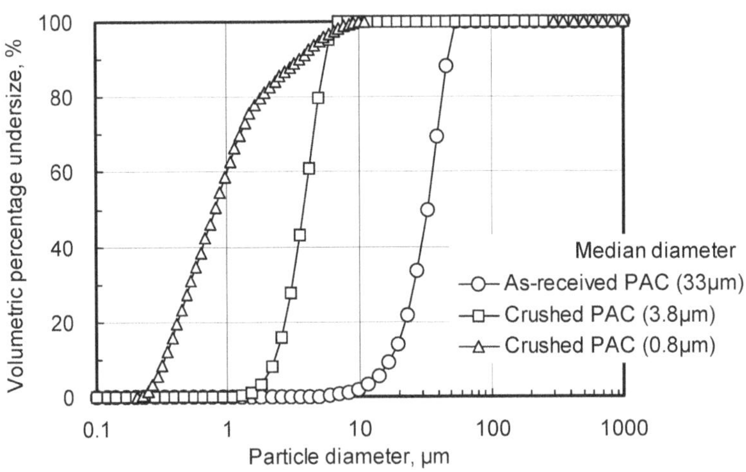

Figure 1 Size distributions of PAC particles.

Water Samples

Water samples from Takkobu Lake (Hokkaido, Japan) and Yodo River (Osaka, Japan) were collected and transported in polyethylene tanks and stored at 4 °C. Total organic carbon (TOC) and ultraviolet absorbance at 260 nm (UV_{260}) were used to quantify bulk NOM concentrations in the samples (TOC measurement: Model 810, Sievers Instruments, Inc., Boulder, CO, USA; UV_{260} absorption: Model UV-160A, Shimadzu Co., Kyoto, Japan). These values were measured after membrane filtration (0.45-μm PTFE membrane filter) for water samples except MF permeates.

Batch Adsorption Tests

Batch adsorption tests were conducted with efficient mixing. Before PAC was added to a beaker containing the water sample (2 L), an aliquot was withdrawn from the beaker to determine the initial NOM concentration. After the addition of a specified amount of PAC, aliquots were withdrawn at intervals and filtered immediately through a 0.45-μm membrane filter for analysis of NOM concentration. The absence of PAC particles in the membrane permeates was confirmed.

Batch Adsorption and Coagulation Tests

Preliminary batch tests were conducted to measure the NOM removal by pretreatment with the PACs and the coagulant. A specified amount of PAC was added to a 500-mL water sample, and the sample was mixed for 1 min. Then the coagulant was added, and the sample was mixed rapidly for 2.5 min and then slowly for 2.5 min. Aliquots withdrawn before and after addition of the coagulant were filtered immediately through a 0.45-μm membrane filter for analysis of the NOM concentration in the liquid phase.

MF Systems

Two systems were used for MF of samples pretreated with PAC and the coagulant (Fig. 2). In System A, PAC was added at the inlet of the tube reactor. After 1 min, the water was transferred to the in-line static mixer (ISM, N40-172, Noritake Co., Ltd., Nagoya, Japan), where the coagulant was added. The residence time in the in-line static mixer was 2.5 s. The membrane was a monolithic ceramic module (multichannel tubular, nominal pore size 0.1 μm, NGK Insulators, Ltd, Nagoya, Japan), which was stored in a stainless steel casing. The flow of feed to the membrane module was maintained at a constant rate (1.5 m/d) inside the module by positive pressure. For each experiment, a new or chemically cleaned membrane was used, and the membrane was operated in a dead-end mode for a predetermined time without backwash, to eliminate any historical fouling effect. In System B, coagulant was added first for comparison of the sequences of PAC and coagulant dosing. Membrane permeates were sampled at intervals, and NOM concentrations were determined in terms of TOC and UV260. At the end of each filtration run, samples were withdrawn from the outlet of the tube reactor (System A) or from the outlet of the in-line static mixer (System B), and NOM concentrations at intermediate treatment stages were analyzed.

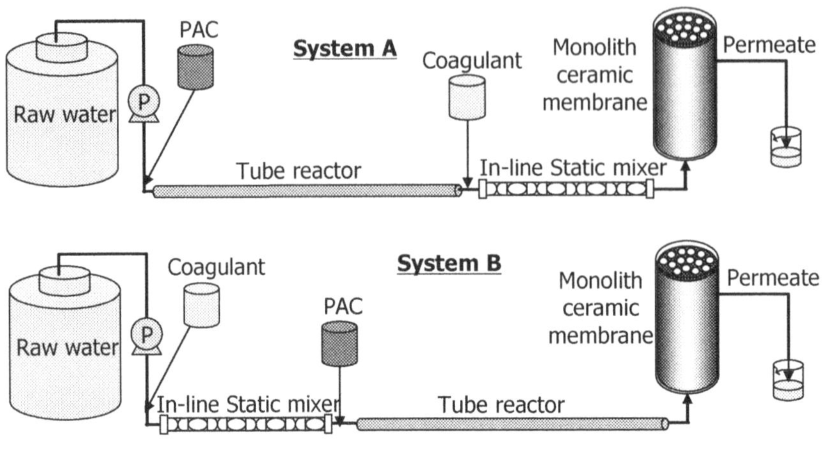

Figure 2 Schematic of the PAC/coagulation-MF systems.

Results and Discussion

We conducted batch experiments to determine the effect of PAC size (Fig. 3). The submicron PAC (0.8-µm PAC) removed more NOM than did the as-received PAC (33-µm PAC): the as-received PAC decreased the NOM concentration by 10% at most after 120 min, whereas the 0.8-µm PAC decreased it by more than 25% in only 10 min. Moreover, in the latter case, the concentration declined rapidly, such that adsorption treatment with a 1-min PAC contact time became feasible.

On the basis of these results, we conducted batch adsorption–coagulation experiments with a fixed PAC contact time of 1 min (Fig. 4). The submicron-sized PAC removed more NOM than an equal dose of the as-received PAC. Use of the submicron-sized PAC permitted a large reduction in the dosage: nearly identical removal percentages were attained with 4 mg/L of 0.8-µm PAC, 8 mg/L of 3.8-µm PAC, and 40 mg/L of 33-µm PAC. Grinding the as-received PAC to form the submicron-sized PAC particles enhanced its ability to adsorb NOM, owing to changes in both adsorption capacity and kinetics (Matsui *et al.*, 2004a). The kinetic effect appeared to be due to an increase in the outer surface area and a decrease in the PAC particle radius. The grinding also changed the NOM adsorptive capacity of the carbon, owing to an increase in mesopore surface area caused by fracture of ink-bottle pore structures. Our results suggest that the use of submicron-sized PAC for water treatment should permit not only lower PAC dosages but also shorter PAC contact times, which are especially important for the use of PAC as a pretreatment before membrane filtration.

We conducted MF experiments with adsorption–coagulation pretreatment using PAC dosages expected to give identical NOM removal, on the basis of the batch test results (Fig. 5). Although we expected a 0.8-µm PAC dose of 5 mg/L to give slightly better NOM removal than a 33-µm PAC dose of 40 mg/L, on the basis of the batch test results (where a 0.8-µm PAC dose of 4 mg/L gave the same NOM removal as a 33-µm PAC dose of 40 mg/L), the opposite occurred. The 33-µm PAC showed a lower effluent concentration, particularly after the adsorption–coagulation–MF treatment. Whereas a 33-µm PAC dose of 40 mg/L yielded a slightly lower NOM concentration than the 0.8-µm PAC dose of 5 mg/L, that difference became amplified after adsorption–coagulation–MF.

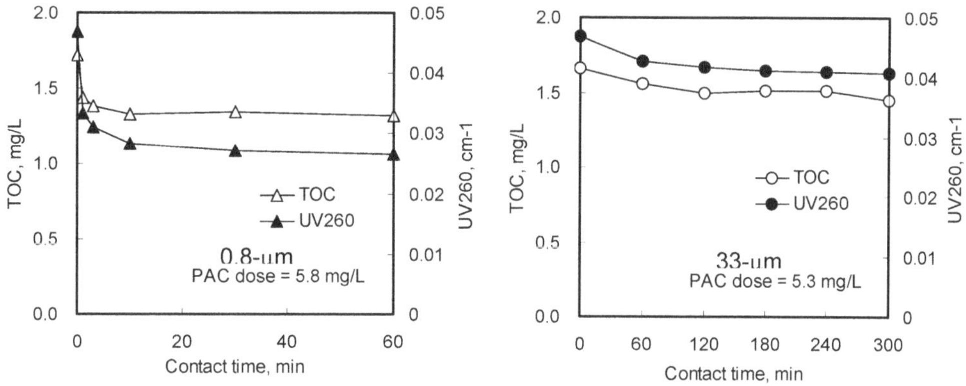

Figure 3 Batch adsorption experiments on Takkobu water.

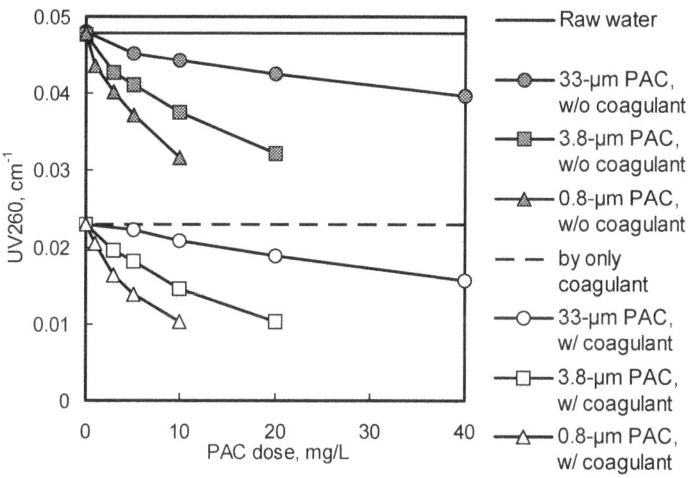

Figure 4 Batch adsorption–coagulation experiments on Takkobu water. PAC was added to the water sample, and then the coagulant PACl was added at 3 mg/L as Al 1 min later. Raw water pH = 7.3; pH after coagulation = 6.8.

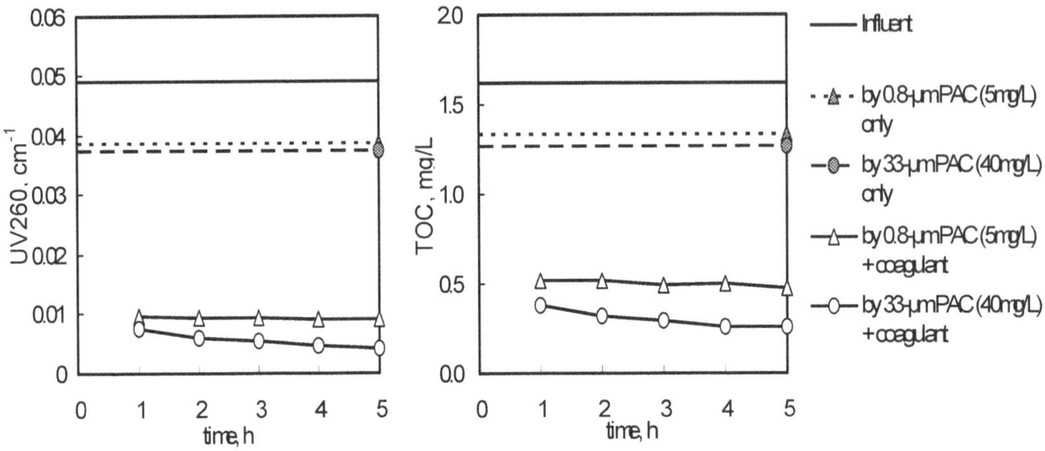

Figure 5 Performance of PAC-coagulation MF of Takkobu Lake water. Coagulant PACl 3 mg/L as Al after PAC dose. Influent pH 7.2, effluent pH 6.7

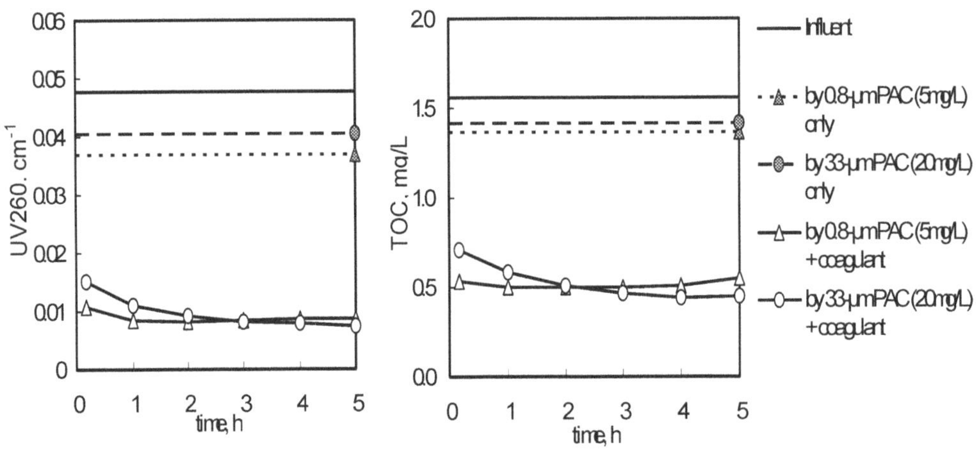

Figure 6 Performance of PAC-coagulation MF of Takkobu Lake water. Coagulant PACl = 3 mg/L as Al after PAC dose. Influent pH = 7.2, effluent pH = 6.7

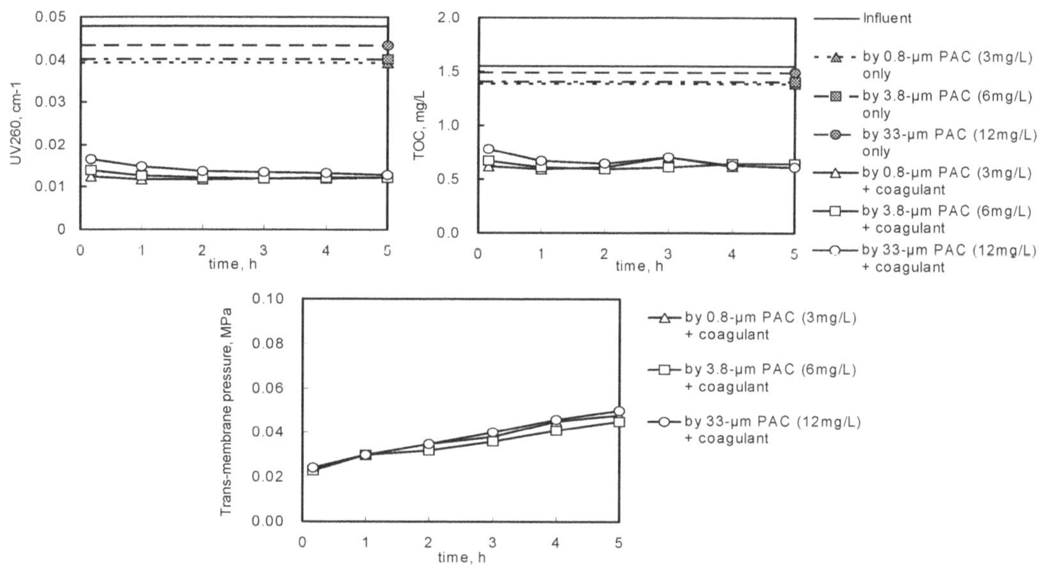

Figure 7 Performance of PAC-coagulation MF of Takkobu Lake water. Coagulant PACl = 3 mg/L as Al after PAC dose. Influent pH = 7.2, effluent pH = 6.7.

To determine equivalent dosages of the 0.8-μm and 33-μm PACs, we conducted further experiments. Although a 33-μm PAC dose of 20 mg/L gave a higher NOM concentration than a 0.8-μm PAC dose of 5 mg/L, the two PACs gave nearly the same NOM removal after the adsorption–coagulation–MF treatment (Fig. 6). During the first half of the experiment, the 0.8-μm PAC performed better than the 33-μm PAC, but in the last half, the 33-μm PAC performed better. On average the two PACs performed equally. The NOM removal performance of the 33-μm PAC improved gradually over the course of filtration, whereas the 0.8-μm PAC yielded high NOM removal from the beginning of filtration, and the removal performance remained constant throughout the filtration. These trends are also evident in Fig. 5. The gradual improvement in the removal efficiency of the 33-μm PAC could be due to adsorption by the PAC that accumulates on the membrane over the course of filtration. For the 33-μm PAC, the contact time of 1 min in the tube reactor was so short that the PAC entered the membrane with much of its adsorptive capacity intact. The remaining capacity could be effective for adsorptive removal on the membrane, particularly as the PAC accumulated. In contrast, the adsorptive capacity of the 0.8-μm PAC was efficiently used during the 1-min contact time, and the PAC that accumulated on the membrane did not remove additional NOM; thus, the NOM removal was high and remained constant over the course of filtration.

The 0.8-μm PAC at a dosage of 3 mg/L showed the same NOM removal as the 33-μm PAC at four times the dosage and the same NOM removal as the 3.8-μm PAC at twice the dosage (Fig. 7). There was a possibility that the submicron-sized PAC might foul the MF membrane and increase the transmembrane pressure (decrease the membrane permeability). The smaller-size fraction of the submicron-sized PACs might penetrate or block the membrane pores, or the submicron-sized PACs might form a less porous cake with smaller pores on the membrane surface than those of the normal PAC. However, the results shown in Fig. 5 indicate that the buildup of transmembrane pressure with the 0.8-μm PAC was the same as with the 33-μm PAC. This result might be due to the lower dosage of the 0.8-μm PAC. We suspect, however, that the aggregation of PACs due to the addition of the coagulant before MF led to a larger effective particle size and that, when deposited in the cake, these larger particles resulted in less filtration resistance. Therefore, submicron-sized PACs should be used in combination with the coagulant because the combination not only enhances NOM removal but also attenuates the transmembrane pressure increase. Finally, the 0.8-μm PAC at one-fourth the dose had the same effect as a full dose of the 33-μm PAC, and in the batch adsorption–coagulation treatment, the 0.8-μm PAC at one-tenth the dose had the same effect as the full dose of the 33-μm PAC.

The results shown in Figs. 5–7 were obtained with System A, where PAC addition was followed by coagulant addition and then MF. This system was designed to prevent incorporation of the PAC particles into the coagulation flocs before adsorption, which slows adsorption. From the viewpoint of competitive adsorption, however, the macromolecular fraction of NOM that adversely affects adsorption of the smaller molecule fraction by blocking PAC pores should be removed by coagulation before PAC is added (System B). Figs. 8 and 9 compare the NOM removal performances of Systems A and B. System A was better at removing NOM than System B. System A gave a longer PAC contact time than System B because the in-line static mixer after the tube PAC contactor also served as a PAC contactor. However, the difference in the PAC contact times of Systems A and B was very small (4%), and therefore the superiority of System A could be due to prevention of coagulation of PAC particles before adsorption. In the experiments using an MF treatment system with an 8.4-min contact time for 2.2-μm PAC, the order of PAC and coagulant additions did not influence the NOM removal (Matsui *et al.*, 2004b). These results suggests that, for a short contact time (1 min) and submicron-sized PACs (0.8 μm), the reduction of adsorption rate by flocculation should be minimized and then PAC should be added before the coagulant.

Conclusions

1. We reduced the particle size of manufacturer-supplied normal PAC to the submicron level (0.8 μm in median diameter) to enhance NOM adsorptive removal with a shorter contact time before MF.
2. With the submicron-sized PAC, we were able to use one-tenth the dose required for the normal PAC.
3. MF with in-line PAC adsorption (1-min contact) and coagulation pretreatment (2.5 s) was conducted. MF of samples pretreated with submicron-sized PAC and the coagulant showed constant NOM removal throughout the filtration, whereas NOM removal by MF with the normal PAC increased with filtration time, probably owing to postadsorption onto the PAC accumulated on the membrane. This postadsorption offset to some extent the advantage of the submicron-sized PAC over the normal PAC.
4. Nonetheless, when the submicron-sized PAC was used to pretreat the samples before MF, we were able to reduce the dosage to one-fourth the dosage required for the normal PAC.
5. The pretreatment with submicron-sized PAC did not influence the transmembrane pressure buildup.
6. The dosing sequence PAC, coagulant, and MF showed slightly better NOM removal than the sequence coagulant, PAC, MF.

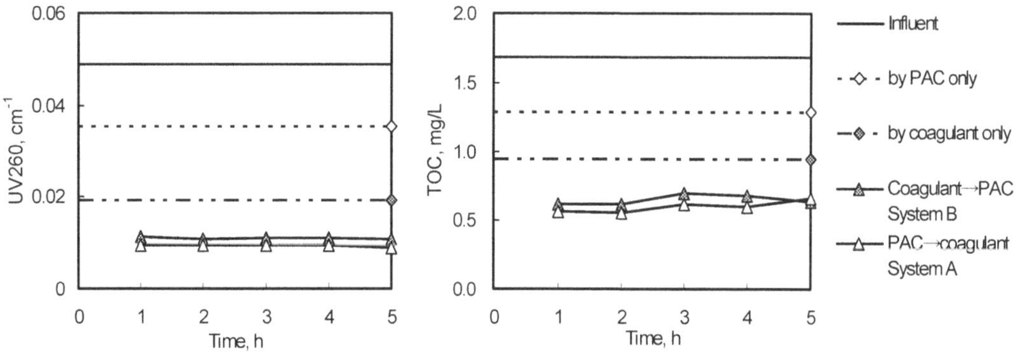

Figure 8 Performances of PAC-coagulation MF and coagulation-PAC MF of Takkobu Lake water. Coagulant PACl 3 mg/L as Al after PAC 5 mg/L dose. Influent pH 7.2, effluent pH 6.7.

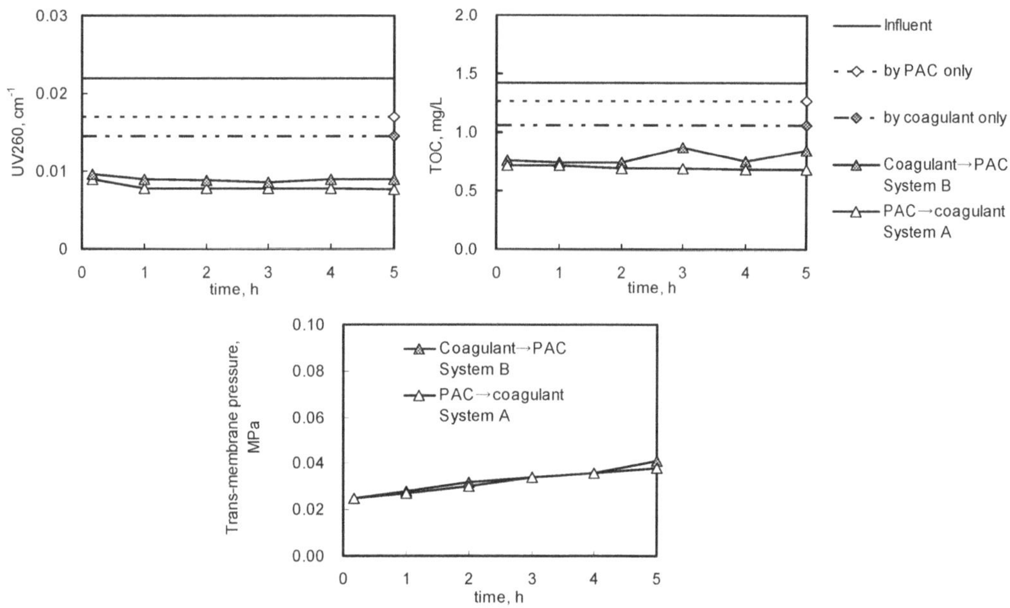

Figure 9 Performances of PAC-coagulation MF and coagulation-PAC MF of Yodo River water. Coagulant PACl 3 mg/L as Al after PAC 5 mg/L dose. Influent pH 7.2, effluent pH 6.7.

Acknowledgements

This study was supported in part by a research grant from the Collaborative Center for Academy/Industry/Government of Gifu University (2002).

References

Clark, M. M., Baudin, I., and Anselme, C. (1996). Membrane-powdered activated carbon reactors. In Water Treatment Membrane Processes, ed. J. Mallevialle, pp. 15.1–15.22. McGraw-Hill, New York, USA.

Letterman, R. D., A. Amirtharajah, and C. R. O'Melia. (1999) Coagulation and Flocculation. In Water Quality and Treatment, ed. R. D. Letterman, pp. 6.1-6.66. McGraw-Hill, Inc., New York., USA.

Matsui, Y., Matsushita, T., Inoue, T., Yamamoto, M., Hayashi, Y., Yonekawa, H., and Tsutsumi, Y. (2003) Virus Removal by Ceramic Membrane Microfiltration with Coagulation Pretreatment. *Water Sci. & Tech.: Water Supply* **3**(5), 93–99.

Matsui, Y., Murase, R., Sanogawa, T., Aoki, N., Mima, S., Inoue, T., and Matsushita, T. (2004a) Micro-ground powdered activated carbon for effective removal of natural organic matter during Water Treatment. *Water Sci. & Technol.* (Submitted).

Matsui, Y., Fukuda. Y., Inoue, T., Matsushita, T., Aoki, N., and Mima, S. (2004b) Enhancing adsorption-membrane hybrid system with micro-grinded activated carbon. Proceedings IWA 4th World Water Congress, Marrakech, 2004. (Submitted).

Najim, I. N., Snoeyink, V. L., Suidan, M. T., Lee, C. H., and Richard, Y. (1990). Effect of particle size and background natural organics on the adsorption efficiency of PAC. *Journal of American Water Works Association* **82**(1), 65–72.

Weber, W. J. Jr., Voice, T. C. and Jodellah, A. (1983) Adsorption of humic substances: the effects of heterogeneity and system characteristics. *Journal of American Water Works Association* **75**(12), 612–619.

Wiesner, M. R. and Laîné, J.-M. (1996) Coagulation and membrane separation. In Water Treatment Membrane Processes, ed. J. Mallevialle, pp. 16.1–16.12. McGraw-Hill, New York, USA.

Air sparging in capillary nanofiltration

J.Q.J.C. Verberk[*], A. Verliefde[**], B. van der Bruggen[**], J.C. van Dijk[*]

* Department of Sanitary Engineering, Delft University of Technology, P.O. BOX 5048, 2600 GA Delft, the Netherlands, j.q.j.c.verberk@citg.tudelft.nl
** Laboratory for Environmental Technology, Department of Chemical Engineering, University of Leuven, W. de Croylaan 46, B-3001Heverlee, Belgium

Abstract Capillary nanofiltration is a new concept in membrane filtration. The technique combines the hydraulic cleaning possibilities of capillary ultrafiltration with the good water quality of nanofiltration. A capillary nanofiltration system is operated in semi dead-end mode. A major drawback of this concept is the high concentration polarization compared to spiral wound membranes. In this study it is shown that water-air two-phase flow can be used to limit the concentration polarization. By air sparging an increase in permeate flux and permeate water quality is observed.

Keywords capillary nanofiltration, air sparging, membrane filtration, concentration polarization

Introduction

Spiral wound nanofiltration is widely used for the production of drinking and process water. A major drawback of spiral wound nanofiltration is the extensive pretreatment needed to overcome fouling problems. Recently a new concept has been developed, capillary nanofiltration. Capillary nanofiltration combines the advantages of nanofiltration membranes (good water quality) and capillary ultrafiltration membranes (good hydraulic cleaning possibility). Surface water can be treated to process water in one single membrane step (van der Bruggen et al 2003).

Capillary nanofiltration is operated in semi dead-end mode with recirculation. The recirculation is necessary to achieve a sufficient cross-flow velocity to limit concentration polarization (Frank et al, 2001). During the filtration run no concentrate is produced. All the feed supplied to the system is withdrawn as permeate (see figure 1 left). During the filtration period salts and suspended matter accumulate in the installation and an increase in concentration in the system is observed. Due to this increase the permeate quality will deteriorate and a decline in permeate flux is observed. When the permeate quality becomes too low, the system is flushed by means of a forward flush, AirFlush® or backwash (Verberk et al, 2001; Futselaar et al, 2002).

© 2004 IWA Publishing. *2nd IWA Leading-Edge Conference on Water and Wastewater Treatment Technologies* edited by Mark Van Loosdrecht & Jonathan Clement. ISBN: 1 84339 508 8.

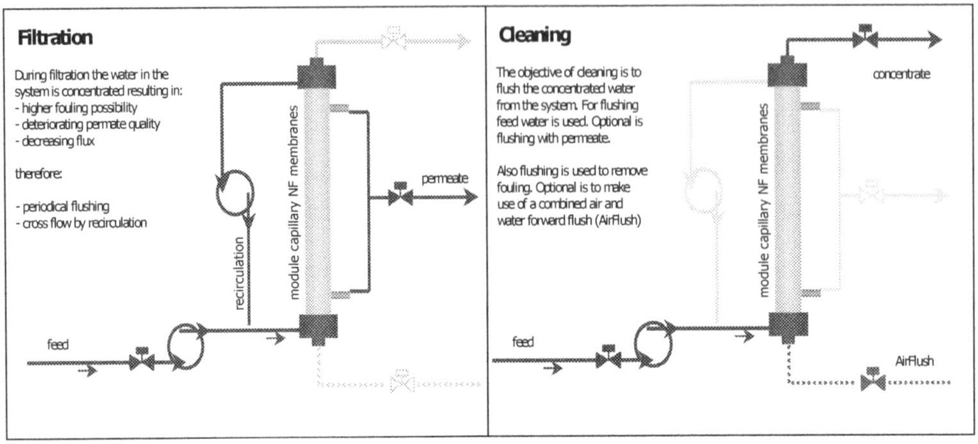

Figure 1 Operation of direct capillary nanofiltration

From operation of ultrafiltration installations it is known that air sparging is an interesting possibility to improve the permeation flux. Air sparging consists of injecting air in the feed flow during filtration. Due to the local increase in turbulence in turbulent wake zones behind an air bubble particle deposition and concentration polarization are limited. Air sparging has shown to be efficient with different membrane geometries like capillary (Cabassud et al, 1997), tubular (Cui and Wright, 1994) and flat sheet membranes (Lee et al, 1993) and different types of flux limiting mechanisms (fouling and concentration polarization).

In this research project it is investigated whether air sparging can be used to limit the concentration polarization of salty solutions. In the Laboratory of Sanitary Engineering at Delft University of Technology a test installation has been built (figure 2). The capillary nanofiltration membrane module is operated in cross-flow mode. The cross-flow operation mode was chosen above the semi dead-end operation mode because no increase in salt concentration during the filtration run occurs. The influence of different operational parameters like superficial water and air velocity, trans membrane pressure (TMP) and salt concentration on the mass transfer coefficient and concentration polarization factor have been investigated more exactly. In the first series of experiments the influence of air sparging on the clean water flux was investigated using demi-water. In the second series of experiments the influence of air sparging at different salt ($MgSO_4$) concentrations was investigated.

Material and methods

An experimental laboratory apparatus was constructed to investigate the effect of the combined water and air flow. A schematic drawing of the installation is given in figure 1, the different components and measurement instruments are assigned with Roman letters and numbers, respectively. A pump (B) was used to feed the water from a storage vessel (A) to the installation. The water flow rate could be regulated by two rota meters (1,2). Air from the laboratory system was controlled by a mass flow controller (3). The water and air flow are mixed in a mixing tee (C). A 1"-membrane module (D) containing 45 capillary nanofiltration membranes with inner diameter of 1,5 mm (resulting in a total membrane area of $0.21m^2$) was placed vertically. Pressure is monitored at the entrance and exit of the membrane module and at the permeate side (4,5,6, respectively) so that the average TMP can be calculated. The electrical conductivity and temperature of the feed and permeate flow was measured with electrical conductivity probes (7,8). The permeate flow was measured with a mass balance

(9). Water and air were separated in a degassifier (E). The concentrate valve (F) was used to regulate the TMP. A data collection system was used to acquire the pressures, electrical conductivities, temperature, air flow rate and permeate mass.

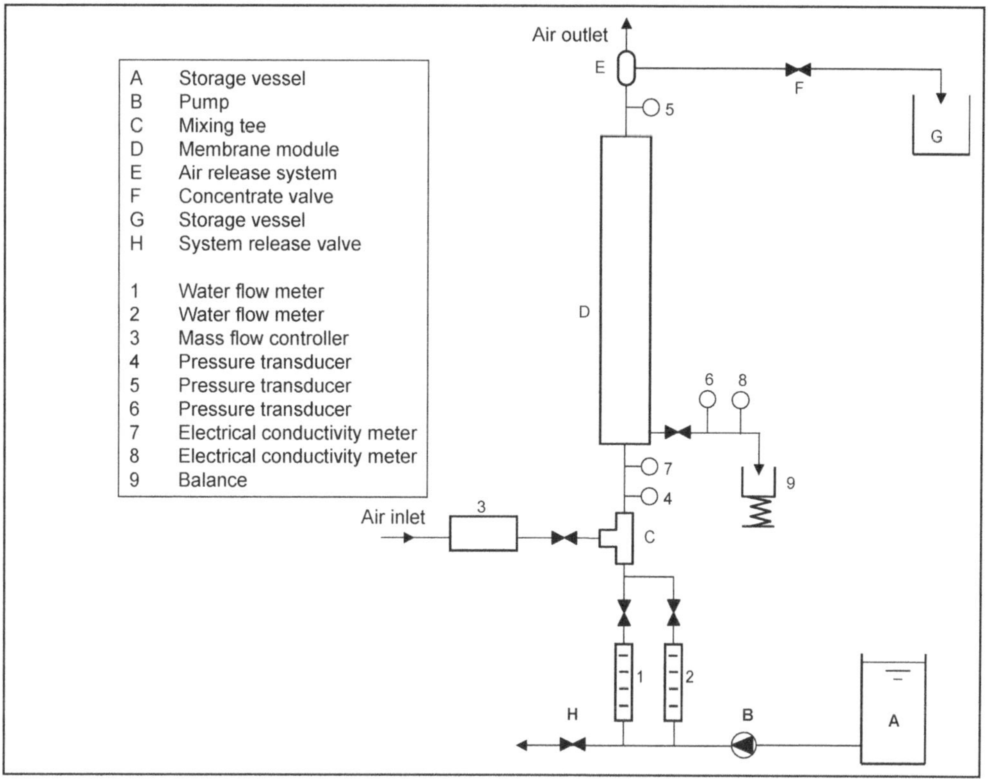

Figure 2 Schematic drawing of test apparatus.

The operating conditions of the experiments were:

- \- TMP: 1 - 4 bar
- \- $u_{L,s}$: 0 - 1,5 m/s
- \- $u_{G,s}$: 0 - 1,5 m/s

The membrane used is a capillary nanofiltration membrane supplied by X-Flow, type NR 015-500, a thin film polyamide/polyethersulfone membrane with high rejections (96%) for bivalent ions and organic molecules with molecular weight above 200 Daltons and a low rejection for monovalent ions (40%). The clean water flux is 12 l/(m^2·h·bar) (X-Flow, 2003). Experiments were done with artificial water at different salt concentrations. 7 and 12 g/l $MgSO_4 \cdot 7H_2O$ were added to demi water, corresponding with 0.34 and 0.59 w% $MgSO_4$. Magnesium sulphate was used because of the high solubililty product. The reference temperature in all experiments was 20 °C.

The results of the different experiments are evaluated by comparing the mass transfer coefficient (MTC) of the different experiments. The MTC is defined as:

$$MTC = \frac{Q_p \cdot TCF}{A_{mem} \cdot NDP} \tag{1}$$

To compare experiments at different temperatures the temperature correction factor (TCF) has to be calculated. The value for TCF is calculated by

$$TCF = \exp^{U\left(\frac{1}{T+273} - \frac{1}{T_{ref}+273}\right)} \tag{2}$$

Experiments have been done with demi-water at a pressure of 2 bar, a cross-flow velocity of 0.35 m/s (Re = 525) and different water temperatures to determine the value of U. From these experiments it follows a value of U equal to 2998.

From the clean water experiments the membrane permeability is calculated. With this value the osmotic pressure difference over the membrane is calculated by

$$J = \frac{K_w}{\mu} \times (TMP - \Delta\pi) \tag{3}$$

After calculating the concentration of salts at the membrane surface the concentration polarization factor β is calculated by

$$\beta = \frac{c_{mem,i} - c_{p,i}}{c_{b,i} - c_{p,i}} \tag{4}$$

Results and discussion

Osmotic pressure limits the performance of membrane filtration. Concentration polarization enhances the performance loss of membrane filtration. Due to concentration polarization the concentration of salts or macro molecules at the membrane surface is higher than in the bulk flow, resulting in a further increase of osmotic pressure. Besides performance loss the solubility product can be exceeded near the membrane surface, resulting in scaling of the membrane module. In the design of a membrane filtration installation much attention is paid to osmotic pressure and concentration polarization. However in operation the only parameter to control the performance is the cross-flow velocity.

Filterability of water at different salt concentrations

The objective of this experiment is to evaluate the filterability without air sparging. The observed flux reduction is significant when salty waters are filtrated, for example up to 75% at a TMP of 2 bar for the 0.59 w% solution (see figure 3 left). A high reduction in permeate flux was found for all other water velocities in the range 0.25 -1.5 m/s. As expected the permeate flux is higher for the 0.34 w% solution than for the 0.59 w% solution. This can be directly related to osmotic pressure. The osmotic pressure in the bulk solution is about 1.0 bar for the 0.34 w% solution and 1.7 bar for the 0.59 w% solution. The permeate quality is expressed by the retention. The retention is influenced by the concentration of the salts and the TMP. From figure 3 right graph it becomes clear that the permeate quality deteriorates when the salt concentration is higher. This is expected because the retention of a membrane is independent of the concentration. The retention increases with the TMP. The values of the retention are in correspondence with the specification of the membrane supplier (X-Flow, 2003).

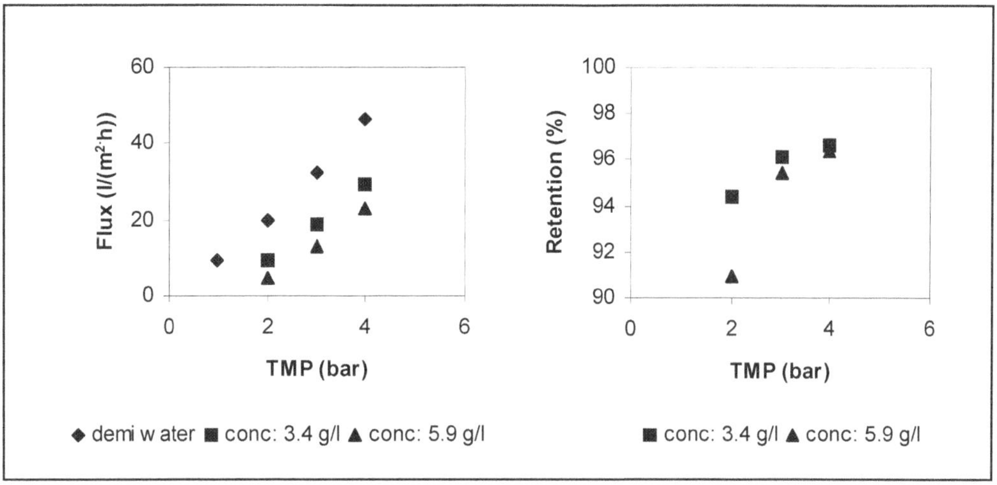

Figure 3 Permeate fluxes and retentions at different TMPs for different salt concentrations at constant cross-flow velocity of 1.5 m/s.

The effect of concentration polarization becomes clear when the fluxes at different cross-flow velocities and constant TMP are observed. At low cross-flow velocities the mass transfer coefficient is considerable lower than at high cross-flow velocities. At a TMP of 3 bar an increase of respectively 56 and 41% was observed when the cross-flow velocity was increased from 0.5 tot 1.5 m/s (see figure 4 left). The higher mass transfer coefficient at high cross-flow velocities is due to the limitation of the concentration polarization. A higher cross-flow velocity results in a larger mass transfer from boundary layer to the bulk solution. With the mass transfer coefficient and retention it is possible to calculate the concentration polarization factor. In figure 4 right the concentration polarization is given for the 0.34 w% salt concentration. It can be observed that the concentration polarization decreases with increasing cross-flow velocity. Also the concentration polarization is dependent on the TMP, high concentration polarization values are found at higher TMP.

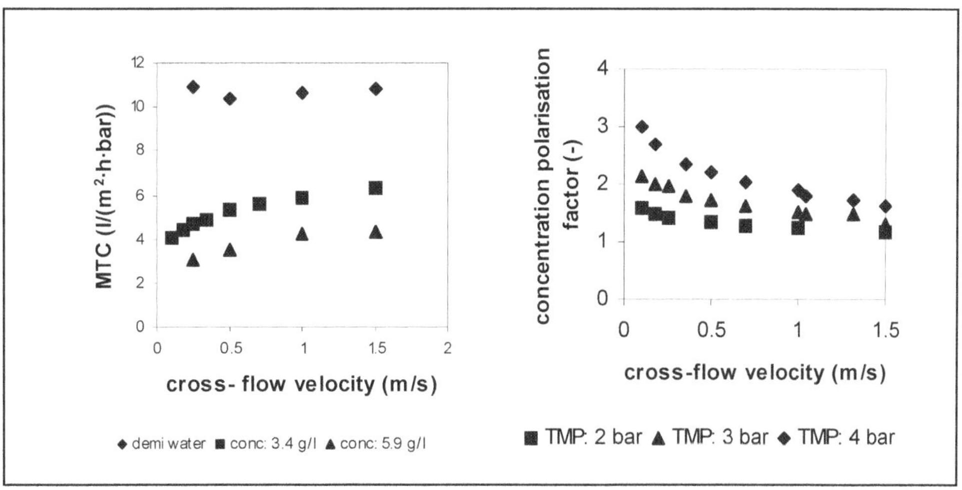

Figure 4 MTC as function of the cross-flow velocity (left), concentration polarization factor for the 0.34 w% solution as function of the cross flow velocity (right).

Effects of air sparging on clean water permeability

Firstly the effect of air sparging on the permeability of demi water was investigated. In figure 5 results of air sparging experiments are presented for a constant liquid cross-flow velocity of 0.5 m/s and several superficial air velocities. It can be observed that the permeate flux at different water-air mixtures is equal to the permeate flux without air sparging. Air sparging does not influence the water permeability. At other water velocities the same result is found. At low values of the liquid hold-up (low ratio of liquid velocity to air velocity) the mass transfer coefficient decreases. This is probably due to the transition in two-phase flow pattern.

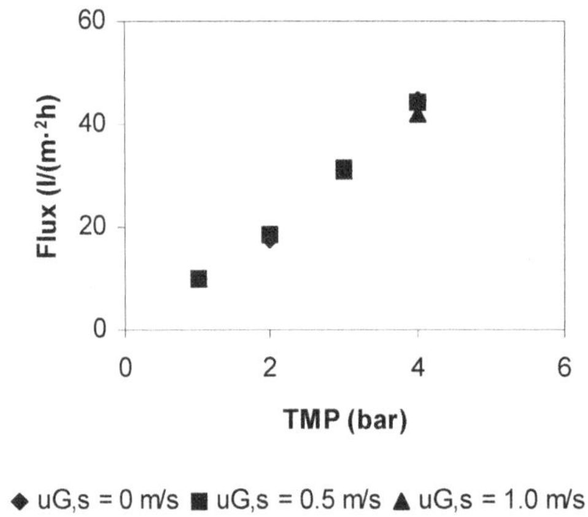

Figure 5 Permeate flux at superficial liquid velocity of 0.5 m/s and different superficial air velocities as function of the TMP.

Effects of air sparging on concentration polarization

Experiments were performed with salty solutions at different TMP and superficial water and air velocities. In figure 6 results of experiments at constant liquid cross-flow velocities of 0.5 m/s and salt concentrations of 0.34 w% are given. Both the permeate flux and the retention at a constant liquid cross-flow velocity of 0.5 m/s are depicted as function of the TMP and air velocity.

Without gas sparging ($u_{G,s}$ = 0 m/s) the permeate flux of the salty solutions is lower than the clean water flux. This is due to the combined effect of osmotic pressure and concentration polarization. When air is injected the permeate flux increases. For the 0.34 w% $MgSO_4$ a permeate flux increase ranging from 12 tot 20% (see table 1) is observed in comparison with the situation without gas sparging. Also an improvement of the retention is observed when air sparging is used. Although expressed in percentage this increase is rather small, it should be considered that an increase in retention from 95 tot 96% corresponds with a 20% (from 5% tot 4%) difference in salt concentration in the permeate. Similar observations were found for liquid cross-flow velocities of 0.25, 1.0 and 1.5 m/s and for the 0.59 w% solution. In general in all experiments a mass transfer coefficient increase is observed when air sparging is used. This increase is relatively insensitive for the air flow rate. This observation is in correspondence with earlier work of Bellara et al (1996). It was reported that for non fouling materials such as dextran the permeate flux enhancement is relatively insensitive to gas flow rates. However when fouling is the flux limiting mechanism the enhancement depends on the gas flow rate.

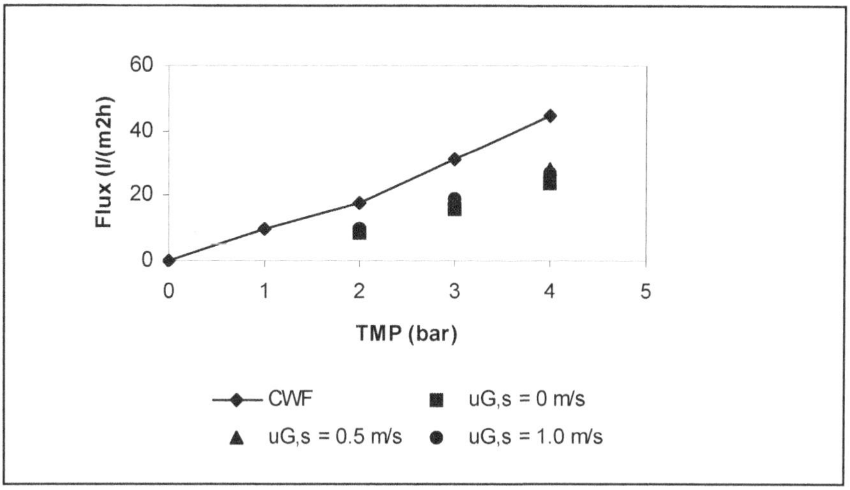

Figure 6 Permeate flux and retention as function of the TMP at constant superficial water velocity of 0.5 m/s and different superficial air velocities; salt concentration: 3.4 w% $MgSO_4$; temperature: 20°C.

Table 1 MTC ($l/(m^2 \cdot h \cdot bar)$ as function of TMPs and superficial water and air velocities and increase of MTC compared to single phase flow ($u_{G,s}$ = 0 m/s).

MgSO4 (%w)	TPM (bar)	$u_{G,s}$ (m/s) $u_{L,s}$ (m/s)	0	0.5	1.0
0.34	2	0.5	4.11	4.64 (13%)	4.83 (17%)
		1.0	4.40	5.02 (14%)	4.96 (13%)
		1.5	4.61	5.11 (11%)	5.15 (11%)
	3	0.5	5.36	6.26 (17%)	6.44 (20%)
		1.0	5.87	6.67 (14%)	6.80 (16%)
		1.5	6.35	6.92 (9%)	6.98 (9%)
	4	0.5	5.95	7.03 (18%)	6.63 (11%)
		1.0	6.68	7.70 (15%)	8.09 (20%)
		1.5	7.30	8.15 (11%)	8.23 (13%)
0.59	2	0.25	1.8	2.27 (26%)	2.37 (32%)
		0.5	2.03	2.16 (6%)	1.79 (-12%)
		1.0	2.22	2.71 (22%)	2.63 (18%)
		1.5	2.34	2.60 (11%)	2.57 (10%)
	3	0.25	3.08	4.16 (35%)	3.68 (19%)
		0.5	3.53	4.04 (14%)	4.07 (15%)
		1.0	4.21	4.72 (12%)	4.85 (15%)
		1.5	4.35	5.00 (14%)	5.02 (15%)
	4	0.25	3.76	3.60 (-4%)	3.68 (-2%)
		0.5	4.34	4.51 (4%)	4.46 (3%)
		1.0	5.08	6.08 (20%)	6.24 (23%)
		1.5	5.68	6.44 (13%)	6.52 (5%)

The mass transfer coefficient increase is higher at low liquid cross-flow velocities. At low liquid cross-flow velocities the flow is still laminar. A liquid cross-flow velocity of 0.5 m/s corresponds with a liquid Reynolds number of only 750. Injection of air will therefore result in more turbulent flow conditions and thus a decrease of the thickness of the boundary layer.

Besides permeate flux enhancement the permeate quality also improves significantly when air is injected into the liquid feed flow. Because the boundary layer is suppressed the concentration of solutes at the membrane surface decreases and thus the concentration of salts in the permeate will become smaller. The combined effect of permeate flux increase and permeate quality improvement is expressed in the concentration polarization factor. Figure 7 shows that the concentration polarization factor is reduced when air sparging is used. The reduction in concentration polarization is large at high values of the TMP.

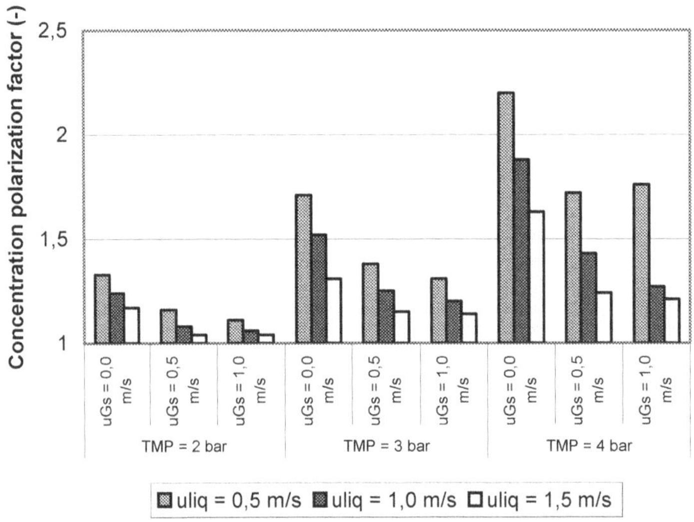

Figure 8 Concentration polarization as function of TMP and different water and air velocities.

The experimental results found in this study are not in agreement with experiments done by Ducom et al (2003). In similar experiments investigating the influence of air sparging to reduce concentration polarization no beneficial effects of air sparging were found. It was concluded that very poor benefits could be expected by air sparging. It has to be noted that even no concentration polarization was found in the experiments, probably due to unfavorable operating conditions. With these conditions no beneficial effects of turbulence increase by air sparging or any other turbulence promoting method could be expected.

Conclusion

In this study air sparging was used in cross-flow operation to limit the concentration polarization in capillary nanofiltration membranes. Two different salt concentrations were used, 0.34 and 0.59 w% $MgSO_4$ respectively. In correspondence with literature the permeate flux decreased when salts were added due to the combined effect of osmotic pressure and concentration polarization.

Air sparging experiments on demi water showed that the water-air two-phase flow does not influence the clean water permeability over a wide range of operating conditions. Only at low liquid hold-ups a decrease of clean water permeability is observed.

When salt/water solutions are used a clear permeate flux enhancement is observed. At the same liquid velocities a permeate flux increase with 30% can be was observed when air sparging was used. Also the retention increased when air was injected. The decrease in concentration polarization factor is up to 35%.

Symbols

A	area (m^2)
MTC	mass transfer coefficient (m/(s·bar))
NDP	net driving pressure (bar)
Q	flow rate (m^3/s)
T	temperature (°C)
TCF	temperature correction factor (-)
TMP	trans membrane pressure (bar)
U	temperature dependent membrane constant (-)

Greek symbols

μ	dynamic viscosity (N·s/m^2)
π	osmotic pressure (bar)

Subscripts

b	bulk
G	gas
i	ion
L	liquid
m	mixture
mem	membrane
p	permeate
ref	reference value
s	superficial

References

Bellara S.R., Cui Z.F., Pepper D.S. (1996). Gas sparging to enhance permeate flux in ultrafiltration using hollow fibre membranes. *Journal of Membrane Science* **121**, 175–184.

Van der Bruggen B., Hawrijk I., Cornelissen E., Vandecasteele C. (2003). Direct nanofiltration of surface water using capillary membranes: comparison with flat sheet membranes. *Separation and Purification Technology* **31**, 193–201.

Cabassud C., Laborie S., Lainé J.M. (1997). How slug flow can improve ultrafiltration flux in organic hollw fibre. *Journal of Membrane Science* **128**, 93–101.

Cui Z.F., Wright K.I.T. (1994). Gas-liquid two-phase cross-flow ultrafiltration of BSA and dextran solutions. *Journal of Membrane Science* 90, 183–189.

Ducom G., Cabassud C. (2003). Possible effects of air sparging for nanofiltration of salted solutions. *Desalination* **156**, 267–274.

Du Pont Technical Information Manual (1980). Du Pont, USA, January 1980.

Frank M., Bargeman G., Zwijnenburg A. Wessling M. (2001). Capillary hollw fiber nanofiltration memebranes. *Separation and Purification Technology* **22–23**, 499–506.

Futselaar H., Schonewille H., Meer W. van der (2002). Direct capillary nanofiltration – a new high grade purification process. *Desalination* **145**, 75–80.

Lee C.K., Chang W.G., Ju Y.H. (1993). Air slugs entrapped cross-flow filtration of bacterial suspensions. *Biotechnology and Bioengineering* **41**, 525–530.

Verberk, J.Q.J.C., Worm,G.I.M., Futselaar, H., Dijk, J.C. van (2001). Combined air-water flush in dead-end ultrafiltration, *Water Science and Technology: Water Supply* **1**, 393–402.

Verberk J.Q.J.C., Post J., Meer W.G.J. van der, Dijk J.C. van (2002). Direct Capillary Nanofiltration for Ground Water and Surface Water Treatment. *Water Science and Technology: Water Supply* **2**, 277–283.

X-Flow (2003). Datasheet capfil nanofiltration membrane NR 015-500. http://www.x-flow.com.

The Behavior of Inorganic and Organic Compounds during Ultrafiltration Treatment of Spent Filter Backwash Water

F. G. Reißmann*, H. Wingrich*, P. Krebs*

*Institute for Urban Water Management, Dresden University of Technology,
D – 01062 Dresden (Florian.Reissmann@mailbox.tu-dresden.de)

Abstract The scope of this study was to investigate possible effects of recycled spent filter backwash water (SFBW) after ultrafiltration treatment on drinking water quality. Backwash water samples from different water treatment plants within the city of Dresden, Germany, were used for batch and pilot plant ultrafiltration experiments. The results indicated an excellent removal of inorganic compounds by more than 96 %. As expected, the removal of organic compounds was much lower. On average, 20 % of DOC removal was analyzed. LC-OCD analysis showed that humic substances almost completely passed the ultrafiltration membrane. In contrast, polysaccharides are rejected by more than 90 %. If the permeate is recycled as an additional raw water source, a deteriorated flocculation process might be possible. Mass balances for an entire WTP demonstrated that most of the contaminants found in the concentrate originate from non-point sources in the watershed. A reduction of these contaminants does not seem to be possible.

Keywords drinking water treatment, spent filter backwash water, ultrafiltration

Introduction

Raw water that is used for drinking water production consists of a broad spectrum of inorganic and organic compounds. In the drinking water treatment process, the concentration of these compounds must significantly be decreased. For example, particles are almost completely removed by coagulation/flocculation and subsequent sand filtration processes. In addition, dissolved compounds like metals or organic compounds partially adsorb onto flocs during the flocculation process.

As a result, in most water treatment plants (WTPs), spent filter backwash water (SFBW) is generated during the water purification process. Depending on source water quality and the treatment process design, between one and ten percent of the treated water volume has to be used for backwashing sand filters. Caused by their removal during the water treatment process, most of the inorganic and organic compounds can be found in water treatment plant residuals. Some of them are known as likely being hazardous elements, for example heavy metals.

In recent years, investigations into the heavy metal content of sludges from groundwater treatment plants have been carried out. Investigations by Wingrich and Wolf (2002) summarized that in Germany significant natural concentrations of arsenic in groundwater occur especially in the northern part of Bavaria, the southern part of Saxony, the eastern part of Lower Saxony and in the north of Hesse. For example, the average arsenic content of

© 2004 IWA Publishing. *2nd IWA Leading-Edge Conference on Water and Wastewater Treatment Technologies* edited by Mark Van Loosdrecht & Jonathan Clement. ISBN: 1 84339 508 8.

Saxon rock is about eight times higher than the average of the entire earth crust. Through the water purification process, these naturally occurring contaminants are transferred to and accumulated in the SFBW.

With regard to their origin, three groups of contaminants in water treatment plant residuals can be distinguished:

(1) contaminants in the raw water that are accumulated through various treatment processes (especially naturally occurring metals like arsenic or cadmium),
(2) contaminants that are generated during the water treatment process (e.g. disinfection by-products), and
(3) trace contaminants in chemicals added during the water treatment process.

Looking at current disposal options for water works sludges in Germany, 36 % of residuals are discharged to wastewater treatment plants and 35 % are deposited in landfills (Wichmann *et al.* 2002). Focussing on the effects of arsenic containing sludges on wastewater treatment plants, Oberacker *et al.* (2003) showed that after the application of sludge containing arsenic to a wastewater treatment plant, the major part of arsenic is firmly bound and hardly mobilized in the sewage sludge. After the incineration of the sewage sludge, almost all arsenic is contained within the ash and removed from the water cycle. In a study published by Eckhardt and Haberer (1986), it was concluded that an acute danger by remobilization of trace elements from deposited sludges even at low pH-values is not expected.

Besides naturally occurring minerals present in the raw water, trace contaminants added by treatment chemicals are of special importance. In a study by Casale (2001), it was shown that impurities in chemicals used for the water treatment process could cause elevated contaminant concentrations in WTP residuals. As an example, arsenic or zinc was found in lime, and arsenic and cadmium was analyzed in ferric coagulants. These examples showed the potential impact of treatment chemicals on the disposal options of WTP residuals.

Over recent decades, in order to reduce the water discharge, SFBW has been returned to the inflow of the WTP after a sedimentation process. Concerns about the recycling of microorganisms such as *Giardia* and *Cryptosporidium*, of heavy metals or precursors for disinfection by-products, have led to a significant reduction of the number of WTPs that directly return filter backwash water to the water treatment process (ASCE and AWWA 1996). Tobiason *et al.* (2003) studied the effects of recycling SFBW on the performance of the clarification and filtration process. It was found that a recycle rate between 10 to 40 % caused a slight increase in clarifier effluent turbidity. Under typical conditions (maximum recycle rate 20 % and pH 7) no impacts were found on the filter effluent quality. In a study by Edzwald *et al.* (2003), the effects of a recycle rate of 10 % on the removal of *Cryptosporidium* were assessed in pilot-scale experiments. For both treatment plants that included a dissolved air flotation step or a plate sedimentation followed by dual media filtration, no oocysts were detected in the effluents of the filtration plant.

However, according to the German Technical and Scientific Association for Gas and Water (DVGW), the reuse of SFBW from surface water treatment is only allowed after the infiltration to groundwater or an equivalent technique. Because of an almost complete recovery of particles and microorganisms, micro- and especially ultrafiltration treatment are an alternative to groundwater recharge of SFBW. Currently, an increasing number of water suppliers are considering the application of membrane filtration such as micro- and ultrafiltration for the treatment and reuse of SFBW. Different pilot and full-scale plants have shown the feasibility of ultrafiltration treatment for the reuse of SFBW (Kötzle et al. 2002, Reissmann et al. 2002).

In order to obtain more information on possible effects of the SFBW-recycling after ultrafiltration treatment, experiments with different ultrafiltration membranes as well as different membrane systems were carried out.

Material and Methods

For the investigations into the removal efficiencies of inorganic and organic compounds through ultrafiltration treatment, three different approaches were applied in this study. A stirred cell ultrafiltration system and two different pilot plants with ultrafiltration membranes were used for the SFBW treatment. An overview of the experimental set-ups is presented in Table 1.

Table 1 Overview of the experiments

	Stirred cell 1	Stirred cell 2	Small-scale pilot plant	Large-scale pilot plant
Abbreviation	Coschütz I	Tolkewitz	Coschütz II	Coschütz III
Experimental set-up	Stirred cell experiments	Stirred cell experiments	hollow fiber modul	submerged membrane modul
Type of SFBW	coag./ flocc. with alum. sulphate and powdered activated carbon	coagulation/ flocculation with ferric chloride	coagulation/ flocculation with aluminum sulphate	coagulation/ flocculation with aluminum sulphate
Time	summer 2001	summer 2002	winter 2002	fall 2002 until spring 2003

Stirred cell ultrafiltration

Stirred cell experiments were conducted in order to get fundamental information about the recovery of specific compounds and contaminants during the ultrafiltration treatment of spent filter backwash water. A detailed description of the stirred cell membrane system is given by Reissmann *et al.* (2002). Approximately 2 000 mL of SFBW were filtered through different types of ultrafiltration and microfiltration membranes. For these experiments, SFBW from two different water treatment plants was used.

The first plant ("Coschütz WTP") with a maximum capacity of 120 000 m^3/d (mean capacity approx. 70 000 m^3/d) is treating surface water. The treatment process includes a coagulation and flocculation step using aluminum sulphate, followed by rapid sand filtration, disinfection, carbon dioxide and lime addition. The plant is equipped with a backwash water recovery system consisting of sedimentation tanks in order to reduce the amount of SFBW directly discharged to the wastewater treatment plant. Caused by an increased microbial contamination of the filtrate, SFBW reuse was stopped for safety reasons in the year 2000 according to German technical standards (DVGW W-221 1999).

The second plant ("Tolkewitz WTP") with a current capacity of approximately 24 000 m^3/d treats infiltrate from the river Elbe within the city of Dresden, Germany. The raw water is treated by a desorption unit followed by a coagulation and flocculation process using ferric chloride, dual media and activated carbon filtration, disinfection, and the addition of sodium hydroxide for pH-adjustment. Approximately 290 m^3 of spent filter backwash water are generated every day and discharged to the wastewater treatment plant.

For all stirred cell experiments, samples from the feed, the permeate, and the concentrate were taken and analyzed for metals such as arsenic, cadmium, aluminum (only "Coschütz WTP"), iron, and manganese. In addition, analysis included organic sum parameters such as adsorbable organic halids (AOX), dissolved organic carbon (DOC), and UV-absorbance (UVA_{254}).

Pilot plants

Results from two ultrafiltration pilot plants were used to obtain additional information about the recovery of organic and inorganic compounds during ultrafiltration treatment as well as information about process parameters, e.g. fluxes, transmembrane pressures, and cleaning intervals. All pilot plant experiments were carried out at "Coschütz WTP". SFBW from the full-scale filtration plant served as a feed for the ultrafiltration pilot plants.

The small-scale pilot plant consisted of a hollow fiber modul with a nominal pore size of 0.03 µm. A maximum flow rate of approximately 1.6 m^3/h deionized water can be achieved with this module. The flow into the hollow fibers with a diameter of 5.2 mm was inside out in dead-end mode. To increase backwash efficiency, an air impulse ("AirLift") at the beginning of the backwash cycle could be applied. After approximately one week of operation, chemical cleaning with citric acid and/or sodium hydroxide was carried out.

A submerged membrane system ZENON ZeeWeed® 500c was used for the large-scale pilot plant with a maximum capacity of approximately 3.5 m^3/h SFBW. The membranes with a membrane area of 60 m^2 are situated in a special storage tank. The negative pressure required for the filtration process through the membranes is generated by a self-priming centrifugal pump. In order to increase the shear strength at the membrane surface, air is continuously directed into the membrane storage tank that ensures a permanent mixing. After the filtration period, membranes are backwashed with permeate from a permeate storage tank. After approximately two weeks of operation, chemical cleaning with hydrochloric acid was carried out.

Samples were taken from the feed (i.e. SFBW), the permeate and the concentrate. Analysis included inorganics (e.g. metals like aluminum, arsenic, iron, manganese, and cadmium) and organic compounds (DOC, UVA_{254}, and AOX). Specific UV-absorbance (SUVA) was calculated from the ratio of UVA_{254} to DOC. In addition, liquid chromatography with online organic carbon detection (LC-OCD) analysis was carried out in order to get more information about the structure of organic compounds passing the membrane. LC-OCD allows the determination and characterization of DOC composition not only as a bulk parameter and provides a more detailed view of the composition of DOC. Based on the data obtained from the pilot plants and the full-scale water treatment plant, mass balances for important metals and AOX were calculated.

Results and Discussion

Recovery of inorganic compounds

Through the application of an ultrafiltration treatment of SFBW, a significant reduction of particulate matter was observed. In all experiments, turbidity decreased from values around 500 FNU to values below 0.05 FNU. Since the majority of metals are adsorbed onto the sludge flocs, an efficient rejection may be expected. As shown in Table 2, for all experiments, a significant reduction of concentrations for aluminum, iron, manganese, and arsenic in a range of 96 to 99.9 % was observed during ultrafiltration treatment of SFBW

Table 2 Recovery rates for different metals during UF treatment of SFBW (aluminum was not analyzed in the "Tolkewitz experiment")

Parameter	Experiment	SFBW (mg/L)	Permeate conc. (mg/L)	Recovery (%)
Aluminum	Coschütz I	75.1	0.10	99.86
	Coschütz II	92.6	0.04	99.96
	Coschütz III	34.7	0.05	99.87
Iron	Tolkewitz	232.0	< 0.02	> 99.99
	Coschütz I	7.61	0.05	99.29
	Coschütz II	2.89	< 0.05	> 98.27
	Coschütz III	1.92	< 0.04	> 98.12
Manganese	Tolkewitz	6.44	0.01	99.84
	Coschütz I	2.48	< 0.01	> 99.52
	Coschütz II	0.94	0.01	98.96
	Coschütz III	0.32	< 0.01	> 96.05
Arsenic	Tolkewitz	0.03	< 0.001	> 96.67
	Coschütz I	0.12	< 0.001	> 99.18
	Coschütz II	0.12	< 0.0005	> 99.59
	Coschütz III	0.07	< 0.001	> 99.08

Even at a very high feed concentration of iron in the "Tolkewitz experiment" that is caused by the use of ferric chloride as a coagulant, permeate iron concentration was below the detection limit (0.02 mg/L). All permeate concentrations for metals were below the German drinking water standard and indicated the almost complete recovery of these compounds by ultrafiltration treatment of SFBW.

Recovery of organic compounds

In contrast to inorganic compounds, a substantially lower removal rate of organic compounds by ultrafiltration treatment could be observed. On average, only a maximum of approximately 20 % DOC recovery was found (see Table 3).

Table 3 Recovery rates for organic sum parameters during UF treatment of SFBW

Parameter	Experiment	SFBW	Permeate conc.	Recovery (%)
UVA_{254} (m^{-1})	Tolkewitz	16.2	5.2	67.9
	Coschütz I	6.1	5.4	12.4
	Coschütz II	7.0	6.4	8.6
	Coschütz III	20.0	11.9	40.5
DOC (mg/L)	Tolkewitz	8.8	8.1	8.0
	Coschütz I	5.1	3.4	33.2
	Coschütz II	3.3	2.5	23.9
	Coschütz III	7.5	6.3	15.5
SUVA (L/(m·mg))	Tolkewitz	1.8	0.6	-
	Coschütz I	1.2	1.6	-
	Coschütz II	2.1	2.6	-
	Coschütz III	2.7	1.9	-
AOX (mg/L)	Tolkewitz	0.08	0.04	56.8
	Coschütz I	0.37	0.01	96.1
	Coschütz II	0.06	0.02	63.2
	Coschütz III	0.13	0.04	65.9

Focusing on the parameters DOC and UVA_{254}, differences between the Tolkewitz experiment and the Coschütz experiments were observed. Although the Tolkewitz-SUVA-values of the SFBW are only slightly below the corresponding Coschütz I and II values, recovery rates for UVA_{254} were very high while recovery rates for DOC were much lower.

The reason may be the use of a different coagulant chemical (i.e. ferric chloride) in the water purification process resulting in higher adsorption rates of complex organic compounds such as humic substances.

Further differences between the Coschütz experiments may be identified. For the first experiment (Coschütz I), the SFBW contained powdered activated carbon added during the water treatment process for taste and odor control. Since complex organic compounds tend to adsorb to activated carbon, lower SUVA values in the SFBW are obtained. The second Coschütz experiment was performed in winter 2002, at a time of low raw water organics concentrations entering the water purification plant. As a result, DOC concentrations in the SFBW were as low as 3.3 mg/L. On the contrary, UVA_{254} values were almost as high as in the first Coschütz experiment indicating a high fraction of complex, humic substances like compounds. By membrane treatment, only low recovery rates for UVA_{254} as well as DOC were achieved. In contrast, the Coschütz III experiment showed high DOC and UVA_{254} concentrations in the SFBW as well as in the permeate. This might have been caused by the disastrous flood event in summer 2002 resulting in destratification and increased turbidity input into the reservoir that provides the raw water for the Coschütz WTP. As a result, the concentration of organic compounds in the raw water increased from approximately 2.5 mg/L up to 3.5 mg/L DOC.

The average recovery rate of adsorbable organic halids (AOX) was estimated to approximately 70 %. For all experiments, the permeate concentration was below 0.04 mg/L. Significant differences were analyzed for AOX concentrations in the SFBW. In the experiment Coschütz I performed in summer 2001 and in the experiment after the flooding in 2002, elevated AOX concentrations were found in the SFBW. Additional investigations showed that most of the AOX contamination of the SFBW originated from diffuse sources in the reservoirs catchment. Therefore, one reason for varying AOX-concentration might be seasonal or climate-related variations in the raw water concentration.

In order to obtain additional information on the behavior of organic compounds passing ultrafiltration membranes, LC-OCD was applied with the source water, the SFBW, and the permeate after membrane treatment at Coschütz WTP. LC-OCD chromatograms show the fraction of DOC depending on the molecular weight. In Figure 1, chromatograms for the OCD and UVA response are presented.

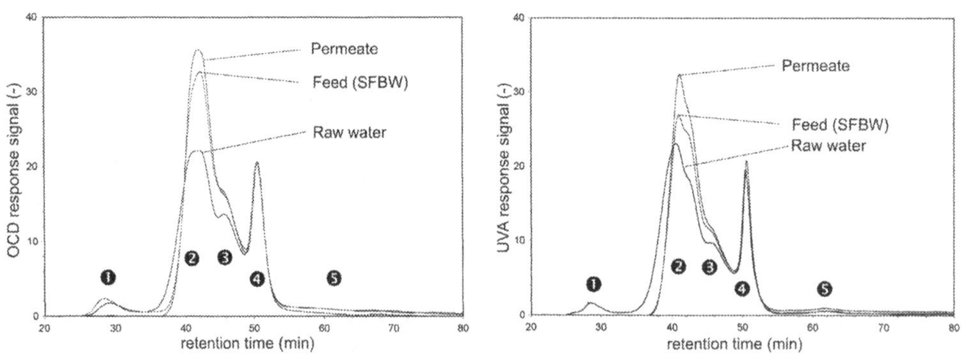

Figure 1 LC-OCD chromatograms of raw water, feed, and permeate

Usually, five major peaks can be distinguished in LC-OCD-chromatograms. The first peak is commonly identified as polysaccharides. The following peaks are recognized as humic substances, followed by the building blocks fraction, low molecular acids, and neutrals and amphiphilic compounds. In the raw water, 55 % of the DOC were humic substances like compounds (symbol '2'). Approximately 4 % of DOC were polysaccharide like compounds (symbol '1'). Comparing the SFBW (feed) with the raw water DOC, a significant increase in humic substances and building blocks fraction can be observed (symbol '2' and '3'). This increase might be caused by elimination of humic substances through the coagulation and flocculation process. Surprisingly, after membrane treatment, a slight increase in the humic substances fraction occurred. In contrast, the permeate concentration of building blocks and amphiphilic and neutral compounds (symbol '5') was lower than in the SFBW sample. Almost no change was observed for low molecular acids (symbol '4').

However, a significant reduction of polysaccharides was observed that could be induced by ultrafiltration treatment of SFBW. Almost no polysaccharides were detected in the permeate sample. In addition, the humic substances concentration was significantly higher in the permeate than the raw water. In contrast, the average molecular mass of the humic substances decreased in the permeate. Consequently, a deterioration of the coagulation and flocculation process might be possible in case of reusing the permeate as an additional raw water source.

Mass balances

Based on the results obtained from ultrafiltration treatment experiments, a more detailed investigation into the origin of contaminants such as arsenic or AOX was carried out. Data from the full-scale water treatment plant were collected and used for the calculation of mass balances including the entire water treatment process. As an example, a mass balance for arsenic is presented in Figure 2.

From the mass balance for Coschütz WTP, it can be concluded that most of the arsenic detected within the water treatment process originates from non-point sources in the watershed and enters the WTP via the raw water. A small amount of arsenic is added as a trace impurity by the coagulant dosage. After the flocculation process, sand filtration, and final disinfection step, arsenic concentrations were found to be below the detection limit. For the balance, a value of 50 % of the detection limit was assumed. However, the arsenic load discharged by drinking water is very low. Based on these results, an arsenic recovery rate of 91 % by the sand filtration process can be calculated. After backwashing the sand filter, the SFBW concentration is estimated to be 0.09 mg/L that corresponds to the measured concentration of 0.07 mg/L. With the application of an ultrafiltration treatment step for the reuse of SFBW, approximately 139 g/d arsenic load are discharged to the Dresden wastewater treatment plant. The amount of arsenic that is directed back to the raw water by the permeate was estimated to be 455 mg/d and is almost negligible. Therefore, the raw water quality should not be affected at all.

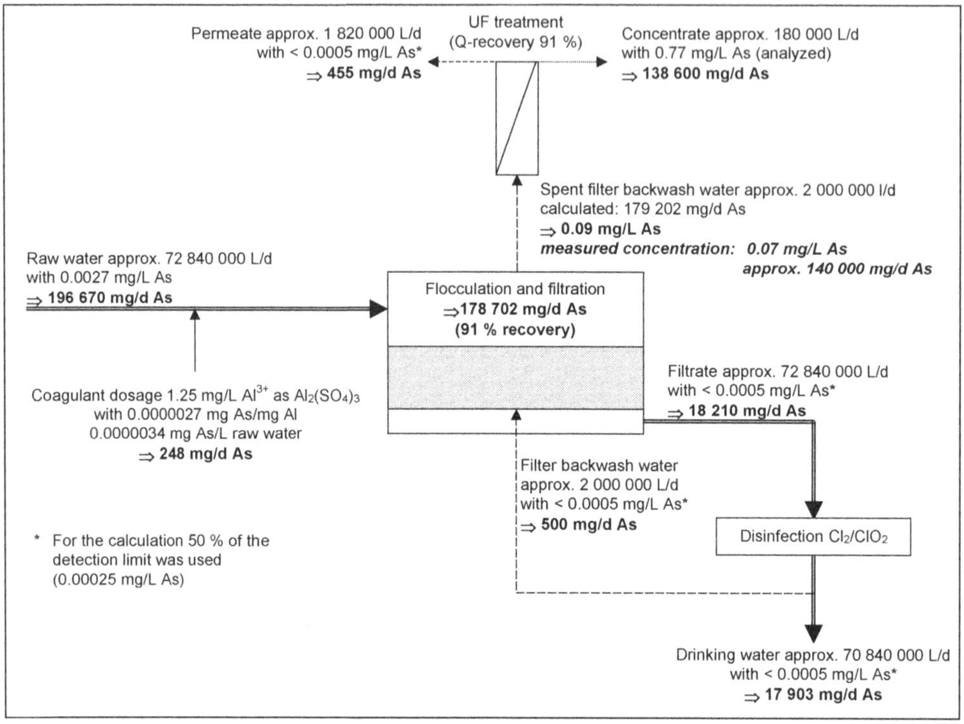

Figure 2 Arsenic mass balance for the Coschütz WTP

However, mass balances provide a good overview on major contaminant sources and sinks within the entire water treatment process. Similar investigations as for arsenic were carried out for AOX. The results pointed out two major sources: AOX concentrations entering the WTP with the raw water, and the formation of AOX during the disinfection process. As a result, most of the AOX found in the SFBW originate from diffuse sources and are recovered by the flocculation and filtration process. AOX formed during the disinfection step are discharged with the drinking water.

Conclusions

The application of ultrafiltration for the reuse of spent filter backwash offers many advantages. Based on batch and pilot-scale studies, recovery rates for inorganic and organic compounds were investigated.

The results confirmed an almost complete recovery of inorganic compounds. In particular, aluminum, iron, manganese, and arsenic were rejected by 96 to 99 %. Permeate turbidity was always below 0.05 FNU. For organic compounds described by organic sum parameters like DOC, much lower recovery rates of around 20 % were obtained. Only adsorbable organic halids (AOX) were rejected by 70 %. Firstly, the structure of organic compounds present in SFBW and secondly, the coagulant used in the water treatment process influenced recovery rates. LC-OCD-analysis showed a significant reduction of the polysaccharide fraction induced by the membrane filtration process. On the other hand, almost no change could be observed for the humic substances fraction in the permeate that was substantially higher than the influent concentration of the full-scale water treatment plant. Therefore, the behavior of organic compounds requires special attention when ultrafiltration treatment is applied, because especially humic-like compounds may cause the flocculation process to deteriorate. Moreover, these compounds are known to be precursors for disinfection by-products.

As a consequence of the accumulation of arsenic or AOX in the concentrate, the respective threshold values of the German federal receiving wastewater utility regulation cannot be met, and either the discharge of concentrate into the sewer will be charged or alternative disposal options must be considered. Mass balances for an entire WTP showed most of these contaminants to originate from diffuse sources in the watershed. Therefore, a reduction of contaminants present in SFBW seems to be very difficult.

Acknowledgements

This work was supported by the "Deutsche Forschungsgemeinschaft (DFG)-Graduiertenkolleg 339 Cross-Media Transport and Reaction Processes at the Compartment Boundaries to Water". The authors also thank the "DREWAG - Stadtwerke Dresden GmbH" (Dresden, Germany) for their excellent support in performing this study.

References

American Society of Civil Engineers (ASCE) and American Waterworks Association (AWWA) (1996) *Technology Transfer Handbook: Management of Water Treatment Plant Residuals*, American Waterworks Association, Denver, New York.

Casale, R.J. (2001) Improving chemical handling procedures can help reduce associated treatment problems. *JAWWA* **9** 95-106.

DVGW 221-1/Technical standard 221-1 (1999) Rückstände und Nebenprodukte aus Wasseraufbereitungsanlagen; Teil 1: Grundsätze und Planungsgrundlagen/Residuals and by-products from water treatment plants; Part 1: principles and fundamentals for planning. *German Technical and Scientific Association for Gas and Water*, Bonn/Germany.

Eckhardt, H. and Haberer, K. (1986) Remobilization of trace elements from water works sludges considering especially the pH-value. *Vom Wasser* **66** 197-210.

Edzwald, J.K., Tobiason, J.E., Udden, C.T., Kaminski, G.S., Dunn, H. J., Galant, P.B., and Kelley, M.B. (2003) Evaluation of the effect of recycle of waste filter backwash water on plant removals of Cryptosporidium. *J Water SRT – Aqua* **52** 243-258.

Kötzle, T., Merkl, G., and Wilderer, P.A. (2002) Large scale reuse of different filter backwash waters applying a submerged microfiltration membrane system. In *Proceedings Management of Wastes from Drinkings Water Treatment*, London, UK.

Oberacker, F., Maier, D., and Maier M. (2003) Arsenic and Drinking Water, Part 2 – A review of the arsenic elimination process for drinking water production and sustainable handling options for arsenic containing water works sludges. *Vom Wasser* **100** 9-48.

Reissmann, F.G., Scherer, M., Weiß, M. and Wingrich, H. (2002) Cost-efficiency of ultrafiltration treatment for the reuse of filter backwash water - a case study. In *Proceedings Membranes in Drinking and Industrial Water Production 2002*, Mühlheim, Germany.

Tobiason, J.E., Edzwald, J.K., Gilani, V., Kaminski, G.S., Dunn, H.J., and Galant, P.B. (2003) Effects of waste filter backwash recycle operation on clarification and filtration. *J Water SRT – Aqua* **52** 259-275.

Wichmann, K., Akkiparambath, A., and Schneider S. (2002) Mengen, Zusammensetzung und Entsorgung von Wasserwerksrückständen im Jahr 1998 sowie zukünftige Datenerfassung – Auswertung der Umfrage von 1999. *gwf Wasser Abwasser* **143**(10) 732 – 737.

Wingrich, H. and Wolf, C. (2002) Research on the removal of arsenic from drinking water by flocculation with iron solutions and adsorption with granulated iron hydroxide (GEH). *J Water SRT – Aqua* **51**(5) 273-281.

UV/H_2O_2 retrofit of PWN's water treatment plant Andijk for primary disinfection and organic contaminant control

Joop C. Kruithof*, Peer C. Kamp*, Mike Belosevic** and Mihaela Stefan***
* NV PWN Water Supply Company North Holland, PO Box 2113, 1990 AC Velserbroek, The Netherlands
(E-mail: joop.kruithof@pwn.nl; peer.kamp@pwn.nl)
** University of Alberta, Department of Biological Sciences, Edmonton, Alberta, Canada T6G 2E9
(E-mail: mike.belosevic@ualberta.ca)
*** Trojan Technologies Inc., 3020 Gore Road, London, Ontario, Canada N5V 4T7
(E-mail: mstefan@trojanuv.com)

Abstract N.V. PWN Water Supply Company North Holland has replaced breakpoint chlorination at their Andijk drinking water treatment plant. Major objectives are restriction of disinfection by-product formation, increased pathogen inactivation and multiple barriers for organic contaminant control. The aim of this study was to show the feasibility of UV/H_2O_2 treatment for primary disinfection and organic contaminant control in direct surface water treatment.

A 2 – 3 log inactivation of Giardia muris and Cryptosporidium parvum was achieved by an UV dose of 20 mJ/cm^2. The highest UV dose was needed for the inactivation of spores of sulphite reducing clostridia. For the required inactivation of all micro-organisms an UV dose up to 105 mJ/cm^2 was needed.

Reactivation of protozoa was established for an UV dose up to 25 mJ/cm2. For doses higher than 60 mJ/cm2 no reactivation was observed.

An 80 % degradation of pesticides, endocrine disruptors and pharmaceuticals can be achieved by a proper combination of electric energy and hydrogen peroxide. At this conversion primary metabolite formation was insignificant, while no bromate was formed.

The electric energy for organic contaminant control is much higher than required for disinfection e.g. 540 mJ/cm2 and 6 mg/L H2O2.

Residual H2O2 removal, AOC and nitrite control are achieved by post GAC filtration.

In view of the very promising results PWN has installed UV/H2O2 at their Andijk plant. Three streets of 4 Trojan Swift 16L30 reactors are in operation since late 2004.

Keywords UV/H_2O_2 treatment, primary disinfection, organic contaminant control

INTRODUCTION

In 1920, when N.V. PWN Water Supply Company North Holland (PWN) was founded, the demand for drinking water was satisfied by ground water extraction . However, caused by the growing drinking water demand PWN was compelled to utilize surface water as an additional source.

Therefore in the 1960's water treatment plant Andijk was constructed for the direct production of drinking water from IJssel Lake (River Rhine) water. Originally the plant consisted of microstraining, breakpoint chlorination, coagulation, sedimentation, rapid

© 2004 IWA Publishing. *2nd IWA Leading-Edge Conference on Water and Wastewater Treatment Technologies* edited by Mark Van Loosdrecht & Jonathan Clement. ISBN: 1 84339 508 8.

filtration and post disinfection. Since 1978 the plant was upgraded with a pseudo moving bed GAC filtration followed by an additional microstraining.

After about 40 years of operation, wtp Andijk still complies with all European Community and Dutch drinking water standards. Nevertheless a retrofit is desired in view of by-product (THM) formation, barriers against pathogenic micro-organisms such as protozoa and control of organic micropollutants such as pesticides, endocrine disruptors and pharmaceuticals.

PWN has pursued the application of UV/H_2O_2 treatment for both primary disinfection and organic contaminant control followed by the already present GAC filtration for removal of residual H_2O_2, AOC and nitrite with very promising results. UV/H_2O_2 is operational since late 2004.

PROCESS SELECTION

PWN's aim was to select an oxidative process able to inactivate pathogenic micro-organisms and to degrade pesticides aselectively with a restricted by-product formation compared to breakpoint chlorination.

Initially the focus was on ozone based technologies. (1) Ozone was able to achieve the required disinfection capacity. However ozone proved to be a selective barrier against pesticides.

In advanced oxidation technologies hydroxyl radicals react 10^6 to 10^9 times faster than ozone, causing a much more aselective pesticide degradation.

O_3/H_2O_2 treatment proved to be a sound barrier against pesticides. In pretreated IJssel Lake water 80 % pesticide degradation could be achieved by an O_3 dose of 2.8 g/m^3 and a H_2O_2/O_3 ratio of 2 g/g. (2) However bromate was found at high levels especially at low water temperature. (3) Feasible treatment options (increase H_2O_2/O_3 ratio, increase pH) could not control bromate formation completely caused by mass transfer limitations. (see figure 1) (3)

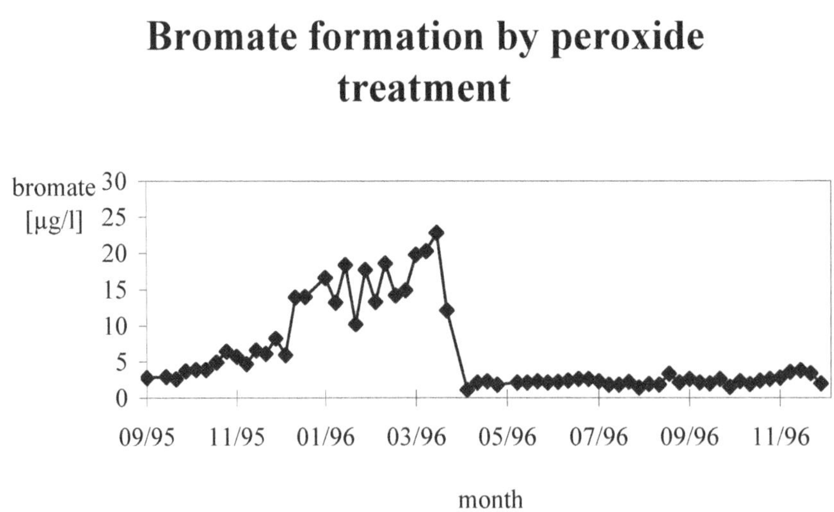

Figure 1 Bromate Formation by O_3/H_2O_2 treatment (O_3/DOC = 1.13 g/g)

before April '96 $H_2O_2/O_3 = 2$ g/g, pH = 7.5
after April '96 $H_2O_2/O_3 = 4$ g/g, pH = 7.8

Under optimal conditions for pesticide and bromate control disinfecting properties of the O_3/H_2O_2 process were very poor. Corresponding with high H_2O_2/O_3 ratios and very low c.t. values for ozone, inactivation of spores of sulphite reducing clostridia was insignificant. (4)

In view of these findings PWN rejected O_3/H_2O_2 treatment for primary disinfection and pesticide control at wtp Andijk and decided to pursue alternative strategies with the focus on UV technologies.

PHOTOLYSIS AND ADVANCED OXIDATION BY UV/H_2O_2 TREATMENT

UV photolysis is based on the absorption of UV photons by contaminants to cause their degradation into smaller products. The absorbed energy must be higher than the energy of the band to be broken to cause photolysis.

In addition to the absorption of UV photons the quantum yield of the degradation is an important parameter. Compounds with a high UV absorbance and a high quantum yield are very susceptible to photodegradation. (5)

However not all substances absorb UV light significantly or have a high quantum yield. In order to promote a more aselective degradation H_2O_2 may be added to produce hydroxyl radicals.

The general expression for the degradation rate of a contaminant is a combination of direct UV photolysis and hydroxyl radical reactions:

$$-\frac{dc}{dt} = Ø_C\, \varepsilon^{\circ}_{abs,C} + k_{C,OH}\,[C]\,[OH^{-}]$$

where $Ø_c$ is the quantum yield, $\varepsilon^{\circ}_{abs}$ is the absorbed light and $k_{C,OH}$ is the hydroxyl radical rate constant.

Not too many $Ø_c$ and $k_{C,OH}$ values for pesticide degradation are given in literature. An electric energy of 0.5 kWh/m^3 is reported for 60 % atrazine photolysis. Under these conditions pesticide degradation varied from 18 % for dicamba to 92 % for mecoprop. (6) Additional research is needed to determine quantum yield and hydroxyl radical rate constants for representative organic micropollutants and to determine the process conditions for UV/H_2O_2 treatment with medium pressure UV lamps.

DISINFECTION BY UV

There is an abundance of literature on the inactivation of fecal coliform bacteria by UV radiation. It has been known for some time that viruses and spore forming bacteria are more resistant to the effect of UV radiation than coliform bacteria. Chang et al (7) showed that viruses, bacterial spores and amoebic cysts required 3 – 4 times, 9 times and 15 times more UV exposure to achieve the same level of inactivation as Escherichia coli.

It had been thought that encysted protozoa were insensitive to UV.(8,9) More recent studies have shown that encysted Cryptosporidium parvum and Giardia muris are relatively sensitive to UV radiation. (10, 11, 12)

Reactivation of pathogenic micro-organisms is a critical parameter with the respect to the treatment of drinking water. It is well known that select pathogenic micro-organisms are capable of reactivating after exposure to UV light. Mechsner et al (13) demonstrated that E coli inactivated by low pressure UV repaired its DNA after several days incubation in the dark.

More work is needed to determine the overall reactivation potential of water-borne pathogenic micro-organisms of concern to the water industry.

The ultimate goal for UV disinfection of drinking water is to damage DNA beyond repair.

Additional research is needed to establish dose-response curves for several pathogenic micro-organisms and to establish the UV dosage where DNA is damaged beyond repair.

OBJECTIVES

Disinfection

Three major objectives were pursued in a collaborative study carried out with the University of Alberta:

to establish UV inactivation by generating UV dose-inactivation curves for MS2 phages, Bacillus subtilis endospores, Giardia muris cysts and Cryptosporidium parvum oocysts using a bench scale collimated beam apparatus;

to determine the ability of one small medium pressure UV reactor and three larger medium pressure UV reactors to inactivate MS2 phages, Bacillus subtilis endospores and Cryptosporidium parvum oocysts;

to examine the reactivation of Giardia muris and Cryptosporidium parvum ex-vivo and in vivo and to determine more precisely the UV dose required for inactivation of these protozoan parasites in drinking water.

The disinfection requirements for the Andijk water treatment plant are based on an acceptable infection risk of 10^{-4} per person per year. Based on measurements in the IJssel Lake water the required inactivations for six priority micro-organisms are calculated. (see Table 1).

Table 1 Concentration of micro-organisms, disinfection requirement and required log removal

Micro-organism	Max. content IJssel Lake	Requirement	Total required inactivation	Inactivation pre-treatment	Inactivation required by UV
	n/L	n/L	log	log	log
E Coli	6000	10^{-2}	5.8	2.0	3.8
Faec. Strept.	3000	10^{-1}	4.5	2.5	2.0
Spores of SRC	4500	10^{-1}	4.7	3.0	1.7
Viruses	0.1	2.5×10^{-7}	5.6	2.0	3.6
Giardia	5.2	7.0×10^{-6}	5.9	2.5	3.4
Cryptosporidium	2.6	3.3×10^{-5}	4.9	2.0	2.9

Organic contaminant control

Three additional major objectives were pursued, partly in collaboration with Trojan Technologies Inc.:

to characterize degradation by UV photolysis and hydroxyl radical reaction for selected priority pollutants (pesticides, endocrine disruptors, pharmaceuticals);

to predict and determine the ability of a medium pressure UV reactor to degrade those priority pollutants;

to design a full scale UV/H_2O_2 system for both disinfection and organic contaminant control.

In the Netherlands around 350 pesticides are used with a great variety in persistence, degradability and toxicity. Monitoring programs have shown the presence of many of these pesticides in drinking water sources such as the IJssel Lake. Priority pollutants such as atrazine, pyrazon, diuron, bentazone, bromacil, methabenzthiaxuon, dicamba, 2,4-D, TCA and trichlorpyr are found in concentrations up to 1 μg/l.

More recently monitoring programs have been focussed on the presence of endocrine disruptors and pharmaceuticals. In the raw water sources up to several hundred nanograms per liter were found for bisphenol A, diethylphtalate, diclofenac, ibuprofen, phenazone, carbamazepine and several antibiotics and X-ray contrast media.

Posttreatment

Three additional objectives were pursued to enable integration of UV/H_2O_2 treatment in the total treatment scheme:

the removal of excess H_2O_2 by GAC filtration;

the removal of AOC by GAC filtration;

the removal of nitrite by GAC filtration.

MATERIAL AND METHODS

Ultraviolet exposure by collimated beam apparatus

The solutions were exposed to UV radiation from a 1-kW medium pressure UV lamp using a collimated beam apparatus. Duration of exposure to UV light and delivered UV dose were controlled by opening and closing a shutter located directly beneath the UV lamp.

The UV irradiance at the surface of the liquid and the center of the UV beam was measured immediately before and immediately after each experiment with a radiometer (International Light, Model IL1400A) equipped with a SED240 UV detector using the procedures described previously. The depth-averaged irradiance was then determined.

Ultraviolet exposure by pilot plants

The small scale pilot plant medium pressure UV reactor was supplied by Berson. Target UV doses were achieved by regulating flow rates, and, therefore, residence times, through the pilot plant. Flow rates within the pilot plant were monitored using calibrated in-line flow meters.

Three medium pressure UV reactors (Manufacturer: Calgon, Berson and Trojan) were tested for their ability to inactivate MS2 phage and Bacillus subtilis endospores. For each trial the power to the UV reactor was turned on and the test water containing spores and phage flushed through the system prior to sampling.

Infectivity analysis

For Giardia muris, the tube and contents were centrifuged at 800xg for 10 min. The supernatant was removed and the concentrated pellet was resuspended in the remaining 3 - 5 ml liquid using a glass Pasteur pipette. For ex vivo experiments, cohorts of five adult male C3H/HeN mice werer inoculated with between 10^4 and 10^5 UV-exposed cysts. Cysts inactivation was determined using the latent period assay described previously for Giarda muris.

For ex vivo experiments with Cryptosporidium parvum, the tubes containing the UV-exposed oocysts were centrifuged at 27,000xg for 10 min. to concentrate the oocysts. The supernatant was aspirated and the cell pellet resuspended in deionized water. Oocysts were counted in quadruplicate using a hemacytometer and an appropriate number of parasites in deionized water was prepared. The procedures for determination of Cryptosporidium parvum infectivity reduction using the neonatal CD-1 mouse model and logistic dose-response model have been described previously.

Pesticide analysis

Pesticides have been analysed by HPLC-UV following solid phase extraction for the water samples.

RESULTS

INACTIVATION OF MICRO-ORGANISMS

Collimated beam experiments

Giardia muris cysts suspended in pre-treated IJssel Lake water were shown to be susceptible to UV inactivation. An UV dose of as low as 20 mJ/cm^2 induced greater than a 2 log-unit inactivation of Giardia muris. However a pronounced tailing effect in the ability of UV to inactivate the infective potential of Giardia muris cysts was observed with UV doses greater than 20 mJ/cm^2. (Figure 2).

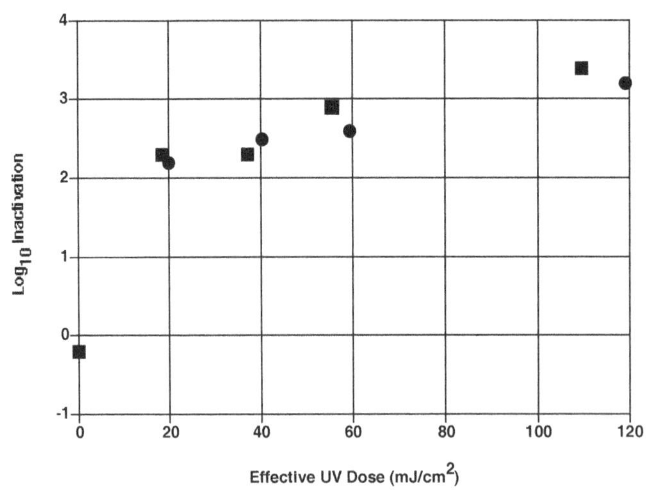

Figure 2 Log Inactivation of Giardia Muris Cysts by UV Disinfection

Exposure of Giardia muris cysts to UV doses as high as 120 mJ/cm^2 resulted in only 1 additional log-unit of inactivation compared to UV doses of 20 mJ/cm^2 only.

Cryptosporidium parvum oocysts suspended in rapid filtrate of IJssel Lake water were extremely susceptible to low doses of UV. Greater than a 3 log-unit reduction in oocyst infectivity was observed in trials where oocysts were exposed to UV doses as low as 20 mJ/cm^2. For Cryptosporidium parvum oocysts the tailing effect of the UV dose-inactivation curve was not as pronounced as that observed for Giardia muris cysts (Figure 3). UV doses of approximately 120 mJ/cm2 induced greater than 4.5 log-units of inactivation.

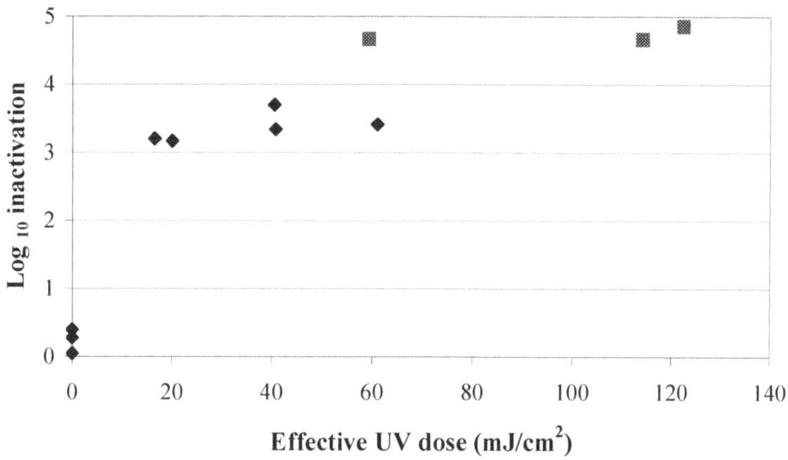

Figure 3 Log Inactivation of Cryptosporidium Parvum Oocysts by UV Disinfection

The presence of the tailing suggests that a UV resistant sub-population of oocysts may be present within any given batch of oocysts, which remains to be confirmed using molecular analysis of the resistant sub-population of the (oo)cysts.

Both Bacillus subtilis endospores and MS2 phages were shown to be more UV-resistant than either G. muris and C. parvum for UV-doses less than 50 to 60 mJ/cm^2. Of all the organisms tested in this study MS2 phages showed the greatest resistance to UV.

For MS2 phages a 3.6 log-unit inactivation was achieved by an UV-dose as high as 105 mJ/cm^2. Spores of sulfite reducing clostridia were most resistant against UV-disinfection. However the required 1.7 log inactivation could be achieved by an UV-dose of 105 mJ/cm2 as well.

Pilot plant experiments

The collimated beam data demonstrate that medium pressure UV radiation effectively inactivated MS2 phages, Bacillus subtilis endospores, Giardia muris cysts and Cryptosporidium parvum oocysts suspended in pre-treated IJssel Lake water. The results were confirmed by data obtained with a small scale Berson UV reactor and with three larger scale reactors from Calgon, Berson and Trojan. Of the three larger scale UV reactors the Berson and Trojan reactors showed inactivation levels equivalent to the collimated beam experiments, while the Calgon unit showed a lower inactivation level (Figure 4).

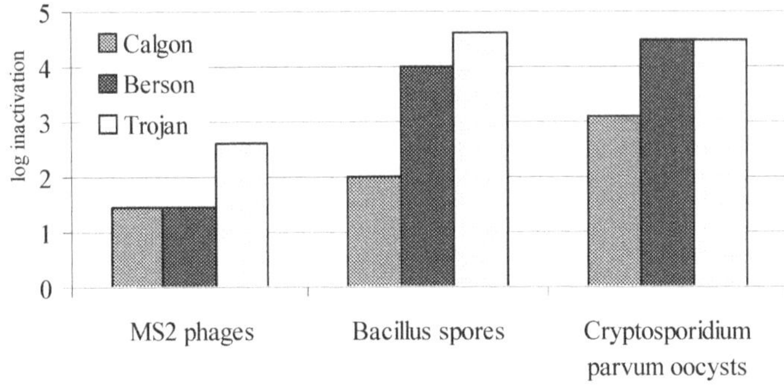

Figure 4 Inactivation of MS2 Phages, Bacillus Subtilis Endospores and Cryptosporidium parvum Oocysts by UV Disinfection with an UV Dose of 105 mJ/cm^2

The findings show that UV disinfection is a very reliable technology for the inactivation of pathogenic micro-organisms in drinking water. An UV dose of 105 mJ/cm^2 gives the required inactivation of phages and (endo)spores and a complete inactivation of protozoan cysts and oocysts.

REACTIVATION OF MICRO-ORGANISMS

It is important to understand that it is debatable as to whether or not UV-irradiation causes permanent or temporary inactivation of micro-organisms. Relatively little is known about the repair of UV-damaged DNA in encysted water-borne protozoan pathogens. Our data indicate that neither Giardia muris cysts nor Cryptosporidium parvum oocysts are capable of in vitro reactivation. The results for in vivo reactivation of Giardia muris cysts are summarized in table 2.

Table 2 In Vivo Reactivation of Giardia Muris Cysts Treated with Medium Pressure UV

Days Post Infection	No. Mice Infected/Total No. Mice			
	< 25 mJ/cm^2		> 60 mJ/cm^2	
5	14/35	0/34	6/20	0/20
10	35/35	1/34	20/20	0/20
15	35/35	1/34	20/20	0/20
20	35/35	5/34	20/20	0/20
25	35/35	6/34	20/20	0/20
30	35/35	7/34	20/20	0/20

From the data it can be concluded that Giardia muris cysts are capable of in vivo reactivation after UV treatment when the UV dose applied is as low as 25 mJ/cm2. This dose is significantly lower than the UV dose of 105 mJ/cm^2 needed for the required inactivation of phages and (endo)spores.

Also Cryptosporidium parvum oocysts are capable of in vitro reactivation for UV doses up to 25 mJ/cm2.

DEGRADATION OF ORGANIC MICROPOLLUTANTS

Pilot plant experiments

Degradation of 10 pesticides atrazine, pyrazon, diuron, bentazone, bromacil, methabenzthiaxuon, dicamba, 2,4-D, TCA, trichlorpyr and the atrazine metabolite desethyldesisopropylatrazine by UV-photolysis was studied in a standard in line Berson reactor. The electric energy was ranging from 0.25 – 2.0 kWh/m^3.

All priority pollutants showed a significant degradation by UV-photolysis. The conversion for an electric energy of 1 kWh/m^3 (about 1000 mJ/cm^2) is summarized in table 3.

Table 3 Pesticide Degradation by UV-Photolysis with 1 kWh/m^3

Pesticide	Degradation (%)	Pesticide	Degradation (%)
Atrazine	70	Dicamba	63
Pyrazon	52	2,4-D	58
Diuron	65	TCA	18
Bentazone	50	Trichlorpyr	52
Bromacil	42	Desethyldesisopropyl-atrazine	30
Methabenzthiaxuon	50		

By UV-photolysis with 1 kWh/m^3 degradation ranged from 18 % for trichloroacetic acid (TCA) to 70 % for atrazine. The criterion of 80 % conversion had to be achieved by an additional H_2O_2-dosage to initiate a supplementary hydroxyl radical reaction. Examples for a compound with a low and a high UV-photolysis conversion are shown in figure 5 and 6.

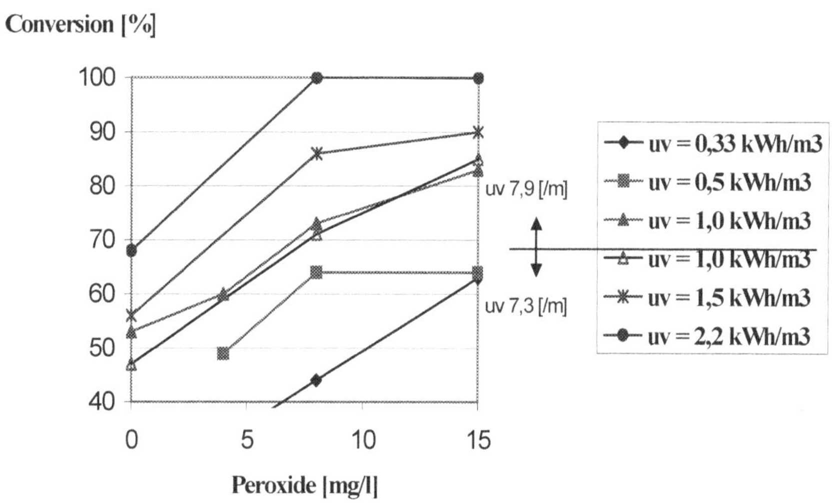

Figure 5 Bromacil Degradation by Combined UV-Photolysis and Hydroxyl Radical Oxidation for UV/H_2O_2-treatment

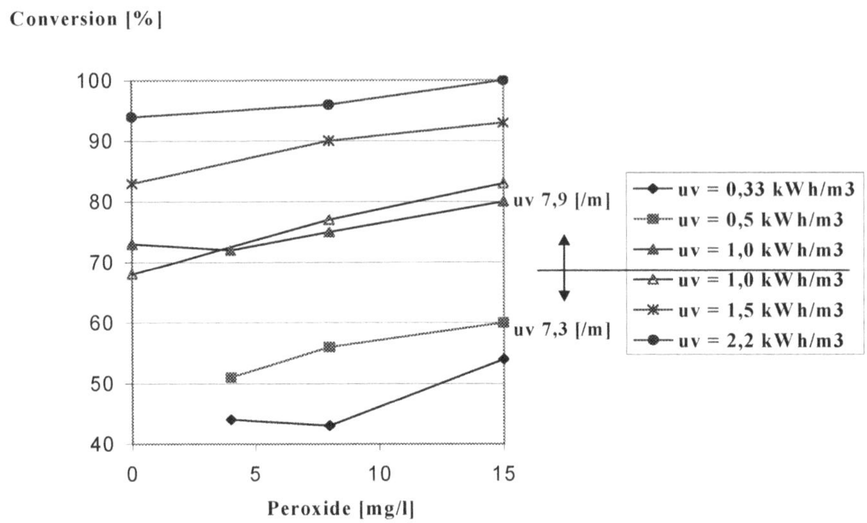

Figure 6 Atrazine Degradation by Combined UV-Photolysis and Hydroxyl Radical Oxidation for UV-H2O2-treatment

Degradation of atrazine by an electric energy of 1 kWh/m^3 amounted 70 %. This degradation was increased to the desired 80 % by adding 13 g/m^3 H_2O_2. Atrazine degradation was caused primarily by photolysis with a polishing effect of hydroxyl radical oxidation.

Degradation of bromacil by an electric energy of 1 kWh/m^3 amounted 42 %. This degradation was increased to the desired 80 % by adding 15 g/m^3 H_2O_2. For bromacil degradation hydroxyl radical oxidation played a much more important part.

For most priority pollutants a degradation of 80 % can be achieved. Depending on UV-absorbance and quantum yield on the one side and chemical structure (double bonds, H-atoms) on the other side photolysis or hydroxyl radical reaction will play a predominant part. Only for TCA and desethyldesisopropylatrazine 80 % degradation may not be achieved completely.

Since these pilot plant runs the reactor design is improved significantly, thereby reducing the electric energy needed for degradation by a factor of about 2. Some results for pesticide, endocrine disruptor and pharmaceutical degradation in a Swift 4L12 reactor are shown in figure 7 and 8.

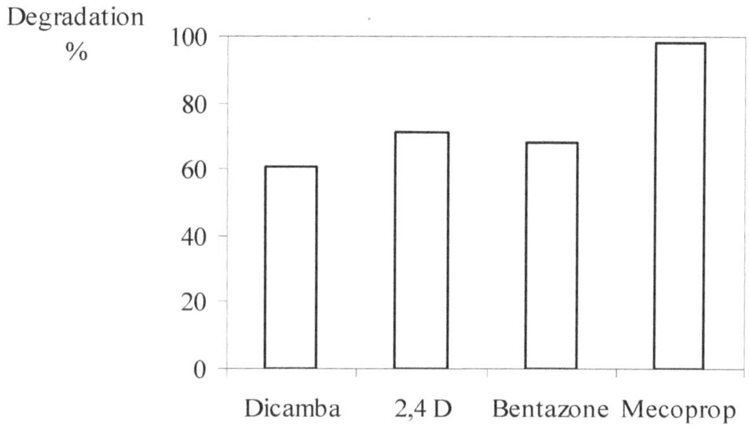

Figure 7 Pesticide Degradation in Pre-treated IJssel Lake Water Swift 4L12 reactor, UV dose 0.54 kWh/m^3 (540 mJ/cm^2), H_2O_2 dose 6 mg/L

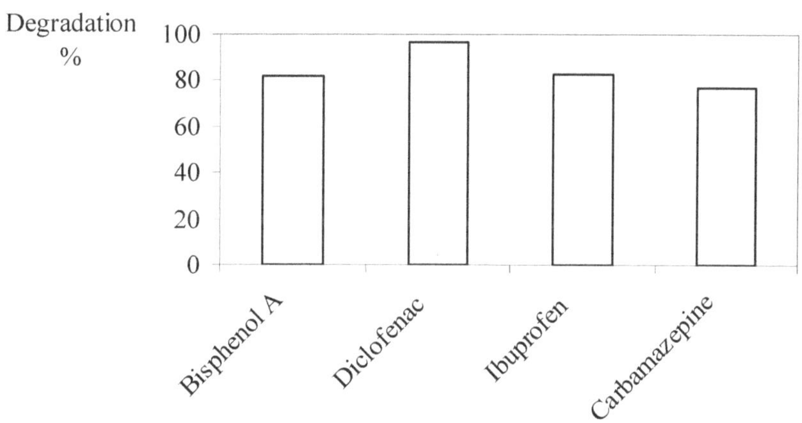

Figure 8 Endocrine Disruptor and Pharmaceutical Degradation in Pre-treated IJssel Lake water Swift 4L12 reactor, UV dose 0.54 kWh/m3 (540 mJ/cm2), H2O2 dose 6 mg/L

With this new reactor design more than 60 % degradation could be achieved at an UV dose of

540 mJ/cm^2 (0.54 kWh/m^3) and 6 mg/L H_2O_2. So the dose for organic contaminant control is much higher than for disinfection.

PERSPECTIVE

At water treatment plant Andijk UV/H_2O_2-treatment will be implemented. For primary disinfection the breakpoint chlorination will be replaced by UV-disinfection. A dose of 105 mJ/cm^2 gives the required inactivation of phages (viruses) and spores and an complete inactivation of Giardia muris and Cryptosporidium oocysts. Although some reactivation of Giardia muris cysts and Cryptosporidium parvum oocysts is observed, this will not take place at UV-doses as high as 105 mJ/cm^2.

Organic contaminant control can be achieved for a broad range of process conditions e.g. 540 mJ/cm^2 (0.54 kWh/m^3) and 6 mg/L H_2O_2. This UV dose is much higher than needed for disinfection. Under these conditions primary metabolite formation is insignificant, while bromate formation is absent irrespective of the process conditions.

PWN decided to implement UV/H_2O_2-treatment in the full scale treatment plant prior to the two step GAC filtration already present. Important issues were removal of residual H_2O_2, biological stability and nitrite formation. Removal of H_2O_2 by GAC filtration is presented in Figure 9.

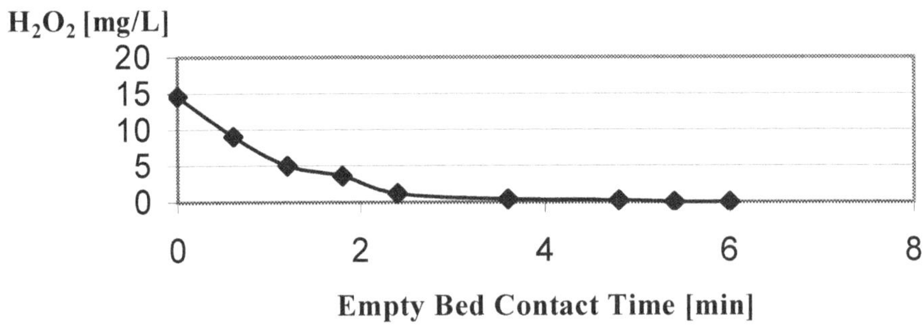

Figure 9 H_2O_2-removal by GAC filtration as a Function of the Empty Bed Contact Time

Removal of 15 g/m^3 H_2O_2 proved to be complete within 5 minutes. After GAC filtration with an empty bed contact time of 30 minutes the biofilm formation rate remained very low (~ 2 pg/cm^2.d) during the total filter run, although AOC increased from 10 to 40 μg Ac C eq/L. Nitrite was completely removed in the GAC step.

Based on these results PWN decided to upgrade water treatment plant Andijk to the following configuration:

microstraining / coagulation / rapid sand filtration / UV-H_2O_2 treatment / two step GAC filtration / microstraining / ClO_2 dosage.

The combination of UV/H_2O_2 treatment / GAC filtration provides a very reliable and flexible process for:

removal / inactivation of micro-organisms;

organic contaminant control.

Three streets of four Trojan Swift 16L30 reactors and H_2O_2 dosage are operational since late 2004.

ACKNOWLEDGEMENT

The authors thank P. Hoogzaad, R. Kooijman (PWN), D.W. Smith, N.F. Neumann, S.Craik (University of Alberta), G. Williams (Trojan Technologies Inc.) and J.R. Bolton (Bolton Enterprises) for their stimulating discussions and their assisting in realizing this study.

REFERENCES

1. Kruithof, J.C., Kamp, P.C.: Proceedings IOA/EA_3G Conference, Moscow p. 331 – 348 (1998)
2. Hoigné, J.: Science of the Total Environment 47, 169 – 185 (1985)
3. Kruithof, J.C., Kamp, Meijers, R.T.: Proceedings 14th Ozone World Congress, Dearborn (1999)
4. Kruithof, J.C., Kamp, P.C.: Proceedings AWWA Annual Conference, Chicago (1999)
5. Sonntag, C. von, Mark, G. Mertens, R., Schuchmann, M.N., Schuchmann, H.P.: J Water SRT – Aqua 43, 201 – 211 (1993)
6. Bourgine, T.P., Chapman, J.I.: Proceedings IOA/EA_3G Conference, Amsterdam (1996)
7. Chang, J.C.H., Ossoff, S.F., Lobe, D.C., Dorfman, M.H., Dumais, C.M., Qualls, R.G.,
8. Johnson, J.D.: Applied and Environmental Microbiology 49, 1361 – 1365 (1985)
9. Ransome, M.E., Whitmore, T.N., Carrington, E.G.: Water Supply 11, 75 – 89 (1993)
10. Rice, E.W., Hoff, J.C.: Applied and Environmental Microbiology 42, 546 – 547 (1981)
11. Bukhari, Z., Hargy, T.M., Bolton, J.R., Dussert, B., Clancy, J.L.: Journal of the American Water Works Association, 91, 86 – 94 (1999)
12. Clancy, J.L., Hargy, T.M., Marshall, M.M., Dyksen, J.E.: Journal of the American Water Works Association 90, (92 – 102 (1998)
13. Craik, S.A., Finch, G.R., Bolton, J.R., Belosevic, M.: Water Research (2000)
14. Mechsner, K., Fleischmann, T., Mason, C.A., Hamer, G.: Water Science and Technology 24, 339 – 342 (1991)

Simulation of Ozone-UV Reactors for Water Treatment

F. Taghipour*, S. Elyasi, and A. Sozzi

Chemical & Biological Engineering Department, University of British Columbia, 2216 Main Mall, Vancouver, B.C., V6T 1Z4, Canada
*E-mail: fariborz@chml.ubc.ca

Abstract A model simulating ozone-UV reactor performance was developed, taking into account parameters such as kinetics, hydrodynamics and fluence rate within the reactor. The fluid flow, calculated by computational fluid dynamics (CFD), was integrated with the radiation and kinetic models to calculate the overall conversion of the target compounds in the reactor. The hydrodynamic model was found to be in good agreement with the particle image velocimetry (PIV) experimental measurements. The results indicated that decomposition of contaminants is influenced by both reactor hydrodynamics and fluence rate.

Keywords UV reactor, advanced oxidation, CFD, ozone

Introduction

Photolysis (direct absorption of UV) and photo-initiated oxidation (activation of strong oxidants such as hydrogen peroxide or ozone by UV to form oxidative hydroxyl radicals) have great potential for the destruction of a wide range of organic contaminants in water. Despite a growing number of studies on the photolysis/photo-initiated oxidation of chemical pollutants, the optimization of these technologies is seldom the topic of reported research and the determination of optimum process and design parameters still remains a challenge. As a result, industrial applications of photo-fluid treatment technologies are limited.

Design optimization of UV systems can best be achieved through the development of sophisticated models that are capable of simulating reactor performance. This is due to the complexity of the UV systems and the interaction of many parameters including hydrodynamics, optics, and kinetics. A limited number of techniques have been employed to simulate UV reactor performance for water treatment. Most of these modelling efforts, however, focused on simulating kinetics, radiation field, or hydrodynamics of the photo-reactors.

Kinetics of ozone-UV systems has been studied by a number of researchers. Staehelin (1982) proposed a chain reaction for the decomposition of ozone in an aqueous solution. This reaction set has been used to develop the mechanistic models of ozone decomposition under various conditions. Peyton et al (1988) modified the proposed mechanism of ozone decomposition by adding the effect of UV on the Ozone/Water reaction that produces hydrogen peroxide and hydroxyl radicals. The proposed mechanism of ozone/hydrogen peroxide decomposition was further developed by including the effect of radical scavengers (Betran et al 1999, Crittenden et al 1999, and Rivas et al 2003). von Gunten (2003) presented the reaction rate constants of ozone and hydroxyl radical with organic and inorganic compounds.

© 2004 IWA Publishing. *2nd IWA Leading-Edge Conference on Water and Wastewater Treatment Technologies* edited by Mark Van Loosdrecht & Jonathan Clement. ISBN: 1 84339 508 8.

Radiation models describing the UV fluence rate in reactors can be developed by solving the radiation transfer equations (RTE). Radiation transfer is defined as the transmission of electro-magnetic radiation through the media. The chemicals and particles in the media can absorb, scatter or re-emit the incident rays. The photon balance can be written for the radiation transfer equation. Romero et al (1997) provided a detailed description of radiation field modelling for an annular photo-catalytic reactor which is adapted from the general transport theory. Cassano et al (1995), Romero et al (1997), and Bolton (2000) presented models for the solution of the RTE. A more accurate solution of the RTE, however, has been proposed by Raithby (1990) and Chui (1993) through the discretization of RTE. The discretization of RTE using the finite volume method presents a general solution but has some limitations, as pointed out by Raithby (1999).

The hydrodynamics of UV water treatment systems have been studied to some extent. Schoenen et al (1993) demonstrated the effect of reactor flow field on the performance of water disinfection systems. Janex et al (1998) used CFD flow modelling to integrate UV-radiation and hydrodynamics to model the performance of a UV-disinfection reactor. Kamimura et al (2002) developed a CFD model for an ozone-UV reactor by combining a fluid dynamics model with a reaction model. Neither the final modelling results nor the CFD flow simulation were, however, verified through conducting relevant experiments. Although CFD has been shown to be successful in simulating single phase flows in complex vessels, turbulent flow models need to be validated through the comparison of the modelling results with flow visualization experiments. Particle Image Velocimetry (PIV) is a non intrusive experimental method, suitable for providing the main characteristics of a given flow configuration with sufficient accuracy for the evaluation of CFD results (Koehlert and Weichert 1998). PIV determines the flow field using a camera to track suspended particles in the flow, which are illuminated by a laser sheet. PIV has been used for the measurement of flow and turbulence intensities in annular reactors (Pruvost et al 2000).

In this work, a computational model simulating ozone-UV reactor performance is presented, taking into account relevant parameters. Hydrodynamics is the main component of this model that is integrated with kinetics and radiation fluence rate. Fluid flow, radiation field, and concentration gradients in an ozone-UV reactor are discussed for atrazine decomposition using the computational model.

Model Development

The model for the simulation of photo-reactors requires consideration of process kinetics, system optics (radiation transfer/absorption), and reactor hydrodynamics.

Kinetics

The kinetics of ozone/UV was integrated from a series of reactions (Elyasi 2004) proposed in the literature (Table 1). The mechanistic kinetic model proposed in this work was found to be in good agreement with the empirical correlation for ozone decomposition suggested by Staehelien (1982), under normal and alkaline conditions in the absence of UV radiation. This model, however, showed some divergence with the empirical correlation under acidic conditions (Elyasi 2004).

The reaction rates for photolysis reactions were derived from Planck's equation for one mole of photons, and quantum yield definition.

$$r_i = \frac{E_a \Theta}{Q_\lambda} \frac{\varepsilon_{\lambda_i} c_i}{\sum \varepsilon_{\lambda_i} c_i} \tag{1}$$

$$Q_\lambda = \frac{h N_p v_c}{\lambda} \qquad (2)$$

where r_i is the rate of photolysis reaction, E_a is the total absorbed energy by a specific point in the media, , Θ is quantum yield, Q_λ is the energy of one mole of photon, $\varepsilon_{\lambda i}$ is molar absorption coefficient of ith component at λ wavelength, c_i is molar concentration of ith component, h is Planck constant (6.62E−34 Js), N_p is Avogadro's constant (6.02E+23), $v_c$is speed of light in vacuum (3.00E+8 ms^{-1}), and λ is the wavelength of UV radiation.

The total absorbed energy can be derived using an infinitesimal sphere of media based on the concentration and total molar absorption coefficient of the chemical components in the media as shown by Elyasi (2004):

$$E_a = \frac{3}{2 Log_{10}(e)} G \sum \varepsilon_{\lambda_i} c_i \qquad (3)$$

where G is UV fluent rate at a specific point in the media.

The reactions of photolysis and oxidation of selected organic compounds were included to the proposed kinetic model in order to study their decomposition in an ozone-UV reactor (see Table 2 for atrazine reactions as an example).

Table 1 The primary reactions in the O_3/UV Advanced Oxidation Process (AOP) system

No.	Reaction	Rate Constant	Reference
	Initiation stage:		
1	$O_3 + OH^- \rightarrow {}^{\cdot}O_2^- + HO_2^{\cdot}$	$k_1 = 70$ $M^{-1} s^{-1}$	Staehelin et al (1982)
2	$O_3 + HO_2^- \rightarrow {}^{\cdot}O_2- + HO^{\cdot} + O_2$	$k_2 = 2.8 \times 10^6$ $M^{-1} s^{-1}$	Staehelin et al (1982)
	Photolysis of the ozone and hydrogen peroxide		
3	$O_3 + H_2O + h\nu \rightarrow 2HO^{\cdot} + O_2$	$\Theta_3 = 0.64$, $\varepsilon_{O3} = 3300$ $M^{-1}cm^{-1}$	Beltran et al (1999)
4	$H_2O_2 + h\nu \rightarrow 2HO^{\cdot}$	$\Theta_4 = 0.5$, $\varepsilon_{H2O2} = 19$ $M^{-1}cm^{-1}$	Beltran et al (1999)
5	$HO_2^- + h\nu \rightarrow HO^{\cdot} + \frac{1}{2} O_2$	$\Theta_5 = 0.5$, $\varepsilon_{HO2} = 228$ $M^{-1}cm^{-1}$	Crittenden et al (1999)
	Propagation stage:		
6	$O_3 + {}^{\cdot}O_2^- \rightarrow {}^{\cdot}O_3^- + O_2$	$k_6 = 1.6 \times 10^9$ $M^{-1} s^{-1}$	Beltran et al (1999)
7	$HO_3^{\cdot} \rightarrow HO^{\cdot} + O_2$	$k_7 = 1.4 \times 10^5$ s^{-1}	Beltran et al (1999)
8	$O_3 + HO^{\cdot} \rightarrow HO_4^{\cdot}$	$k_8 = 2.0 \times 10^9$ $M^{-1} s^{-1}$	Rivas et al (2003)
9	$HO_4^{\cdot} \rightarrow HO_2^{\cdot} + O_2$	$k_9 = 2.8 \times 10^4$ s^{-1}	Beltran et al (1999)
10	$H_2O_2 + HO^{\cdot} \rightarrow HO_2^{\cdot} + H_2O$	$k_{10} = 2.7 \times 10^7$ $M^{-1} s^{-1}$	Beltran et al (1999)
11	$HO_2^- + HO^{\cdot} \rightarrow HO_2^{\cdot} + OH^-$	$k_{11} = 7.5 \times 10^9$ $M^{-1} s^{-1}$	Beltran et al (1999)
12	$H^+ + {}^{\cdot}O_3^- \rightarrow HO_3^{\cdot}$	$k_{12} = 5 \times 10^{10}$ $M^{-1} s^{-1}$	von Gunten (2003)
13	$HO_3^{\cdot} \rightarrow H^+ + {}^{\cdot}O_3^-$	$k_{13} = 3.3 \times 10^2$ s^{-1}	von Gunten (2003)
14	${}^{\cdot}O_3^- \rightarrow O_2 + {}^{\cdot}O^-$	$k_{14} = 2.1 \times 10^3$ s^{-1}	von Gunten (2003)
15	$O_2 + {}^{\cdot}O^- \rightarrow {}^{\cdot}O_3^-$	$k_{15} = 3.3 \times 10^9 M^{-1} s^{-1}$	von Gunten (2003)
16	$H_2O + {}^{\cdot}O^- \rightarrow HO^- + HO^{\cdot}$	$k_{16} = 1 \times 10^8$ $M^{-1} s^{-1}$	von Gunten (2003)
17	$HO_4^{\cdot} \rightarrow HO^{\cdot} + O_3$	$k_{17} = 1 \times 10^2$ $M^{-1} s^{-1}$	Rivas et al (2003)
18	$HO_2^{\cdot} \rightarrow H^+ + {}^{\cdot}O_2^-$	$k_{18} = 3.2 \times 10^5 s^{-1}$	Rivas et al (2003)
19	$H^+ + {}^{\cdot}O_2^- \rightarrow HO_2^{\cdot}$	$k_{19} = 2 \times 10^{10}$ $M^{-1} s^{-1}$	Rivas et al (2003)
	Termination stage:		
20	$HO_4^{\cdot} + HO_2^{\cdot} \rightarrow O_2 + O_3 + H_2O$	$k_{20} = 10^{10}$ $M^{-1} s^{-1}$	Beltran et al (1999)
21	$HO^{\cdot} + HO^{\cdot} \rightarrow H_2O_2$	$k_{21} = 5 \times 10^9$ $M^{-1} s^{-1}$	Beltran et al (1999)
22	$HO^{\cdot} + {}^{\cdot}O_2^- \rightarrow OH^- + O_2$	$k_{22} = 10^{10}$ $M^{-1} s^{-1}$	Beltran et al (1999)

23	$HO^{\cdot} + HO_3^{\cdot} \longrightarrow H_2O_2 + O_2$	$k_{23} = 5 \times 10^9\ M^{-1}\ s^{-1}$	Beltran et al (1999)
24	$HO_3^{\cdot} + {}^{\cdot}O_2^{-} \longrightarrow OH^{-} + 2O_2$	$k_{24} = 10^{10}\ M^{-1}\ s^{-1}$	Beltran et al (1999)
25	$HO_3^{\cdot} + HO_3^{\cdot} \longrightarrow H_2O_2 + 2O_2$	$k_{25} = 5 \times 10^9\ M^{-1}\ s^{-1}$	Rivas et al (2003)
26	$HO_4^{\cdot} + HO_4^{\cdot} \longrightarrow H_2O_2 + 2O_3$	$k_{26} = 5 \times 10^9\ M^{-1}\ s^{-1}$	Beltran et al (1999)
27	$HO_4^{\cdot} + HO_3^{\cdot} \longrightarrow H_2O_2 + O_3 + O_2$	$k_{27} = 5 \times 10^9\ M^{-1}\ s^{-1}$	Beltran et al (1999)
28	$HO_4^{\cdot} + HO^{\cdot} \longrightarrow H_2O_2 + O_3$	$k_{28} = 5 \times 10^9\ M^{-1}\ s^{-1}$	Beltran et al (1999)
29	$HO_4^{\cdot} + {}^{\cdot}O_2^{-} \longrightarrow OH^{-} + O_2 + O_3$	$k_{29} = 10^{10}\ M^{-1}\ s^{-1}$	Beltran et al (1999)
30	$HO_2^{\cdot} + O_2^{\cdot -} \longrightarrow HO_2^{-} + O_2$	$k_{30} = 9.7 \times 10^7\ M^{-1}\ s^{-1}$	Peyton et al (1988)
	Equilibrium and dissociation in water:		
31	$H_2O_2 \rightleftarrows H^{+} + HO_2^{-}$	$pk_{31} = 11.7$	Beltran et al (1999)
32	$H_2O \rightleftarrows H^{+} + HO^{-}$	$pk_{32} = 14$	

Radiation

The fluence rate distribution inside photo-reactors was calculated by solving the radiation transport equations (RTE). A discrete ordinate (DO) model, based on a spherical discretization of space and numerical solution of the radiation transfer equation in this space, was used for the solution of the radiation transport equations (Fluent, 2003). Algorithms that can take into account refraction, reflection, and adsorption can be used with DO radiation model.

Hydrodynamics

The fluid flow within UV reactors was characterized through the treatment of mass and momentum conservation equations along with various turbulence models. With an impinging round jet at the inlet of typical UV reactors and a relatively high Reynolds number, the choice of the appropriate turbulence model could significantly impact the accuracy of the results. Different turbulence models including, standard k-ε, realizable k-ε, and Reynolds stress transport were applied to determine the model that best captures the turbulent characteristics at reasonable CPU time for UV reactors.

The fluid flow patterns and turbulence within the UV reactors was characterized experimentally using PIV flow visualization techniques. A dual-head frequency doubled Nd:YAG laser was used to illuminate seeding particles with a mean diameter of 20μm and a specific weight of 1.02. The digital CCD camera (1280×1024 pixel) was operated at 4Hz to capture the instantaneous flow fields at the reactor inlet and at several points along the reactor. 500 pairs of images were taken for each field of view and the adaptive correlation method with 32×32 pixel interrogation windows was employed for the PIV velocity determination. A Plexiglas annular reactor was used for the PIV experiments. The reactor was placed into a rectangular Plexiglas box filled with water to reduce the optical distortion caused by the refractive index of the cylinder material. The time averaged data was used to assess the accuracy of the CFD simulations by comparing the overall flow patterns, as well as the radial and axial liquid velocity fields. The details of the experimental and computational models as well as the results obtained are discussed in the following sections.

Reactor performance calculation

Taking into account the aforementioned parameters, a model for simulating UV reactor performance was developed using Fluent 6.18 computational fluid dynamics (CFD) software. The kinetic model was included by writing the related subprograms and making use of the user-defined functionality of commercial CFD codes. The kinetic subprograms incorporate the mechanistic reaction model and solve the model based on a balance in steady state concentrations of radicals in each cell. For the direct photolysis and photo-initiated oxidation processes, the species transport (from the hydrodynamic model), species generation/consumption (from the kinetic model), and the fluence rate (from the radiation model) was considered. The overall conversion (degradation) of the target compounds in the reactors was calculated by the numerical solutions of the aforementioned governing equations within the reactors.

Results and Discussions

Evaluation of hydrodynamic model

The hydrodynamics component of the model was evaluated through comparison of the CFD model predictions with PIV experimental measurements for a model UV reactor. The Plexiglas annular reactor of 0.09 m in diameter and 1m in length with a concentric Quartz tube of 0.02 m in diameter was used for PIV experiments. Both inlet and outlet had an ID of 0.019 m and were placed centred on the top of the reactor 0.016 m from each respective end. The water flow rate was set at 6.9×10^{-4} m^3/s (11GPM), which corresponds to a mean axial velocity in the annulus of 0.11 m/s and a Reynolds number of 8000. PIV measurements were performed at several locations perpendicular and parallel to the main body of the reactor but concentrated around the inlet jet, since this is the zone of the highest interest.

The flow field inside the abovementioned reactor was simulated using Fluent CFD software. The fluid flow was shown to be best represented by the Realizable κ-ε turbulence model. This is in accordance with the fact that this model is best suited to describe round jets as seen at the inlet and impinging flows as present in the flow separation on the UV-lamp.

Figure 1 shows the position of the laser sheet for the PIV measurements (presented in Figure 2); the two lines indicate the position of the extracted Z velocity component data (presented in Figure 3). A qualitative comparison of the measured and simulated velocity vectors showed a good agreement for all major flow structures (Figure 2). Comparison of the z-velocity vectors on 2 vertical axes along the reactor also showed a reasonable agreement between PIV experimental data and CFD simulated results (Figure 3).

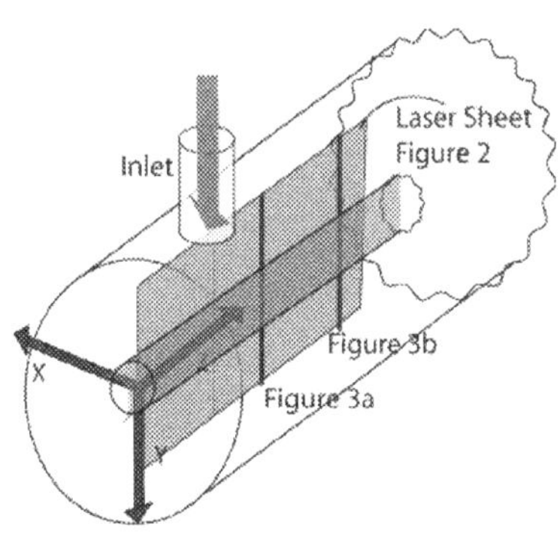

Figure 1 Position of laser sheet for PIV measurements in UV reactor.

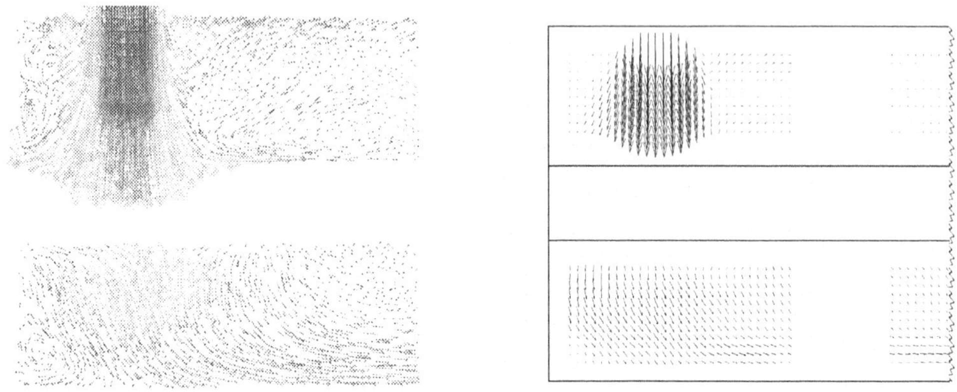

A B

Figure 2 Velocity Magnitude vectors from CFD simulation (A) and PIV measurements (B). The solid lines in Figure B indicate the reactor and lamp boundaries; areas where no velocity vector is shown was inaccessible for PIV measurements.

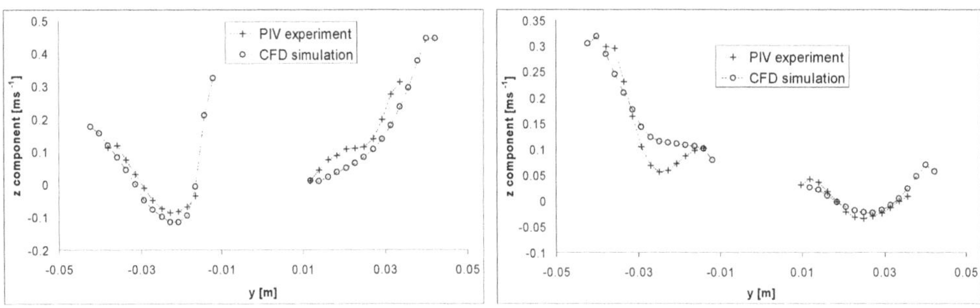

A B

Figure 3 Z-Velocity components at 6 cm (A) and 12 cm (B) from the reactor inlet end.

Modelling UV advanced oxidation reactor

An ozone-UV reactor was simulated using the aforementioned model of integrating reactor kinetics, fluence rate, and hydrodynamics. Atrazine, one of the contaminants in drinking water was selected as a model micro-pollutant. The reaction rates of photolysis and oxidation of atrazine are provided in Table 2.

Table 2 The primary reactions of atrazine photolysis and oxidation

No.	Reaction	Rate Constant	Reference
1	Atrazine + $h\nu \longrightarrow$ Products	$\Theta_{Atr} = 0.1$, $\varepsilon_{Atr} = 3560\ M^{-1}cm^{-1}$	Arantegui (1995)
2	Atrazine + $HO^{\cdot} \rightarrow$ Products + H_2O	$k = 3 \times 10^9\ M^{-1}s^{-1}$	von Gunten (2003)
3	Atrazine + $O_3 \longrightarrow$ Products + H_2O	$k = 6\ M^{-1}s^{-1}$	von Gunten (2003)

A two dimensional reactor was used to simulate the atrazine degradation. The configurations and dimensions of the reactor are provided in Figure 4. The Discrete Ordinate radiation model was used with UV irradiation of 2000 W/m^2 from the lamp surface. The fluid flow rate was 10^{-3} m^3/s. The mass concentration of dissolved ozone and atrazine in reactor influent were 2.7 E–4 and 1.2 E–4, respectively; the pH of the influent was set at 10.

The velocity vectors at the inlet of the reactor are shown in Figure 5. After hitting the UV lamp, the flow direction changes and swirl flow is formed above the lamp. These swirls help to improve the mixing conditions inside the reactor, which will affect the profiles of concentrations (e.g. ozone, Figure 5) in the reactor, and ultimately the reactor performance.

Distribution of UV fluence rate shows that the incident radiation declines as distance from the lamp increases (Figure 6). The fluence rate distribution is not quite uniform alongside of the lamp due to a non-uniform concentration gradient of radiation absorbance species (e.g. ozone) in the reactor.

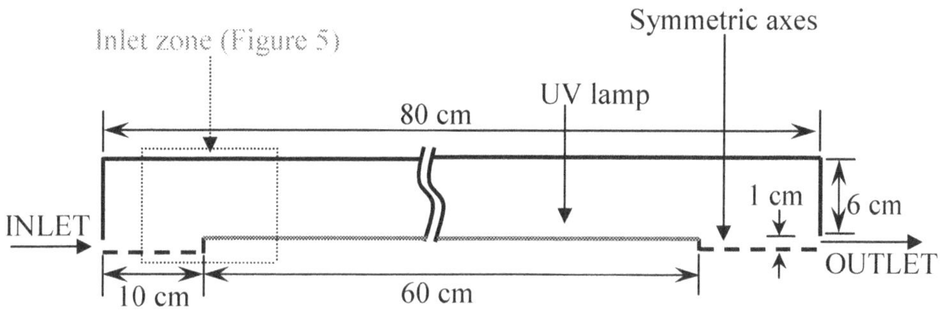

Figure 4 Configurations and dimensions of the two dimensional UV reactor.

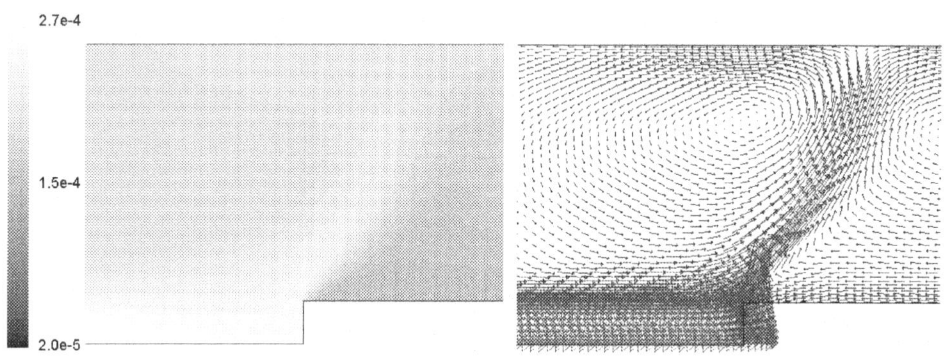

Figure 5 Velocity vectors (right) and ozone mass fraction (left) at the inlet zone of the UV reactor (see Figure 4).

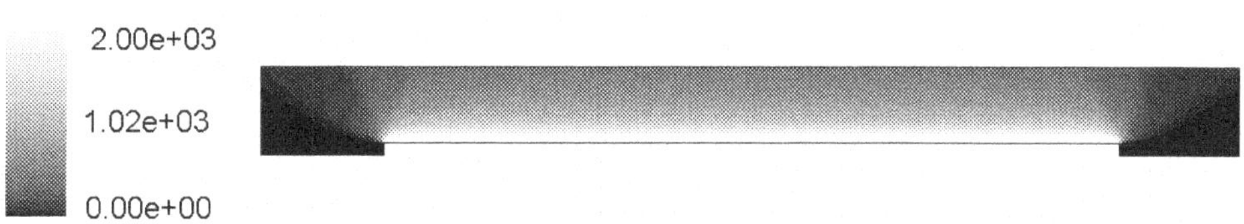

Figure 6 Incident radiation (W/m^2) in the UV reactor.

Mass fractions of hydroxyl radical (·OH), ozone, and atrazine are provided in Figure 7. Ozone concentration decreases alongside of the reactor. This is due to the ozone decomposition by UV radiation. The concentration of ·OH is higher near the UV lamp, where the UV fluence rate is higher. This is expected as ·OH formation depends on the fluence rate. Atrazine concentration reduces along the reactor due to direct photolysis and oxidation with ·OH. The radial atrazine concentration gradient shows lower concentration near the UV lamp, where higher UV fluence rate and ·OH present. Relatively high atrazine and ozone concentration gradients at the inlet zone of the reactor is the result of back mixing and sharp changes in flow direction. It is evident that atrazine decomposition is influenced by both reactor hydrodynamics and fluence rate.

Figure 7 Mass fractions of hydroxyl radical (A), Ozone (B), and atrazine (C) in the UV reactor.

Conclusions

Simulation of the ozone-UV reactors through integration of kinetics, hydrodynamics and fluence rate can provide insight into detailed concentration gradient of the species in the reactor. The results indicate that the decomposition of organic contaminants depends both on UV fluence rate and reactor hydrodynamics. Once validated, these models can be used for parametric studies to investigate the effect of various related factors such as concentrations of chemical reactants, UV intensity, and flow rate on reactor performance. The information provided by modeling can also be used for reactor design optimization and improving the reactor performance.

References

Arantegui, J., Prado, J., Chamarro, E., and Esplugas, S. (1995) Kinetics of the UV degradation of atrazine in aqueous solution in the presence of hydrogen peroxide. *J. Photochemistry and Photobiology A: Chemistry*, **88** (1), 65–74.

Beltran, F.J., Rivas, J, Alvarez, P.M., Alonso, M.A. and Acedo, B. (1999) A kinetic model for advanced oxidation processes of aromatic hydrocarbons in water: Application to phenanthrene and Nitrobenzene. *Ind. Eng. Chem. Res.* **38** (11), 4189–4199.

Bolton, J. R. (2000) Calculation of ultraviolet fluence rate distributions in an annular reactor: significance of refraction and reflection. *Wat. Res.* **34** (13), 3315–3324.

Cassano, A.E., Martin, C.A., Brandi, R.J. and Alfano, O.M. (1995) Photoreactor analysis and design: Fundamentals and application. *Ind. Eng. Chem. Res.* **34** (7), 2155–2201.

Chui, E.H. and Raithby, G.D. (1993) Computation of radiant heat transfer on a nonorthogonal mesh using the finite-volume method. *Num. Heat Transfer B* **23**, 269–288.

Crittenden, J.C., Hu, S., David W. Hand, D.W., and Green, S.A. (1999) A kinetic model for H_2O_2/UV process in a completely mixed batch reactor. Wat. Res. **33** (10), 2315–2328.

Elyasi, S. (2004) CFD modelling of UV photo-reactor for water treatment and water microbial disinfection. M.Sc. thesis, Dept Chem. Eng., Univ. of Chalmers/ Univ. of British Columbia, SE-412 96 Göteborg, Sweden/ V6T 1Z4 Vancouver, Canada.

Fluent User's Manual (2003), Fluent Inc., Lebanon, NH.

Janex, M.L. (1998) Impact of water quality and reactor hydrodynamics on wastewater disinfection by UV, use of CFD modelling for performance optimization. *Wat. Sci. Tech.* **38** (06), 71–78.

Kamimura, M., Furukawa, S., and Hirotsuji, J. (2002) Development of a simulator for ozone-UV reactors based on CFD analysis, *Water Sci.e Technol.*, **36** (11-12), 13-19

Koehlert, U., Weichert, R (1998) Three dimensional particle image velocimetry in turbulent flows *7th Eur Symp on Particle Characterization* 255–263.

Peyton, G.A. and Glaze, W.H. (1988) Destruction of pollutants in eater with ozone in combination with ultraviolet radiation. 3. Photolysis of aqueous ozone. *Environ. Scl. Technol.* **22** (7), 761–767

Romer, R.L., Alfano, O.M. and Cassano, A.E. (1997) Cylindrical photocatalytic reactors. Radiation absorption and scattering effects produced by suspended fine particles in an annular space. *Ind. Eng. Chem. Res.* **36** (8), 3094–3109.

Pruvost, J., Legrand, J., and Legentilhomme, P. (2000) Particle Image Velocimetry investigation of the flow field of a 3D turbulent annular swirling decaying flow induced by means of a tangential inlet. *Experiments in Fluids* **29**, 291–301.

Raithby, G. D. and Chui, E. H. (1990) A finite-volume method for predicting a radiant heat transfer in enclosures with participating media. *J. Heat Transfer.* **112**, 415–423

Raithby, G.D., (1999) Evaluation of discretization errors in finite-volume radiant heat transfer predication. *Num. Heat Transfer* **36** (4), 241–264.

Rivas, F.J., Beltran, F.J., Carbajo, M., and Gimeno, O. (2003) Homogeneous catalyzed ozone decomposition in the presence of Co(II). *Ozone Sci. Eng.* **25** (4), 261–271

Staehelin, J. and Hoigné, J. (1982) Decomposition of ozone in water: Rate of initiation by hydroxide ions and hydrogen peroxide. *Environ. Scl. Technol.* **16** (10), 676–681.

Schoenen, D.; Kolch, A.; and Gebel, J. (1993) Influence of geometrical parameters in different irradiation vessels on UV disinfection rate *Int. J. of hygiene and env. Med.* **194** (3), 313–20.

von Gunten, U. (2003) Ozonation of drinking water: Part I. Oxidation kinetics and product formation. *Wat. Res.* **37** (7), 1443–1467.

Nitrite Formation, a useful Tool in Medium Pressure UV Fluence Control

A.J. van der Veer[1], G.F. IJpelaar[2], G.J. Medema[3], J.C. Kruithof[4]

[1]Evides, P.O. Box 59999, 3008 RA Rotterdam, the Netherlands
[2,3]Kiwa N.V. Water Research, P.O. Box 1072, 3430 BB Nieuwegein, the Netherlands
[4] N.V. PWN Water Supply Company North Holland, P.O. Box 2113, 1990 AC Velserbroek, the Netherlands

Abstract Water Company Evides (formerly Water Company Europoort) started testing UV disinfection with polychromatic medium pressure mercury lamps in 2000. In 2003 a full scale UV test facility, consisting of two serial Trojan Swift™ 10L30 reactors, capacity 2500 – 3000 m3/h at 70mJ/cm2 (UVT_{254}: 87%) became operational for fluence testing and validation with challenge tests. Because of experienced problems with disinfection by-products from chlorine and ozone a lot of attention was paid to potential by-product formation from the application of UV. At reduction equivalent UV fluences (REF's) of 70 – 90 mJ/cm^2 and nitrate levels of 7-16 mg/l nitrite appeared to be formed in small quantities, below the Dutch standard of 0.1 mg NO_2^-/L. In case flow and lamp power were changed to adjust the fluence, the amount of nitrite formed in the continuous flow UV reactor(s) showed a linear relation with the fluence. Since additional methods of fluence monitoring in UV-reactors are welcome, nitrite formation was tested for this purpose. UVT-scans demonstrated that UVT's in the wavelength range below 240 nm, where the absorption by nitrate is the strongest, are much lower compared to UVT's at wavelengths, which are responsible for disinfection. Because of the steeper dropping intensity with increasing distance from the lamps, formation of nitrite takes place in a smaller volume around the lamps compared with disinfection. Inactivation results, and for this reason also the reduction equivalent fluence, are much more influenced by changes in UVT_{254} than nitrite formation. The effects of the changed UVT on the reduction equivalent fluence itself primarily explain the changes in the relation nitrite formation versus REF caused by different UVT's. Switching off lamps in a 10 lamps UV reactor resulted in a reduction of nitrite formation, proportional to the percentage of lamps switched off, whereas the levels of inactivation deteriorated much more. Nitrite formation can be utilised as a fluence monitor system for disinfection in a well-designed and validated UV reactor, but is not likely to be suitable for assessing reactor designs because of its underestimation of the effects of (hydraulic) shortcuts and lamp spacing.

Keywords: by-product formation, nitrite formation, UV fluence control, full-scale testing

Introduction

Water Company Evides (formerly Water Company Europoort) provides drinking and industrial water to households and industries in the city of Rotterdam and adjacent areas. Drinking water is produced by three Water Treatment Plants, of which Berenplaat is the largest with an annual capacity of 100 Mm^3. From the start in 1967, Water Treatment Plant Berenplaat utilises chlorine (hypochlorite) for the primary and post disinfection. In recent

© 2004 IWA Publishing. *2nd IWA Leading-Edge Conference on Water and Wastewater Treatment Technologies* edited by Mark Van Loosdrecht & Jonathan Clement. ISBN: 1 84339 508 8.

years the Berenplaat process results in a TTHM formation of 30 – 50 μg/L and the treated water is subjected to customer taste and odour complaints. In 2006 Dutch Regulations will restrict TTHM's to 25 μg/l on a 90- percentile base. Within the current disinfection regime, the Berenplaat process can not meet this standard, apart from the taste and odour problems. So Evides decided to change the disinfection process of Water Treatment Plant Berenplaat, resulting in a study of alternative disinfection processes that runs for several years already. Since 1991 pilot plant research was started, using ozone as an alternative primary disinfectant. After recognising bromate as a suspected genotoxic compound a Dutch standard for bromate in drinking water was pending and became operational in 2000. Although in compliance with the bromate standard, Evides was not satisfied with the ozone disinfection process because of the tight margins between by-product formation and disinfection requirements. Because of its disinfection potential (Clancy et al, 2000), Evides started investigating UV at pilot-plant scale in 2000. In 2002 after gaining enough confidence in UV disinfection, a full-scale test facility was built with a hydraulic capacity of 3000 m^3/h. Due to the experiences with disinfection by-products formation by chlorination and ozonation attention was paid to by-product formation during UV treatment. This paper focuses on nitrite formation. It is well known that bromate formation does not occur during UV treatment (IJpelaar et al., 2001) and by replacing chlorine THM formation is avoided as well. Nitrite was formed at levels below the Dutch standard of 0.1 mg NO_2^-/L even at UV fluences from 70 to 90 mJ/cm^2. Orientating experiments nitrite showed nitrite formation was proportional to the UV fluence. Since the precursor (nitrate) is already present in the raw water, the idea was born to test nitrite formation as a fluence monitoring system in a WTP like Berenplaat. Mark and Von Sonntag (1996), founding nitrite formation dependent of pH, nitrate content, oxidizable organic compounds and radiation intensity, have performed fundamental research on the photochemistry of nitrate ion at 254 nm.

Nitrite formation as a fluence-monitoring tool

UV fluence can be approached by modelling flow and intensity distributions, e.g. with Computational Flow Dynamics, CFD. The water flow, UVT, lamp power and reactor design are the main input parameters for these calculations. For the moment, the general accepted way of verification and validation of fluence in flow-through UV reactors consists of performing biodosimetry challenging bacterial spores or bacteriophages with a calibrated UV-inactivation potential. Since spores or bacteriophages can not be dosed to a WTP this method is limited to test installations. Alternative fluence monitoring methods, applicable during drinking water production are welcome.

The research into the suitability of nitrite formation as fluence meter is carried out in close co-operation with Kiwa Water Research. Kiwa studies the fundamental aspects of this research, which a.o. consists of the impact of water quality matrix effects.

Evides examines the practical application, relating nitrite formation to (undesired) changes in process parameters, like lamp failures. Evides's UV test work started in 2000 with two Berson 1000InLine™ (cross flow) reactors in series, which were integrated in a pilot plant process in conformity with the future Berenplaat process. The UV feed water consisted of coagulated and filtered water. Nitrite formation and the inactivation of SSRC were measured as a function of the UV fluence at flows from 100 – 180 m^3/h and lamp power settings (60, 80, 100%). The calculated fluences were verified by challenge tests with Bacillus subtilis as well as MS2-fages. The challenge tests were performed in close co-operation with Kiwa Water Research (Hijnen et al). Tests were continued in 2003 The water quality of the UV feed water was rather constant (see table 1).

The first tests using the Berson UV-reactors were performed at rather constant UVT_{254} (10mm) levels, see table 1. Nitrite was formed proportional to the UV fluence. The inactivation of SSRC showed a linear relation with the nitrite formation, in other words: disinfection could be related to the photochemical reduction of nitrate to nitrite. A relevant question was born: Can nitrite formation be used as a tool for fluence control in the future Berenplaat UV disinfection process? This might be interesting since:

- Nitrite formation can be measured accurately at very low levels below the EC/Dutch standard of 0.1 μg/L
- The alternative, performing biodosimetry tests, is complicated and expensive and can not be used in a drinking water production plant
- The levels of nitrate in the raw water enable the detection of nitrite throughout the year, which avoids spiking nitrate salts.

From a disinfection point of view, Evides had gained enough confidence in UV for primary disinfection. That is why a larger experiment set-up has been scheduled for the full scale cross flow Trojan Swift™10L30-reactor test street, which includes challenge tests for fluence control, reactor validation and options of using nitrite formation as a fluence monitoring tool.

Table 1: Characteristics and water quality of the pilot plant (Berson), the full-scale units and future plant (both Trojan) using cross flow polychromatic medium-pressure mercury lamps.

	Biesbosch water	Reservoir water	Pilot plant "after filtration	Future process "after filtration"
UV-installation	Trojan SWIFT™ 10L30	Trojan SWIFT™ 10L30	Berson 1000 InLine™	Trojan SWIFT™ 10L30
Unit scale	1:1	1:1	1:12.5	1:1
No. reactors/street	2	2	2	2
No. lamps/reactor	10	10	4	10
Water quality	Biesbosch water	Reservoir water	Pilot plant "after filtration	Future process "after filtration"
UVT_{254}	85 – 87%	81 – 83%	86-88%	87 – 90 %
DOC (mg/L)	3.0 -3.5	2.4 – 2.8	2.4 – 2.8	2.4 – 2.8
pH	8.6 – 9.1	8.2 – 8.4	7.6- 8.0	7.8 – 8.0
Turb.(FTU)	1.0 – 2.5	1.0 – 2.5	0.08- 0.1	0.08 0.1
nitrate	13 - 16	7 – 10	7–10 & 13–16	13 - 16
HCO_3^- (mg/L)	100-120	100-120	110 - 130	110 - 130

Experimental set-up

The Trojan Swift™ 10L30 reactor was selected for the full-scale test facility and the final plant. A scheme of the full-scale test facilities is presented in fig. 1. The system consists of two serial reactors, with 10 lamps each. Only the first one is used for the experiments in this article. The flow rate through the UV reactors was continuously monitored, using an Altoflux magnetic flow meter. The required flow range for the test facility, up to 3000m3/h water, free of residual chlorine, could only be realised by using raw water. Differences in water quality parameters between the microstrained raw water and the future filtered water are found in turbidity, UVT, DOC, nitrate, bicarbonate and pH. The Berenplaat facility has two available water sources: the regular water source, Biesbosch water and local reservoir water as a backup water source. The latter often shows a UVT_{254} of 87% (1 cm) which is equal to the worst-case future filtered water UVT. The following water qualities are characteristic for the test and future plant facilities, see table 1.

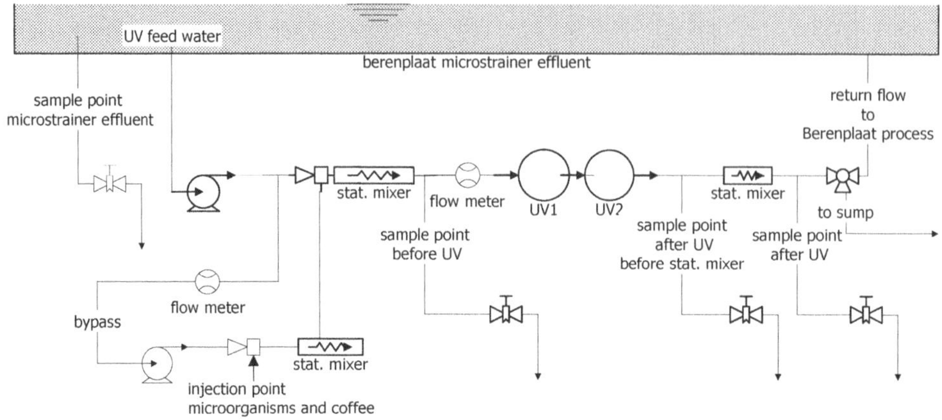

Figure 1 scheme of the full-scale test facilities, 2x Trojan Swift™ 10L30 in series

The full test street has been fitted with static mixers before and after the UV reactors, respectively in order to homogenise the dosed organisms in the main stream and to be able to take a representative sample of the water after passing the UV reactor. During the validation tests of the Trojan Swift™ 10L30 the static mixers were verified for their mixing capacities. These tests were performed with coffee (Nescafe Gold instant coffee, in Experimental) as a dye tracer and with micro-organisms. The results of the samples of organisms and UVT measurements show complete mixing over both static mixers. MS2-phages (from GAP, EnviroMicrobial Services) were used as a challenge organism in the bioassay tests. Each test day a batch was prepared by dispersing and diluting high-concentration stock solutions of MS2-phages into a stirred stainless steel container.

From this container the challenge organisms were injected into a bypass stream mixed, and again injected just before a static mixer into the main stream. To determine the reduction equivalent fluences and nitrite formation were taken simultaneously samples of the MS2-phages were taken before and after the UV-reactor. The MS2-phages were analysed by GAP EnviroMicrobial Services, Canada; the other water quality parameters by Aqualab, Netherlands. Nitrite was analysed using the NEN-ISO 6777-method. This method is based on measuring the absorbency (10 mm) of the formed azo dye at 540 nm, after adding sulphanilamide and N-(1-naphtyl) ethyleendiamine dihydrochloride to the acidified sample.

Two relevant objectives with respect to nitrite formation as a fluence-monitoring tool are:

1. The impact of water quality differences of both raw waters on nitrite formation Both water qualities differ in UVT and nitrate level, see table 1. In which way is the amount of nitrite formed influenced by differences in water quality. Fluences will be varied in this test by changing lamp power and flow through the reactor. By dosing coffee to the higher UVT reservoir water, UVT will decrease without influencing the nitrite level.
2. The effect of switching off lamps in a 10-lamps reactor on the fluence and nitrite formation.

   To simulate the effect of lamp failure, lamps in this test will be switched off in this test. As well the effect on fluence as the amount of nitrite are will be determined simultaneously. CFD calculations show a lamp position dependant effect on fluence, caused by corresponding differences in the amount of shortcut. The questions is: Will nitrite formation will be influenced the same way and be indicative for the fluence left.

Results and discussion.

1. The influence of changes in UV feed water nitrate concentrations and UVT on nitrite formation

As mentioned Evides has two water qualities available: the regular Biesbosch water and local reservoir water. The relevant water quality parameters with respect to disinfection and nitrate formation are presented in table 1. To see how the relation REF against nitrate formation is affected by the differences in both water qualities, a test program was run. The amount of nitrite formed at several settings of flow and lamp power was compared for three waters:

- Reservoir water
- Biesbosch water
- Reservoir water spiked with instant coffee to level the UVT254 (10 mm) to Biesbosch water

Since instant coffee can be considered a source of DOC both the UVT_{254} and the DOC levels were adjusted, while bicarbonate, pH and nitrate levels were not affected. Nitrite and MS2-phages (to determine the REF) were sampled simultaneously before and after the UV reactors. The rather small changes of DOC and bicarbonate/pH were assumed to have a minor role in the amount of nitrite formed.

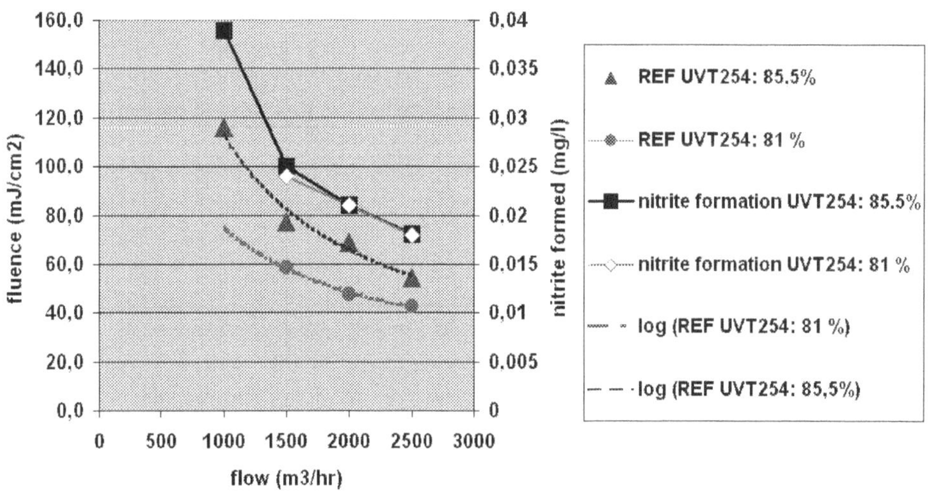

Figure 2 REF's and nitrite formed versus flow at UVT 85.5 and 81%; Trojan Swift™ 10L30 one reactor, 100 % lamp power.

Figure 2 shows the effects of variation of flow on nitrite formation and REF at 100% lamp power. Both REF and the amount of nitrite formed are reciprocal functions of the flow through the reactor. In contrast to the level of nitrite formed, the obtained fluences are influenced by the changes in UVT_{254}. So at the same conditions of flow and lamp power the amount of nitrite formed is hardly influenced by (small) changes in UVT_{254}. This phenomenon was confirmed by the 60 and 80% lamp power tests results.

In figure 3 nitrite formation is plotted against the REF for MS2inctivation and shows a linear relationship. The plotted points are obtained by varying flow from 1000 to 2500 m^3/hr and lamp power from 60 – 100%.

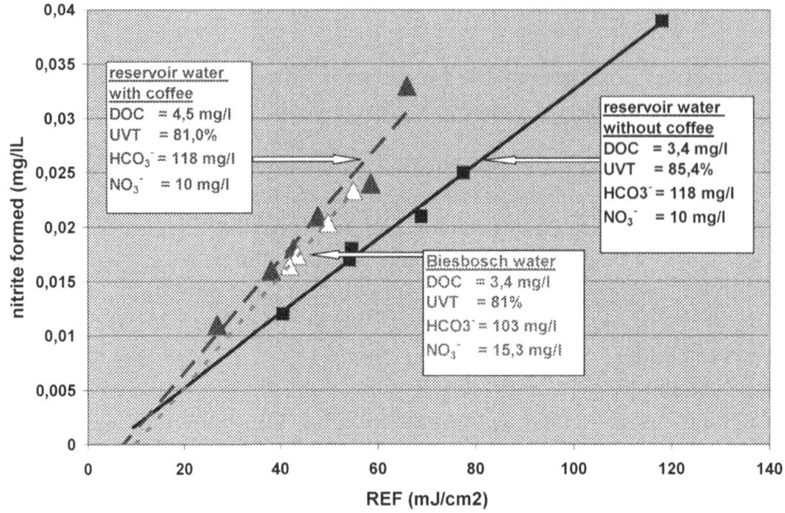

Figure 3 Nitrite formation versus REF at three, partly comparable water qualities; Trojan Swift™ 10L30, flow 1000- 2500 m^3/hr, lamp power 60, 80 and 100%

Nitrite formation as a function of the fluence differs if the UVT254 was changed by dosing coffee to the UVT level of the UV feed water; the nitrate level was kept unchanged. The Biesbosch water however, which has a UVT_{254} comparable to coffee spiked reservoir water, but with a higher nitrate level seemed to form equal amounts of nitrite as coffee spiked reservoir water. The results presented here indicate that nitrite formation at medium pressure lamp UV disinfection conditions is determined by flow and lamp power, whereas the fluence is determined by flow, lamp power and UVT.

Adding UVT-scans (figure 4) to these findings shows what happened. The UVT-scans of the involved waters show changes in UVT over the total spectral range, for 200 to 300nm. Nitrate, the precursor for nitrite formation, and DNA, linked with disinfection, absorb UV radiation effectively at different wave lengths, 200 – 230 nm and 240-300 nm respectively. Coffee addition only decreases UVT at 254 nm (disinfection) range, but not in the 200-230 nm (nitrate) range. Because of the high absorption coefficient of nitrate in the 200 – 230 range, the effect on $UVT_{200\text{-}230}$, as found in the scan, might well be caused by the changes in nitrate concentration. Reservoir water with a lower nitrate level showed not only a higher UVT in the 254 nm range but also in the 200 – 230 nm range. The combination of higher nitrate - lower $UVT_{200\text{-}230}$ or lower nitrate - higher $UVT_{200\text{-}230}$ seems to result in equal amounts of nitrite formed. This is in contrast with changes in UVT_{254} which strongly affects the amount of nitrite formed at a certain fluence. This is hardly caused by diversity in the formation of nitrite but by the dominant way fluence is affected by changes in UVT_{254}.

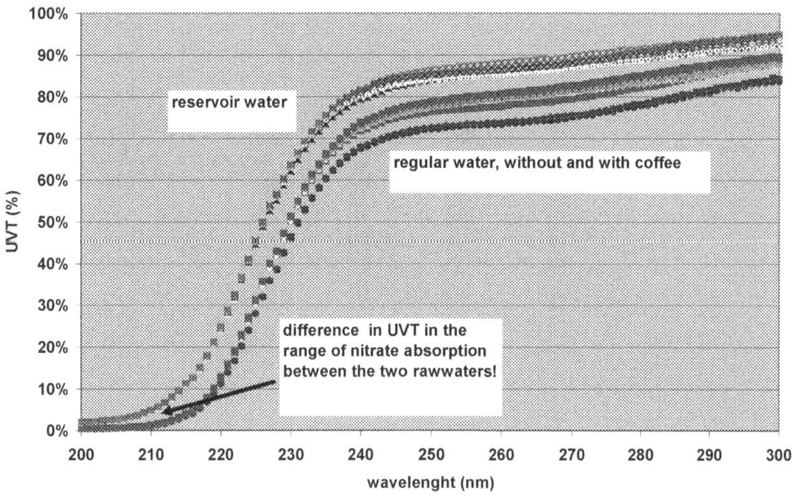

Figure 4 Representative spectral scans of Reservoir water, Biesbosch water, Biesbosch water with added coffee.

2. The effect of switching off lamps in a 10-lamps reactor on nitrite formation and fluence

Challenge tests with a Trojan Swift™ 10L30 reactor were performed on Biesbosch water. During the test changes in lamp configuration were made by switching off lamps.

Flow was maintained at 2000 m^3/h; lamp power settings were either 80 or 100% of the total energy of 13,6 kW/lamp. The results are presented in figure 5.

Resulting REF's in lamp-off situations depend on which lamp(s) is (are) switched off, since the amount of under-irradiated water depends on the position of the non burning lamps. The plot presents the average reduction of both nitrite and REF for 80 and 100% lamp power and combinations of one and two lamps off at different positions. The level of nitrite formation and REF with all 10 lamps switched on is assigned as '100%'. The reduction in nitrite formation was proportional to the number of lamps switched off whereas the reduction in the REF's was higher. This implies that a fluence calculated from a linear relationship with nitrite formation is not the same as the REF. This can be explained as follows: Shortcuts in a well-designed medium-pressure polychromatic multilamp UV reactor is limited as a result of an optimal reactor design and lamp geometry.

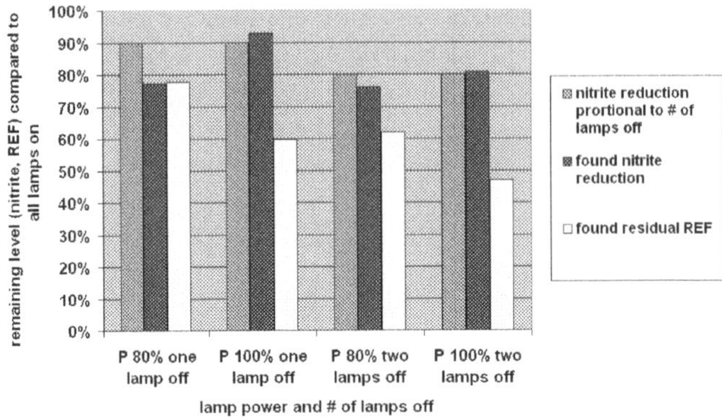

Figure 5 Average nitrite reduction and REF during lamp switch off situations

In figure 6 the decreasing intensity of the UV radiation with increasing distance from the lamp sleeve is plotted, based on the different relevant UVT's for nitrite formation and disinfection. Since the penetration depth of 200 – 230 nm radiation is low compared to the 240 – 300 nm radiation for disinfection, nitrite formation seems to be located within a cylinder surrounding the lamp having a limited radius. So the range of effective UV radiation for nitrite formation is smaller than for disinfection. This means that well-designed disinfection reactors will show a higher amount of "shortcut" (non irradiated water) towards nitrate irradiation resulting in a nitrite-fluence relation that differs from the relation between disinfection and the fluence. This implies that the nitrate fluence meter can be used for monitoring the fluence in a well-characterised UV reactor within a given set of process parameters. The existence of a relation between the 'nitrite' fluence and the 'disinfection' fluence is currently under investigation.

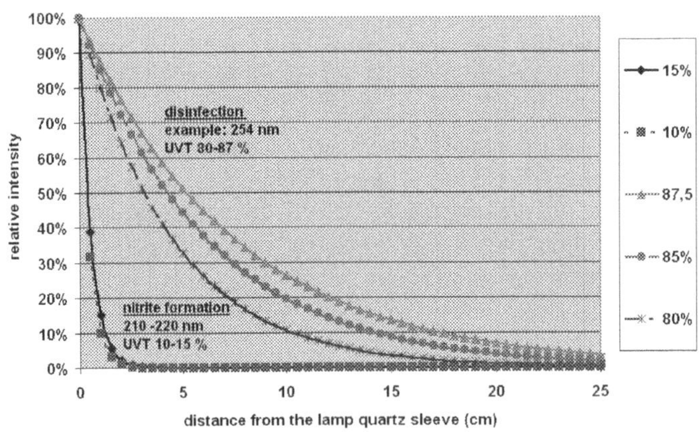

Figure 6 Intensity distribution around a medium pressure UV-lamp, comparison between disinfection (254 nm, UVT 85%) and nitrite formation (215 nm, UVT 15%)

From the above mentioned results can be concluded that the effect of lamp aging and lamp fouling on fluence might be detected by monitoring nitrite formation in time, although probably liable to wavelength dependent absorption aspects as well.

Conclusions

Nitrite is formed in small quantities as a by-product of UV polychromatic lamp disinfection processes. Nitrite formation can be used quantitatively as a fluence monitoring tool for already validated well designed reactors, showing a linear relation with process variations limited to flow and lamp power. It does not look promising as an alternative for biodosimetry tests.

Lamp failures can be detected by following nitrite formation, although the remaining fluence is overestimated by the measured effect on nitrite formation. Effects of water quality changes on nitrite formation, especially UVT and nitrate, are much more complicated. Fluence is more pronouncedly influenced by UVT_{254} -changes in comparison with nitrite formation. Flow through reactor testing and challenge test fluences are required to observe effects of UVT on fluence. More research is needed to fully understand the complex and linked role of UVT, as well in the disinfection action spectra as in the nitrate/nitrite action spectra. Attention should be paid to the feed water nitrate concentration, the role of DOC, bicarbonate and pH.

References

Clancy J.L., Bukhari Z., Hargy T.M., Bolton J.R., Dussert B., Marshall M.M. (2000). Using UV to inactivate Cryptosporidium *JAWWA* **92** (9); 97–104.

Mark G., Korth H.G., Schuchmann H.P. , Sonntag C. von (1996), The photochemistry of aqueous nitrate ion revisited, J. *Photochem. Photobiol. A: Chem.* 101, 89103.

Hijnen W.A.M., van der Veer A.J., Beerendonk E.F. , and Medema G.J. (2004) Increased resistance of environmental anaerobic spores to inactivation by UV *Wat. Sci.Tech: Water Supply* **4**(2) 54-61.

IJpelaar G.F., Kruithof J.C. (2001) Advanced oxidation processes for the degradation of pesticides in ground water and surface water. *Proceedings IOA* - 15th World Congress, 2001, London.

Recent Advances in NOM Removal

M Drikas

Cooperative Research Centre for Water Quality and Treatment, Australian Water Quality Centre, Private Mail Bag 3 Salisbury, South Australia, Australia

Abstract Natural organic matter (NOM) occurring naturally in surface waters has been shown to be one of the major characteristics that affects drinking water quality and treatment processes. Considerable emphasis has been placed in recent years on increasing the removal of NOM both by optimising current processes and by developing new processes. This has resulted in the realisation that it is not only the amount but also the character of the NOM that is important and impacts on the effectiveness of treatment processes. This paper provides an overview of possible methods for NOM removal, highlighting more recent novel developments.

Keywords: NOM removal, coagulation, MIEX, adsorption

Introduction

Organic matter of microbial, animal and/or vegetative origin occurs naturally in surface waters and has been shown to be one of the major characteristics that affects drinking water quality and treatment processes. It reduces the effectiveness of water treatment by interfering with the flocculation process, reacting with disinfectants and generating disinfectant by-products (DBP). Removal of problem contaminants by adsorption with activated carbon is reduced due to competition for adsorption capacity and membrane fouling is increased reducing membrane life and effectiveness. Furthermore, the biodegradable component of natural organic matter (NOM) acts as a food source for micro-organisms resulting in bacterial regrowth in distribution systems. These concerns have resulted in the removal of NOM from raw water being of prime concern for water authorities. This has been acknowledged by the United States in their proposed Disinfectant and Disinfection By-product Rule. This requires removal of NOM, as measured by total organic carbon (TOC), to minimise disinfection by-product (DBP) formation; the extent of removal required is dependent on starting concentrations and alkalinity. Considerable emphasis has been placed in recent years on increasing the removal of NOM both by optimising current processes and by developing new processes. This has resulted in the realisation that it is not only the amount of NOM that is important but that the character of the NOM varies and that this impacts on the effectiveness of treatment processes due to the different NOM reactions. This paper provides a summary of recent developments in NOM removal.

Enhanced Coagulation

Conventional treatment, consisting of coagulation/flocculation followed by sedimentation/filtration (or flotation) is still the major process used for water treatment. Acceptable turbidity removal can be achieved at reasonably low alum doses with increased alum doses providing only minimal decrease in treated water turbidity. However a detailed

© 2004 IWA Publishing. *2nd IWA Leading-Edge Conference on Water and Wastewater Treatment Technologies* edited by Mark Van Loosdrecht & Jonathan Clement. ISBN: 1 84339 508 8.

evaluation of coagulation using aluminium sulphate at a range of doses and pH has shown that removal of NOM as measured by dissolved organic carbon (DOC) can be maximised at an acidic pH around 5 (Figure 1). However when this is tempered with the need to achieve aluminium residuals of less than 0.2 mg/L the optimum pH is around 6 (van Leeuwen et al, 1999). This can be achieved in treatment plants either by the use of pH correction or by increased alum doses, whichever is the most economical and practical, and is generally referred to as enhanced coagulation (Krasner and Amy, 1995, White et al, 1997).

Alternative coagulants have been assessed by many researchers in attempts to improve the removal efficiency with coagulation which varies between 30 – 70%, depending on the NOM character. The organics removed by coagulation have been shown to be higher molecular weight, conjugated compounds as represented by high UV 254 nm absorbance and high specific UV absorbance (SUVA). In addition it appears that lower molecular weight, neutral hydrophilic compounds are not well removed by coagulation, hence the range in removal efficiencies for different waters.

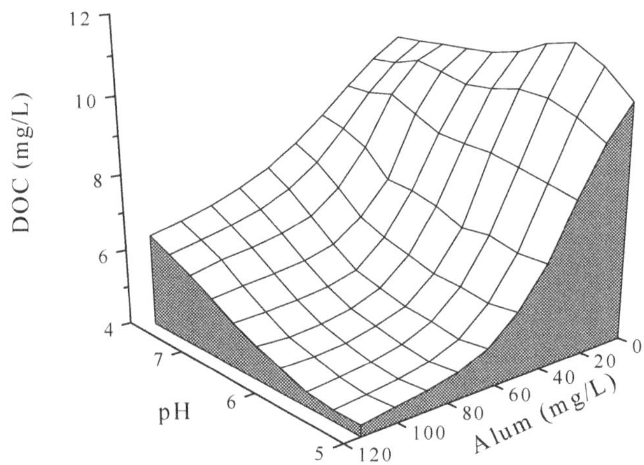

Figure 1 Removal of DOC from Myponga Reservoir water

Fabris et al, 2003, showed that alum performed best for DOC removal at pH between 5 and 6 whilst poly-aluminium chlorohydrate was better at pH 6. Ferric sulphate and ferric chloride at equivalent molar concentrations of the metal ion showed their best DOC removal between pH 4 and 5. Ferric chloride was slightly better than ferric sulphate at optimal pH and dose and resulted in 10 - 20% higher DOC removal in the waters tested than treatment with alum due to the improved removal of DOC possible at low pH with ferric salts. Poly-aluminium chlorohydrate was clearly inferior for NOM removal to the other coagulants in all conditions. Although poly-aluminium chlorohydrate did not perform as well as the other coagulants in this study, the fact that it can be used at high doses without greatly affecting the pH would make it ideal for treatment of low alkalinity, low DOC waters where turbidity removal is the main reason for treatment.

New ferric polymeric coagulants have been developed and appear promising for removal of humic acids (Cheng, 2002) whilst poly-aluminium ferric chloride appeared effective for removal of synthetic dyes (Gao et al, 2002).

Oxidation/ Biodegradation

Pre-oxidation with ozone has been successfully used to increase the removal of NOM by coagulation in some waters, resulting in micro-flocculation and reduced coagulant doses. However this is not a consistent effect. Recently potassium ferrate was shown to enhance coagulation of lake water with high organic content (Ma and Lui, 2002). Other oxidants such as Fenton's reagent, photo – Fenton's reagent and titanium dioxide photocatalysis show promise in laboratory studies in reduction of NOM, particularly the lower molecular weight material (Murray and Parsons, 2004).

Ozone and ultraviolet irradiation have also been used to directly oxidise NOM although this is not economical or practical in most situations. An alternative has been to use these oxidative processes to increase the biodegradability of the water and pass it through a biologically active medium such as granular activated carbon or slow sand filtration. This results in long term removal of NOM but usually only at low levels of around <30% DOC (Graham, 1999).

Similarly the use of bank filtration has effectively treated groundwater for many years in Europe for removal of microbial contaminants and NOM but the NOM levels in these supplies have always been low (Kuehn and Mueller, 2000). A recent US study (Wang et al, 2002) assessed riverbank filtration for the treatment of surface water with TOC 2-3 mg/L. The overall TOC reduction, including the impact of groundwater dilution, was more than 50%. The TOC data indicated that both biodegradation and physical removal were responsible. The ability of biological processes of this type to effectively remove high concentrations of NOM either with or without pre-oxidation remains a challenge for the future.

Magnetic Ion Exchange Resin (MIEX®)

There would appear to be clear benefits for the stability of process performance and water quality outcomes, if an effective method of NOM removal could be employed as an initial step in the treatment process, rather than relying on enhanced coagulation.

Anion exchange has been used previously for the effective removal of organic matter but the incorporation of magnetic iron oxide particles enables the resin to be applied as the first step prior to any further treatment. The MIEX® process is based on a magnetic anion exchange resin with a high adsorptive capacity for NOM developed by the Commonwealth Scientific and Industrial Research Organisation (CSIRO) in collaboration with the Australian Water Quality Centre (AWQC) and is manufactured under licence by Orica (Morran et al, 1996). The patented application developed by AWQC and Orica is shown in Figure 2.

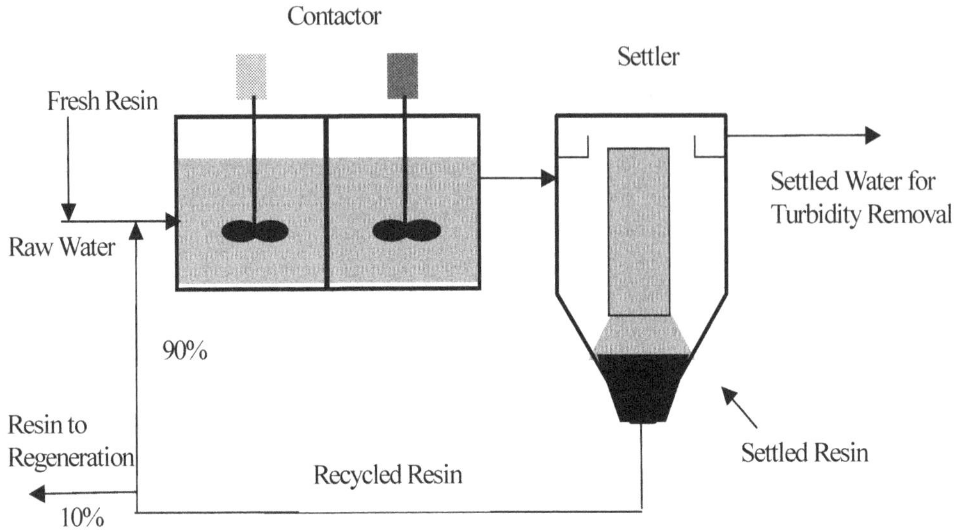

Figure 2 MIEX® Process Schematic

The process is designed so that it can be retro-fitted into a conventional water filtration plant. The resin can be added to the raw water to achieve the desired removal of NOM, without affecting or being affected by turbidity. The small particle size (approx. 150 μm) provides a high surface area for adsorption and removal proceeds rapidly with moderate stirring. Contact times of between 10 and 20 minutes can achieve NOM reductions as high as 85–90% in some waters. The magnetic properties of the resin promote aggregation of the resin particles under quiescent conditions in a separation vessel, thereby facilitating resin recovery. The recovered resin is returned to the head of the plant as the resin is not heavily loaded with NOM and can be effectively recirculated a number of times. A small proportion (5–15%) of the resin flow is diverted for regeneration, which is achieved by brief contact with a brine solution, which can also be re-used many times. The MIEX® treated water containing low organics must then be processed to remove any suspended matter present (Drikas et al, 2002).

In July 2001 the SA Water Corporation commissioned the first plant incorporating MIEX® at the 2.5 ML/day Mt Pleasant Water Treatment Plant in the northern Adelaide Hills. The plant has been divided into two streams, each of 1.25 ML/d capacity, each incorporating the MIEX® DOC removal process. Stream 1 is designed to structurally simulate the retrofit of MIEX® into a conventional plant without any major infrastructure modifications or additions and then followed by coagulation, flocculation, sedimentation and filtration. This means that the plant is more complex than would be required for a greenfield MIEX®/ coagulation/ filtration plant to enable the many advantages of MIEX® to be fully evaluated. Stream 2 comprises the greenfield process utilising MIEX® treated water directly onto Veolia Water CMF-S (submerged) microfiltration with PVdF membranes. The plant also incorporates an associated MIEX® regeneration plant that uses brine as the regenerant, with the ability to assess different regeneration mechanisms.

Table 1 Raw water quality data November 2001 – Feb 2004

	Turbidity NTU	Colour HU	DOC mg/L
Minimum	28	2	3.5
Maximum	99	8	10.1
Mean	60	5	6.1

The raw water source is the River Murray via the Mannum-Adelaide pipeline, a variable water source with a 5 year average turbidity of 55 NTU, and peaks commonly above 200 NTU. The raw water quality since plant commissioning steadily improved with a minimum of 28 NTU early July 2003 and then deteriorated in October 2003 when the turbidity increased to a peak of 99 NTU in January 2004 (Table 1). DOC reached a maximum of 10.1 mg/L in May 2002, its minimum in July 2003 and stabilised around 6 mg/L in February 2004. The water entering the plant from the Mannum-Adelaide pipeline has been chlorinated at Mannum and so the colour is always low. Whilst the raw water turbidity has varied, the treated water turbidity resulting from the combined treatment streams has always been well below 0.3 NTU, generally below 0.1 NTU.

Operational data indicating the effectiveness of the treatment processes incorporating MIEX® provides substantiation of the benefits of using the MIEX® process that to date have only been proven in laboratory and small scale demonstration plants. DOC removal has averaged 30-40% at MIEX® resin doses between 5-15 ml/L. Further DOC removal of ~15% was achieved using conventional coagulation or direct filtration with ~10 mg/L alum and 0.2 mg/L of a cationic polymer. Coagulation also reduced turbidity to <0.5 NTU from an average of 60 NTU. Coagulation without MIEX® pre-treatment required alum doses of ~40 mg/L and resulted in only 20% DOC reduction.

Table 2 Comparison of MIEX® and Alum Treatment

	DOC mg/L	DOC % removal	UVabs @ 254 nm	UVabs @ 254 nm % removal	THMFP* ug/L
Raw	4.5 (TOC)				128
40 mg/L alum	2.56	21.3	0.028	34.9	69
80 mg/L alum	2.13	34.5	0.022	48.8	55
8 mL/L MIEX®	1.35	58.5	0.010	76.7	34
8 ml/L MIEX® + 10 mg/L alum + 0.2 mg/L cationic poly	1.53	52.9	0.010	76.7	42

*THMFP: THM formation using 20 mg/L chlorine, 35°C, 4 hours contact time

Table 2 compares the removal of DOC using MIEX® (20 minute contact time) with that obtained using alum at the required dose (40 mg/L) and also under enhanced coagulation conditions (80 mg/L alum at pH 6). The water quality at this time was quite difficult to treat and only 35% DOC removal could be achieved even with enhanced coagulation. MIEX® achieved significantly more DOC removal. The improved DOC removal results in reduced disinfection byproduct formation as illustrated by the trihalomethane formation potential (THMFP), which is reduced by 40% from that which is formed with alum at the required dose (42 ug/L compared with 69 ug/L). This improvement in THMFP results not only from reduced DOC but also because MIEX® removes bromide. However in this prechlorinated water, with little bromide present, this is not relevant.

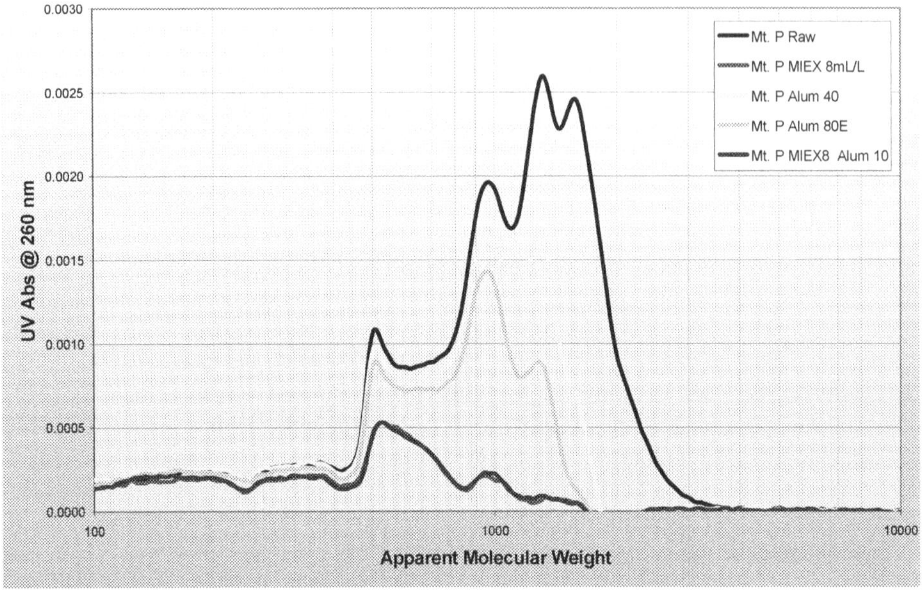

Figure 3 Molecular weight for different treatments.

Alum treatment removes the higher molecular weight organics during coagulation while MIEX® removes a broader range of organics, enabling higher removal of DOC than is possible with coagulation alone. This is illustrated in Figure 3 which compares the molecular weight scans for the different treatments. It is apparent that increasing alum dose removes organics predominantly above 1000 Daltons while MIEX® removes those organics as well as the ones below 1000. In some waters the alum used to remove turbidity remaining after MIEX® treatment also removes additional higher molecular weight organics (Chow et al, 2002).

The MIEX® process has been evaluated in a number of locations in the US (Budd et al, 2003) and although DOC removal efficiencies varied depending on water quality, removal exceeded that possible with alum alone and results have generally confirmed the findings in Australian waters. Further work (Singer and Johnson, 2003) has confirmed that MIEX® removes UV absorbing organics and bromide substantially reducing the ozone demand of waters and the formation of bromate.

Adsorption

Adsorption of NOM by activated carbon has always been considered more of a hindrance than a treatment process for 2 reasons – NOM interferes with the primary role of activated carbon which is to remove problem contaminants such as taste and odour compounds or manmade contamination such as pesticides and secondly due to the poor economics of using granular activated carbon (GAC) to remove NOM. GAC filters, once past the initial adsorptive stage are well known to remove small amounts of NOM (around 10%) which is generally attributed to biodegradation. Pre-ozonation can sometimes improve this removal up to 20%.

An operating plant that has successfully used GAC filters to control problem contaminants and also to remove NOM is the Cincinnati Water Works, Cincinnati, Ohio,

USA (Metz et al, 2004). This plant consists of coagulation and sedimentation followed by secondary sedimentation and rapid sand filtration prior to entering 12 post-filter GAC adsorbers containing Calgon Filtrasorb 400 and two multiple hearth furnaces for on-site reactivation. The GAC process was installed to remove synthetic organic chemicals from industrial contamination of the Ohio River and to reduce DBP precursors. TOC input to the GAC filters has been between 1-3 mg/L with a treated water goal of 1 mg/L TOC. Recent comparison of virgin carbon to five and twelve times reactivated granular activated carbon indicated a reduction in micropore and an increase in mesopore volume. There was minimal deterioration in performance of the reactivated carbons for removal of TOC and DBP precursors with 91 days to reach breakthrough of 1 mg/L TOC for virgin carbon compared with 81 days for 12 times reactivated carbon to reach the same.

Matsui et al (2004) described the improved removal of NOM using thermally activated wood based powdered activated carbon (PAC) that had been ground to 0.8 um compared with as-received PAC (33um). The ground carbon achieved the same NOM removal as the as-received PAC at 10 times less dose. The time to reach 80% of adsorption equilibrium was also reduced to less than 1 minute contact time. The micro ground PAC showed a larger surface area for pore sizes >1.5 nm suggesting that grinding exposed areas of mesopore surface not previously accessible. The micro ground PAC preferentially adsorbed NOM having high UV absorbance and whilst improving NOM adsorption of waters with high SUVA, little effect of PAC size was seen on NOM adsorption in water with a low SUVA (1.22 L/g/cm). Nevertheless this micro grinding process, which is currently being patented, shows potential as a means of not only reducing PAC dosages but increasing reaction rate enabling application of PAC in locations where short contact times previously precluded its usefulness, for example prior to micro or ultrafiltration membranes for NOM removal.

A new adsorption technology for NOM removal has been developed using fixed-bed adsorbers with granular ferric hydroxide (GFH) as an adsorbent (Genz and Jekel, 2004). GFH was developed at the Technical University Berlin, Germany, primarily for adsorptive removal of arsenic from drinking water (Jekel and Seith, 2000). GFH is a 100 % synthesized Akaganeite (ß-FeOOH) which is a poorly crystallized but highly active modification of iron hydroxide with a specific surface area of 280 m^2/g. The treatment is designed in two stages: in a first step, the GFH is loaded with NOM until the maximum adsorption capacity is reached; and in a second step, the material is oxidatively regenerated by hydrogen peroxide or sodium hydroxide. Sorption of higher molecular weight, UV absorbing material appears preferentially adsorbed and adsorption was significantly improved at lower pH. The minimum empty bed contact time was determined to be 5 minutes. This technology appears promising and its long term viability and benefits compared with other available processes needs to be determined.

Nanofiltration

Pressure driven membrane processes are becoming more economically viable as an alternative to conventional water treatment techniques for removal of NOM. Micro and ultra filtration are not able to remove NOM without pre-treatment but combining these membrane systems with pretreatment is an effective way of both minimising fouling and removing NOM (Zhang et al, 2003). The effectiveness of NOM removal is thus dependent on the pre-treatment process applied.

Nanofiltration (NF) is a membrane process that targets the removal of NOM, multivalent salt ions (eg. those responsible for hardness) and to a lesser degree, simple salts (eg. fluoride and chloride). NF can also remove some trace organic micropollutants, such as pesticides.

NF is therefore a potential alternative process for removal of NOM. It has the advantage of being a simple physical process, allowing a high degree of automated control and reduced operator requirements. However, it is necessary to install pre-treatment prior to NF to remove particulate and colloidal material that may otherwise result in blockage of the feed channels and cause membrane fouling. High energy requirements are a feature of the process and regular chemical cleaning to ensure acceptable productivity of the membranes is required. In NF, the waste flow can represent 10-25% of the raw water flow, an inefficiency that makes its acceptance difficult in countries without an abundance of water resources.

Recent studies have shown that high molecular weight, high UV absorbing, high SUVA NOM was preferentially rejected through nanofiltration membranes studied whilst lower molecular weight material, particularly that which was derived from algae, fouled the membrane (Her et al, 2003). Other studies have confirmed that the extent of NOM removal is dependent on membrane type (Reiss et al. 1999).

Nanofiltration is effectively employed for NOM removal in a number of operating treatment plants such as the Mery-sur-Oise plant in Paris France. This plant treats water from the polluted River Oise, which experiences chemical spills from industry each year and is high in dissolved organic material. The 140 ML/day plant utilises nanofiltration (NF) membranes. Pre-treatment prior to NF is by conventional treatment using the patented (Veolia) Actiflo® process (compact flocculation/sedimentation), followed by rapid sand filtration and ozonation. The plant also has an older treatment train (200 ML/day capacity), comprising conventional sedimentation tanks and filtration through rapid sand filters. This is followed by ozone/GAC for pesticide removal (and also taste/odour and algal toxin removal). The NF stage essentially replaces the ozone/GAC stage in the older stream but also achieves higher removal of NOM, along with some softening. The water from the two streams is blended prior to distribution. The plant operates the NF train at design capacity, with the old process train providing the make-up requirement to meet demand.

Conclusion

Of the processes discussed only MIEX® and, to a lesser extent, GFH have been developed specifically to remove NOM. The other processes discussed have been developed, or used, for other purposes and, coincidentally, NOM has also been found to be removed and/or to interfere with the desired purpose. This does not lessen their effectiveness or importance but highlights the need for processes such as MIEX® and GFH to be combined with other processes to provide an improvement in the total water quality. This is of importance as none of the described processes are mutually exclusive but are chosen in various combinations to address particular water quality issues.

The realization of the contamination of water sources with manmade products, such as pesticides, has resulted in conventional treatment plants originally considered to produce acceptable treated water quality being declared inadequate, and ozone and/or GAC has been introduced. Concern with DBP has resulted in an emphasis on removal of NOM with the introduction of alternative disinfectants and enhanced coagulation. Concern with protozoa such as *Cryptosporidium* has increased the usage of membrane treatment and UV irradiation.

The processes identified in this paper are possible solutions for the removal of NOM. The most appropriate solution will depend on effectiveness of the process, infrastructure already in place, other water quality issues to be solved (the total treatment objective) and of course capital, operating and life cycle costs.

References

Budd G.C., Singer P.C., Edwards J.C., Long B.W., Meisch M.E., Rexing D. J., Owen C.A., Cook J.B. and Joslyn B.L., (2003) Magnetized Ion Exchange (MIEX): Evaluation of process impacts in different sources, *Proceedings of the American Water Works Association Water Quality Technology Conference*, Philadelphia, November

Cheng W.P., (2002) Comparison of hydrolysis/coagulation behaviour of polymeric and monomeric iron coagulants in humic acid solution, *Chemosphere*, 47, 963-969

Chow C., Cook D. and Drikas M. (2002) Evaluation of Magnetic Ion Exchange Resin (MIEX®) and alum treatment for formation of disinfection by-products and bacterial regrowth, *Water Science & Technology: Water Supply*, 2(3), 267–274

Drikas M., Morran J.Y., Pelekani C., Hepplewhite C. and Bursill D.B. (2002) Removal of natural organic matter - a fresh approach, *Water Science & Technology: Water Supply*, 2(1), 71–79

Fabris R., Chow C. and Drikas M., (2003) The impact of coagulant type on NOM removal, *Proceedings of the AWA 20th Federal Convention*, Perth, Australia, April 2003

Gao B., Yue Q. and Miao J., (2002) Evaluation of polyaluminium ferric chloride (PAFC) as a composite coagulant for water and wastewater treatment, *Water Sci. Tech.* 47(1), 127–132.

Genz A. and Jekel M. (2004) A new treatment process for removal of NOM by adsorption and regeneration of granular ferric hydroxide (GFH). *Proceedings of the International Conference on Natural Organic Matter Research*, Victor Harbor, Australia, March

Graham N.J.D., (1999) Removal of humic substances by oxidation/biofiltration – a review, *Water Science & Technology*, 40(9) 141 - 148

Her N, Amy G. Yoon J. and Song M. (2003) Novel methods of characterizing algogenic organic matter and associated nanofiltration membrane fouling, *Water Science & Technology: Water Supply*, 3(1), 71–79

Jekel, M. and Seith R., (2000) The removal of arsenic: Comparison of conventional and new techniques for the removal of arsenic in a full scale water treatment plant, *Water Supply*, 18(1): 628-631.

Krasner S. and Amy G. (1995) Jar-test evaluations of enhanced coagulation. *J. AWWA*. 87(10), 93-107.

Kuehn W. and Mueller U. (2000) Riverbank filtration: an overview, JAWWA, 92(12), 60

Ma J. and Lui W., (2002) Effectiveness of ferrate (VI) preoxidation in enhancing the coagulation of surface waters, *Water Research*, 36 (20), 4959-4962

Matsui Y., Murase R., Sanogawa T., Aoki N., Mima S., Inoue T. and Matsushita T., (2004) Micro-ground powdered activated carbon for effective removal of natural organic matter during water treatment. *Proceedings of the International Conference on Natural Organic Matter Research*, Victor Harbor, Australia, March

Metz D.H., DeMarco J. Pohlman R., Canon F.S. and Moore B.C, (2004) Affect of multiple GAC reactivations on disinfection by-product precursor removal. *Proceedings of the International Conference on Natural Organic Matter Research*, Victor Harbor, Australia, March

Morran J., Bursill D., Drikas M. and Nguyen H., (1996) A new technique for the removal of natural organic matter. *Proceedings of the AWWA WaterTECH conference*. Sydney, Australia, May

Murray C.A. and Parsons S.A., (2004) Comparison of AOPs for the removal of natural organic matter: performance and economic assessment. *Water Science & Technology* 49(4), 267 - 272

Reiss C.R., Taylor J.S. and Robert C., (1999) Surface water treatment using nanofiltration – pilot testing results and design considerations. *Desalination*, 125(1-3), 97-112

Singer P.C. and Johnson C.J. (2003) Reduction in ozone demand and bromate formation using a magnetic ion exchange resin. *Proceedings of the American Water Works Association Water Quality Technology Conference*, Philadelphia, November

van Leeuwen J. A., Chow C., Fabris R. Drikas M. and Spark K., (1999) Enhanced coagulation for dissolved organic carbon removal in conventional treatment with alum. *Proceedings of the 18th Federal Convention*, Adelaide, Australia. April

Wang J.Z., Hubbs S.A. and Song R., (2002) Evaluation of riverbank filtration as a drinking water treatment process. AWWARF Report 90922

White M.C., Thompson J.D., Harrington G.W. and Singer P.C. (1997). Evaluating criteria for enhanced coagulation compliance. *J. AWWA*, 89(5), 64-77.

Zhang M.M., Li C., Benjamin M. M. and Chang Y.J. (2003) Fouling and natural organic matter removal in adsorbent/membrane systems for drinking water treatment. *Environmental Science & Technology* **37** (8), 663-1669

Characterization of Natural Organic Matter in Advanced Water Treatment Processes for DBPs Control

H.C. Kim*, M.J. Yu*, G.N. Myung*, J.Y. Koo* and Y.H. Kim**

*Department of Environmental Engineering, University of Seoul, 90 Junnondong, Dongdaemungu, Seoul, 130-743, Korea (E-mail: animaplus@hanmail.net; myong@uos.ac.kr; gnmyung@hotmail.com; jykoo@uos.ac.kr)
**Department of Environmental Management, Shinheung College, Howondong, Euijungbusi, Kyungido, 480-701, Korea (E-mail: yhkim@shinheung.ac.kr)

Abstract A demonstration plant consisted of granular activated carbon (GAC) and ozone/GAC process was operated using filtered water of a conventional water treatment plant as influent to study feasibility of introducing advanced water treatment. Natural organic matter (NOM) from process waters at the demonstration plant was isolated into hydrophobic and hydrophilic fractions by physicochemical fractionation methods to investigate its characteristics. In the case of hydrophobic fraction (i.e. humic substances; HS), the structural and chemical characteristics were investigated using various spectroscopic methods. Through ozonation and carbon adsorption, the hydrophobic fraction and hydrophilic fraction gradually decreased from 22.0 % to 14.7 % and from 38.6 % to 20.8 % based on the content in raw water, respectively. Formation of trihalomethanes (THMs) was highly influenced by hydrophobic fraction, whereas haloacetic acids formation potential (HAAFP) depended more on the hydrophilic fraction. Concentration of phenolic group in the HS gradually decreased from 60.5 % to 15.8 % through the water treatment and a higher yield of THMs resulted from chlorination of NOM with a higher phenolic content. The structural changes of HS identified from FT-IR and ^{1}H-NMR were consistent with the results from the isolation of functional groups in the HS fractionated using A-21 resin. Decreases of ratio of UV absorbance at 253nm and 203nm, respectively (A_{253}/A_{203}) and DBPFPs/DOC showed consistent trends, therefore, the A_{253}/A_{203} ratio may be a good indicator of tendency for the DBPs formation.

Keywords DBPs/DBPFPs; FT-IR; ^{1}H-NMR; humic substances (HS); hydrophobicity; natural organic matter (NOM)

Introduction

Humic substances (HS), which are described as heterogeneous polyfunctional polymers formed through the breakdown of plant and animal tissues and/or synthesis of products by chemical and biological processes (Thurman, 1985). Because of their complex polymeric properties, HS are still among the least understood and characterized components in the environment. To understand the role of HS in water chemistry, it is often necessary to fractionate natural organic matter (NOM). Various methods have been used to isolate NOM from natural water and the XAD resin method has been reported in many applications for fractionation of NOM (Aiken *et al.*, 1979; Leenheer, 1981; Yu *et al.*, 2002).

One of the major problems for NOM including HS is the production of disinfection by-products (DBPs) such as trihalomethanes (THMs) and haloacetic acids (HAAs), and NOM is

© 2004 IWA Publishing. *2nd IWA Leading-Edge Conference on Water and Wastewater Treatment Technologies* edited by Mark Van Loosdrecht & Jonathan Clement. ISBN: 1 84339 508 8.

widely known as precursors of DBPs (Bellar *et al.*, 1974; Rook, 1974). Ozonation followed by biological activated carbon adsorption is one of the promising processes to remove NOM from raw water. Some of NOM is removed in the conventional drinking water treatment process system such as coagulation/sedimentation and sand filtration. The rest should be removed before releasing the water to the distribution system by oxidation and activated carbon adsorption. It is generally known that formation of DBPs highly depends on the organic matter content, but there are many other factors such as organic matter composition and water treatment methods. Although many studies on the DBPs formation have been carried out, the reduction of the DBPs concentration in distribution is still an important issue.

The alterations in structural and chemical characteristics of NOM by physicochemical water treatment processes have been shown to cause changes in reactivity with disinfectants (Schnoor *et al.*, 1979; Vartiainen *et al.*, 1988). The main purposes of this paper were to compare the structural and chemical characteristics of NOM fractionated from process waters in the advanced water treatment, and to investigate the relationships between the formation of DBPs and various characteristics of NOM i.e. hydrophobicity, functionality and different chromatography. Knowledge about the interactions of chlorine with NOM considering characteristics of the NOM is essential to establish optimal treatment strategy for DBPs control. Therefore, we attempted to estimate the characteristics of DBPs generated by the different properties of the NOM in aqueous environments, and to characterize behavior of NOM as DBPs precursors through advanced water treatment.

Materials and Methods

The first phase of this study was to fractionate NOM from process waters in 10 months operation into hydrophobic and hydrophilic fractions with XAD resin at the demonstration plant. On the other hand, using a weak-base secondary amine resin, hydrophobic fraction (i.e. humic substances; HS) in NOM was isolated into phenolic and carboxylic functional groups. In the second phase study, these fractions obtained in the first phase were analyzed with following analytical techniques: dissolved organic carbon (DOC), UV-abs, specific ultra-violet absorbance (SUVA) and disinfection by-product formation potentials (DBPFPs). And then, HS samples were selectively extracted and analyzed by Fourier transform infrared (FT-IR) and proton nuclear magnetic resonance (^{1}H-NMR) spectroscopy.

Demonstration plant

The demonstration plant is located at the Gueui Water Treatment Plant (GWTP) in Seoul, Korea. GWTP adopts conventional water treatment consisting of pre-chlorination, coagulation/sedimentation, sand filtration and chlorination, and uses Han River water as a water source. The demonstration plant that consisted of two trains, granular activated carbon (GAC) and Ozone/GAC respectively, was operated using filtered water of GWTP as influent to study feasibility for introduction of advanced water treatment. Table 1 outlines the operational parameters for the demonstration plant. Each train was optimized based on total organic carbon (TOC) removal efficiency and economical aspect, and designed to treat $40m^3$/day of sand filtered water.

Table 1 Operating parameters of the advanced water treatment demonstration plant

Process	Parameters	Design values
Ozonation	Output	40 g/hr
	Contactor	L×W×H: 1.7 m×0.3 m×5.5 m
	Contact time	10 min
	Ozone dose	1~1.5 mg/L
	Residual ozone	0.05~0.1 mg/L
GAC adsorption	Raw material	Bituminous coal type
	Particle size	0.9~1.1 mm (12×40 mesh)
	Effective size	0.55~0.75 mm
	Uniformity coefficient	1.9 (Max.)
	Apparent density	0.44 g/mL (Min.)
	EBCT[a]	15 min
	LV[b]	11 m/hr
	Bed volume	ID×H: 450 mm×6 m

[a]Empty bed contact time, [b]Line velocity

Fractionation and extraction of NOM

NOM from various water samples was fractionated and extracted using the methods conducted by Thurman and Malcolm (Thurman and Malcolm, 1981). A glass column (ID: 3 cm) was filled with XAD resin (Amberite XAD-7HP, Rohm & Haas Co., France) up to 40 cm. 50 L of the water filtered with 0.45 μm membrane filter was passed through the glass column with flow rate of 10~15 mL/min, and the effluent collected is termed the hydrophilic fraction. 0.1N NaOH was used to extract HS (i.e. hydrophobic fraction) adsorbed in the resin bed. And then, Na^+ ions in the solution were substituted with H^+ using cationic exchange resin (Amberite IRC-50, Rohm & Haas Co., France). Each fraction collected was analyzed for various physicochemical analyses.

Isolation of phenolic and carboxylic groups

A secondary amine weak base [~$N(CH_3)_2$] resin (Amberlyst A-21, Rohm & Haas Co., France) was used for separating phenolic and carboxylic groups present in the HS (Lin *et al.*, 2001). The HS solution after passing through the resin column contains little phenolic functional group by strong interaction between neutral nitrogen of the resin and the phenolic functional group present in the HS solution at neutral pH. The effluent collected is termed the carboxylic fraction.

FT-IR and ^{1}H-NMR

HS powder obtained through freeze-drying HS solution substituted with H^+ was analyzed for its structural and chemical characteristics. IR spectra were obtained by scanning HS extracted from process waters. KBr (FT-IR Grade, Aldrich Co., USA) was mixed with HS in the ratio of 100 to 1 and the IR spectra of the mixture were scanned by IR spectrophotometer (Infinity Gold 60AR, Thermo Mattson, USA). A NMR analyzer (Avance 400, Bruker, Germany) was used to obtain the ^{1}H-NMR spectra of HS. Approximately 50 mg of HS powder was added to 0.5 mL D_2O in a 10 mm NMR tube. The signal for D_2O was used as reference and set to 4.8 ppm chemical shift.

Determination of physicochemical characteristics

DOC was determined with TOC analyzer (DC-180, Dohrmann, USA) after filtration of the samples with 0.45 μm membrane filters. A UV-spectrophotometer (UV-2101PC, Shimadzu, Japan) was used to determine the absorbance at various wavelengths. For DBPFPs test, solutions were chlorinated at pH 8 and 20 °C for 48 hr. In case of THMs, using the headspace method, volatile compounds in samples were concentrated and injected into GC (HP-5890, Hewlett Packard, USA). GC/MSD (3800CP, Varian, USA) was used for determining HAAs by slightly modifying EPA Methods 552.2 (USEPA, 1995).

Results and Discussion

Removal of NOM and DBPs in water treatment processes

The removal trends of TOC, SUVA and DBPs in process waters at the demonstration plant are shown in Figure 1. TOC decreased from 2.67 to 1.23 mg/L during conventional water treatment and decreased to 0.72 mg/L through ozonation and carbon adsorption. Also, SUVA values gradually lowered with decrease of UV absorbance. In case of carbon adsorption process alone, TOC decreased from 1.23 to 0.70 mg/L. However, DBPs formed during pre-chlorination were not effectively removed by ozonation, and THMs even increased in the range of 29.6~31.1 μg/L after carbon adsorption. On the other hand, HAAs were effectively removed by carbon adsorption with/without pre-ozonation, indicating that the adsorption capacity of HAAs by activated carbon was relatively higher than those of THMs.

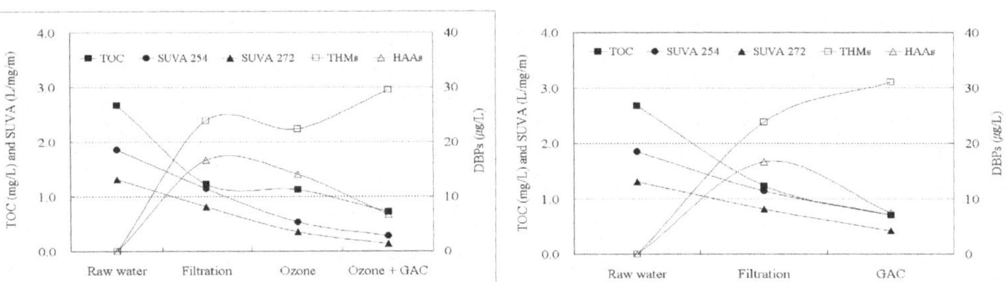

(a) Ozone/GAC (b) Only GAC

Figure 1. Variation in treated water quality at the demonstration plant

The variation in distribution of DOC concentration of hydrophobic and hydrophilic fractions in NOM during water treatment is shown in Figure 2. Hydrophobic fraction and hydrophilic fraction decreased from 46.1 % to 22.0 % and 53.9 % to 38.6 % through conventional water treatment based on the content of raw water, respectively. Through ozonation and carbon adsorption, the hydrophobic fraction and hydrophilic fraction gradually decreased from 22.0 % to 14.7 % (from 0.45 to 0.30 mgDOC/L) and from 38.6 % to 20.8 % (from 0.78 to 0.42 mgDOC/L), respectively. In case of carbon adsorption process alone without pre-ozonation, both fractions decreased from 22.0 % to 15.1 % (from 0.45 to 0.31 mgDOC/L) and from 38.6 % to 19.6 % (from 0.78 to 0.42 mgDOC/L), respectively as shown in Figure 2.

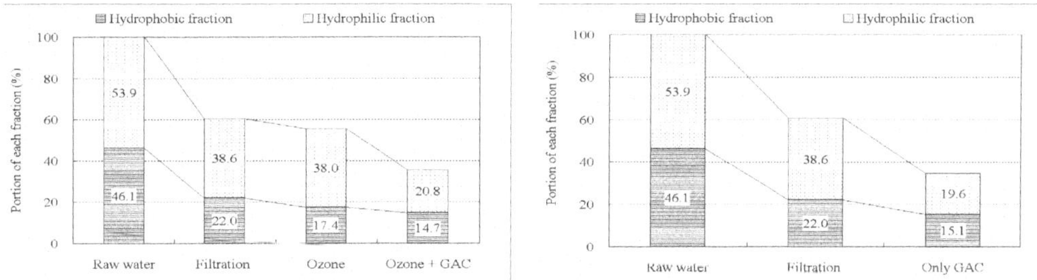

(a) Ozone/GAC (b) Only GAC

Figure 2. Distribution of NOM fractionated from process waters at the demonstration plant

These results are indicating that hydrophobicity of the carbon surface decreased in 10 months operation of the demonstration plant. That is, the hydrophobicity of activated carbon surface is reduced by hydrophilic oxygenated functional groups increased by interaction with oxidants such as dissolved oxygen, free chlorine and residual ozone in water. Although not all of the surface functional groups are hydrophilic, the adsorption capacity of hydrophilic compounds was enhanced by increase in hydrophilic sites. As a result, the activated carbon adsorption resulted in an early breakthrough of hydrophobic compounds including THMs. Therefore the adsorption capacity of hydrophilic compounds by activated carbon was relatively higher than those of hydrophobic compounds. And, ozonation and carbon adsorption processes mainly contributed to reduction of HAAs precursor rather than decrease of THMs precursor.

DBPFPs characteristics of NOM fractionated into hydrophobic and hydrophilic fractions from process waters are shown in Table 2. THMFP/DOC values of hydrophobic and hydrophilic fractions of raw water were 85.6 μg/mg and 43.3 μg/mg, and HAAFP/DOC values of each fraction were 20.2 μg/mg and 33.7 μg/mg. Through ozonation and carbon adsorption, THMFP/DOC value of hydrophobic fraction was gradually decreased from 38.4 μg/mg to 24.8 μg/mg and HAAFP/DOC value of hydrophilic fraction was decreased from 19.9 μg/mg to 16.7 μg/mg. The yield of THMs formed by hydrophobic fraction as compared to the HAAs yield by the same fraction was significantly high. That is, the formation of THMs was mainly influenced by hydrophobic fraction, while HAAFP depends more on the hydrophilic fraction.

Table 2. DBPs formation characteristic of NOM fractionated from process waters at the demonstration plant

Item	Raw water (TP^a/HP^b)	Filtration (TP^a/HP^b)	Ozone (TP^a/HP^b)	Ozone/GAC (TP^a/HP^b)	Only GAC (TP^a/HP^b)
Unfractionated water	55.6/32.6	35.5/27.7	35.0/21.2	29.7/13.5	31.7/15.3
Hydrophobic fraction (A_{253}/A_{203})[c]	85.6/20.2 (0.160)	38.4/13.8 (0.102)	30.3/10.3 (0.073)	24.8/9.0 (0.055)	25.6/9.7 (0.062)
Hydrophilic fraction	43.3/33.7	18.4/19.9	15.4/19.1	16.0/16.7	21.6/14.1

[a]THMFP/DOC (μg/mg) [b]HAAFP/DOC (μg/mg)
[c]The ratio of UV absorbance at 253 nm and 203 nm, respectively

A_{253}/A_{203} ratio of the hydrophobic fraction (i.e. humic substances) was analyzed and the results are shown in Table 2. A_{253}/A_{203} ratio decreased from 0.160 to 0.102 by conventional water treatment system. As the A_{253}/A_{203} ratio gradually decreased from 0.102 to 0.055 through ozonation and carbon adsorption, DBPs formation potential decreased. The changes in A_{253}/A_{203} ratio of HS suggest that aromatic rings substituted with various functional groups in the HS molecules are structurally altered by physicochemical water treatment. The A_{253}/A_{203} ratio is low for unsubstituted aromatic ring structures and increases for HS in which the aromatic rings are highly substituted with OH, carbonyl, ester and carboxylic groups. These groups are also considered to participate preferentially in reactions generating DBPs. Decreases of A_{253}/A_{203} ratio and DBPFPs/DOC showed consistent trends, therefore, the A_{253}/A_{203} ratio may be a good indicator of the tendency for the DBPs formation.

Structural and chemical characteristics of HS

Determination of functional groups in HS

HS was isolated from process waters in the water treatment and then fractionated into the phenolic and carboxylic groups using A-21 resin. The variation in distribution of molar concentration of each group through treatment process is shown in Figure 3.

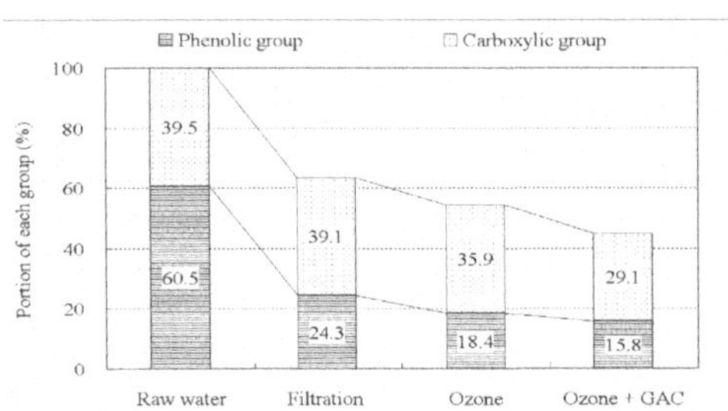

(a) Ozone/GAC

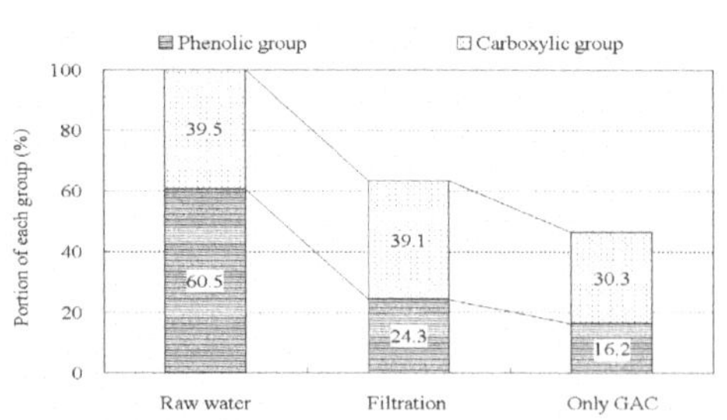

(b) Only GAC

Figure 3. Functional group distribution of HS in NOM fractionated from process waters at the demonstration plant

The concentration of phenolic group and carboxylic group decreased from 60.5 % to 24.3 % and 39.5 % to 39.1 % through conventional water treatment, respectively. The phenolic group decreased by 36.2 % through conventional treatment while the carboxylic group only changed by 0.4 %. This suggests the reduction in A253/A203 ratio from 0.16 to 0.102 after conventional treatment was solely due to the removal of phenolic group. That is, A253/A203 ratio depended more on the phenolic group content than carboxylic group content in the HS. In addition, it was recognized that the content and substitution type of phenolic ring structure in the HS molecule highly influenced DBP formation. Through ozonation and carbon adsorption, the concentration of phenolic group and carboxylic group decreased from 24.3 % to 15.8 % (from 4.9 to 3.2 μM/L as phenolic-OH) and from 39.1 % to 29.1 % (from 7.8 to 5.8 μM/L as COOH), respectively. The content of phenolic group decreased from 24.3 % to 18.4 % by pre-ozonation, however adsorption capacity of phenolic group was decreased by structural alteration through pre-ozonation. In the case of carbon adsorption process alone, both groups decreased from 24.3 % to 16.2 % (from 4.9 to 3.3 μM/L as phenolic-OH) and from 39.1 % to 30.3 % (from 7.8 to 6.1 μM/L as COOH), respectively as shown in Figure 3. In advanced processes such as ozone/GAC and GAC alone, A_{253}/A_{203} ratio decreased as phenolic group decreased.

FT-IR spectra of HS

The IR spectra of HS extracted from ozonated and carbon adsorbed waters in advanced water treatment system are shown in Figure 4. The band at around 3400cm^{-1} is generally attributed to OH groups and bands at 2995~2965 cm^{-1} are assigned to C-H, C-H_2 and C-H_3 stretching of the aliphatic groups. The bands at 1646~1640 cm^{-1} and 1560~1551 cm^{-1} are attributed to C=O stretching vibration of carboxylic acids and ketones/quinones, respectively. The bands at around 1450 and 1410 cm^{-1} are attributed to C-H deformation of aliphatic and CH_3 groups, respectively. Also, bands in the 1280~1137 cm^{-1} region are attributed to C-O stretching of esters, ethers and phenols, and band at around 830 cm^{-1} can be assigned to OH stretching vibration of carboxylic groups.

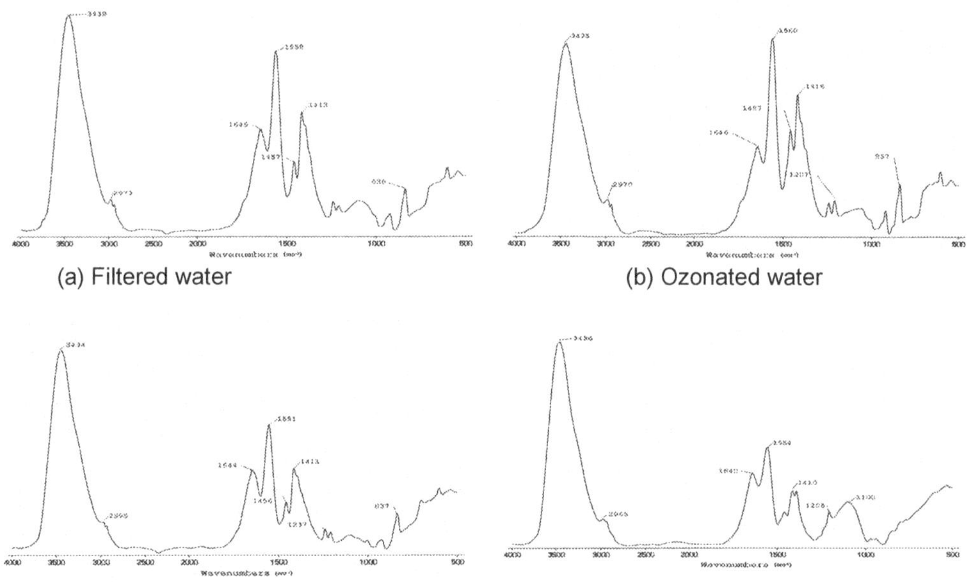

(a) Filtered water (b) Ozonated water

(c) Ozonated and GAC adsorbed water (d) GAC adsorbed water

Figure 4 FT-IR spectra of HS extracted from process waters at the demonstration plant

From this result, the peak area of 1646~1640 cm^{-1} was not changed much by ozonation, while this peak area decreased by carbon adsorption after pre-ozonation. In case of carbon adsorption process alone without pre-ozonation, the peak area also decreased. It suggested that hydrophobicity of carboxylic acids in the HS was decreased by ozonation, the hydrophilic compounds were selectively adsorbed by activated carbon. The structural changes of HS through water treatment process from FT-IR were consistent with the result from the isolation of phenolic group and carboxylic group using A-21 resin.

^{1}H-NMR spectra of HS

^{1}H-NMR spectra were obtained for basic structural information of HS extracted from process waters at the demonstration plant. For a more effective comparison, in accordance with literature results of assignments for the chemical shift in ^{1}H-NMR spectra, each spectrum was quantitatively analyzed and the results are shown in Table 3 (Ma *et al.*, 2001).

Table 3. ^{1}H-NMR chemical shift regions and their relative contributions

Chemical shift region (ppm)	Relative contributions (%)				
	Raw water	Filtration	Ozone	Ozone/GAC	Only GAC
I (0.0~1.6)[a]	38.6	41.3	46.6	44.7	40.3
II (1.6~3.2)[b]	37.0	39.0	38.7	38.7	38.0
III (3.2~4.3)[c]	11.3	10.9	6.1	6.3	10.1
IV (6.0~8.5)[d]	13.1	8.8	8.7	10.3	11.6
P_{Al}/P_{Ar} ratio[e]	5.8	9.1	9.8	8.1	6.8

[a]Aliphatic methyl and methylene

[b]Protons of the methyl and methylene groups α to aromatic rings, protons on carbons in α position to carbonyl, carboxylic acid, ester, or amino acid

[c]Protons on carbon of hydroxyl, ester and ether, and protons on methyl, methylene, and methyne carbons directly bonded to oxygen and nitrogen

[d]Aromatic protons including quinones, phenols, oxygen containing hetero-aromatics

[e]The ratio of aliphatic protons to aromatic protons (Region I and II/Region IV)

Percentage of the region III (3.2~4.3ppm) including olefin and acetylene compounds decreased more than region IV (6.0~8.5ppm) including aromatic compounds by ozonation. It was caused by the reduction of specific reaction sites. That is, since specific reaction sites such as aromatic ring structures in the HS molecule were already attacked by chlorine through pre-chlorination in conventional water treatment process, specific reaction sites to react with ozone or OH radical through ozonation were reduced. Therefore, other unsaturated sites such as olefin and acetylene bonds were directly or indirectly oxidized by ozonation. On the other hand, the ratio of aliphatic protons to aromatic protons (P_{Al}/P_{Ar}) was increased from 5.8 to 9.1 by conventional water treatment system as shown in Table 3. This ratio increased by ozonation, while it was relatively decreased by carbon adsorption. And percentage of the region IV increased by carbon adsorption with/without pre-ozonation. It was because that aliphatic group in the HS was selectively adsorbed to activated carbon, and portion of aromatic protons relatively increased. These results were mostly consistent with the result from FT-IR spectra of the HS.

Summary and Conclusions

The formation of THMs was highly influenced by hydrophobic fraction, while HAAFP depended more on the hydrophilic fraction. Also, a higher yield of THMs resulted from chlorination of NOM with a higher phenolic content. Through conventional water treatment, the hydrophobic fraction was reduced more than the hydrophilic fraction and the phenolic group was mainly removed compared to the carboxylic group. It was found carbon adsorption with/without pre-ozonation decreased the hydrophilic fraction than the hydrophobic fraction. Therefore residual hydrophilic NOM after conventional treatment could to be removed by advanced treatment such as ozone/GAC and GAC alone to reduce HAAs formation. GAC alone appeared virtually as effective as ozone/GAC in the results. However one of these can be suggested considering the necessity of selective oxidation and adsorption of each NOM constituent, and reactivation frequency of GAC.

Amount and species of DBPs produced were more influenced by chemical and structural characteristics such as hydrophobicity and functionality rather than merely molecular size of organic matter. The structural alteration of the HS caused the change in A_{253}/A_{203} ratio, therefore the A_{253}/A_{203} ratio may be a good indicator for changes in reactivity of the NOM with chlorine. The formation of DBPs may be predicted by monitoring the A_{253}/A_{203} ratio, a relatively simple measurement.

References

Thurman E.M. (1985). *Organic Geochemical of Natural Waters*, Kluwer Academic, Boston, MA, USA, 497.

Aiken G.R., Thurman E.M., Malcolm R.L. and Walton H.F. (1979). Comparison of XAD macroporous resins for the concentration of fulvic acid from aqueous solution, *Anal. Chem.* **51**, 1799-1803.

Leenheer J.A. (1981). Comprehensive approach to preparative isolation and fractionation of dissolved organic carbon from natural waters and wastewaters, *Environ. Sci. Technol.* **15**, 578-587.

Yu M.J., Kim Y.H., Han I. and Kim H.C. (2002). Ozonation of Han River humic substances, *Water Sci. Technol.* **46**(11-12), 21-26.

Bellar T.A., Lichtenberg J.J. and Korner R.C. (1974). The occurrence of organohalide in chlorinated drinking water, *J. AWWA* **66**, 703-706.

Rook J.J. (1974). Formation of haloform during chlorination of natural waters, *Water Treat. Exam.* **23**, 234-243.

Schnoor J.L., Nitzschke J.L., Lucas R.D. and Veenstra J.N. (1979). Trihalomethanes yields as a function of precursor molecular weight, *Environ. Sci. Technol.* **13**, 1134-1138.

Vartiainen T., Liimatainen A., Kauranen P. and Hiisvirta L. (1988). Relations between drinking water mutagenicity and water quality parameters, *Chemosphere* **17**, 189-202.

Thurman E.M. and Malcolm R.L. (1981). Preparative isolation of aquatic humic substances, *Environ. Sci. Technol.* **15**, 463-466.

Lin C-F., Liu S-H. and Hao O.J. (2001). Effect of functional groups of humic substances on UF performance, *Water Res.* **35**(10), 2395-2402.

USEPA. (1995). Methods for the determination of organic compounds in drinking water: supplement 3. EPA600R95131.

Ma H., Allen H.E. and Yin Y. (2001). Characterization of isolated fractions of dissolved organic matter from natural waters and a wastewater effluent, *Water Res.* **35**(4), 985-996.

A New Treatment Process for Removal of NOM by Adsorption and Regeneration of Granular Ferric Hydroxide (GFH)

A. Genz*, B. Baumgarten*, M. Jekel*

*Department of Water Quality Control, Technical University Berlin, Sekr. KF 4, Strasse des 17. Juni 135, 10623 Berlin, Germany

Abstract Adsorption onto Granular Ferric Hydroxide (GFH) with subsequent regeneration was studied as a new removal process for natural organic matter (NOM). Berlin groundwater and Suwannee River NOM were studied in terms of adsorption equilibrium, filter dynamics and regeneration of the adsorbent. With regard to the NOM structure, prerequisites for adsorption onto GFH are both a minimum amount of functional groups and molecular size small enough to enter the GFH pores. Regeneration was successfully performed in lab-scale using 0.1 M sodium hydroxide and 3% hydrogen peroxide as regenerants.

Keywords DOC isotherms, Fixed-bed adsorption filters, Breakthrough curves, Oxidative regeneration

1 Introduction

1.1 The treatment process

Aquatic natural organic matter (NOM) may cause several health-related and technical problems during drinking water purification and distribution. Direct issues are associated with the presence of naturally colored groundwater and surface waters containing higher amounts of fulvic and humic substances. The latter might additionally influence other treatment processes that do not aim directly on NOM removal, e.g. particle separation or membrane filtration. Humic substances are well known for their potential to enhance the transport of heavy metals or organic micro pollutants due to complexation in the aquatic environment. Indirect issues arise as a consequence of a potential disinfection step, like the formation of potentially harmful disinfection by-products or biological re-growth in the distribution system.

The removal of NOM by conventional and advanced treatment processes (flocculation, adsorption on activated carbon, oxidation and membrane processes) is not always adequate with regard to removal efficiency, chemical and energy consumption, and management of residuals.

A new adsorption technology for NOM removal is developed using fixed-bed adsorbers with granular ferric hydroxide (GFH) as an adsorbent. In a first step, the adsorbent is loaded with NOM until the maximum loading capacity is reached. In a second step, the adsorbent is regenerated either by NaOH or oxidatively by H_2O_2.

GFH was developed at the Department of Water Quality Control (Technical University Berlin) primarily for adsorptive removal of arsenic from drinking water (Jekel and Seith

© 2004 IWA Publishing. *2nd IWA Leading-Edge Conference on Water and Wastewater Treatment Technologies* edited by Mark Van Loosdrecht & Jonathan Clement. ISBN: 1 84339 508 8.

2000). GFH is 100 % synthesized akaganeite (ß-FeOOH) which is a poorly crystallized but highly active modification of iron hydroxide with a specific surface area of 280 m^2/g. Table 1.1 gives a summary of its chemical properties. It was found that the loading capacity of GFH for arsenic is significantly reduced by the presence of fulvic acids indicating a competitive adsorption. This was the starting point for further studies regarding the adsorption of organic compounds onto GFH which are now expanded to the adsorption of NOM.

Table 1.1 Properties of Granular Ferric Hydroxide

grain size	0.32 - 2 mm
BET surface area	51 - 292 m^2/g
porosity (grains)	0.72 – 0.77
bulk density (wet/dry)	1250 g/L / 560 g/L
water content	54 - 60 %
pH_{PZC} (point of zero charge)	pH 7.0 - 8.2
iron content	0.62 – 0.64 g/g dry matter

1.2 Actual state of knowledge

Numerous authors investigated the adsorption of humic acids onto iron oxides/hydroxides (Tipping 1980; Gu et al. 1994; Edwards et al. 1996). The most important mechanism was found to be ligand exchange between the acidic carboxylic and phenolic groups of the humic substances and the hydroxyl groups of the hydrated iron oxide surface. Additionally, coulombic and hydrophobic interactions were discussed involving aromatic and aliphatic parts of the molecules.

Concerning the adsorption of fulvic and humic acids (Sigma Aldrich, Roth) and single organic acids onto GFH, loadings of 30-60 mg DOC per g were achieved (Teermann and Jekel 1999). The pH dependence of adsorption was relatively low for these fulvic acids, and a co-adsorption of calcium at pH values above the point of zero charge (pH_{PZC}) was assumed. At small adsorbent/solvent ratios a steep increase of the fulvic acid isotherms was observed and interpreted as multilayer adsorption and/or as formation of calcium fulvates.

With regard to regeneration, complete desorption of fulvic acids and reactive dyes was achieved by use of H_2O_2 or NaOH (Teermann and Jekel 1999; Kornmüller et al. 2002). NaOH increases the pH leading to poor adsorption conditions and thus to desorption of the organic substances. The regenerate is alkaline and requires subsequent disposal. Regeneration with H_2O_2 should lead to an oxidation and partial mineralization of the adsorbed organic molecules. H_2O_2 is catalytically activated at the GFH surface following a radical chain mechanism proposed by Lin and Gurol (1998). In this case the regenerate should be easily biodegradable.

2 Methods

2.1 Experimental procedures

GFH was dried at room temperature for 16h, ground in a mill and sieved to a fraction <63 μm (water content 22 %). For determination of adsorption isotherms, the adsorbent mass was varied (0.01 to 1 g dry matter) at constant solution volume (200 mL), initial DOC concentration and pH. To develop pH-loading curves, a constant mass to volume ratio of 0.5 g/L and an initial DOC of 5 mg/L were used while the pH was varied between 4 and 11.

The flasks were shaken for 48 h (a pre-determined equilibrium time) at 20 °C and filtered by 0.45 μm cellulose nitrate filters.

As adsorbates, Suwannee River NOM (SR NOM, purchased from the IHSS, isolated by reverse osmosis) and Berlin groundwater (BGW) were studied.

For filter studies, small columns (diameter 2.5 cm, height 20 cm) filled with a medium size GFH fraction (0.56-1 mm) were continuously fed upstream with BGW without interim back flushing (filter velocity 1.5 m/h).

For regeneration studies a differential bed reactor (DBR) was used as described by Sontheimer et al. (1988). From a well-mixed storage tank (V=0.5 L) the solution was cycled for 48h over a bed of glass beads containing a thin layer of adsorbent (m=0.5 g dry matter). The loaded columns were regenerated after the first loading by cycling NaOH or H_2O_2 solutions for 2h. For reconditioning, columns were rinsed with deionized water and the pH was adjusted before reloading.

2.2 Analytics

UV absorption (254 nm) and color (436 nm) were determined using a Lambda 12 UV/VIS-spectrophotometer (Perkin-Elmer) with cuvettes of 1 cm length. Dissolved organic carbon (DOC) was analyzed by thermal-catalytic oxidation using a highTOC analyzer (elementar Analysensysteme, Germany). The pH was measured by a WTW pH 340-meter using a Sentix 41-electrode (WTW) calibrated with technical standard solutions.

Advanced NOM characterization was carried out using SEC-OCD (size exclusion chromatography – organic carbon detection) following the method of Huber und Frimmel (1996). This procedure allows a separation by molecular size with an online UV and OC detection dividing up the NOM fractions into polysaccharides, humic substances, hydrolysates (building blocks), low molecular acids and amphiphylics. The analyzer was equipped with a column (2x25 cm) filled with Toyopearl Gel HW 50 S with a separation size of 105 g/mol. Oxidation was performed in a thin film reactor (by Alfred Graentzel) with subsequent IR detection (WellChrom K 200 UV-detector).

3 Results

3.1 Adsorption equilibrium

Adsorption characteristics of Berlin groundwater and Suwannee River NOM

Despite the different origin of BGW and SR NOM, comparable loadings of up to 35 mg DOC/g dry matter were observed at pH 6.5 (Figure 3.1). The SR NOM isotherm shows a non-adsorbable DOC fraction of 0.4 mg/L and a smaller slope compared to the BGW NOM that exhibits a non-adsorbable portion of 1.1 mg/L.

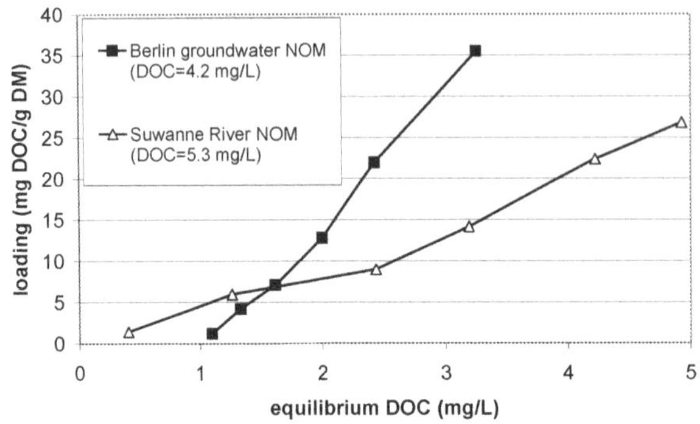

Figure 3.1 DOC isotherms at pH 6.5

Results from SEC-OCD revealed that, in particular, the smaller molecular fractions of BGW NOM (building blocks, low molecular weight acids) are poorly or not adsorbable whereas the larger and UV-active fractions (mainly fulvic acids) are well adsorbable (see Figure 3.5). In comparison, the SR NOM revealed a different behaviour. In this case, the larger and UV-active fulvic acid fraction is poorly adsorbable compared to the fulvic acids in the medium size range of the SR NOM. This finding might arise from a higher content of larger molecules of the SR NOM. The two NOM samples exhibited a different molecule size distribution (Figure 3.2) especially in the range of smaller retention times. The GFH pores were determined with a maximum size of 10 nm (Teermann and Jekel 1999). Consequently, part of the pores might be too small for the larger SR NOM molecules to diffuse through. The size of SR fulvic acids were determined by Aiken et al. (1994) with 7-11 Å at pH 6–9.

Influence of pH, ionic strength and presence of ions on adsorption equilibrium

The adsorption of SR NOM shows a quite strong pH influence, but only a low influence of ionic strength (NaCl), indicating a ligand exchange mechanism rather than an electrostatic mechanism (Figure 3.3).

A concentration of 1 mmol/L phosphate suppresses NOM adsorption completely at pH>6 due to competition, while with decreasing pH the adsorption is improved again, probably due to an increased overall number of adsorption sites. Phosphate is known for its ability of specific adsorption onto iron oxides even above the pH_{PZC} (Geelhoed et al. 1998). The presence of sulfate (10 mmol/L) leads to an equal decrease in NOM loading over the whole pH range, although no adsorption of sulfate was expected at pH values above the point of zero charge (pH_{PZC}) due to a negative charge of the GFH surface. The decrease in loading is smaller compared to phosphate. The NOM adsorption is improved by the presence of calcium at pH>7 due to co-adsorption. Below pH 7, no effects are observed.

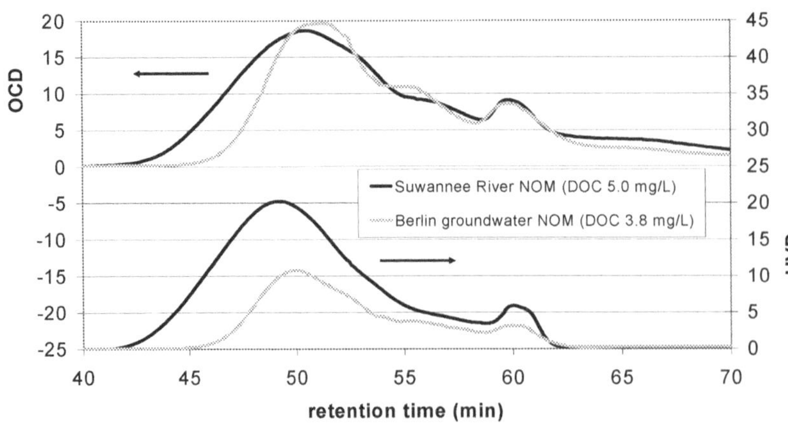

Figure 3.2 SEC-OCD diagram of Berlin GW NOM (black) and SR NOM (grey)

The mechanisms for adsorption of NOM appear to be similar to those found by other authors for the adsorption of small organic acids onto iron oxides (Filius et al. 1997). With decreasing pH an increasing number of positively charged surface groups ($-OH_2^+$) are formed which can be easier exchanged by negatively charged NOM acidic functional groups.

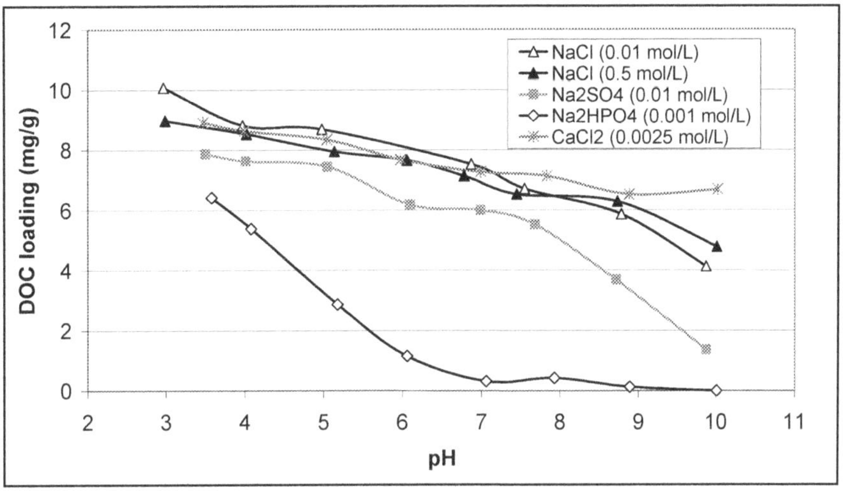

Figure 3.3 pH-loading curves for SR NOM (m/L=0.5 g/L, DOC_0=5 mg/L)

3.2 Filter dynamics

Studies of filter dynamics (Figure 3.4) for BGW NOM showed a relative rapid 50%-breakthrough of DOC after 1000 bed volumes (BV) while UV active compounds were removed preferentially (50%-breakthrough after 4000 BV). The non-adsorbable DOC fraction (1.5 mg/L at pH 7.8) breaks through instantaneously. SEC-OCD measurements during the filter run (Figure 3.5) reveal a preferential removal of the larger fulvic acid fraction. At the beginning of the filter run, almost the entire DOC can be removed with exception of the low molecular weight acids.

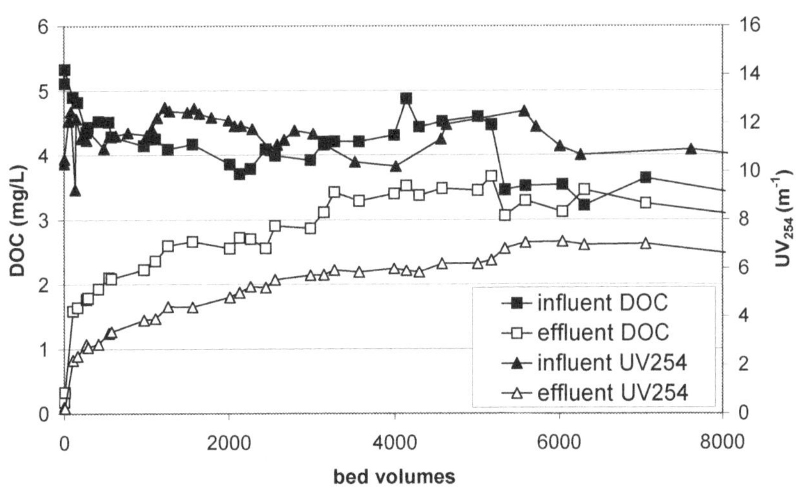

Figure 3.4 Filter breakthrough curves for Berlin GW (pH 7.8, EBCT=10 min, v_F=1 m/h)

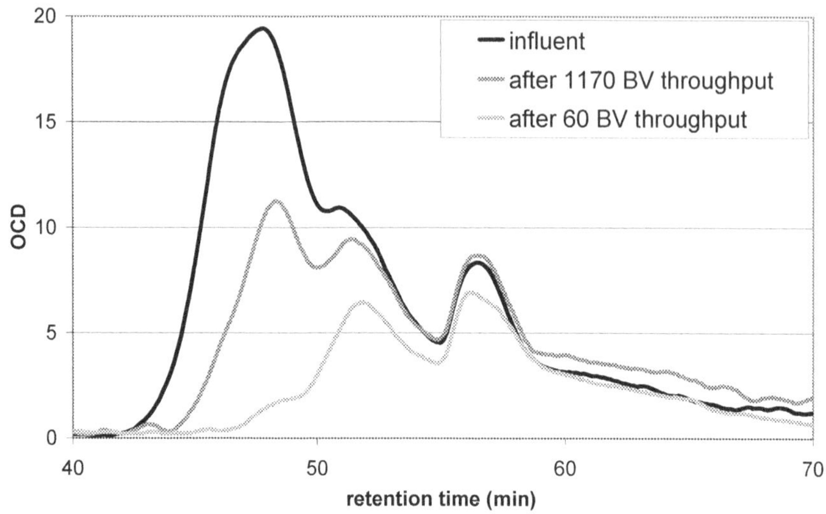

Figure 3.5 SEC-OCD diagram for Berlin groundwater and GFH filter effluents

The minimum empty bed contact time required for the mass transfer from the liquid to the solid phase was determined to be 5 minutes by sampling the column over height corresponding to increasing EBCT. For the used filtration rate of 1 m/h, the mass transfer was dominated by external mass transfer (film diffusion).

3.3 Regeneration

Effects of regeneration are shown in Figure 3.6. The black curve represents the original BGW before adsorption onto GFH showing a typical groundwater DOC distribution. The DOC desorbed from GFH by NaOH (dark grey curve) is nearly unaltered in its structure, whereas the DOC after H_2O_2 regeneration (light grey curve) is shifted mainly to smaller and more oxidized molecules that do not exhibit UV activity.

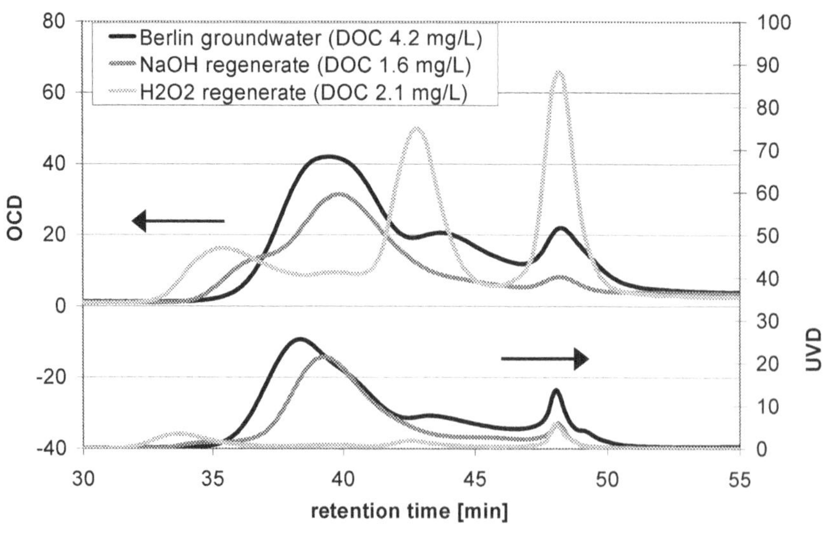

Figure 3.6 SEC-OCD diagram for BGW NOM (black curve), 0.1M NaOH (dark grey) and 3% H_2O_2 (light grey) regenerates

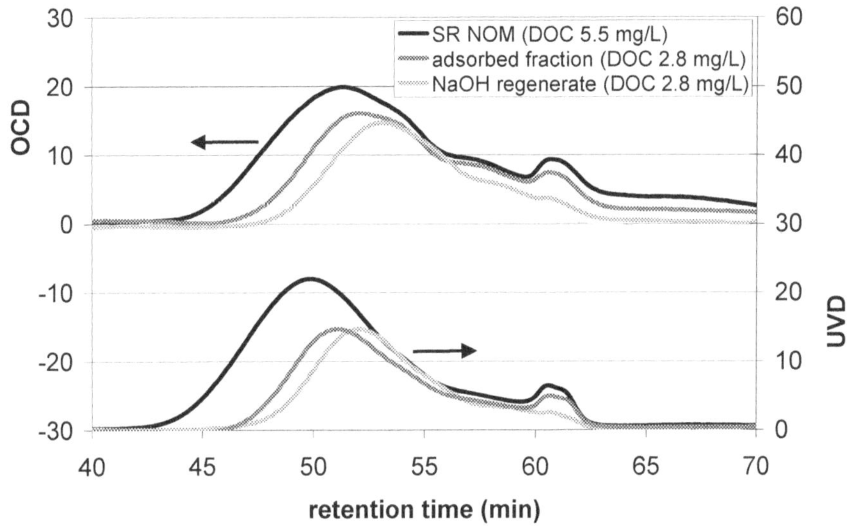

Figure 3.7 SEC-OCD and UVD diagram for SR NOM (black curve), the adsorbed fraction (dark grey) and 0.1M NaOH (light grey) regenerate

For SR NOM, Figure 3.7 illustrates regeneration of GFH loaded with SR NOM. Additionally given is the adsorbed fraction of DOC calculated as the difference between initial DOC and residual DOC in solution (dark grey curve). In contrast to Berlin GW, the smaller fulvic acid fraction is preferentially adsorbed, while the UV-active and coloring substances are poorly adsorbed. The DOC desorbed by NaOH almost equals the adsorbed DOC with a slight shift to smaller molecular sizes, indicating that some of the larger adsorbed molecules remain on the solid phase. In the H2O2 regenerate, no DOC was detectable pointing on an enhanced biodegradability of the desorbed SR NOM.

Loading-regeneration cycles performed in DBRs with Berlin GW NOM (Figure 3.8) showed a constant loss in reloading capacity with increasing number of reloading cycles for NaOH (0.1 M), while for H_2O_2 an initial decrease compared to the first loading is observed followed by an almost constant reloading capacity.

Carbon mass balances exhibited that a part of the adsorbed NOM can neither be desorbed from GFH with NaOH nor oxidized with H_2O_2. For BGW, this fraction amounted to approximately 10%, and from SEC-OCD it is concluded that this fraction mainly consists of the larger fulvic acid molecules. For SR NOM, this fraction is significantly smaller (<5%) and consists mainly of larger humic substances and, to a lesser extent, building blocks.

Regarding the required concentration of regenerants, it was found that 0.01M NaOH should be the preferred concentration in practical use. By using 0.1M or 1M NaOH, no benefit is obtained for the reloading capacity with SR and BGW NOM. By use of 0.3% H_2O_2, no acceptable regeneration could be achieved. Therefore, 3% should be the preferred concentration for H_2O_2 regeneration.

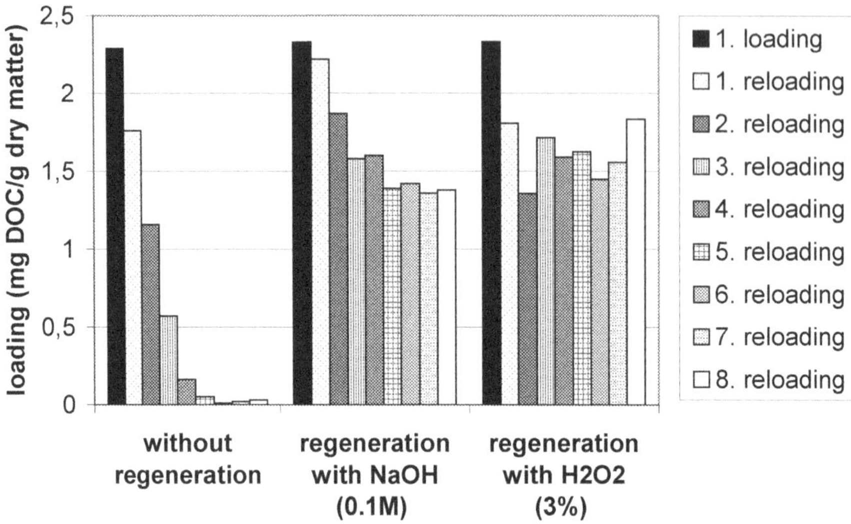

Figure 3.8 Initial loading with Berlin GW NOM and reloadings 1-8 without and after regeneration with NaOH and H_2O_2 in differential bed reactors (pH 7.5, m/L=1 g/L, regeneration time 2h)

4 Evaluation of the treatment process

Based on results from batch tests and lab-scaled filter tests, GFH proved to be an applicable and hydraulically stable adsorbent for NOM removal. In general, an adsorptive treatment may provide advantages in comparison to other conventional water treatment techniques, e.g. no adding of chemicals, no sludge production and low maintenance and operation costs. However, GFH must compete with other suitable adsorbents like activated carbon, macro porous anion exchange resins or other metal oxides (Fettig 1999). A potential advantage of GFH may arise from the selectivity with regard to a certain target substance group and from the possibility of an in-situ regeneration.

However, the DOC loadings achievable with GFH are significantly lower than those known from activated carbon. Cost calculations need to be performed considering production and regeneration costs, of which the latter depend not only on the efficiency of regeneration but also on the number of regeneration cycles.

In principle, a regeneration of GFH with both NaOH and H_2O_2 is possible in terms of stability of adsorbents. The reloading capacity was determined to be 70 - 90% of the initial capacity. H_2O_2 provides advantages concerning disposal (biodegradability of regenerants), but the handling proved to be more difficult due to gas production during the decay of H_2O_2 on the iron oxide surface, which might be a problem in up-scaling. The use of an fluidized-bed reactor might be advantageous in this view. Regeneration using NaOH was successful even after 12 reloading cycles in lab scale. Due to buffering of the iron oxide surface, a conditioning step is necessary before reloading. The disposal is more difficult (regenerant is alkaline). Since the reloading efficiency (and not the desorbed mass) is the decisive parameter for an effective regeneration, long-term scaled tests need to be performed to evaluate the regeneration process.

Furthermore it has to be emphasized that GFH is only capable of removing certain NOM fractions and that there is always a non-adsorbable fraction that cannot be removed in a fixed-bed adsorber. Condition for a practical application of GFH is the presence of well adsorbable fractions. These structural NOM characteristics are 1) a maximum NOM size which enables passing through the GFH pores, and 2) a minimum number of functional groups for the ligand exchange. For BGW and SR NOM, especially the medium and larger UV-active fulvic acid fractions were well adsorbable.

References

Aiken, G., Brown, P., Noyes, T. and Pinckney, D. (1994). Molecular size and weight of fulvic and humic acids from the Suwannee River. In: Humic substances in the Suwannee River, Georgia, U.S. Geological Survey Water Supply Paper. 2373.

Edwards, M., Benjamin, M. M. and Ryan, J. N. (1996) Role of organic acidity in sorption of natural organic matter (NOM) to oxide surfaces. Colloids and Surfaces A 107, 297-307.

Fettig, J. (1999) Removal of humic substances by adsorption/ion exchange. Wat. Sci. Tech. 40(9), 173-182.

Filius, J., Hiemstra, T. and Van Riemsdijk, W. H. (1997) Adsorption of small weak organic acids on goethite: Modeling of mechanisms. J. Colloid Interface Sci. 195, 368-380.

Geelhoed, J. S., Hiemstra, T. and Van Riemsdijk, W. H. (1998) Competitive interaction between phosphate and citrate on goethite. Environ. Sci. Technol. 32, 2119-2123.

Gu, B., Schmitt, J., Chen, Z., Liang, L. and McCarthy, J. (1994) Adsorption and desorption of natural organic matter on iron oxide: Mechanisms and models. Environ. Sci. Technol. 28, 38-46.

Huber, S. and Frimmel, F. (1996) Size-exclusion chromatography with organic carbon detection (LC-OCD): A fast and reliable method for the characterization of hydrophilic organic matter in natural waters. Vom Wasser 86, 277-290.

Jekel, M. and Seith, R. (2000) The removal of arsenic: Comparison of conventional and new techniques for the removal of arsenic in a full-scale water treatment plant. Water Supply 18(1), 628-631.

Kornmüller, A., Karcher, S. and Jekel, M. (2002) Adsorption of reactive dyes to granulated iron hydroxide and its oxidative regeneration. Wat. Sci. Tech. 46(4-5), 43-51.

Lin, S. and Gurol, M. (1998) Catalytic decomposition of hydrogen peroxide on iron oxide: Kinetics, mechanism and implications. Environ. Sci. Technol. 32, 1417-1423.

Sontheimer, H., Crittenden, J. C. and Summers, R. S.(1988) Activated carbon for water treatment. Karlsruhe, Denver:DVGW-Forschungsstelle am Engler-Bunte-Institut der Universität Karlsruhe (TH), AWWA Research Foundation.

Teermann, I. and Jekel, M. (1999) Adsorption of humic substances onto ß-FeOOH and its chemical regeneration. Wat. Sci. Tech. 40(9), 199-206.

Tipping, E. (1980) The adsorption of aquatic humic substances by iron oxides. Geochim. Cosmochim. Acta 45, 191-199.

Removal of AOC and NOM from water with Nanofiltration

Hans S. Vrouwenvelder*, Erwin F. Beerendonk*, Jil Verdouw* and Dick van der Kooij*

*Kiwa Water Research, Groningenhaven 7, 3433 PE Nieuwegein, the Netherlands

Abstract Assessment of the utility of nanofiltration (NF) for reducing the microbial growth potential of water requires information about the removal efficiency of low molecular-weight biodegradable organic compounds. For this purpose, the effect of NF on easily assimilable organic carbon (AOC) and NOM was tested using three types of membranes, one with a molecular-weight cut-off (MWCO) of 1000 D and two with MWCO of 200 D, in a single-element test unit (SETU) at different pH values without and with dosage of low concentrations of acetate and ethanol to the feed water. NOM was effectively removed by these membranes at pH 8, but the removal of acetate ranged from very limited (at MWCO of 1000 D) to >90% (at MWCO of 200 D). The membranes with an MWCO of 200 D showed a poor removal of acetate and ethanol at pH 5. One of the membrane types with MWCO of 200 D was also used in a SETU fed with ozonated water with an elevated AOC concentration at pH 8. AOC concentrations were reduced from 67 to 1.8 µg AOC-C/l (site 1) and from 40 to 1.0 µg C/l (site 2), respectively. Also, NOM was largely removed (DOC > 85%, UV absorbance > 92%). Biofouling may however limit the direct application of spiral-wound NF membrane systems on water with elevated AOC concentrations. Consequently, NF may be used after biological filtration to achieve water with a high degree of biological stability.

Keywords AOC, biological stability, nanofiltration, water treatment

Introduction

Continuing discussions about further improvement of drinking water quality and the decreasing costs of membrane technology are stimulating research into the application of membranes in water treatment. In the Netherlands, the water supply companies aim at distributing biologically-stable drinking water without a disinfectant residual (Van der Kooij et al. 1999). Increased concerns about the multiplication of *Legionella* in plumbing systems in buildings and households have raised questions about the effects of water quality on biofilm formation and growth of *Legionella* in such systems. Nanofiltration (NF) removes natural organic matter (NOM) and inorganic compounds, organic and inorganic contaminants and also is a barrier against microorganisms. Hence, a considerable improvement of water quality may be achieved with NF. However, limited information is available on the removal efficiency of NF for biodegradable compounds. Escobar and Randall (1999) reported that NF did not remove AOC from acidified groundwater, although BDOC and DOC concentrations were greatly reduced. Sibille et al. (1997) also observed a considerable BDOC reduction by NF, but still some regrowth occurred in the system supplied with this water.

© 2004 IWA Publishing. *2nd IWA Leading-Edge Conference on Water and Wastewater Treatment Technologies* edited by Mark Van Loosdrecht & Jonathan Clement. ISBN: 1 84339 508 8.

To elucidate the effects of NF on concentrations of AOC and NOM three types of commercially available NF membranes with different pore sizes were used in a single element test unit (SETU) with and without dosage of selected compounds to the NF feed water at different pH values. Additionally, the SETU with a selected membrane was also supplied with ozonated water with an elevated AOC concentration in two water treatment plants.

Methods

The AOC test as applied on the feed and product water, is based on determining the maximum colony counts of two selected bacterial cultures growing in a sample of pasteurised water, contained in thoroughly cleaned glass-stoppered Erlenmeyer incubated flasks (in duplicate) at 15°C. The selected bacterial cultures, which had been isolated from drinking water, are *Pseudomonas fluorescens* strain (P17) which can utilise a wide range of organic compounds and *Spirillum* sp. strain NOX, which only utilises organic acids (Van der Kooij and Hijnen, 1984). The AOC concentration is calculated as μg of acetate-C equivalents/l from the maximum colony counts of the individual strains and their yield factor for acetate (Van der Kooij, 1992). Additional water quality parameters included: DOC, UV absorbance (at 254 nm) conductivity and pH. These analyses were conducted with standardised procedures.

The investigations were carried out with three different types (M1, M2 and M3) of spiral-wound NF membranes (Ø 4 inch) in a SETU at a recovery of ± 10% and a flux of ± 20 $l/(m^2.h)$. Water temperature ranged from 12 to 22°C. M1 is a Hydranautics CoRe membrane with a molecular-weight cut-off (MWCO) of 1000 D, M2 is a Trisep TS-80 membrane and M3 is a Koch TFC-S membrane, both with an MWCO of 200 D. The feed water (treated ground water; DOC = 1.9 mg/l) was supplemented with 100 μg C/l of either acetate or ethanol from stock solutions. Before and after 2 hours of dosage, samples were taken of the feed and product water for analysis. The SETU with a selected membrane (M2) was also supplied with ozonated water with an elevated AOC concentration at two water treatment plants. The pre-treatment at one treatment plant (site 1) using river Rhine water was: coagulation, sedimentation, rapid sand filtration, transport, dune passage, abstraction, aeration, rapid sand filtration and ozonation. The ozone dosage was 0.5 to 2.0 g O_3/m^3 at a minimum contact time of 15 minutes. Pre-treatment at site 2 with river Meuse water was: reservoir storage, coagulation, sedimentation, ozonation and dual media filtration (DMF, anthracite and sand). The DMF effluent was fed to the SETU. The ozone dosage was 1 to 2 g O_3/m^3 at a minimum contact time of 7 minutes.

Results

Removal of added biodegradable compounds and NOM

Figure 1 shows the results of the study with dosages of about 100 μg C/l to the feed water of a SETU. The membrane types were all tested with dosages of acetate, but the M2 membrane was also supplied with 100 μg ethanol-C/l. Filtration with the relatively 'open' M1 membrane showed a limited reduction of conductivity (<10%), a reduction of UV absorbance of 70 to 80% and a DOC removal of 77%, respectively The removal of AOC with the M1 membrane also was <50% (Fig. 1). The M2 and M3 membrane types both demonstrated a good conductivity reduction (>92%) and effective removals of UV (>95%), DOC (88% and 91%) and also AOC (90%). Lowering the pH of the feed water from 8 to 5 with the M3 membrane did not affect NOM removal (UV, DOC), but decreased the conductivity reduction to 82% and clearly limited AOC removal (Fig. 1 D). The M2 membrane supplied with ethanol at pH 5 also showed a limited AOC removal but effective removals of DOC, UV, and conductivity, respectively,. Experiments with ethanol at pH 8 were not performed.

Comparison of the feed and product water characteristics prior and during addition of 100 μg C/l showed similar values for DOC, conductivity and UV-absorbance. AOC concentrations of the feed water prior to dosage ranged from 3 to 14 μg C/l and concentrations in the product water ranged from 0.7 to 2 μg C/l, respectively. Dosage of acetate and ethanol gave elevated concentrations of AOC (≥6.7 μg C/l) in the product, demonstrating that part of the dosed compounds passed the membrane.

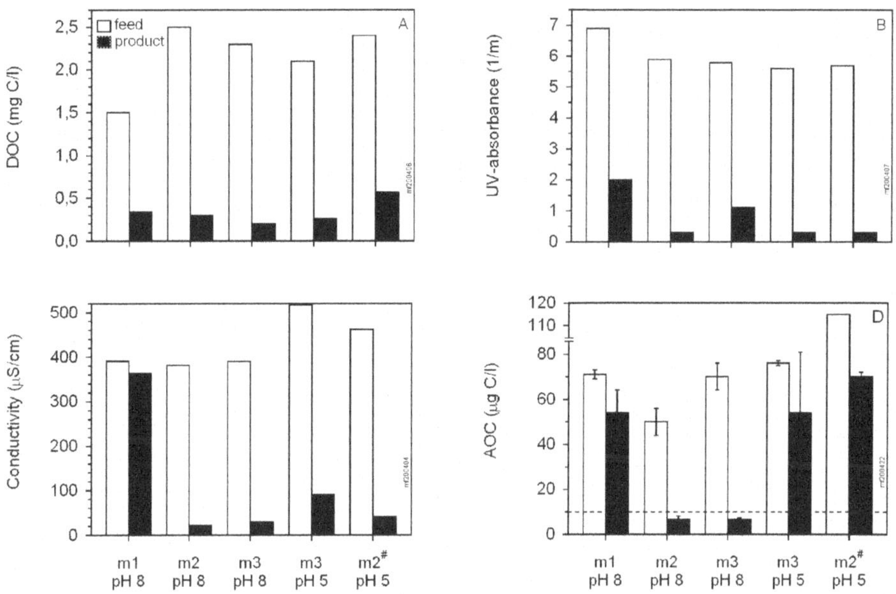

Figure 1: Effects of three NF membrane types (M1, M2 and M3) on selected water quality parameters at pH 8 and 5. (A) Dissolved organic carbon (DOC); (B) UV absorbance at 254 nm; (C) conductivity, and (D) assimilable organic carbon (AOC). # dosage of 100 μg ethanol C/l.

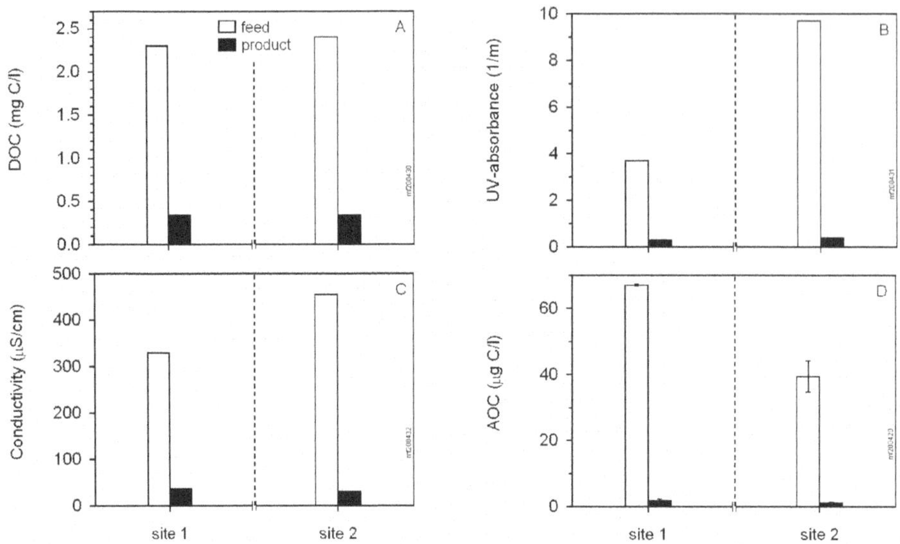

Figure 2: Effect of membrane type M2 on selected quality characteristics of ozonated water (pH 8) at two water treatment plants. (A) Dissolved organic carbon (DOC); (B) UV absorbance at 254 nm; (C) conductivity, and (D) assimilable organic carbon (AOC).

Removal of AOC and NOM from ozonated water

Figure 2 depicts the results of the study with the M2 membrane type in a SETU supplied with water after ozonation at pH 8 (no addition of biodegradable compounds). AOC concentrations were reduced from 67 to 1.8 μg AOC-C/l (site 1) and from 40 to 1.0 μg C/l (site 2), respectively. Also clear reductions of the conductivity (> 90%), the concentration of DOC (> 85%) and UV absorbance (> 92%) were observed.

Discussion

SETU experiments with and without added low molecular weight biodegradable compounds, (acetate, MW 82; ethanol, MW 46) proved to be well suited for studying the removal of AOC and NOM by NF. The obtained results clearly show that the removal capacity of NF membranes for the selected water quality parameters is affected by (i) the MWCO of the membrane used and (ii) the pH of the feed water. The conductivity reduction is related to the MWCO characteristic of the membrane, with only a limited effect at MWCO = 1000 D. This membrane type (M1) also gave a limited removal of AOC, although most NOM was removed. Lowering the pH of the feed water clearly reduced the acetate removal by the M3 membrane with a MWCO value of 200 D. Obviously, undissociated acetic acid molecules (MW 60) pass this NF membrane. A change in membrane charge at lower pH may also affect AOC removal. AOC removal from ozonated water at pH 8 was very effective (>95%) with MWCO of 200 D.

The presented results are in agreement with the observations of Escobar and Randall (1999) who showed that NF did not remove AOC from acidified groundwater (pH not reported), but DOC and BDOC concentrations were clearly reduced (> 90%). In their study at two plants without ozonation the AOC concentrations of the feed water ranged from 60 to 250 μg C/l and the MWCO value (200 D) of the Fluid Systems TFCS membrane was similar to the MWCO of the M2 and M3 membrane types. In another study on NF (MWCO and pH not reported) the AOC-concentration of pre-treated water was reduced from 25 to 4 µg C/l (Yeh et al. 2000). Sibille et al (1997) also observed that NF (pH 8, MWCO not specified) considerably reduced the concentrations of DOC (85%) and BDOC (>70%) of water after ozonation and GAC filtration. However, this treatment only gave a limited reduction of the growth of microorganisms in an experimental distribution system.

A study on integrated membrane systems for evaluation of conventional pre-treatment processes followed by NF or reverse osmosis (RO) showed that at locations without biofouling AOC values of feed water ranged from 3 to 10 µg C/l and AOC values of the product were ≤ 1 µg C/l (Schippers et al., 2004). Biofouling of membranes was observed at an AOC concentration of about 27 µg C/l (Vrouwenvelder, 1998). Dosage of certain antiscalants can increase the growth potential of the feed water and cause biofouling (Vrouwenvelder 2000). Reverse osmosis gave a 94% reduction of an AOC concentration of about 1000 µg C/l in a test period of 3 hours (Ng et al., 2001), but biofouling is inevitable with such a high AOC concentration in the feed water. Consequently, the application of spiral-wound (NF) membranes for the removal of elevated AOC concentrations may be limited because AOC present in the feed water may cause biofouling, resulting in operational problems. Hence, pre-treatment is needed to achieve relatively low AOC concentrations in the feed water and NF will further reduce these AOC concentrations to a very low level. Another option is to improve cleaning procedures for periodic removal of biomass from the membrane.

Further studies will be conducted to define the conditions required for optimal use of NF filtration to improve biostability. These studies will include (i) SETU experiments with dosages of selected nutrients at different pH values and (ii) investigations with model distribution systems supplied with NF treated water to assess the impact on regrowth.

Conclusions

- NF membranes cause significant reductions of NOM concentrations, but the reduction of conductivity and the removal of AOC is limited at low pH-values and/or MWCO of 1000 D;
- Biofouling may hamper the use of NF for direct AOC removal thus requiring biological pre-treatment and/or effective cleaning techniques;
- Further studies are needed to demonstrate the effects of NF on limiting regrowth in distribution systems and water installations at different water types and pretreatment scenarios.

Acknowledgements

This study was conducted within the framework of the Joint Research Program (BTO) of the water supply companies in The Netherlands. The authors acknowledge the technical support of Amsterdam Water Supply and Evides to study the effect of NF on ozonated water.

References

Escobar I. C. and A. A. Randall, 1999. Influence of NF on distribution systems biostability, J. Am. Water Works Assoc., 91 (6): 76-89.

Ng W.J., S.L. Ong and J.Y. Hu. 2001. The effects of water reclamation technologies on biological stability of industrial water. Water Science and Technology. 43 (10), 327 – 334.

Schippers, J.C., J.C. Kruithof, M.M. Nederlof and J.A.M.H. Hofman (2004). Integrated Membrane Systems. Report Awwa Research Foundation and American Water Works Association.

Sibille, I., L. Matthieu, J.L. Paquin, D. Gatel and J.C. Block, 1997. Microbial characteristics of a distribution system fed with nanofiltered drinking water. Wat Res. 31(1997): 2318-2326.

Van der Kooij, D. and W.A.M. Hijnen, 1984. Substrate utilisation by an oxalate-consuming *Spirillum* species in relation to its growth in ozonated water. Appl. Environ. Microbiol. 47:551-559.

Van der Kooij, D., 1992. Assimilable organic carbon as an indicator of bacterial regrowth. J. Am. Water Works Assoc., 84 (2): 57-65.

Van der Kooij, D., J.H.M van Lieverloo, J. Schellart and P. Hiemstra. 1999. Maintaining quality without a disinfectant residual. J. Am. Water Works Assoc. 91(1):55-64.

Vrouwenvelder, J.S., J.A.M. van Paassen, H.C. Folmer, J.A.M.H. Hofman, M.M. Nederlof and D. van der Kooij. 1998. Biofouling of membranes for drinking water production. Desalination 188, 157-166.

Vrouwenvelder, J.S., S.A. Manolarakis, H.R. Veenendaal and D. van der Kooij. 2000. Biofouling potential of chemicals used for scale control in RO and NF membranes. Desalination 132, 1-10.

Yeh H.H., I.C. Tseng, S.J. Kao, W.L Lai, J.J. Chen, G.T. Wang and S.H. Lin. 2000. Comparison of the finished water quality among an integrated membrane process, conventional and other advanced treatment processes. Desalination 131, 237 – 244.

Measuring Discoloration Risk: Resuspention Potential Method

J.H.G. Vreeburg (University of Technology, Delft), P. Schaap (Kiwa Water Research), J.C. van Dijk (University of Technology Delft)

PO BOX 1072, 3430 BB NIEUWEGEIN, The Netherlands
Jan.vreeburg@kiwa.nl; Peter.schaap@kiwa.nl; j.c.vandijk@citg.tudelft.nl

Abstract: Customer complaints are often used as an indicator for the discoloration risk. Many disadvantages are attached to this method of which the most important are that it is neither reproducible nor reliable. The Resuspention Potential Method introduced in this paper is based on a standardised protocol to quantify the mobility of loose deposits in the network and the associated discoloration risk. Within a length of pipe a moderate flow disturbance (0,35 m/s) is invoked during a short time (15 minutes). The reaction of the continuous monitored turbidity gives a good ranking of the discoloration risk. An important application of the RPM is found in decision support and prioritisation of cleaning programs. Wide implementation of the method within the Dutch water companies allowed them to optimise the cleaning programs and actually reduce the number of discoloration complaints.
Keywords: Discoloration, Risk assessment, Water Quality, Drinking Water Networks, Deposits

Introduction

Discoloration of the water is a major cause for customers to complain about the quality of the supplied water. 'Dirty water' may account for 60 to 80% of all water quality related complaints (Polychronopolous 2003, et al.). An objective and reproducible method to quantify the discoloration risk is described in this article. Most important use of such a method is in quantifying the discoloration risk in itself and in assessing the effect of actions to manage this risk.

Discoloration complaints are mostly associated with the presence of iron and manganese, colouring the water red, brown or even black. As source for these materials, initially the corrosion of ferrous materials such as cast iron pipes, valves and fire hydrants are often held responsible for releasing corrosion products. More recently suspended particles in finished drinking water are recognised as a source for loose deposits and sediment in pipes and reservoirs (Gauthier,et al. 2001a and 2001b). In distribution systems suspended particles may be transported and deposited as sedimentary deposits, depending on flow conditions (Slaats, 2003, et al.).

Discoloration events, leading to customer complaints, are mostly connected to 'hydraulic events' causing a disturbance in the normal situation in the network. In a number of cases the relation with hydraulic events is not obvious, but closer study of the event mostly reveals the underlying hydraulic disturbance (Vreeburg, 1996). Typical 'hydraulic incidents' are pipe burst, hydrant use and valve exercise that result in velocity and flow direction changes in the pipes.

© 2004 IWA Publishing. *2nd IWA Leading-Edge Conference on Water and Wastewater Treatment Technologies* edited by Mark Van Loosdrecht & Jonathan Clement. ISBN: 1 84339 508 8.

Discoloration events and customer complaints

Customer complaints are an important identifier for discoloration risk. As a measuring value to quantify the discoloration risk they are not suitable: they are not reproducible or reliable and moreover they are too late. When customers complain the discoloration risk is already too high. Another significant problem with complaints as identifier for discoloration risk is in most cases a lack of historic records. Increase or decrease in number of complaints then is difficult to analyse. Lastly customers tend to get used to a certain level of turbidity and don't report it anymore to the water company. Positive or negative effects of remedial actions cannot be monitored accurately, let alone be reproduced. For these reasons a method that allows for an objective and reproducible assessment of the discoloration risk is needed.

In this study the discoloration is connected to the turbidity of the water and hydraulically induced turbidity changes to discoloration risk. Turbidity is recognised by the customer as an organoleptic deviation of the normal water quality that may lead him or her to notify the water company. The complaint risk increases with an increasing level of turbidity. Also the time an increased turbidity lasts influences the discoloration risk or better the complaint risk: a customer has more chance to notice the discoloration. Turbidity is better noticed in a white bathtub than during an activity as washing hands. Not only the absolute turbidity level, but especially sudden changes in turbidity cause complaints. A high yet constant turbidity level (for instance in the order of 1-3 FTU) seldom causes complaints whereas a changing turbidity as a sudden rise from 0,3 to 5 FTU does.

In case of a discoloration complaint water samples can be taken and analysed for parameters such as iron, manganese, turbidity and colour. If one succeeds in taking a sample of the actually discoloured water, then all the parameters will show high levels. If samples are taken that do not represent the discoloration event, because the sediment resettled again, the analysis will not be conclusive.

Figure 1 shows the turbidity in a distribution net during a typical discoloration event.

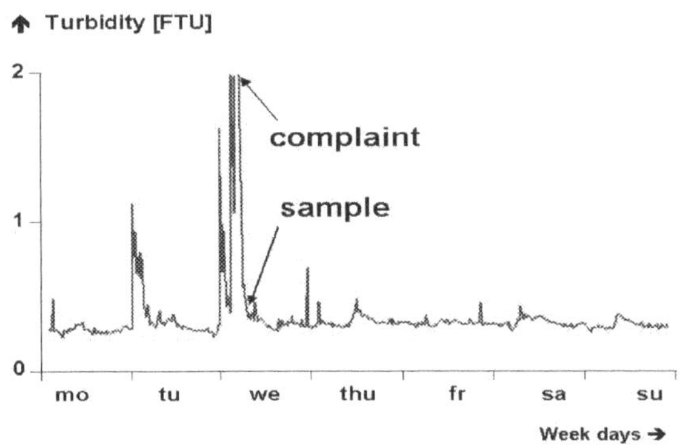

Figure 1 Typical turbidity pattern during a discoloration event in a network.

On Wednesday morning turbidity rises as result of a hydraulic disturbance (use of a fire hydrant). The complaint arises at the peak of the turbidity rise, while the first opportunity to take a sample is after the peak occurred. Turbidity in the rest of the network was not affected (not shown).

Conclusion is that a discoloration event will take place if two restrictions are met: A disturbance of the normal flow velocity in a pipe together with some resuspendable sediment. This makes the presence of loose and resuspendable deposits one of the major risk factors causing discoloration.

Loose sedimentary deposits in the network

Resuspendable deposits in combination with changes in the hydraulic regime lead to discoloration events. The origin of the deposits can be multiple. As often-assumed corrosion of ferrous materials in contact with drinking water can produce solid materials that either is transported to the customer or can deposit in the pipes. Particulate matter in finished water is also an important source for loose deposits (Slaats 2003, et al.). In case of filtered water (Colton,1996 et al.) found that 40% of all particles that pass into supply during a 48-hour filter run do so in the first hour of operation. This agrees with findings that the turbidity of finished water evaluated with grab samples can be quite low, while the fouling in a network can be quite considerable. Irregularities in turbidity, which are not detected with the daily grab samples, do contribute significantly to the sediment load of a network.

Once deposited loose sediment may cause discoloration when velocity increases. Apart from the amount of sediment present in the pipes, the mobility is of importance in determining the discoloration risk. Mobility is partly determined by the type and nature of the sediment: relative heavy particles as sand grains will settle quickly. Lighter particles, mostly of organic origin and iron flocks will take more time to deposit and are easily resuspended. Also the roughness of the pipe wall influences the mobility of the sediment and may vary over the network. Gauthier (2001, et al.) found that the loose deposits primarily consist of inorganic material (90%).

The Resuspention Potential Method (RPM)

Irrespective of the origin, the presence and mobility of loose deposits determines the discoloration risk. The Resuspention Potential Method (RPM) as introduced in this paper is based on measuring the capability of the sediment in a network to resuspend and create visually noticeable turbidity levels.

The Resuspention Potential Method (RPM) is based on a continuous monitoring of turbidity during a controlled and normalised increase of the velocity in a pipe. The hydraulic shear stress as result of the increased velocity causes sediment to resuspend. Practical restriction in this procedure is of course that the change in hydraulic regime only causes a moderate increase in shear stress to prevent too much increase in turbidity levels leading to actual complaints. Figure 2 illustrates the protocol of the method:

- Isolate the pipe for which the discoloration risk is assessed by closing the valves that connect other feeding pipes to the pipe to be evaluated;
- Induce acceleration of the flow by opening a fire hydrant in such a way that the velocity in the pipe is increased by an additional 0,35 m/s on top of the normal velocity and maintain the higher velocity during fifteen minutes.

  The velocity of 0,35 m/s is empirically determined. For a 100 mm. internal diameter pipe this requires an extra volume flow of 2,78 l/s (10 m^3/h), which leads to a moderate disturbance. The increased levels of turbidity are very well measurable, and in most cases don't exceed unacceptable levels.
- Monitor the turbidity in the pipe
- After fifteen minutes, reduce the velocity to normal and monitor the turbidity until it is back at the starting or initial level.

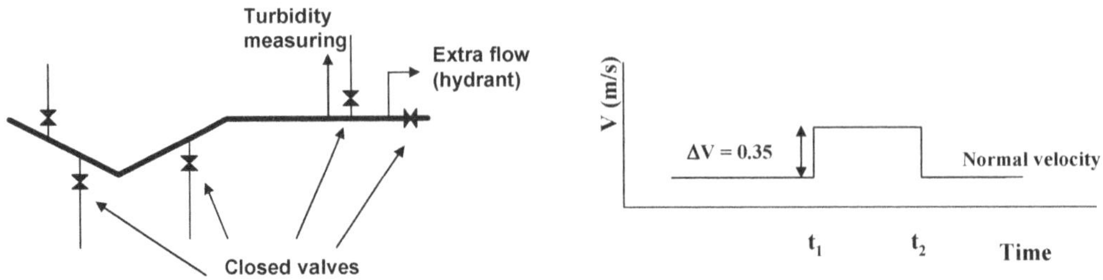

Figure 2 Principle Resuspention Potential Method: Increase velocity in isolated pipe

Materials and methods

The RPM is developed for smaller water mains ranging from 80 to 200 mm internal diameter. The velocity needed to induce the increase of the velocity is dependant of the internal diameter of the pipe. This ranges from 1,76 l/s (6,3 m^3/h) for a 80 mm pipe tot 11,0 l/s (40 m^3/h) for a 200 mm pipe. Typical values for a 100 mm pipe are 2,78 l/s (10 m^3/h).

Opening a hydrant creates the extra flow needed for the increased velocity. Figure 3 shows the standpipe developed for this purpose. The flow is restricted using calibrated plates in the standpipe. Applying this pipe with the plates limits the flow to the desired one without the risk of extra volume flow when opening the hydrant. In normal practice a hydrant with a standpipe is 'cracked' with opened orifices. This introduces an instant disturbing flow that in most cases is higher than the desired flow for the RPM causing an extra velocity and an extra resuspention of the sediment. To avoid this the flow is limited to the desired flow, which is enough to clear the hydrant from pollution.

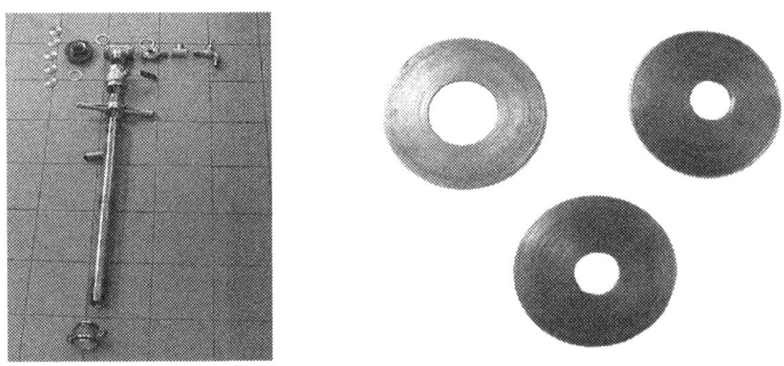

Figure 3 Standpipe and calibre plates for standardized disturbances

For measuring the turbidity several flow through turbidity meters can be used. In the Netherlands the Sigrist KT65 and the Dr. Lange Ultraturb are used, which use different measuring principles. Comparison of the two led to the conclusion that the meters measure comparable turbidity patterns in turbulent solutions, but that the absolute values of turbidity are at different levels. (Slaats 2000, et al.) This should be carefully taken into account when interpreting the results of the RPM measurements.

The flow through the standpipe, responsible for the disturbance in the pipe can be monitored using a flow meter or a water meter. These flows should be recorded because any deviation from the standard disturbance will cause a different sheer stress on the sediment and consequently a different effect on the turbidity. Using well-calibrated plates in the standpipe can eliminate the need for flow measurement.

The RPM-curve

The result of a RPM experiment is the turbidity pattern as observed in the pipe. A typical turbidity curve of the RPM is shown in figure 4. For rating the discoloration risk four elements of the curve are important

These elements are:

- Base turbidity level
- Initial increase in turbidity at the start of the hydraulic disturbance
- Development of turbidity during the hydraulic disturbance
- Resettling time and pattern to base (initial) turbidity level

Base turbidity level

The base turbidity level is the level preceding the hydraulic disturbance. Base line turbidity can be linked to the turbidity of the source water and give some insight in the source of the sediment. Base line turbidity is also needed to judge the time needed for the turbidity to resettle after the increased velocity stopped.

Initial increase in turbidity

Following the actual disturbance the turbidity will rise immediately to a certain level. This initial increase gives the instantaneous mobility of the sediment, resulting in peak turbidity. A loose layer that is immediately available causes the initial turbidity.

The initial increase is an indication of the maximum turbidity caused by a hydraulic incident. A high initial turbidity increases the chance discoloration is noticed.

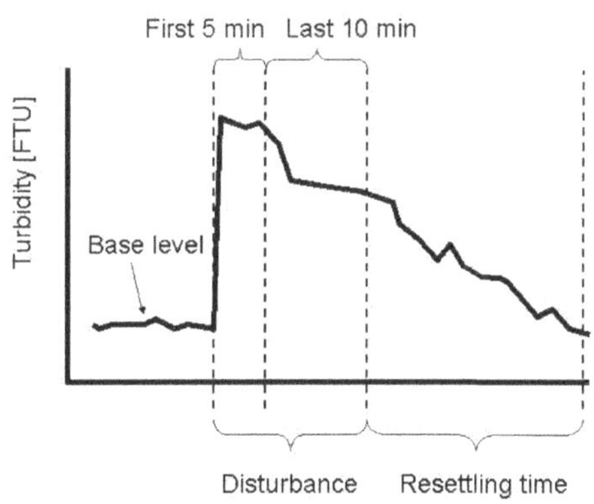

Figure 4 typical RPM-curve

Development of turbidity during the hydraulic disturbance

The hydraulic disturbance is kept for 15 minutes, allowing the turbidity to develop to a stable level. If the turbidity stays at almost the first five minutes level, than the amount of sediment is considerable and composition of the sediment is homogeneous. A quickly reached and during 15 minutes stable level indicates a high discoloration risk.

In many cases the turbidity drops during the 15 minutes disturbance time. Three phenomena can explain this:

- A relatively small amount of heavy sediment is present in the pipe. The extra forces in accelerating the flow also promote the initial whirling up of this heavy sediment. (Blokker, 2003, et al.)The significance of this sediment is limited because it is settling even during the deviating hydraulic circumstances, making the discoloration risk smaller. The chance the initial discoloured water actually reaches a tapping point in which it can be visually identified is small.
- A too short length of isolated pipe can be the second reason for a declining turbidity during the 15 minutes disturbance. If the length is less than 315 meter, water is draw from pipes upstream the isolated pipe. This water originates from the looped network or from pipes with larger diameters and is less disturbed than the 0,35 m/s.
- A non-homogeneous deposit over the length of the tested pipe, for instance in a hilly area with concentrations of sediment in the depression of a pipe.

In all cases however the level of turbidity following the first peak determines the continued discoloration risk. This level will be present during a longer time, allowing customers more time to see it.

Resettling to base level

After closing the hydrant, it takes a certain time for the turbidity to resettle again to base level. The time needed is important for the discoloration risk or actually the complaint risk. If the turbidity stays high during a longer period, the risk of noticing the turbidity in an application as filling a white basin (bathtub, washing bin, bathroom sink, etc) is larger.

RPM and discoloration risk

The RPM is interpreted on five aspects items that are rated equally:

- Absolute maximum value of turbidity during first five minutes of disturbance (20%)
- Average value of turbidity during first five minutes of disturbance (20%)
- Absolute maximum value of turbidity during last ten minutes of disturbance (20%)
- Average value of turbidity during last ten minutes of disturbance (20%)
- Time needed to resettle again to initial turbidity level. (20%)

For each aspect a validation on a scale of 3 is made: 0 is the lowest or best rating and 3 the highest or worst rating. The lowest value equivalent with 'no discoloration risk' is 0 (zero) and the highest value or 'maximum discoloration risk' is 15.

For the rating per aspect a scale must be made, that is calibrated to the turbidity equipment used. Also site-specific elements can be taken into account. If for instance the intuitive feeling of a network is that the discoloration risk is moderate, then the rating scale can be adjusted to this level. With the RPM the intuitive feeling is objectified. Changes in the discoloration risk that may occur when for instance the treatment is improved or a cleaning program is started can be related to the objectified level. The discoloration risk established in this way is a relative figure that can be company-specific or even area specific, allowing for assessing effects of changed operation of the network. For every situation and measuring equipment a ranking table can be made, depending on the type of turbidity measuring equipment and local circumstances.

Tables 1 and 2 give the values for discoloration risk for the Sigrist turbidity measure and the Dr Lange equipment.

Table 1 Example of ranking RPM for discoloration risk using the Sigrist equipment

	0	1	2	3
Absolute max first 5 min	<0,3 FTU	0,3-1,0 FTU	1,0-2,4 FTU	>2,4 FTU
Average first 5 min	<0,3 FTU	0,3-1,0 FTU	1,0-2,4 FTU	>2,4 FTU
Absolute max last 10 min	<0,3 FTU	0,3-1,0 FTU	1,0-2,4 FTU	>2,4 FTU
Average max last 10 min	<0,3 FTU	0,3-1,0 FTU	1,0-2,4 FTU	>2,4 FTU
Time to clear	< 5 min.	5-15 min	15-60 min	>60 min

Table 2 Example of ranking RPM for discoloration using the Dr Lange equipment

	0	1	2	3
Absolute max first 5 min	<3 FTU	3 – 10 FTU	10-40 FTU	>40 FTU
Average first 5 min	<3 FTU	3 – 10 FTU	10-40 FTU	>40 FTU
Absolute max last 10 min	<3 FTU	3 – 10 FTU	10-40 FTU	>40 FTU
Average max last 10 min	<3 FTU	3 – 10 FTU	10-40 FTU	>40 FTU
Time to clear	< 5 min.	5-15 min	15-60 min	>60 min

The ranking tables give a certain flexibility to tailor the results for a specific application. If the RPM is used to prioritise areas for an initial cleaning program, the maximum values can be adjusted to the outcomes of the measurements. For instance most of the measurements end up with a total value larger than 12, because the average turbidity over the whole period is larger than 40 FTU. By adjusting the ranges it is possible to distinguish very high and high degrees of discoloration risk enabling the desired prioritisation.

Typical RPM-curve

Figure 5 gives an example of a RPM with a high discoloration risk. The measurement has been done with Sigrist-equipment, so table 1 is applicable. The base turbidity is rather constant and low (0,22 FTU). The initial turbidity during the first five minutes as well as during the following 10 minutes is high and above 2,4 FTU, which is the maximum turbidity that can be measured in this setting. The time to clear is several hours, so on all the items the maximum score is reached, resulting in a maximum score of 15. Despite the fact that the base line turbidity is very constant and low, this location will experience a high turbidity with any hydraulic event.

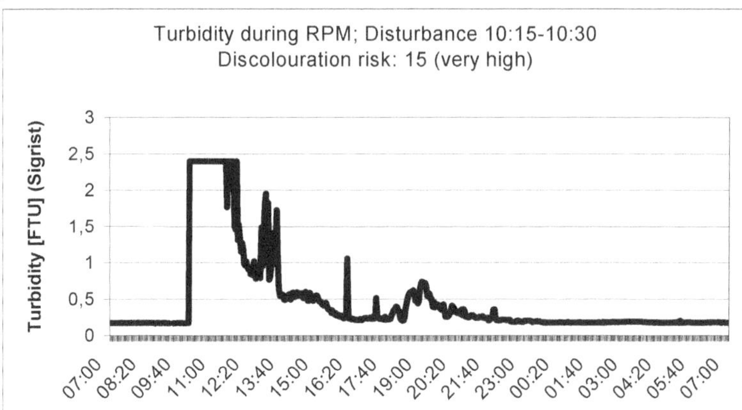

Figure 5 Example of RPM, measured with Sigrist-equipment, indicating a high discoloration risk.

Experience with RPM and discoloration risk in the Netherlands

The RPM has been developed within the joint research program of the Dutch water companies (Bedrijfstakonderzoek BTO) and has been applied for more than a decade by the water companies. At present the method is used mostly for analysis of need for cleaning pipes and networks. Following a cleaning action the method is used to evaluate the effectiveness of the cleaning. Regular assessment of the resuspention potential in a cleaned network gives information on the frequency of cleaning. Figure 6 gives the result of a RPM before and after cleaning by flushing a pipe.

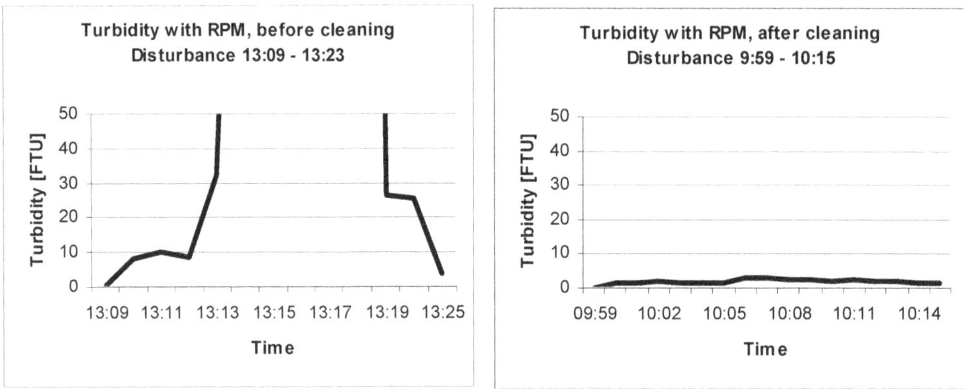

Figure 6 Left: pre-cleaning, Discoloration risk is high (>12); post cleaning discoloration risk is very low (0)

Alternative analysis of the RPM

A full evaluation of the RPM requires a measuring time of at least a few hours, mostly consumed by determining the initial base level and pattern and the resettling time. It also requires turbidity measuring equipment that can be costly and requires skill to operate. The implementation of the principle of the RPM during the last years led to some alternatives in the analysis methods. The first adjustment is to limit the measuring time of the base level turbidity to some minutes prior to the disturbance or the time needed to isolate the pipe. Limiting the time used to measure the resettling time to a specific time, for instance 30 minutes is another adjustment. The examples of figure 6 use this abbreviated version of the RPM. As seen the resettling time is not clear in the first graph. However the overall impression is that the discoloration risk is so high (>12) action is required.

Limitation of the measuring time allows for more measurements in a working day. It is possible to do 4 measurements an 8-hour working day. Replacing the continuous monitoring by taking 5 to 8 grab samples during and after the disturbance limits the total measuring time even further. All the adjustments however cause loss of information, specifically on the base level and the resettling time. For the assessment of the trigger level towards the discoloration risk, this is less important as the examples of figure 6 show. The increase in number of measurements and the decrease of labour costs however can compensate for this.

Conclusion

The RPM is a powerful tool to assess the discoloration risk and to evaluate objectively the effect of counter measures. The method is operational within many Dutch water companies allowing them to optimise the cleaning programs and actually reduce the number of discoloration events.

References

Blokker, E.J.M., P. Schaap (2003), Pulserend spuien, Theorie en praktijk, Kiwa report BTO 2003.050, Kiwa NV Nieuwegein, The Netherlands (in Dutch)

Colton-JF; Hillis-P; Fitzpatrick-CSB (1996) Filter backwash and start-up strategies for enhanced particulate removal Water Research, 30/10 (2502-2507)

Gauthier-V; Portal-JM; Yvon-J; Rosin-C; Block-JC; Lahoussine-V; Benabdallah-S; Cavard-J; Gatel-D; Fass-S (2001a) Characterization of suspended particles and deposits in drinking water reservoirs Water-Science-and-Technology:-Water-Supply. 2001; 1(4): 89-94

Gauthier-V; Barbeau-B; Millette-R; Block-J-C; Prevost-M (2001b) Suspended particles in the drinking water of two distribution systems Water-Science-and-Technology:-Water-Supply. 2001; 1(4): 237-245

Polychronopolous, M, Dudley K, Ryan, G, Hearn J. (2003) Investigation of factors contributing to dirty water events in reticulation systems and evaluating of flushing methods to remove deposited particles, Water Science and technology, Water Supply, 2003; 3(1-2): 295-306

Slaats, PGG, L.P.M. Rosenthal, W.G. Siegers, M. van den Boomen, R.H.S. Beuken, J.H.G. Vreeburg (2003) Processes involved in the generation of discolored water, AwwaRF, Denver

Slaats, PGG, (2000), Comparison of turbidity meters for quick measurement of resuspention potential, Kiwa report SWI 00.122, Kiwa NV, Nieuwegein, The Netherlands (in Dutch)

Vreeburg J.H.G. (1996). Red water, cleaning, cause and consequences, Kiwa report SWE 96.008, Kiwa NV, Nieuwegein, The Netherlands (in Dutch)

Physical Behaviour of GAC Filter Beds Under Warm, Eutrophic Conditions

Johannes Haarhoff

Department of Civil and Urban Engineering, Rand Afrikaans University, PO Box 524, Auckland Park, Johannesburg, 2006 South Africa. (E-mail jh@ing.rau.ac.za)

Abstract This paper reports on practical experience gained with GAC filtration (preceded by conventional treatment without ozone) at the Rietvlei water treatment plant near Pretoria, which treats highly eutrophic water in a warm climate. Observation of the GAC filter beds after the first four years of operation indicated that these beds are unusually difficult to clean properly during backwash and that backwash losses of 10% per regeneration cycle were more than expected. A simple test was developed to quantify and characterise the specific deposit on the media and this showed that about 70% of the deposit is organic in nature. The specific deposit is not removed efficiently during backwash and steadily increases with time. The backwash expansion of the GAC in the filters is almost double that of the same GAC after being oven-dried. This phenomenon also steadily increases with time and offers a partial explanation for the unexpectedly high media loss. It is, therefore, necessary to adapt the conventional freeboard calculation methods to allow for higher bed expansion and also to consider more aggressive backwash methods than conventionally used.

Keywords Granular activated carbon (GAC), specific deposit, bed expansion, media loss

Introduction

The Rietvlei water treatment plant near Pretoria, South Africa, had been treating water from a surface impoundment with conventional coagulation, hydraulic flocculation, sedimentation, sand filtration and chlorination since 1934. Increasing eutrophication necessitated a switch from sedimentation to dissolved air flotation in 1988. Soon thereafter, however, further source deterioration required more advanced treatment processes. Pilot testing of about 200 days (Van Staden, 1996) pointed to the insertion of a GAC filtration step with an empty bed contact time of 10 minutes between sand filtration and chlorination, which was commissioned towards the end of 1998. Since then, the GAC plant has been closely monitored in terms of its adsorption efficiency (Olivier & Haarhoff, 2003), its operational performance (De Kloe, 2000) and the practical aspects of managing the GAC inventory and curbing its losses (Clements & Haarhoff, 2004a). Table 1 provides a water quality summary for the period January 2000 to September 2000. During these studies, two aspects were noted which stimulated the work reported in this paper. The first was that the GAC was visibly dirty and sticky after the usual backwash routine, even when the backwash was preceded by air scouring (an operational option allowed for in the design). The second was that the average GAC washout of 0.05% per backwash was higher than expected. This loss will remove 20% of the bed after 400 backwash cycles. The hypothesis was that the warm, eutrophic conditions stimulated the development of biofilm on the GAC, which in turn rendered the GAC difficult to clean using standard backwash procedures and also affected the bed expansion during backwash. This paper summarises the work done to support this hypothesis.

© 2004 IWA Publishing. *2nd IWA Leading-Edge Conference on Water and Wastewater Treatment Technologies* edited by Mark Van Loosdrecht & Jonathan Clement. ISBN: 1 84339 508 8.

Table 1 Water quality summary for Rietvlei water treatment plant for the period January 2000 to September 2000 (from De Kloe, 2000)

	raw water			after sand filtration			after GAC filtration		
	X_{10}	median	X_{90}	X_{10}	median	X_{90}	X_{10}	median	X_{90}
Colour (Hazen)	17	23	41	6	12	18	5	6	12
PH	6.9	8.2	8.7	6.7	7.4	7.7	6.7	7.4	7.7
Conductivity (mS/m)	26	32	41	25	32	39	20	28	37
Temperature (°C)	12.5	15.5	20.5	-	-	-	-	-	-
Turbidity (NTU)	1.2	1.7	5.0	0.22	0.36	1.00	0.10	0.18	0.44
UV absorbance (/m)	4.5	6.6	7.6	3.0	3.8	5.3	2.5	3.1	4.3

Methods

The Rietvlei water treatment plant

There are 20 GAC filters at the Rietvlei water treatment plant, each measuring 4.4 m by 2.4 m and filled with GAC to a depth of 1.4 m. This allows for an EBCT of 10 minutes at the design flow rate of 40000 m^3/day with 18 filters operating. (Two filters are held in reserve to allow for GAC regeneration, which is done off-site by a commercial contractor.) The GAC filters are washed on reaching the maximum allowable head loss, which happens on the average two to three times per week. The computer-controlled backwash programme allows for 2.5 minutes of air scour at a rate of 16 mm/s followed by 7.5 minutes of water backwash at a rate of 5.6 mm/s. The original intention was that the air scour would normally be disabled and only used when the operator asks for it, but the problem with dirty filters dictated the permanent use of air.

The GAC is procured by tender. Two GAC gradings have been used up to present. At the time of the tests reported on in this paper, 18 of the filters were in operation, of which 14 were filled with a grading between approximately 0.4 mm and 1.6 mm and the remaining four filters with a grading between approximately 0.6 mm and 2.4 mm.

Samples were taken through the full depth of the GAC beds with a core sampler, from a minimum of five different positions evenly distributed over each filter bed. The samples from each filter were pooled and reduced with a standard riffle splitter to obtain a representative sample of convenient mass for laboratory work.

Measurement and characterisation of specific deposit

A newly developed method was used for the analysis of the specific deposit (Van Staden & Haarhoff, 2004a). This method is loosely patterned after the "floc retention test" suggested by the American Water Works Association (AWWA, 2000). Firstly, the deposit is stripped from the media by inverting about 50g media sample with 100 ml water 20 times in a measuring cylinder before decantation. (The AWWA test calls for stripping by vigorous shaking, but this method was found to be less reproducible). This process is repeated five times and the decant pooled to get a 500 ml sample containing the stripped deposit. Secondly, the solids thus stripped are filtered, dried and measured gravimetrically. (The AWWA method measures this nephelometrically). Finally, the solids are split into volatile (by burning for 15 minutes at 550°C) and acid-soluble (in a 0.185M HCl solution for a minimum of 2 minutes) fractions, which are loosely interpreted as follows:

- Non-volatile and non-soluble – the mineral fraction from the raw water
- Non-volatile and soluble – the chemical precipitates formed upon coagulation
- Volatile – the biological component of the specific deposit

All tests are performed on triplicate samples and the averaged results are expressed on a mass/mass basis in terms of oven-dried media.

Measurement of bed expansion

A representative GAC sample was transferred from the filter bed to a laboratory column with an internal diameter of 70 mm. The sample, which filled the column to a depth of approximately 250 mm, was first washed at 75% bed expansion for 4 minutes. The aim of this step was twofold; firstly to expose all the GAC samples to the same backwash intensity under controlled laboratory conditions, and secondly to simulate the "maximum cleanliness" achievable by a backwash system using water only, using the optimum backwashing guidelines established by Amirtharajah (1978). The bed was then allowed to settle gently before the bed depth and water temperature were noted. A series of four different backwash rates were then applied, namely at 3.6, 7.2, 10.8 and 14.4 mm/s. Each flow rate was maintained until the bed expansion reached equilibrium before the expanded bed depth and headloss were noted.

The GAC sample was then dried at 105°C for 24 hours before sieving and weighing. The sieving process (also used to obtain the GAC grading for mathematical modelling, not reported on in this paper) probably also removed any dried residue of the specific deposit that may have remained on the GAC grains after drying. From a mechanical point of view, the outside surface of the GAC after sieving was comparable to newly regenerated GAC; also confirmed by visual comparison under a stereoscope.

The bed expansion test was then repeated in an identical fashion with the resulting GAC. There were thus two bed expansion tests carried out on each sample – first with the GAC as removed from the filter, and second with the same sample after drying and sieving. It should, therefore, be a simple matter to establish whether the specific deposit had any effect on the expansion behaviour of the media.

Results and discussion

Total specific deposit

An analysis of variance of the triplicate measurements on each sample showed that the reproducibility of the test was excellent and that the differences measured between the different GAC filters were highly significant at an alpha level of less than 0.001. Values could thus be averaged for the rest of the analysis that follows.

The total specific deposit measurements for each filter are shown on Figure 1 as a function of the number of days of operation since its GAC had last been regenerated. The top line indicates the specific deposit immediately after each filter had been backwashed at the plant, and the bottom line after the GAC had been further backwashed in the laboratory at 75% bed expansion.

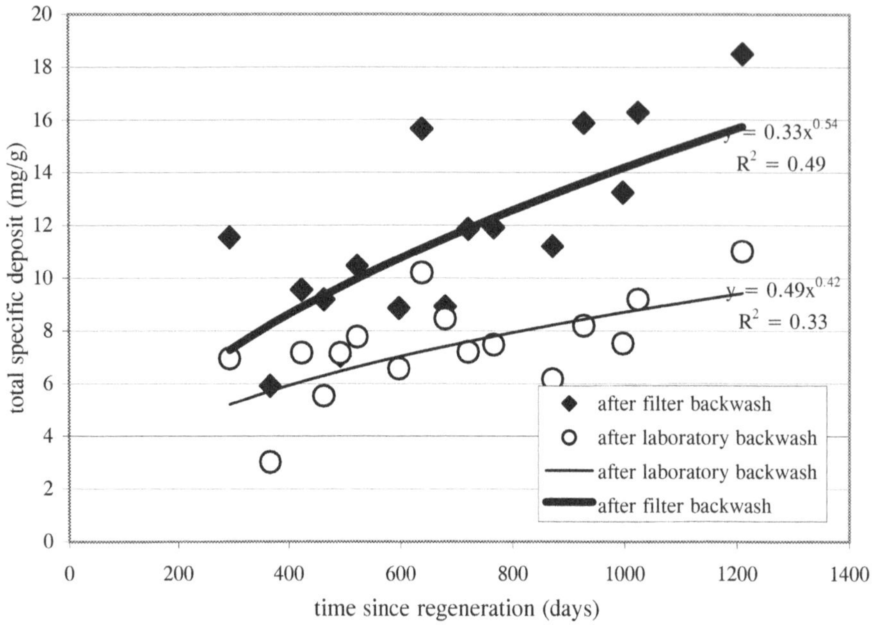

Figure 1 Total specific deposit on the GAC of individual filters, after a) plant backwash and b) laboratory backwash

The specific deposit on the GAC within the filters does not reach equilibrium as would be expected. After even more than three years of operation, the specific deposit seems to be still increasing fairly rapidly. Moreover, measurement of the specific deposit at a number of South African sand filtration plants indicated that values of the specific deposit of between 6 and 10 mg/g (Van Staden & Haarhoff, 2004b) indicate a threshold above which sand filters are considered to be unacceptably dirty. Correcting for the lower density of GAC, this threshold roughly translates into 10 to 16 mg/g indicating that the GAC filters become unacceptably dirty. It is unlikely that this can simply be attributed to an inefficient backwash system. The plant is fairly new, the backwash rates comply with the best practice and the GAC suppliers' recommendation, and the plant is substantially free from the usual filter ailments of dead spots, cracked beds and broken or blocked nozzles.

Further evidence of a tenacious specific deposit is provided by the values measured after laboratory backwash. The laboratory backwash procedure provides a measure of the best possible result that can be obtained by backwashing with water only. But even this backwash process (albeit only once) is not able to return the different GAC samples to the same state of cleanliness – the increasing trend with time remains. Neither is it able to reduce the specific deposit to levels appreciably below 10 mg/g, indicating GAC that is still relatively dirty. It is, of course, not possible to know what the effect would have been if a different backwash regime had been used in the plant since its commissioning – this can only be a matter of speculation.

The nature of the specific deposit

Further understanding was sought by analysis of the different fractions of the specific deposit. Figure 2 shows a plot of the three components of the specific deposit after plant backwash, shown as a function of the time since regeneration. The data for the specific deposit after laboratory backwash shows essentially the same pattern and are not represented here.

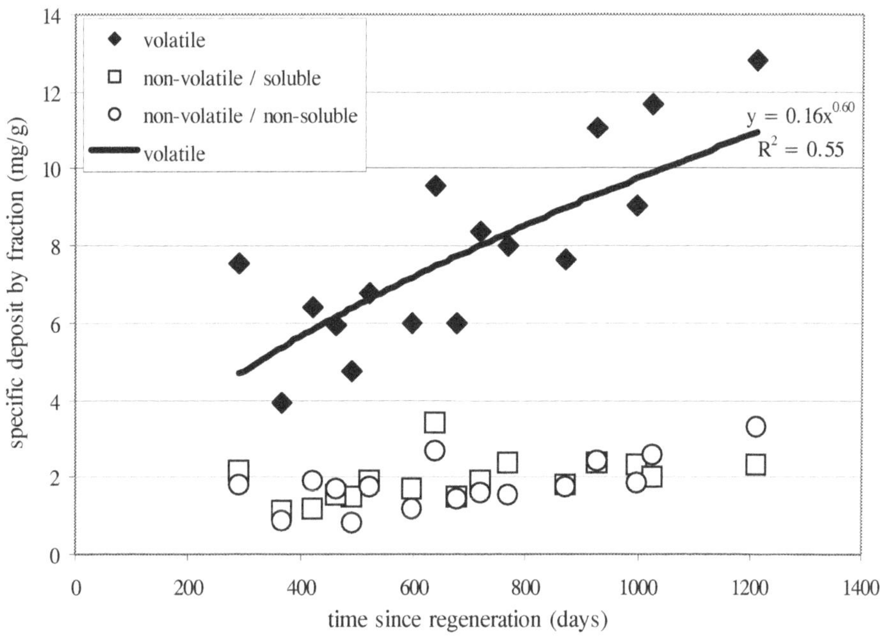

Figure 2 The specific deposit of individual GAC filters after plant backwash, broken into its three constituent fractions

It is clear that the specific deposit is dominated by the volatile fraction, which probably is an organic component with biological origin. On the average, 67% of the total specific deposit is volatile; this percentage is very consistent (61% to 72%) irrespective of the time since regeneration. For the GAC samples after laboratory backwash, the results are the same (69%, varying between 59% and 71%). Further confirmation of the biological origin came from a number of semi-quantitative proof-of-presence tests for extracellular polymeric substances (EPS) tests which indicated EPS presence of approximately 4 mg/g.

Bed expansion

To have a consistent base for comparison amongst the many bed expansion experiments performed, the percentage bed expansion at an arbitrary, constant backwash rate of 10.8 mm/s was extracted for each experiment. This backwash rate was selected to attain a fairly large, easily measurable bed expansion, and also to ensure complete fluidisation of the bed. (During the initial stages of bed expansion in the laboratory column, the media grains form a fixed matrix and the media bed expands erratically as a plug. As the flow rate increases, the matrix structure breaks down, which allows the individually suspended media grains to move freely. For reproducible bed expansion results, it is important to use flow rates that are high enough.) Figure 3 shows these values for the GAC samples immediately after being withdrawn from the filters, as well as after the same had been oven-dried and sieved. Only the filters with 12x40 GAC are included; these filters were filled with carbon from the same supplier.

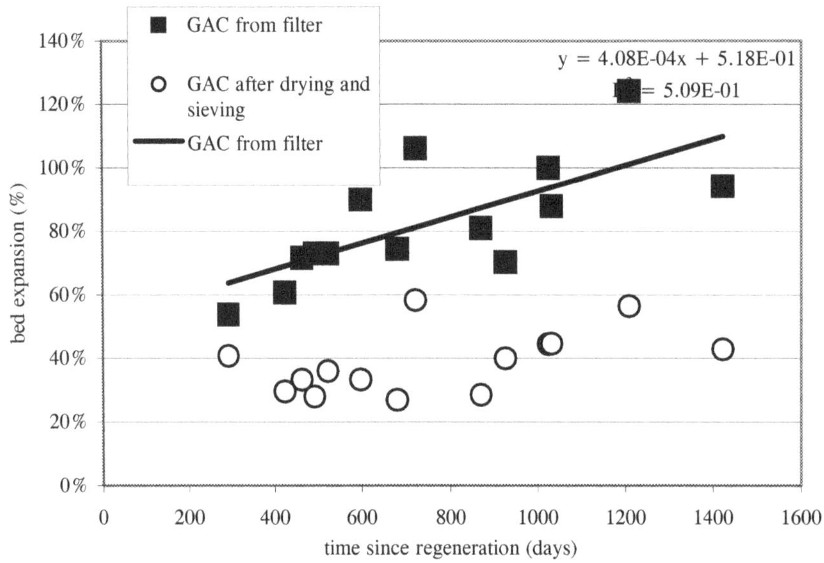

Figure 3 Percentage bed expansion of 12x40 GAC from individual filters for a) as taken from the filter and b) after drying and sieving

The first point to note is that the expansion of the in-situ GAC is approximately double than the expansion of the same media after drying and sieving for the full range of data points shown. Immediately after regeneration, there should be no difference. The second point is that the bed expansion of the in-situ GAC increases sharply with time since regeneration, a phenomenon apparently related to the increase in specific deposit earlier shown. The third point is that there is no significant trend (at α = 5%) in the bed expansion with respect to time since regeneration for the GAC after drying and sieving, as would be expected.

The first two of the above observations have serious consequences for freeboard design. Even for GAC beds of only 1000m deep, the spread of 80% in Figure 3 (from 30% to 110%) translates into an "uncertainty" of 800mm when mathematically predicting the freeboard with, for example, the well-established bed expansion model of Dharmarajah & Cleasby (1986).

The same phenomenon has also been observed and measured in conventional South African sand filters where warm, eutrophic water is treated. In that case, the analysis has advanced to the point where a preliminary design guideline could be proposed, namely that the media sphericity term in the Dharmarajah model could be artificially reduced by a maximum of 30% (from the value obtained from laboratory calibration on dry, clean media) to compensate for the higher bed expansion eventually obtained in the filters (Clements & Haarhoff, 2004b). A similar guideline for GAC is not available yet.

Summary and conclusions

1. Empirical observation of the GAC filter beds at the Rietvlei water treatment plant near Pretoria during the first four years of operation showed that the GAC was insufficiently cleaned by the backwash system and that the GAC losses were unexpectedly high. The plant treats warm, eutrophic surface water.
2. A new test method was standardised to quantitatively measure the specific deposit on the GAC. This measure showed that the level of media of cleanliness after backwash was indeed inadequate, and that the media steadily got dirtier with time.
3. By rewashing samples from the same filters at 75% bed expansion in the laboratory, the GAC became appreciably cleaner, but still on the borderline of being acceptable. The trend of increasing specific deposit with time was also still evident.
4. The test method further fractionated the specific deposit into soluble and volatile fractions. This showed 70% of the specific deposit on the GAC is volatile (therefore organic or biological in nature, confirmed by the presence of EPS).
5. The expansion of GAC samples taken directly from the filters was significantly more for samples that had been in operation for a longer time.
6. When these same samples were oven-dried and sieved (thus getting the media in approximately the same mechanical condition as newly regenerated GAC), the bed expansion was less. There was no significant trend of increasing bed expansion with time for the GAC treated by drying and sieving.
7. The overall conclusion is that there is significant development of biofilm on the GAC, similar to the results recently reported for slow sand filters (Mauclaire *et al*, 2004). This biofilm seems to be at the root of both the tenacious nature of the specific deposit and the high media loss – serious practical problems that demand further research.

Acknowledgments

Experimentation at a full-scale, working treatment plant requires unusual patience on the side of the plant managers, for which Carel Taljaard and Leanne Coetzee have to be thanked. It also hinges on persistence and dedication of the researchers – Samantha van Staden and Michele Clements who cheerfully and unreservedly gave their best.

References

American Water Works Association (2000) Filter surveillance techniques for water utilities. Video, catalog number 65160, Denver, Colorado.

Amirtharajah, A (1978) Optimum backwashing of sand filters. Journal of the Environmental Engineering Division, ASCE, 104(EE5), Proc. Paper 14078, 917-931.

Clements, M and Haarhoff, J (2004a) Practical experiences with granular activated carbon (GAC) at the Rietvlei water treatment plant. Water SA 30(1), 89-95.

Clements, M and Haarhoff, J (2004b) Filter media expansion during backwash – the effect of biological activity. Presentation to the Ninth Biennial Conference of the Water Institute of Southern Africa, Cape Town, 2-8 May 2004.

De Kloe, H (2000) Die prestasie van geaktiveerde koolstof by Rietvlei watersuiweringsaanleg. (The performance of activated carbon at the Rietvlei water treatment plant). Undergraduate thesis presented to the Department of Civil Engineering at the Rand Afrikaans University, Johannesburg.

Dharmarajah, AH and Cleasby, JL (1986) Predicting the expansion of filter media. Journal of the American Water Works Association, 78(12), 66-76.

Haarhoff, J and Olivier, J (2003) GAC performance at three Southern African water treatment plants. Water SA (special edition from WISA proceedings), 37-42.

Mauclaire, L, Schürmann, A, Thullner, M, Gammeter, A and Zeyer, J (2004) Sand filtration in a water treatment plant: biological parameters responsible for clogging. AQUA, 53(2), 93-108.

Van Staden, AL (1996) Activated carbon and ozone as supplementary water treatment options at Rietvlei Dam. Masters thesis presented to the Department of Civil Engineering at the Rand Afrikaans University, Johannesburg, 1996.

Van Staden, SJ and Haarhoff, J (2004a) A standard test for filter media cleanliness. Water SA 30(1), 81-88.

Van Staden, SJ and Haarhoff, J (2004b) What simple media tests can tell you about filter problems and rehabilitation. Presentation to the Ninth Biennial Conference of the Water Institute of Southern Africa, Cape Town, 2-8 May 2004.

MTBE removal by biofiltration in a water works

E. Arvin, L. K. Nielsen, A.G. Tully, H.-J. Albrechtsen and H. Mosbæk

Environment and Resources. Technical University of Denmark, Building 115, DK-2800 Kgs. Lyngby, Denmark

Abstract MTBE is a relatively "new" contaminant in groundwater that is expected to affect water works wells in the future because of the widespread application of MTBE as an octane booster. This work is a field study of the MTBE removal in a conventional water works where MTBE removal was observed in the biological active sand filters. This study shows for the first time that in a conventional water works rapid sand filter MTBE can be biodegraded from 10-65 μg/L to below 5 μg/L. Although the filter velocity is normally as low as 1.1 m/h, MTBE was effectively biodegraded even at a filter velocity of 3 m/h after an adaptation time of 3 weeks. Studies of the effect of backwashing duration and intensity indicated that the MTBE degrading biomass adhere firmly to the sand particles in the filter.

Keywords MTBE, methyl tert-butyl ether, water works, biofiltration, biodegradation, rapid sand filters, residence time, backwashing.

Introduction

Methyl tert-butyl ether (MTBE) has substituted lead in gasoline since the beginning of the 1980'es and due to leaking underground gasoline storage tanks, MTBE is found widely in groundwater in the US, (Squillace et al. 1997 and 1999). However, a recent review indicates that the frequency of detection of MTBE contamination is decreasing (Thomsen et al., 2003). A regional study from Denmark showed that MTBE was found in 86% of contaminations downstream of gasoline service stations. In 53% of the contaminated sites, the MTBE concentration exceeded 30 μg/L, and in 40% of these contaminated sites, the concentration exceeded 1000 μg/L, Munch (2001).

Salanitro et al. (1994) showed for the first time that MTBE can be biodegraded under aerobic conditions. Salanitro et al. (1994) and several other researchers since then have worked with remediation of groundwater at relatively high concentrations, for example 10-100 mg/L. However, in water works wells, significant contaminations occur at much lower concentrations in the μg/L range and very little work has been carried out concerning biodegradation at such low concentrations.

This study is a field study at the Grubbemølle water works in the city of Svendborg, Funen, Denmark, where for the first time it has been demonstrated that MTBE can be biodegraded in full-scale bioactive rapid sand filters even in the μg/L concentration range. This is important information because the Danish drinking water standard for MTBE is 5 μg/L, the lowest in the world. The purposes of the study were to investigate: 1. The MTBE removal efficiency at different loadings of the aeration plant and the biofilters, 2. The effect of residence time in the sand filters, and 3. The effect of various backwashing strategies on the MTBE-degrading biomass and thereby the MTBE removal efficiency.

© 2004 IWA Publishing. *2nd IWA Leading-Edge Conference on Water and Wastewater Treatment Technologies* edited by Mark Van Loosdrecht & Jonathan Clement. ISBN: 1 84339 508 8.

Materials and methods

The Grubbemølle water works is a traditional water works treating groundwater from various wells. Some of the wells have been contaminated by MTBE from a nearby gasoline station. The concentration of MTBE varied between 10-65 µg/L, depending on the actual production pattern. Aromatic compounds like benzene, toluene, ethyl benzene and xylenes (BTEX) could not be detected in the inlet water to the water works (< 0.1 µg/L). The ground water is anaerobic, and typical for Danish groundwater's, with iron concentrations of 0.5-1 mg/L and an ammonium concentration about 0.5 mg/L. The water is relatively hard with a calcium concentration of 100-120 mg/L and a bicarbonate concentration of 270-350 mg/L. The concentration of non-volatile organic carbon (NVOC) was 2 mg/L.

The Grubbemølle water works has a diffused air aeration plant with three tanks connected in series with a total residence time of 12-47 min. and an air:water-ratio in the range 5:1 - 9:1. Afterwards, the water is treated in six parallel one-stage rapid sand filters with a total filter area of 80 m^3 and with a height of the filtration layer of 120 cm. The upper part consists of a 30 cm anthracite layer (2.5-4 mm) and the lower layer consists of 90 cm sand (0.8-1 mm). The filter velocity (Darcy velocity) is low, 1.1 m/h. The filters are backwashed every 7'th day. First, the filters are backwashed with air (67.5 m/h) for 10 min, followed by a combined air (67.5 m/h) and water (2 m/h) backwash for 12 min, and finally a water backwash for 2 min at a velocity of 26 m/h. The filters are idle for 2-4 hours per day.

Results and discussion

MTBE removal in the water works

Table 1 shows that MTBE is removed from the raw water in the aeration tanks and in the bioactive rapid sand filters at different flow rates and at different initial MTBE concentrations. The removal due to stripping in the aeration tanks was 20-30 % of the MTBE in the raw water, followed by a nearly complete removal of MTBE in the biofilters. The effluent MTBE concentration was well below the Danish drinking water standard of 5 µg/L.

No traces of the metabolites tert-butyl alcohol, isopropanol, tert-butyl formate, and acetone could be detected (detection limit of 0.5 µg/L).

Table. 1. Changes in MTBE concentration (µg/L) through the water works at different flows per filter and at different influent concentrations.

	MTBE concentration (µg/L)		
Flow (m^3/h)	Influent	After aeration	After sandfilter
4	66	55	2
13	16	14	2
17	11	8	2
28	9	9	3

Samples of filter sand were collected from a filter and incubated under aerobic conditions in 1 L bottles with MTBE concentrations from 100 to 1000 µg/L. MTBE was removed in all the bottles, but not in a reference bottle where sodium azide (1 g/L) was added to inactivate the biomass. This indicated that the MTBE removal was biotic.

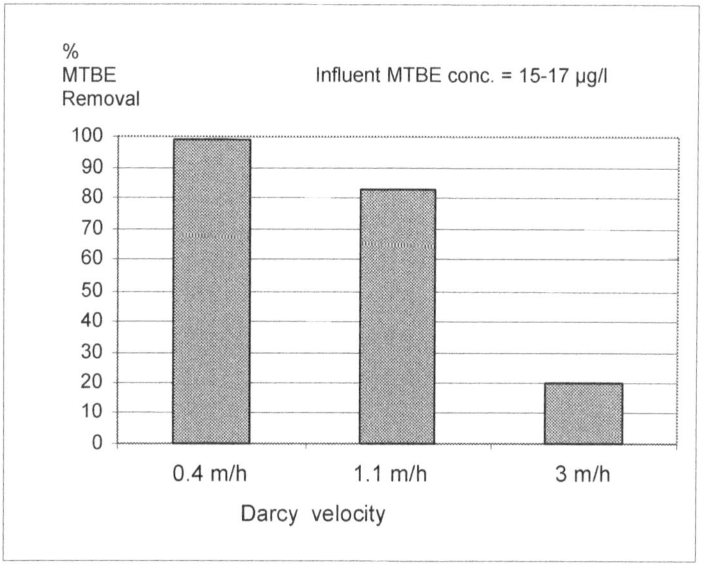

Figure 1 Percent removal of MTBE at different water velocities in the filter. The removal was measured before backwashing.

Effect of residence time in the sand filters

The normal Darcy velocity in the filters is 1.1 m/h, which is a relatively low velocity. Short-term (hours) changes in this rate strongly influenced the MTBE removal, which decreased substantially when the rate was increased to 3 m/h (Fig. 1).

In a long term experiment, the filter velocity was kept at 3 m/h in one of the filters. The MTBE removal slowly increased from 30% to nearly 95% during four weeks. This indicates that very slowly growing bacteria are responsible for the MTBE removal. In this experiment, backwashing had a significant effect on the MTBE removal.

Effect of backwashing

The effect of backwashing, including the effect of intensity and duration of the backwashing were tested. The effect varied, but in most cases the removal of MTBE was not significantly different before and after backwashing. This is illustrated in Figure 2. However, by increasing the duration of air flow (67.5 m/h) from 10 to 19 minutes and at the same time increasing the duration of the final, high water flow (26 m/h) from 2 to 5 minutes, the MTBE removal was significantly reduced from about 95 % before backwashing to 67-81% after backwashing.

Conclusions

This study showed for the first time that MTBE can be removed efficiently in bioactive rapid sand filters in a conventional water works treating groundwater with concentrations of MTBE as low as 10-65 µg/L. A short-term increase of the filter velocity from 1.1 m/h to 3 m/h significantly reduced the MTBE removal efficiency. However, after approximately one month of adaptation, the filter had regained much of its removal capability. This indicates, that the MTBE degrading bacteria are growing very slowly. Various backwashing strategies with different water and air intensities and treatment durations had little influence on the MTBE removal. Only a combination of relatively long term air and water backwashing reduced the MTBE removal significantly. This indicates that the MTBE degrading bacteria adhere firmly to the sand grains.

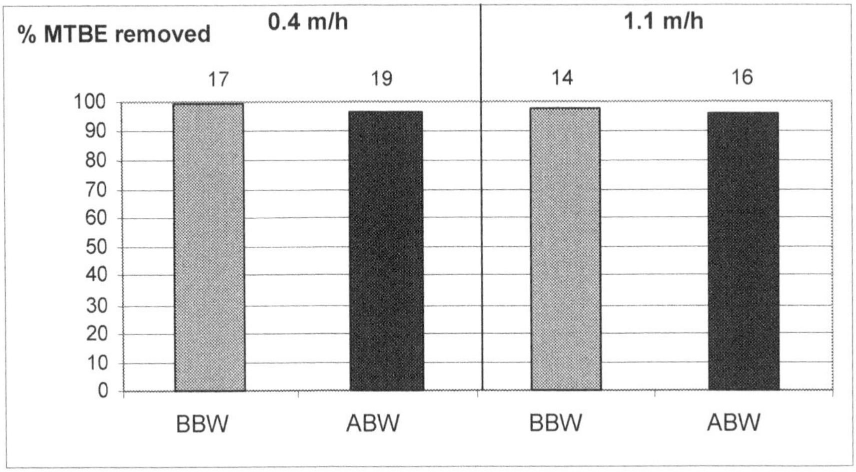

Figure 2 MTBE removal before backwash (BBW) and after backwash (ABW). Inlet MTBE concentrations in µg/L are shown above the bars. Filter velocities were 0.4 m/h and 1.1 m/h.

References

Munch, S. K. (2001) How often is MTBE found and in which concentrations? – *A review of results from Fyn's county*. (In Danish) ATV Meeting, Winter meeting on soil- and groundwater contamination. Vingsted Centre, 6. – 7. Marts, 2001. p. 209-215.

Salanitro, J.P., Diaz, L.A., Williams, M.P., og Wisniewski, H.L. 1994. Isolation of a bacterial culture that degrades methyl t-butyl ether. *Applied and Environmental Microbiology*. ***60*** *(7), 2593-2596.*

Squillace, P.J., Pankow, J.F., Korte, N.E., and Zorgorsky, J.S. 1997. Review of the environmental bahavior and fate of methyl tert-butyl ether. *Environmental Toxicology and Chemistry*. ***16****(9), 1836-1844.*

Squillace, P.J., Moran, M., Lapam, W., Price, C., Clawges, R., and Zogorski, J.S. 1999. Volatile organic compounds in untreated ambient groundwater in the United States, 1985-1995. *Environmental Science and Technology*. ***33,*** *4176-4187.*

Thomsen, J.A.M., McKinley, J.W., Harris, R.C., Hart, A.J., Hicks, P., Ramsden, D.K., Wilson, B. In: MTBE remediation handbook. Chaptor 4. Editors: Moyer, E. og Kostecki, P. T.Amherst Scietific Publishers. Amhurst. Massachusetts. ISBN 1-884940-29-3.

Assessing Riverbank Filtration Removal Mechanisms

V. Partinoudi and M.R. Collins

Water Treatment Technology Assistant Center, Department of Civil Engineering, University of New Hampshire, USA

Abstract The paper focus was to quantify subsurface filtration (or underground passage) and groundwater (GW) dilution contributions to overall riverbank filtration (RBF) removals of selected microorganisms and dissolved organic carbon (DOC) precursors. Four field-scale RBF sites were evaluated. The microbes quantified included aerobic spore forming bacteria (ASFB), *Total coliforms*, *E. coli* and male-specific bacteriophage. Both GW dilution and subsurface filtration were important removal processes. The relative importance of subsurface filtration may be a function of RBF site characteristics and especially travel time/distance between the river and the RBF extraction well. Complete removals were observed for all microorganisms except viruses which were considered the most problematic. Significant DOC reductions were observed with adsorption being more influential than biodegradation.

Keywords riverbank filtration, removal mechanisms, groundwater dilution, subsurface filtration, microorganisms, dissolved organic carbon.

Introduction

Riverbank filtration (RBF) can be an effective drinking water treatment process for a variety of pollutants including pathogens and organic precursor material (Kuehn and Mueller, 2000; Wang et al., 2002; Ray et al., 2002a and 2002b; and Tufenkji et al., 2002). Although sometimes viewed as a simple process similar to slow sand filtration, RBF can be complex to understand as biological, biogeochemical, and hydrogeological factors may control the removal of both dissolved and particulate contaminants.

Riverbanks filtrate is typically presented in the literature as surface water that "percolates" through the banks or bed of a river to an extraction well by induced filtration due to pumping. During its underground passage the water changes due to biological, chemical, and physical processes and by dilution with groundwater. These processes are schematically shown in Figure 1.1. Most of the RBF treatment literature has rightly focused on the removal mechanisms associated with the surface water underground passage to the RBF extraction well. This natural treatment process has been identified by a variety of descriptors including underground passage, soil passage, soil aquifer treatment, subsurface filtration, or simply riverbank filtration. This paper will use subsurface filtration to note the removals achieved by underground passage of the river water to the RBF extraction well. The contribution of groundwater dilution to the overall quality of the RBF extracted water or filtrate is generally ignored or minimized.

© 2004 IWA Publishing. *2nd IWA Leading-Edge Conference on Water and Wastewater Treatment Technologies* edited by Mark Van Loosdrecht & Jonathan Clement. ISBN: 1 84339 508 8.

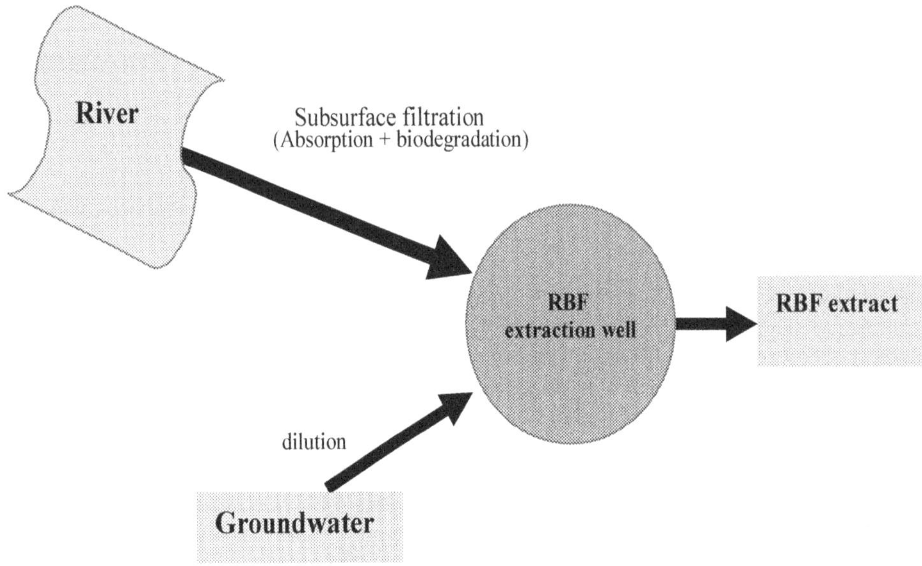

Figure 1.1 Selected removal processes typically associated with RBF

There are several questions that should be addressed before RBF's removal capabilities can be fully assessed for a specific contaminant. Three specific questions are listed below:

What is the travel time from the river to the RBF extraction well?

What are the removals due to subsurface filtration?

What are the removals due to groundwater dilution?

The focus of this paper is to address these questions at selected RBF sites. The main objective is to quantify subsurface filtration and groundwater dilution contributions to overall RBF removals of selected microorganisms and organic precursors as quantified by dissolved organic carbon (DOC).

Methodology

RBF study sites

Four RBF sites were chosen for this phase of a more comprehensive study (Partinoudi, 2004). A brief description of each RBF site is summarized in Table 1.1. The four sites provided different RBF characteristics to consider including degrees of hydraulic connection and distances between the river and RBF extraction well, source water quality variations, different well-constructions and pumping rates, varying hydrogeologies, and different subsurface materials. The two sites in New Hampshire were chosen for their convenience while the study sites in Louisville, KY and Cedar Rapids, IA were chosen because these sites were considered well characterized from previous studies.

Analytical Procedures

Each sampling event involved sample collections from the river water, the background groundwater and the RBF extracted water or filtrate. All analytical measurements on these waters followed the procedures outlined in Standard Methods (1998) unless specified otherwise (Partinoudi, 2004).

Table 1.1 Summary of extraction well characteristics for RBF study sites

Sampling Sites	Source Water	Sampling Dates	# Sampling Events	Extraction Well Info			
				Type	Distance m	Depth m	Diameter
Pembroke, NH	Soucook River	8/01-11/02	19	V	55	17	30 cm
Milford, NH	Souhegan River	11/01-11/02	13	V	23	19.8	60 cm
Louisville, KY	Ohio River	9/01-5/03	11	H/V	12.2	12.2	6.1 m caisson
Cedar Rapids, IA	Cedar River	9/02-4/03	5	V	19.5	17.4	10.7 cm

A variety of microbial analyses were performed during the course of this overall study. More detailed descriptions may be found elsewhere (Partinoudi, 2004). Briefly, *Total coliforms and E. coli* were determined using MPN from the IDEXX Quanti-Tray/2000 method. Aerobic spore forming bacteria (ASFB) determinations followed the procedures developed by Ballester and Margolin (2000).

Procedures for Estimating Travel Time

The travel time between the river and the RBF extraction well needs to be estimated in order to quantify river water quality changes from underground/soil passage. The quality of the RBF filtrate needs to be compared with the quality of river water at the commencement of its underground passage since surface water quality can change dramatically over short timeframes. The variability in the characteristics of the RBF study sights necessitated the use of various techniques to estimate travel time as noted in Table 1.2. The travel time estimates reflect values that were determined to be fairly constant throughout the course of this study.

Table 1.2 Estimated travel times for the RBF study sites

Sampling Site	Travel Time, d (From river to RBF well)	Evaluation of Travel Time
Pembroke, NH	4.5-5.5 (5.0 avg)	Darcy's Law in terms of seepage velocity
Milford, NH	0.75-1.25 (1.0 avg)	Darcy's Law in terms of seepage velocity
Louisville, KY	1 avg	Estimated from information provided by the Louisville Water Company (initial pumping test data)
Cedar Rapids, IA	5	Estimated from information provided by the City of Cedar Rapids (Schulmeyer, 1995)

Source contributions to RBF Filtrate

A fundamental distinction that first needs to be made in assessing RBF removal mechanisms is to quantify the contributions of river water and groundwater to the RBF extracted water/filtrate. Knowledge of the mixing ratio between these two sources would be necessary to determine contaminant removals achieved solely by subsurface filtration and groundwater dilution. The degree of mixing depends on numerous factors which, although are well understood, are difficult to easily ascertain in the field especially when taking into account the subsurface heterogeneity, the hydraulic grade line variability, and the transmissivity of the infiltrating surface water at a changing interface.

For purposes of this study a more simplistic approach was taken to quantify source water contribution to the RBF filtrate. Basically mass balances were performed on a variety of water quality parameters (see Table 1.3) at each RBF study site. The source contribution estimates were made on a water quality parameter or modeling result that best represented the majority trend. The varying characteristics of each RBF study site resulted in different parameters upon which the mixing ratio was determined as noted in Table 1.4. These contribution ratios remained consistent throughout the study period.

Table 1.3 Water parameters used to assess mixing ratios in the RBF filtrate

Sulfate	Color (True)	Hardness
Chloride	UV_{254} absorbance	Alkalinity
pH	Particle counts	Redox Potential
Temperature	Selected radioisotopes (222Ra)	Conductivity

Table 1.4 Average contribution of river water and groundwater to the RBF filtrate

	% river water in *RBF* well	% Groundwater in *RBF* well	Parameter upon which ratio is based
Pembroke, NH	40.7±3.7	59.3±3.7	Conductivity
Milford, NH	40.8±6.4	59.2±6.4	Sulfate
Louisville, KY	78.1±4.4	21.9±4.4	Hardness
Cedar Rapids, IA	70	30	Based on groundwater flow modeling (Schulmeyer 2002)

Results and Discussion

A summary of overall RBF removals observed during the course of this study is presented in Table 1.5. A perusal of the data highlights the ability of a RBF system to produce a filtrate with low microbial counts and significant organic precursor reductions compared to the river water. The RBF literature contains numerous references to excellent removals of similar contaminants (Kuehn and Mueller, 2000; Ray et al., 2002b; Tufenkji et al., 2002; and Wang et al., 2002). However, removals achieved by the RBF removal processes, e.g. subsurface filtration and groundwater dilution, are not generally quantified. The focus here is to first assess microbial and second organic precursor (DOC) removals achieved by the various RBF process mechanisms.

Microbial Process Removals

The contributions of subsurface filtration and groundwater dilution to overall microbial removals by RBF were difficult to assess during this study primarily because no detectable colony forming units (CFU) were found in either the RBF filtrate or background groundwater (see Table 1.6). In this study, consequently, no clear distinctions can be made between overall RBF removals of *Total coliforms*, *E. coli*, and ASFB and removals achieved by subsurface filtration or groundwater dilution. However, subsurface filtration does provide significant microbial reductions with log removals limited more by river water concentrations than the RBF characteristics noted in this study.

Table 1.5 Typical RBF removals of selected analytes during this study

Parameter /Site	**Pembroke, NH**		**Milford, NH**		Louisville, KY		Cedar Rapids, IA	
Sampling Dates	**8/01-11/03**		**11/01-11/02**		**9/01-12/02**		**9/02-1/03**	
	River	***RBF***	**River**	***RBF***	**River**	***RBF***	**River**	***RBF***
pH	6.3-7.3	6.1-6.6	6.3-7.6	5.8-6.6	NA	NA	8.2-8.5	7.6-7.7
DOC (mg/L)	1.7-6.6	0.3-1.2	1.6-6.4	0.5-1.3	3.0-5.0	1.7-2.0	2.8-4.3	1.5-2.5
Temperature (degrees C)	0.3-25.7	8.5-13.8	0.4-26.1	7.3-16.2	3.6-29.7	13.3-26.4	0.4-20.6	8-22.9
Turdidity (ntu)	0.7-2.0	0.1-0.2	0.8-1.4	0.1-0.2	3-500	<0.7	2.1-23	0.4-1.0
Conductivity (μS/cm)	42-150	193-269	119-214	107-152	NA	NA	454-666	473-624
Total Coliforms (CFU/100mL)	141-1399	BDL	423-2431	BDL*	10-24192	BDL	75-2572	BDL
E.coli (CFU/100mL)	4-108	BDL	6-119	BDL	2-1031	BDL	23-56	BDL
Aerobic spore forming bacteria (CFU/100mL)	84-1997	BDL	124-1093	BDL	3500-145000	BDL	39-217	BDL

- BDL- Below detection limits (< 1cfu/100ml)

Table 1.6 Average microbial reductions by RBF processes

Location	# of sampling events	River	*RBF*	Background well	Average Log Removals: Total	Subsurface Filtration	GW Dilution
AEROBIC SPORE FORMING BACTERIA (CFU/100mL)							
Pembroke, NH	19	502±11	BDL	BDL	≥1.9	≥1.51	≥0.39
Milford, NH	13	673±55	BDL	BDL	≥2.1	≥1.52	≥0.58
Louisville, KY	11	33404±5672	BDL	BDL	≥3.5	≥3.39	≥0.11
Cedar Rapids, IA	5	101±8	BDL	BDL	≥2.6	≥2.45	≥0.15
TOTAL COLIFORMS (CFU/100mL)							
Pembroke, NH	19	521±63	BDL	BDL	≥2.1	≥1.71	≥0.39
Milford, NH	13	1091±114	BDL	BDL	≥2.6	≥2.02	≥0.58
Louisville, KY	11	3921±182	BDL	BDL	≥1.0	≥0.89	≥0.11
Cedar Rapids, IA	5	1391± 25	BDL	BDL	≥1.4	≥1.25	≥0.15
E.coli (CFU/100mL)							
Pembroke, NH	19	27±4	BDL	BDL	≥0.6	≥0.21	≥0.39
Milford, NH	13	55±10	BDL	BDL	≥0.8	≥0.22	≥0.58
Louisville, KY	11	5±2	BDL	BDL	≥0.3	≥0.19	≥0.11
Cedar Rapids, IA	5	16±1	BDL	BDL	≥0.7	≥0.55	≥0.15
VIRUS INDICATORS (MS COLIPHAGE) (PFU/100mL)							
Louisville, KY	4	4342±24	3703±22	3402±18	≥0.1	≥0.06	≥0.01
Cedar Rapids, IA	5	3438±21	753±9	BDL	≥0.7	≥0.55	≥0.15

Although not enough samples were collected to be statistically significant, the virus indicators (e.g. male specific bacteriophage) used in this study were found in two selected RBF filtrates. The background lateral used as the groundwater source at the Louisville study site may contain over ten percent surface water (Wang et al., 2002) which could account for the levels found in the groundwater background source. In any case, viruses seem to be the most mobile microorganisms assessed under the RBF conditions of this study.

Dissolved Organic Carbon (DOC) Removals

Organic precursor removals as quantified by DOC reductions were evaluated at all RBF sites throughout the course of this study (Partinoudi, 2004). Total DOC removals range from 18-92% with a weighted average calculated at 63% which is similar to what has been found in the literature (Drewes and Summers, 2002). Of this 63% weighted average, 29% was calculated to be due to groundwater dilution and 34% was due to subsurface filtration. These DOC subsurface filtration removals are higher than the 5-20% typically observed from conventional slow sand filters (Collins et al., 1992).

Two RBF study sites (Pembroke and Louisville) were selected for a more detailed assessment of the DOC removal processes associated with RBF. The total removals as well as the average removals due to groundwater dilution and subsurface filtration for the two sites are summarized in Table 1.7. The significantly higher reductions noted for Pembroke could be directly related to the extraction well's distance from the river water which would be expected to enhance both groundwater dilution and subsurface filtration removals.

Table 1.7 Average DOC removals achieved at selected RBF sites

Site	% Total Removal	% Removal due to:		# of Samples
		Dilution	Subsurface Filtration	
Louisville, KY	34±18.0	6±4.0	28±4.0	11
Pembroke, NH	86±4.19	26±8.6	60±8.6	19

The subsurface filtration DOC removals were further grouped into reductions achieved by biodegradation and absorption as noted in Table 1.8. The biodegradation removals were determined by quantifying the biodegradable dissolved organic carbon (BDOC) content of the various waters using a 5-day batch-shaker method (Algeier et al., 1996). The resulting significant reductions achieved by the absorption process emphasized the importance of subsurface material such as metal oxides in RBF removals of DOC.

Table 1.8 Assessment of subsurface filtration DOC removal processes

Site	% Average Removal by Subsurface Filtration		
	Total	Biodegradation	(Bio)adsorption
Louisville, KY	28±4.0	9±9.6	19±9.6
Pembroke, NH	60±8.6	7±0.7	53±0.7

Summary

Both groundwater dilution and subsurface filtration were important RBF removal processes. For microbial removals under the conditions of this study, subsurface filtration was an effective removal mechanism/process with overall estimated removals more a function of river water concentration than groundwater dilution. No *Total coliforms*, *E. coli,* and aerobic spore forming bacteria were observed in either the RBF filtrate or background groundwater. The most problematic microbes evaluated in this RBF study were viruses. For DOC removals, the importance of subsurface filtration increased while groundwater dilution decreased as the travel time/distance to the RBF extraction well increased. Significant DOC reductions were more a function of adsorption than biodegradation for the conditions found in this study.

Acknowledgments

- EPA for funding this project through the UNH WTTAC
- S. Hefron, N. Ballester & J. Fontaine, UNH
- The Cedar Rapids Water Department, IA
- The Louisville Water Company, KY
- The Milford, NH Fish Hatchery personnel
- The Pembroke, NH Waterworks
- Drs L. Brannaka & A. Margolin, UNH

References

Algeier, S.S. et al. (1996) A simplified and rapid method for biodegradable dissolved organic carbon measurement. In Proc. of the WQTC, AWWA, Denver, CO.

Ballester, N.A. and Margolin, A.A. (2000) Detection of Bacillus Spores in Environmental Samples, Univ. of New Hampshire waterborne disease laboratory SOP manual.

Collins, M.R., Eighmy, T.T., Fenstermacher, J.M. and Spanos, S.K. (1992) Removing natural organic matter by conventional slow sand filtration. *Jour. AWWA* **84**(5) 80-90.

Drewes, J.E. and Summers, R.S. (2002) Natural organic matter removal during riverbank filtration: current knowledge and research needs. *In Riverbank Filtration: Improving Source Water Quality.* Edited by C. Ray, G. Mour and R.B. Lindsay, Kluwer Academic Publishers, Dorsrecht.

Kuehn, W. and Mueller, U. (2000) Riverbank filtration: An overview. *Journ. AWWA,* 92, (12: 60-69.

Partinoudi, V. (2004) Riverbank filtration as a viable pretreatment method. M.S. thesis, Univ. of New Hampshire, Durham, NH.

Ray, C., Grischek, T., Schubert, J., Wang, J.Z., and Speth, T.F. (2002a) A perspective of riverbank filtration. *Jour. AWWA,* **94**(4): 149-160.

Ray, C., Melin, G., and Linsky, R.B., editors (2002b) *Riverbank Filtration: Improving Source Water Quality,* Kluwer, Academic Publishers, Dordrecht.

Schulmeyer, P.M. (1995) *Effect of the CedarRiver on the quality of the groundwater supply for Cedar Rapids, Iowa,* U.S. Geological Survey Water Resources Investigations Report 94-4211

Tufenkji, N., Ryan, J.N. and Elimelech, M. (2002) The promise of bank filtration. *Enviromental Science and Technology,* **36**(21): 423A-428A.

Wang, J.Z., Hubbs, S.A. and Song, R. (2002) *Evaluation of riverbank filtration as a drinking water treatment process,*American Water Works Association Research Foundation and American Water Works Association, Denver, CO.

Part Two: Wastewater

Simulation of full scale plants: Benefits and drawbacks

J. Ferrer*, A. Seco **, F. García-Usach*, A. Bouzas and R. Barat***

* Dpto. Ingeniería Hidráulica y Medio Ambiente. Universidad Politécnica de Valencia. Camino de Vera s/n. 46022. Valencia. Spain
(E-mail: jferrer@hma.upv.es; magarus@hma.upv.es; rababa@hma.upv.es)
** Dpto. Ingeniería Química. Universidad de Valencia. Doctor Moliner 50. 46100. Burjassot. Valencia. Spain
(E-mail: aurora.seco@uv.es; alberto.bouzas@uv.es)

Abstract In the last years simulation of WWTPs has become an important tool to increase the knowledge on the process and system behaviour. Simulation tools have been developed making use of mathematical models which include the information related to the processes considered. A successful control strategy must include all the processes taking place in a WWTP: organic and nutrient removal, digestion processes and physico-chemical processes. This fact points out the usefulness of a mathematical model capable of comprising all these processes. The complexity of these models involves the use of simulation platforms.

The application of models to real systems requires calibration techniques in order to obtain the kinetic and stoichiometric parameters. On-line, off-line or fitting calibration techniques can be used. Wastewater characterization is an important feature in modelling but hydraulics, plant layout and operational parameters must be taken into account. Some model applications focused on design, optimization or control systems are included in this paper to emphasize the usefulness of mathematical modelling and simulation.

Keywords Calibration; full scale plants; modelling; simulation applications

Introduction

The more strict standards for treatment plant effluents, the demand of cost effective plants and the need of finding future sustainable solutions have made the monitoring, control and automation of wastewater treatment processes a top priority. The more stringent environmental regulations not only affect the wastewater plant effluents but also the sludge production. Moreover, in terms of operational costs, the treatment of sludge is often higher than the treatment of water phase and therefore cannot be neglected.

Permanent increase in efficiency of existing WWTPs is required. For this, it is necessary to have a deep knowledge of the processes involved, as well as the analysis tools capable of representing all of them. Mathematical models have been found as very successful tools for research, development and optimization since they include all the information related to the processes. Models have to include the following aspects:

- Hydraulic model. Mass balances have to be applied according to the plant layout. Reactor configurations (discontinuous, continuous stirred or plugflow tanks), recirculation and waste flows must be well defined in order to obtain the better description of the real system.

© 2004 IWA Publishing. *2nd IWA Leading-Edge Conference on Water and Wastewater Treatment Technologies* edited by Mark Van Loosdrecht & Jonathan Clement. ISBN: 1 84339 508 8.

- Biological model. Mathematical definition of the different biochemical reactions between the system components (microorganisms) have to be described. At this level, the wastewater treatment modelling can be discussed from two points of view:
- The use of single biological models to describe each process, which implies specific wastewater characterization for each model.
- The use of global biological models able to represent the global plant behaviour.
- Settling model. Suspended solids behaviour linked to biological reactions taking place inside settlers have to be defined in order to achieve reliable results.

In the recent years, several biological and physico-chemical models have been proposed to simulate different processes taking place in WWTPs. Most of them have as objective the processes related to the water treatment line. Besides of these, other models represent the anaerobic processes that occur in prefermentation or digestion systems and the solids precipitation processes.

- The Activated Sludge Models proposed by the IWA (ASM1, ASM2, ASM2d, ASM3) are the most accepted for the simulation of activated sludge systems, including biological nitrogen and phosphorus removal. These models can also be used to describe the aerobic digestion process. Others models, especially focused on metabolic modelling of biological phosphorus removal have been described as the TUDP model (Murnleitner et al., 1997) or the model proposed in Filipe et al. (1998).
- Different models have been proposed to describe anaerobic processes. Among them can be quoted the models proposed by Mosey (1983) and Negri et al. (1993) modified later by Costello et al. (1991) and Münch et al. (1999), respectively. A more detailed model was proposed by Siegrist et al. (1993) and finally, in 2002, the IWA published the Anaerobic Digestion Model No. 1 (ADM1) following the biochemical kinetic format used in ASM series.
- Physico-chemical processes, as precipitation or gas interchange, have a great importance when anaerobic digestion, anaerobic stage in biological phosphorus removal or sludge thickening are simulated. Several models have been proposed to consider these processes (Loewenthal et al., 1994; Maurer et al., 1999; Musvoto et al., 2000a,b).
- A new trend in WWTP simulation consists in considering all the processes involved in a WWTP from a global point of view. BNRM1 (Seco et al., 2004a) is one attempt to achieve this objective. In these models, a unique set of variables are proposed to favour the interconnection between all the streams represented.

A successful control strategy must include all the processes that take place in a WWTP. This fact points out the usefulness of a mathematical model capable of comprising all the processes involved in the wastewater treatment: organic matter and nutrients removal, digestion processes and physico-chemical processes.

The complexity of these models requires their implementation in simulation platforms. There are several simulation platforms which include some of the models stated before. The main characteristics of these simulators are summarized in Table 1. Most of them use the ASM series or ASM1 modifications to represent the biological processes and the Takács model to describe the settling process (Takács et al., 1991). Others allow the user to introduce the model that better represents the process.

Table 1 Main simulator characteristics.

Simulator	Biological Model	Settling Model
BioWin®	Extended version of ASM1 with biological phosphorus removal	Based on Vesilind (1968) model
EFOR®	Various predefined models, including ASM1	Dupont & Dahl (1995) model.
GPS-X®	ASM1	Based on Tákacs model
Matlab/Simulink®	User-defined models	User-defined models
SIMBA®	ASM series, TUDP, Siegrist digestion model and user-defined models	Takács model
STOAT®	ASM1	Takács model
WEST®	ASM series and user-defined models	Tákacs model
DESASS®	BNRM1	Based on Tákacs model

To assure the reliability of simulation strategies, these should be based on rigorous methodology including the definition of a comprehensive simulation model of the plant, plant layout, influent load characterization, model parameters calibration, controllers and operational data. Thus, the acceptance of these strategies by designers and plant operators should be improved. Several guidelines have been presented in order to standardize simulation studies (Hulsbeek et al., 2002; Langergraber et al., 2003).

Measurement campaign

In general, the design procedure based on process modelling requires an exhaustive influent wastewater characterization. Experience on simulation strategies has shown that the wastewater characterization is one of the most important aspects before the application of a model to a specific case. An appropriate influent characterization will include the following analysis: total and volatile suspended solids, COD, BOD, volatile fatty acids, NO_3-N, NH_4-N, total nitrogen, total phosphorus, PO_4-P and alkalinity. However, the number of measured parameters will depend on the requirements of the model chosen for the simulation.

When simulation is used to upgrade plants, not only influent wastewater characterization but also the effluent and mixed liquor characterization is required. In these characterizations, influent, effluent and mixed liquor parameters are analysed from samples collected during an appropriate period of time. The mixed liquor characterization can be performed analysing the total and volatile suspended solids, COD, NO_3-N, NH_4-N, total phosphorus, PO_4-P and alkalinity. This sludge characterization should be carried out for the different reactors, especially when the studied system is a biological nutrient removal process. "On-line" recordings (flowrates, pH, and dissolved oxygen concentration) are also needed for calculations. When control strategies have to be evaluated, oxidation-reduction potential and conductivity monitoring would be also required. Hydraulics must be checked considering the flow characteristics through line reactors and settlers. Frequently, the influent flow distribution differs from the ideal situation causing simulation deviations from experimental values.

During both, designing and upgrading, the sample campaign must be performed under pseudo-stationary conditions, covering a period of time long enough to consider the possible influent fluctuations. Furthermore, dry weather conditions are required for the sample campaign. The obtained data should be compared to the historical system data in order to confirm the stationary conditions. Mass balances must be applied.

Numerous efforts to describe wastewater components have been carried out throughout the last years. In fact, there is a large variation in different methods proposed. The estimation of the different fractions of wastewater is usually carried out by two different ways. One is the respirometric method as has been reported by several authors (Ekama et al. 1986; Sollfrank and Gujer, 1991; Spanjers et al., 1995), and the other one, frequently used, consist in physico-chemical separation techniques (Mamais et al., 1993; Roeleveld and van Loosdrecht, 2002).

Calibration

Calibration consists in determining all parameters of the selected model capable of reproducing the process behaviour. Given the high number of parameters to be determined, a selective calibration should be done in order to facilitate the study. High influence kinetic and stoichiometric parameters have to be distinguished from low influence parameters. Low variations in a high influence parameter have a great influence on simulation results. In the other hand, high variations in a low influence parameter produce low changes on simulation results. For that reason, both types of parameters undergo different mathematical treatment. Sensitivity analysis, changing the parameters manually or automatically with software tools, reveals the major influence parameters (Langergraber et al., 2003).

The low influence parameter values can be taken from published parameter sets. Opposite, the high influence parameters show a great variability in literature, so these parameters must be obtained using different calibration techniques:

- *On-line* calibration technique. This calibration procedure consists in on-line measurements carried out at the WWTP in fixed intervals (Larrea et al., 1992). The measurement results are introduced in a mathematical algorithm (e.g. Kalman filter) which finds the best parameters to reproduce the experimental data. As the different measurements take into account the system variations, the model finds the best set of parameters to reproduce the total period of time studied. Therefore, the calibration is valid for the considered conditions.
- *Off-line* calibration techniques. Laboratory measurements allow the analyst to obtain a parameter or a set of parameters from individual experiments using pilot or full scale plant biomass (Penya-roja et al., 2002). Once the kinetic and stoichiometric parameters are determined, the plant behaviour is simulated in order to reproduce the experimental data. In these experiments, a wide range of conditions are tested since these assays require a great number of experimental points. The off-line calibration is a dynamic calibration where the experiment conditions change throughout the assay. Hence, the values obtained applying this calibration methodology can be considered as representative for the studied system. The main problem of this methodology focus on two points: the fact that all parameters can not be calibrated in this way and the laborious experiments needed.
- *Manual or automatic fitting using software tools*. In that case, simulation tools are used to fit the simulation results with the measured data.

The verification of the calibrated parameters should be carried out with an independent set of performance data from an additional measurement campaign (Hulsbeek et al., 2002; Langergraber et al., 2003).

Besides the wastewater quality, other characteristics of the system have an important influence on calibrated parameters:

- The plant layout becomes a very important factor when either denitrification or prefermentation processes take place in the calibrated system. The percentage of denitrification zone directly affects the reduction factor for anoxic activity. The existence of a prefermentation zone affects the volatile fatty acids concentration and composition, two variables with a high influence on biological phosphorus removal.
- It has been stated (García-Usach, 2002) that the sludge retention time affects the phosphorus-glycogen accumulating organisms ratio. If glycogen accumulating organisms are not included in the calibrated model the phosphorus accumulating parameters differ from the default values (Ferrer et al., 2004).

Applications

Mathematical modelling, besides the required simulation software, has improved design, operation efficiency and many other aspects related to wastewater treatment processes. The main applications of mathematical models are described next.

- *Design of new WWTPs*. Process design studies can be carried out for building new plants or enlarging existing ones. Normally, for plants that have to be constructed no measurement data are available and therefore no model calibration and validation can be carried out. This lack of data must be compensated with a flexible treatment scheme. In the case of large WWTP, to build a pilot plant that provides the kinetic and stoichiometric parameters could be a worthwhile possibility. In any case, modelling can help to bridge the gap between lab and full-scale application.
- *Diagnosis and optimization of existing WWTPs*. Once the model has been calibrated, it can be used to optimize the plants detecting non optimal performance and proposing modifications focused on the plant layout and the operation strategies.
- *Teaching and staff training*. Models implemented in simulation tools are very useful for studying the whole plant performance taking into account the interactions between main stream and side streams. The use of simulators can help plant operators to know the effects of operation decisions. Furthermore, the didactic capability of models must be highlighted. The use of models in simulation enhances the comprehension of the global process and the influence of the interactions between the different variables.
- *Test of new control systems by simulation studies*. Mathematical modelling is considered a time and cost-saving tool for testing and developing new control systems.

Application examples

Some of the applications stated above have been tested using the simulation software DESASS developed by the CALAGUA research group. The model included in DESASS is the BNRM1 (Seco et al. 2004a). BNRM1 is a global model capable of simulating the most important physical, chemical and biological processes taking place in a WWTP. This fact allows simulating primary settlers; activated sludge system for organic matter, nitrogen and phosphorus removal; gravity thickeners; sludge digesters (aerobic and anaerobic) and primary sludge prefermentig systems. Thus, it can be considered, on the same simulation platform, the linked effects of the different process units in a WWTP. Some examples of modelling applications are described below:

Inhibition problems

The calibration and simulation of Elche WWTP (Alicante) showed problems in the nitrification process (Ferrer et al., 2004). The historical data of the plant indicated that nitrification occurred randomly. Besides the parameter values obtained showed a low maximum nitrifying growth rate (0.55 d^{-1}) and a high rate constant lysis for the nitrifying organisms (0.26 d^{-1}). These values pointed out the existence of inhibition problems probably as a result of an industrial component in wastewater (footwear and textile contributions). These facts confirmed the important dependence on wastewater characteristics of the model parameters. The inhibition problems were considered during the plant upgrading, increasing the sludge retention time and modifying the suitable operational parameters.

Biological processes in settlers

As can be observed in Table 2, the consideration of the biological processes occurring in settlers improves the reliability of the simulated results. This situation is stressed for the main components implicated in that processes, such as nitrogen and phosphorus.

Table 2 Simulation and experimental settler effluent data.

Parameter (mg l^{-1})	Experimental	Simulation	
		Without biological processes	With biological processes
DQO total	65	63	62
N-NO_3	26.1	48.3	25.9
N-NH_4	2.5	1.44	2.6
P-PO4	5.2	4.9	5.1

Optimization studies

The biological model was calibrated for Elda (Alicante) full scale biological nitrogen removal plant (Ferrer et al., 2004). The activated sludge and effluent characteristics were simulated using the software DESASS. Due to the lack of an internal recirculation flow, the anoxic reactor was working as anaerobic one. As consequence, the presence of biological phosphorus removal was detected. Moreover, the presence of glycogen accumulating organisms was observed. The model predictions reproduced the plant performance accurately and were used to upgrade it. The upgrading consisted in introducing an internal recirculation flow preserving the anaerobic zone. Thus, nitrogen and phosphorus removal were improved.

New designs

Several designs, based on process modelling, have been carried out using DESASS.

- Seco et al. (2004b) proposed the Peñíscola WWTP (Castellón) design considering two treatment schemes in order to optimize the plant performance for two extremely different seasonal operations:
    - Low-load influent conditions. The plant is operated under extended aeration conditions without further stabilization of excess sludge. The possibility of working with one or two secondary settlers has been tested by simulation. As can be seen in Figure 1, the settler sludge blanket height and wastage sludge concentration are similar in both cases, but suspended solids concentration in the hindered-settling zone of the clarifier obtained working with only one settler is higher than working

with two settlers. This fact could cause excessive carry-over of solids during short periods of high-load influent conditions. Transition between low and high load conditions was simulated in order to assess the settler performance (Figure 2).

- High-load influent conditions. Extended aeration can not be achieved and the plant is operated under conventional process conditions. In this operation mode, excess sludge is aerobically digested. As a consequence of the high amount of phosphorus released during the aerobic digestion, the phosphorus concentration in the supernatant from dewatering process will increase. Phosphorus precipitation had to be considered in this stream in order to avoid the circulation of phosphorus from the sludge treatment system back to the main stream system.

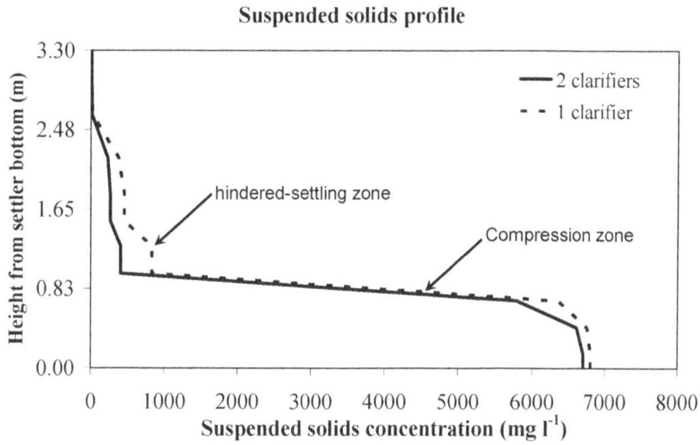

Figure 1 Suspended solids profile

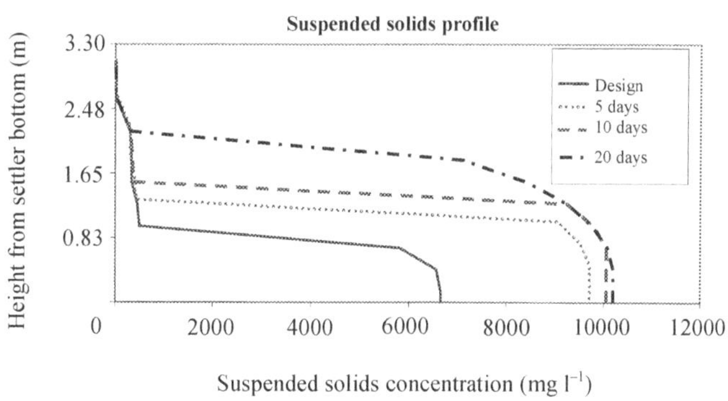

Figure 2 Suspended solids profile in time

Test of new control systems

DESASS has been used to developed a supervisory control system for optimizing nitrogen removal and aeration energy consumption in wastewater treatment plants (Serralta et al., 2002). Nitrogen removal is optimized by adjusting the internal recirculation flow rate. Aeration energy consumption is minimized by adjusting control valves opening and rotational speed of the blowers. The supervisory control system modifies the dissolved oxygen set point according to ammonia concentration in order to optimize the process performance. Energy consumption saving of up to 10% in the blowers and 50% in the recirculation pumps.

Training and teaching

DESASS has been used at the University of Valencia and the Polytechnic University of Valencia for the last years. It has been very useful to facilitate the understanding of the theoretical concepts previously introduced. Students analyse the operation of different WWTPs and proposed alternatives based on their knowledge. On the other hand, plant operators training are being carried out.

Conclusions

Further research must aim at developing and applying analysis tools to assess future sustainable solutions. Mathematical models capable of representing all the processes involved in a WWTP will be of great importance. Application of mathematical models to full scale plants requires exhaustive wastewater characterization, model parameters calibration, operational data and hydraulics knowledge.

An exhaustive characterization implies a huge number of analyses which mean a costly measurement campaign, not only in terms of money but also in time. Calibration methodologies usually include laborious experimental determinations and complex mathematical resolutions. Nevertheless, the model parameters determination is required to obtain an accurate simulation of the process. Parameter values give important information about the system performance as information about possible inhibitions or competitive microorganism populations.

In this paper it have been stated that simulation is not only an important tool for the scientific community but also can be a useful tool in the general wastewater treatment practice. Different control strategies and operation scenarios can be studied. Moreover, simulation tools enable environmental engineering students to understand specific details about the implemented models or the numerical solution procedures concerned. On the other hand, the use of simulation tools without understanding the processes involved can lead students or plant operators to obtain wrong solutions.

References

Batstone, D.J.; Keller, P.; Angelidaki, I.; Kalyuzhnyi, S.V.; Pavlostathis, S.G.; Rozzi, A.; Sanders, W.T.M.; Siegrist, H.; Vavilin, V.A. (2002). *Anaerobic Digestion Model No. 1, IWA Scientific and Technical Report No. 13*, IWA, London.

Dupont, R. and Dahl. (1995). A one-dimensional model for a secondary settling tank including density current and short-circuiting. *Wat. Sci. Tech.*, **31**(2), 2151-224.

Ekama G.A., Dold P.L. and Marais G.v.R. (1986). Procedures for determining influent COD fractions and the maximum specific growth rate of heterotrophs in activated sludge systems. *Wat. Sci. Tech.*, **18**(6), 91-114.

Ferrer J., Morenilla J.J., Bouzas A. and Garcia-Usach F. (2004). Calibration and simulation of two large wastewater treatment plants operated for nutrient removal. Proceedings of the International conference on wastewater treatment for nutrient removal and reuse 2004 (ICWNR ´04). Thailand.

Filipe C. D. M. and Daigger G. T. (1998). Development of a revised metabolic model for the growth of phosphorus-accumulating organisms. *Wat. Environ. Res.*, **70**(1), 67-79.

García-Usach F. (2002). *Influence of sludge retention time and temperature on biological matter and nutrient removal of wastewaters*. PhD thesis, Department of Hydraulic Engineering and Environment, Universitat Politécnica of Valencia.

Henze M., Gujer W., Mino T. and van Loosdrecht M. C. M. (2000). *Activated Sludge Models ASM1, ASM2, ASM2d and ASM3.* IWA Scientific and Technical Report No. 9. IWA Publishing, London, UK.

Hulsbeek J. J. W., Kruit J., Roeleveld P. J., van Loosdrecht M. C. M. (2002). A practical protocol for dynamic modelling of activated sludge systems. *Wat. Sci. Tech.*, **45**(6), 127-136.

Langergraber G., Rieger L., Winkler S., Alex J., Wiese J., Owerdieck C., Ahnert M., Simon J. and Maurer M. (2003). A guideline for simulation studies of wastewater treatment plants. Proceedings of the 9th IWA Specialised Conference Design, Operation and Economics of Large Wastewater Treatment Plants. Praha.

Larrea L., García-Heras J.L., Ayesa E. and Florez J. (1992). Designing experiments to determine the coeficients of activated-sludge models by identification algorithms. *Wat. Sci. Tech.*, **25**(6), 149-165.

Loewenthal R. E., Kornmüller U. R. C. and van Heerden E. P. (1994). Modelling struvite precipitation in anaerobic treatment systems. *Wat. Sci. Tech.*, **30**(12), 107-116.

Mamais D., Jenkins D. and Prrr P. (1993). A rapid physical-chemical method for the determination of readily biodegradable soluble COD in municipal wastewater. *Wat. Res.*, **27**(1), 195-197.

Maurer M., Abramovich D., Siegrist H. and Gujer W. (1999). Kinetics of biologically induced phosphorus precipitation in waste-water treatment. *Wat. Res.* **33**(2), 484-493.

Mosey, F. (1983). Mathematical modelling of the anaerobic digestion process: regulatory mechanisms for the formation of short-chain volatile acids from glucose. *Wat. Sci. Tech.* **15**, 209-232.

Münch, E.v.; Keller, J.; Lant, P.; Newell, R. (1999). Mathematical modelling of prefermenters - I. Model development and verification. *Wat. Res.* **33** (12), 2757-2768.

Murnleitner E., Kuba T., van Loosdrecht M. C. M. and Heijnen J. J. (1997). An integrated model for the aerobic and denitrifying biological phosphorus process. *Biotechnol. Bioeng.*, **54**, 434-450.

Musvoto E. V., Wentzel M. C., Lowenthal R. E. and Ekama G. A. (2000a). Integrated chemical-physical processes modelling-I. Development of kinetic-based model for mixed weak acid/base systems. *Wat. Res.*, **34**(6), 1857-1867.

Musvoto E. V., Wentzel M. C. and Ekama G. A. (2000b). Integrated chemical-physical processes modelling-II. Simulating aeration treatment of anaerobic digester supernatants. *Wat. Res.*, **34**(6), 1857-1867.

Negri, E.D.; Mata-Álvarez, J.; Sans, C.; Cecchi, F. (1993). A mathematical model of volatile fatty acids (VFA) production in a plug-flow reactor treating the organic fraction of municipal solid waste. *Wat. Sci. Tech.*, **27** (2), 201-208.

Penya-Roja J.M., Seco A., Ferrer J. and Serralta J. (2002). Calibration and validation of Activated Sludge Model No. 2d for Spanish municipal wastewater. *Environ. Technol.*, **23**, 849-862.

Roeleveld P. J. and van Loosdrecht M.C.M. (2002). Experience with guidelines for wastewater characterisation in The Netherlands. *Wat. Sci. Tech.*, **45**(6), 77-87.

Seco A., Ribes J., Serralta J. and Ferrer J. (2004a). Biological nutrient removal model No.1 (BNRM1). Proceedings of the International conference on wastewater treatment for nutrient removal and reuse 2004 (ICWNR ´04). Thailand.

Seco A., Ferrer J., Serralta J., Ribes J., Barat, R. and Bouzas A. (2004b). Use of biological and sedimentation models for designing Peñíscola WWTP. *Environ. Technol.* (in press).

Serralta J., Ribes J., Seco A. and Ferrer J. (2002). A supervisory control system for optimising nitrogen removal and aeration energy consumption in wastewater treatment plants. *Wat. Sci. Tech.*, **45**(4-5), 309-316.

Siegrist, H.; Renggli, D.; Gujer, W. (1993). Mathematical modelling of anaerobic mesophilic sewage sludge treatment. *Wat. Sci. Tech.* **27** (2), 25-36.

Sollfrank U. and Gujer W. (1991). Characterization of domestic wastewater for mathematical modelling of the activated sludge process. *Wat. Sci. Tech.*, **23**(4/6), 1057-1066.

Spanjers H. and Vanrolleghem P. (1995). Respirometry as a tool for rapid characterization of wastewater and activated sludge. *Wat. Sci. Tech.*, **31**(2), 105-114.

Takács I., Patry G.G. and Nolasco D. (1991). A dynamic model of the clarification-thickening process. *Wat. Res.* **25**(10), 1263-1271.

Vesilind P.A. (1968). Design of prototype thickeners from batch settling tests. *Wat. Sewage Wks.* 115, 302-307.

Advanced CFD tool for Wastewater: today complex Modelling & tomorrow easy-to-use Interface

Guimet V., Savoye P., Audic J.-M., Do-Quang Z.

SUEZ ENVIRONNEMENT, 38 rue du Président Wilson, 78230 Le Pecq, France
Tel: +33 1 34 80 23 36 – Fax: +33 1 30 53 62 07 - Email: valerie.guimet@suez-env.com

Abstract CFD modeling applies for full-scale wastewater treatment processes to improve water quality, to optimize existing reactors or test new designs. Single and two-phase flow patterns are modeled for three applications: secondary clarifier, aeration basin and UV. As a review of CFD techniques, each process and the respective CFD model are detailed. The objective is the assessment of the process mechanism, in particular the sludge transport in the solid-liquid separation, the oxygen transfer in the aeration process and eventually the inactivation dose received in a UV process.
Keywords Wastewater, Computational Fluid Dynamics CFD, Modelling

Introduction

Since the late 90's, Computational Fluid Dynamics (CFD) modelling became popular in drinking water and wastewater industry (Do-Quang, 1999; Lyn, 1999; Craig 2002) as an alternative or a support tool to pilot studies, reducing realization cost and time. Moreover, process or quality optimisation with CFD has lead to important cost savings up to 50% on capital expenditures and 20% on operating expenditures (Guimet, 2003). However, operators, regulators and sometimes engineers are still reluctant to use it because of the lack of simplicity and accessibility of the CFD tools and results.

The aim of the work presented herein is, first, to review the actual CFD models available for wastewater, to contribute to CFD enhancement on complex phenomena modelling, to facilitate the understanding of CFD results and eventually, to study the feasibility of accessible CFD and automatic interfaces.

CFD modelling in the wastewater context

In the wastewater context (figure 1), processes of the clarification step, biological removal step (aeration ditch) and disinfection step (pathogens inactivation by UV) can be modelled by CFD. For each process, CFD study gives the assessment of the internal physical mechanism. Then, optimisation, retrofit or qualification can follow.

In particular, modelling leads to the representation of the sludge transport in the solid-liquid separation by settling or flotation, the oxygen concentration distribution in water in the aeration basin, and eventually the inactivation dose received in a UV process. Those parameters control the corresponding process efficiency.

© 2004 IWA Publishing. *2nd IWA Leading-Edge Conference on Water and Wastewater Treatment Technologies* edited by Mark Van Loosdrecht & Jonathan Clement. ISBN: 1 84339 508 8.

CFD methodology

At first, water flow is supposed to be incompressible, Newtonian and turbulent (because of high local velocity in the process or large scale facility). The flow is governed by Reynolds Averaged Navier-Stokes equations, RANS modeling equations and theory can be found for example in (Mohammadi 1994). The turbulence model of the Reynolds stresses associated with the RANS approach is described by the standard k-epsilon model. Whenever impacts (on baffles for example) arise, other models may be considered to limit the production of turbulent kinetic energy near the walls.

Following numerical rules, established by CFD publishers (Freitas 1993), models are at least 2nd order accurate in space, convergence is checked carefully (on residuals and main variable levels), boundary and initial conditions are described. Furthermore, any new application is preceded by a feasibility study. Eventually, any user-defined patches or numerical models are validated on reliable experimental data.

Applications

In the following sections, three applications will be presented together with the corresponding CFD procedure and possible enhancement. Those applications concern one secondary clarifier for the liquid-solid separation problem, an aeration basin for the biological removal step and one UV system for disinfection.

Secondary clarifier

Problem description

In a secondary clarifier or in any settling tank, to reproduce the internal phenomena, the model has to take into account the sludge and the water or at least the interaction between the two phases encountered.

A multiphase model can model sludge transport: a Lagrange-Euler model for flow-particles interaction or an Euler-Euler model for two interacting phases. Because the interaction between phases is not well known, one may consider as an alternative by adding a scalar transport equation representing the sludge concentration to a single-phase flow representing then the mixing between water and sludge (Zhou 1992). So far, this last approach gives the more representative results. That's the reason why this two-phase flow (sludge and water) phenomenon is described by a single-phase flow coupled with an interacting scalar equation.

Continuity and momentum equations for the mixture of water and floc are derived from the liquid (phase f) and solid (phase p) set of equations. Then, under Boussinesq assumption that considers gravity only governs density gradients, a single-phase flow with variable density models the mixture. Sludge concentration is transported by the flow, modifying the fluid density, equation (1), and viscosity.

$$\rho = \rho_f + C(1 - \frac{\rho_f}{\rho_p}) \tag{1}$$

Notation: ρ is the mixture density, ρ_f the density of the water, ρ_p the density of the floc particles and C the sludge concentration.

Velocities of water and floc (U_f and U_p) are assumed to slip at the settling velocity ($V_s = U_p - U_f$). Settling velocity formula is given by a double exponential to take into account both accumulation and compression in settling of the characterized sludge (Takacs, 1991). A settling velocity source term is added to the sludge concentration equation, equation (2), describing the fact that settling is progressively reduced by the sludge accumulation.

$$\frac{\partial C}{\partial t} + \nabla\left(\underline{u}.C\right) = \nabla.\left(D.\nabla C\right) + \left(\frac{\partial\left(V_S.C\right)}{\partial z}\right) \tag{2}$$

The RANS system is closed by k-epsilon model including buoyancy effects (Gibson, 1976), which are non-negligible in high concentrated sludge zones, in particular in the sludge bed (Lakehal 1999, Honoré 2004).

CFD procedure

1. Local hydrodynamics representing the mixture (water+floc) is coupled to a scalar equation of sludge concentration that changes in time the viscosity and density of the mixture. Therefore, a time dependent algorithm is compulsory to solve the system.
2. Checking the global convergence (e.g. the numerical residuals of the equations to be solved) and the local convergence (e.g. sensors which assess the main variable levels, like concentration) achieves a stable and meaningful representation of the mixture.
3. At the end of an operating cycle (time needed before sludge removal), mean velocity field, pressure, turbulence and sludge concentration are represented in the reactor. They give the flow patterns and in particular the zones of high sludge concentration. It is then possible to locate precisely the possible zones of poor water quality and if so to propose a better design or operating condition.

Example

The sludge blanket was computed in plant A (figures 2, 3) for different flow rates to assess the limits at which the operating system may run without any discrepancies. Flow instabilities and/or a too high sludge blanket could have occurred for one specific flow regime.

Aeration basin

Problem description

Physics in aeration basin is one of the most complex problems encountered in water industry. Three-phase flow governs the phenomena: water, activated sludge and air injected. Aeration and biological phenomena are key issues to assess the process efficacy.

Oxygen transfer from the dispersed to the continuum phase insures bacteria growth. This is important to maintain the biological pollutants removal. Therefore, oxygen concentration in the liquid phase is one of the important CFD criterions assessing the process efficiency.

The oxygen transfer to water is ensured through extra source terms (Cockx, 2000) at the second hand term of the transport equation (3).

$$\frac{\partial S_O}{\partial t} + \nabla\left(\underline{u}_f.S_O\right) = \nabla.\left(E_L.\nabla S_O\right) + OUR + k_L a(S_{O,sat} - S_O)$$

$$OUR = -k_D X_{BH} = cte$$

S_O represents the dissolved oxygen concentration; u_f : the water velocity field; E_L : the overall oxygen diffusivity in the water; OUR: oxygen uptake rate or biomass consummation, supposed constant at a value of 20 kg/m3/h; $k_L a$: overall transfer coefficient (takes the local hydrodynamics into account); $S_{O,sat}$: the saturation oxygen concentration.

CFD procedure

1. A two-phase flow is considered coupling the water and the air injected. Oxygen concentration is resolved through a transport equation, containing extra source terms for the fact the oxygen is consumed over time (by the bacteria) and transferred to the continuum phase. A time dependent algorithm is compulsory to solve the system.
2. Checking the global convergence over time.
3. At the end of an operating cycle, the time needed before an anaerobic cycle for example, the main variables (mean gas-liquid velocity field, pressure and turbulence) are represented in the reactor. Moreover, the oxygen concentration assesses the process efficacy. In particular, the dead zones give the regions where bacteria will die; those regions decrease the process quality.

Example

Figure 4 represents an example of application for the aeration basin problem. Regrouping the air injection was considered as an alternative in plant B, motivated by CAPEX savings. The aim of the CFD study was to check if there were no dead oxygen zones resulting in poor water quality. CFD results show that the new configuration doesn't enhance the oxygen transfer: regions of low oxygen concentration in water were found. Therefore the change in the aeration inlet wasn't recommended in order to maintain water quality at a certain level.

UV system

For UV systems the criteria used to compare or optimize designs is the UV dose received by the pathogens. A homogeneous UV dose enhances the efficacy of the disinfection.

Spatial distribution and irradiation of the UV lamps, local hydrodynamics and transmittance contribute to the assessment of the dose and the process efficacy (Craig 2002). The dose is calculated by coupling the intensity of the lamps together with the local hydrodynamics. Without loss of generality, the intensity of the lamps is supposed to be constant in time for a fixed percentage of UV transmittance. Combining the local kinetics of the physical and chemical phenomena together with the corresponding microorganisms residence time assesses the quality of the disinfection process.

CFD procedure

1. Once the hydrodynamics is computed (see previous sections), the UV dose is assessed through a Lagrangian post-treatment combining the local intensity of the lamp viewed by the particle through its traveling in the system and the time spent by the particle in the system (Lyn 1999, Do-Quang 1997, 2002).
2. The dose represents the integration of the intensity along the streamlines generated by the local hydrodynamics.
3. Particle tracks along the continuous phase give a representation of particles travel from their injection to the outlet zone in the process.

Example

The model three steps are recalled on figure 6 (hydrodynamics, lamp radiation and dose). Recirculating zones behind the lamps combined with the lamp intensity in the same region were leading to over irradiation in these zones (waste of energy). Moreover, for the reactor presented, the zones near the walls revealed low UV dose: the velocities were high while the UV intensity was low in these regions (figure 7). Short baffles were introduced to increase the turbulent energy and to block the short-circuiting along the walls (last in figure 6). The homogeneity of the dose was improved, especially at the outlet of the reactor.

Automatic interface

Besides the existing reviews and validations, engineers, operators and regulators are still reluctant to use CFD complex modelling. In order to enhance accessibility, automatic interfaces can be developed so non-experts could conduct CFD computations.

Feasibility studies are possible onto single phase-flow before dealing with the complex flows encountered in clarification wastewater processes.

One methodology is first to establish CFD best practices: check all the numerical and physical parameters describing the problem onto academic test cases then onto industrial data. Space and time discretisations, turbulence modelling, solution algorithm procedure need to be validated. Those best practices define the reference for automatic macros or at least guide the user through a friendly user-interface. An automatic interface prototype version on chlorination tanks is in development (Guimet, 2003).

Conclusion

Single or two-phase flow models are presented for the wastewater industry. The list of model and applications is not exhaustive but gives the actual state of the art in CFD computations run for the operator needs.

More complex models can be used for multiphase flows. For example, combining different families of dispersed phase (e.g.: bubble characteristics) leads to a better understanding of on-site full-scale process. Besides, simpler single phase-model can be used for a complex phenomena or geometry to give an a priori internal hydrodynamics result before pilot or process implantation.

On one side, CFD simulations are about to model the complete process and on the other side they are about to be conducted by non-experts. Recalling that CFD studies could lead to important cost reductions and better control of the reliability of the waste water treatment plants, the tool automation will allow to gain more and more productivity in the future.

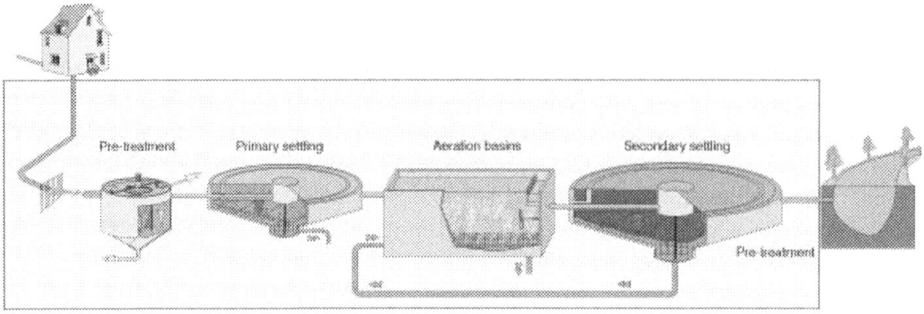

Figure 1 Wastewater context and CFD applications

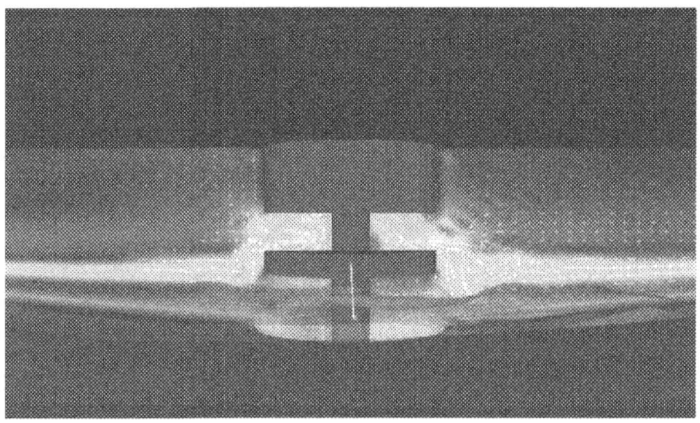

Figure 2 3D view of the sludge bed formation

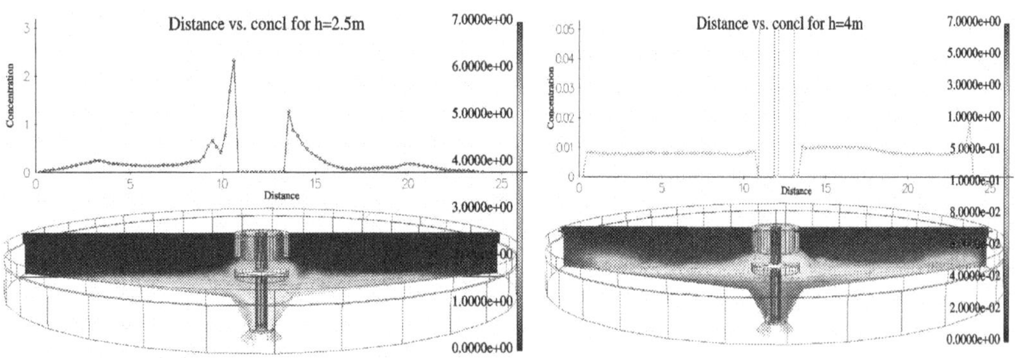

Figure 3 Sludge concentration in the water (mg/l) for two different regimes

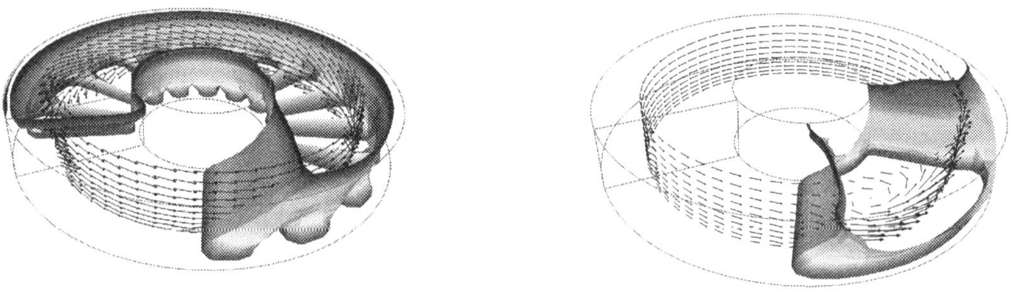

Figure 4 Gas hold-up for dispersed air injection (left) and grouped air injection (right)

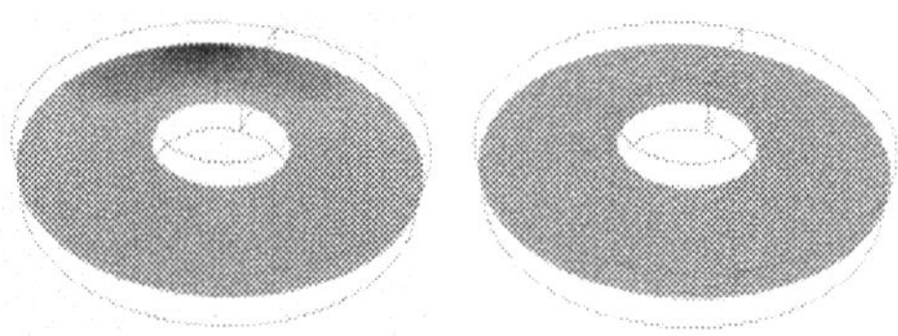

Figure 5.1 Dispersed air injection

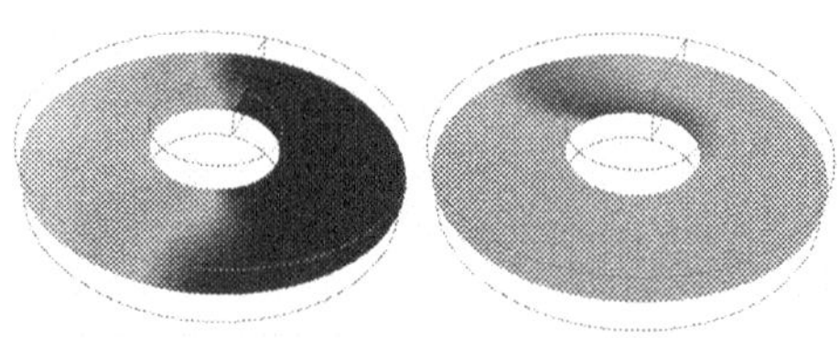

Figure 5.2 Grouped air injection

Figure 5 Oxygen rate in the basin for two inlet regimes (0 m/s and 0.25 m/s)

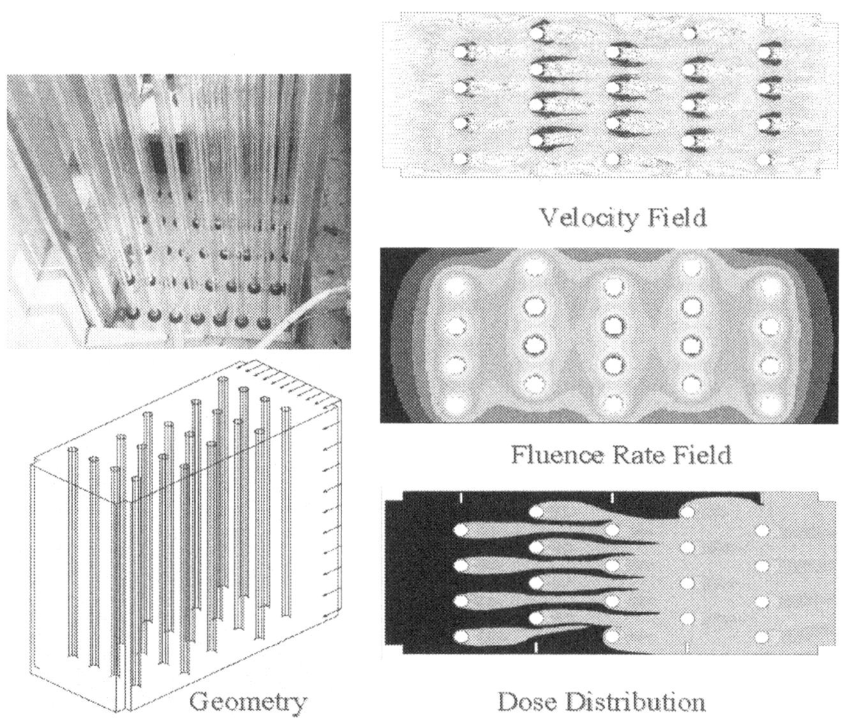

Figure 6 CFD modeling for one UV process pilot (open channel configuration)

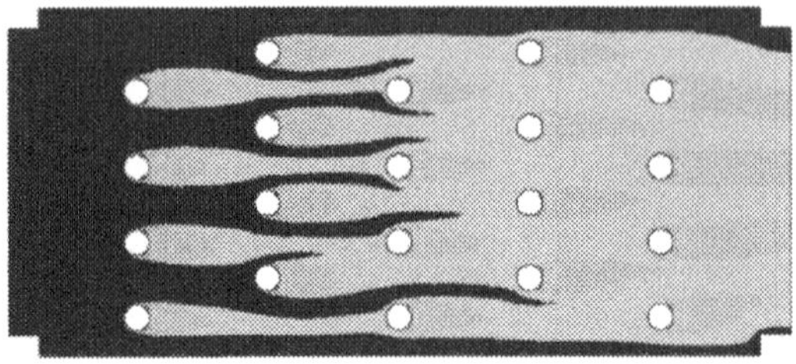

Figure 7 UV dose for the initial configuration

References

Craig K., De Traversay C., Bowen B., Essemiani K., Levecq C., Naylor R. (2002). Hydraulic study and optimisation of water treatment processes using numerical simulation, Water Supply **2**(5-6), 135–142.

Do-Quang Z., Cockx A., Line A., Roustan M. (1999). Computational Fluid Dynamics applied to water and wastewater treatment facility modeling, Journal of Environmental Engineering and Policy **1**, 137-147.

Do-Quang, Z., Djebbar, R., Blatchley III, E.R., Laine, J.M. (1997). Computational Fluid Dynamics (CFD) Modeling of UV Disinfection Reactor Performances: Optimization of the Flow in Vertical Lamps Open Channel. Conference of the ASCE, Edmonton.

Do-Quang, Z., Janex M.L., Perrin D. (2002). Predictive tool for UV dose distribution assessment: validation of CFD model by bioassays. AWWA Annual Conference, New Orleans.

Freitas C.J. (1993). Journal of Fluids Engineering Editorial Policy Statement on the Control of Numerical Accuracy. Journal of Fluids Engineering Transactions of the ASME **115**, 339-340.

Gibson M.M. & Launder B.E., (1976). On the calculation of horizontal, turbulent, free shear flows under gravitational influence, Journal of Heat Transfer, **2**, 81-87.

Guimet V., Do-Quang Z. (2003). Virtual Plant™: CFD Tool for full-scale water treatment train modelling. *2003 Water Quality and Technology Conference Proceedings*.

Honoré T (2004). A general CFD model for activated sludge reactors including biological and hydrodynamics interactions, DEA Research Report.

Lakehal D., Krebs P., Krijgsman J., Rodi W. (1999). Computing shear flow and sludge blanket in secondary clarifiers, Journal of Hydraulic Engineering, Vol. **125**, n°3, 253-262.

Lyn D.A., Chiu K. and Blatchley III E.R. (1999). Numerical modeling of flow and UV disinfection channels". *Journal of Environmental Engineering ASCE* **125(1)**, 17-26.

Mohammadi B., Pironneau O. (1994). Analysis of the k-epsilon turbulence model. Research in Applied Mathematics, J. Wiley & Sons, Masson.

I. Takacs, G. Patry, D. Nolasco (1991). A dynamic model of the clarification-thickening process, *Water Research*, **25**, pp 1263-1271.

Zhou S., McCorquodale J.A. (1992). Influence of skirt radius on performance of circular clarifier with density stratification, Int. J. Numer. Methods Fluids, **4**, 919-934.

Instrumentation and Control in Anaerobic Digestion

Damien J. Batstone*, Krist V. Gernaey and Jean-Philippe Steyer*****

*Environment & Resources, DTU, Bygningstorvet Building 113, Lyngby 2800DK, Denmark
djb@er.dtu.dk
** CAPEC, Department of Chemical Engineering, DTU, Building 229, Lyngby 2800DK, Denmark
*** Laboratoire de Biotechnologie de l'Environnement, Institut National de la Recherche Agronomique. LBE-INRA, Avenue des Étangs, 11100 Narbonne, France

Abstract

Anaerobic digestion is a multistep process, and is most applied to solids destruction and wastewater treatment for energy production. Despite wide application, and long-term industrial proof of application, some industries are still reluctant to apply this technology. One of the classical reasons has been lack of process control handles, instruments, and developed control algorithms. This has improved dramatically in the past 10 years, and all of these areas have now been addressed. The main gap in instrumentation technology has been a rapid intermediate sensor to detect overload conditions. There are now a number of sensors that can measure total or individual organic acids, and some are in commercial production. Control has also been widely applied, with a wide variety of controllers, for direct beneficial results, and one application has been shown in this paper. Available control handles are still a limitation, but this is being partly addressed by the increased complexity of digestion processes. Methods for control benchmarking have also been improved, as there is now an industry standard model (the IWA ADM1), and this is being applied in an improved whole wastewater treatment plant benchmark. There has therefore been, overall, a quantum advance in application and sophistication of instrumentation and control in anaerobic digestion, and it is an effective option for improved process loading rate and conversion efficiency.

Keywords anaerobic digestion, benchmarking, control, instrumentation, modeling, ADM1

Introduction

Anaerobic digestion (AD) is a multistep process in which complex organic material is converted to simpler compounds without an external electron acceptor such as oxygen or nitrate. The main steps are: (a) Extracellular hydrolysis of complex organic particulates to simple sugars and amino acids; (b) Acidogenesis of these sugars and amino acids to produce organic acids and alcohols; (c) Acetogenesis, in which the organic acids and alcohols are converted to hydrogen and acetic acid; (d) Hydrogenotrophic methanogenesis, in which the hydrogen in step (c) is converted to methane; and (e) Aceticlastic methanogenesis, in which the acetate in step (c) is converted to methane. Normally the process as applied is complete digestion to carbon-dioxide and methane, but partial digesters are also used in specific applications (e.g., prefermenters). AD is an established technology for wastewater treatment, and is diversely applied. Depending on the application, different control and instrumentation

© 2004 IWA Publishing. *2nd IWA Leading-Edge Conference on Water and Wastewater Treatment Technologies* edited by Mark Van Loosdrecht & Jonathan Clement. ISBN: 1 84339 508 8.

objectives, and different control variables are practically available. Some common examples are:

(1) Primary and activated sludge digesters in municipal wastewater treatment plants (WWTPs) for sludge destruction. Control objectives are normally stable operation, and minimisation of the effect on the activated sludge plant of disturbances caused by the recycle of nutrients (nitrogen and phosphorus) via the treated sludge reject water stream. Maximisation of energy production (via the biogas) can also result in important operating cost reductions for the entire municipal WWTP.

(2) Solids digesters; in which a solid or semi solid stream (3-30% VS) is converted to methane, for renewable energy production, and solids destruction. Control objectives are stable operation, and maximum energy production.

(3) High-rate digesters treating industrial streams with mostly soluble organics, to minimise downstream treatment charges and avoid environmental pollution. Control objectives are stable operation, and to avoid damage to, or inhibition of the reactor sludge, which is normally retained.

(4) Partial fermenters and prefermenters, which produce a specific product. These may produce organic acids for enhanced biological phosphorous removal, ethanol, hydrogen, or specific acids for industrial use. Control objective is maximum product yield.

It is immediately obvious that the main control objective in methanogenic treatment plants is stability, normally as measured by effluent soluble COD or organic acids. However, the causes of instability may be varied. In manure digesters, the most common cause of instability is free ammonia inhibition, which stops aceticlastic methanogenesis. In primary and activated sludge digesters, it may be overloading, ammonia, or long chain fatty acid inhibition. In high-rate digesters, the cause may be pH, or organic acid inhibition. Because of different hydraulic retention times, dynamics are much faster in high-rate reactors than semi-solid digesters. Therefore, the instrumentation to be selected, manipulated variable, and control strategy can depend on application.

Advances in Instrumentation, Control and Benchmarking

Advances in Instrumentation

Standard instrumentation in anaerobic digesters has been pH, gasflow, and temperature. pH is an indirect indicator of organic acid accumulation. It is relatively slow in response to overloads, as the system is buffered against pH changes by a number of compounds, including bicarbonate and ammonia. Gasflow is an indicator of both reactor loading, and overload conditions. During increased loads, gasflow will increase, though response is generally slow. In extreme overload conditions, gas production will decrease or cease. Normally by this time, the reactor has been damaged.

There is an obvious need for a direct, rapid indicator of overload. In the liquid phase, organic acids (total or individual) have been regarded as an excellent indicator because accumulation during overload is rapid. There are also some indications that individual organic acids are useful for indicating overload classes (Siegrist *et al.*, 2002). However on-line measurement of organic acids has been traditionally difficult because the main offline measurement method has been by gas chromatography (GC), which is expensive, and difficult to apply on-line, as sample filtration is required. Recently, a number of alternative methods have been developed, including:

(1) Titrimetric methods, which have been implemented as on-line sensors (e.g., Feitkenhauer *et al.*, 2002; Van de Steene *et al.*, 2002) for measuring partial and total alkalinities, bicarbonate and total VFAs concentrations (see Figure 1). These methods do not measure individual organic acids, as all organic acids have similar acidity coefficients. Other weak acids and bases may also interfere with measurement according to the methods used for the titration curves analysis. Nevertheless, these sensors can be implemented relatively cheaply and are presently commercially available.

(2) Mid infra-red spectrometry (Steyer *et al.*, 2002a) can measure on-line a broad variety of compounds, including COD, TOC, total VFA and acetate, dissolved CO_2, partial and total alkalinities, but it is rather expensive and may require a filtered sample.

(3) On-line gas chromatography; Pind *et al.*, (2002) automated GC analysis with sample preparation using a rotating membrane. Individual organic acids are quantified separately (including isomers), but the sample must be pretreated, and the GC automated. The overall equipment cost is high, and no commercial option has yet been offered. Measurement of organic acids by gas phase analysis is under development, and has promise, as there is no need for sample pretreatment.

(4) Biofilm-type biosensors, often relying on respirometry (Liu and Mattiasson, 2002), are effective for measuring in-reactor soluble COD (which in some conditions approximates VFA concentrations). These have not yet been commercially developed, and may suffer from long term drift as well as sensor poisoning. Nevertheless, the sensors are very promising since they can detect a wide variety of compounds.

A number of other rapid sensor types exist, apart from organic acid sensors (Vanrolleghem, 1995). Gas phase composition changes can indicate overload. A drop in methane composition can indicate an overload in most reactors, and is easily measured by IR absorption. In the gas phase, hydrogen is often recognised as very useful (Pauss *et al.*, 1990, Moletta *et al.*, 1994), especially for carbohydrate fed reactors, even though there is occasionally a delay in overload detection. It can be measured by on-line GC-TCD, or a palladium sensor in the gas phase. Björnsson *et al.*, (2001) has used a flexible silicon tube with nitrogen carrier gas to directly measure in-reactor hydrogen concentrations. Respirometric, pH-based activity sensors have also been increasingly applied to anaerobic digestion. This has been via pH stat systems (Rozzi *et al.*, 2002, Mösche and Meyer, 2002), that measure aceticlastic methanogenic activity in a specified culture. The pH drops when acetate is fed to the respirometer. As the acetate is removed, pH rises, and more acetic acid may be added to maintain a stable pH. Therefore, the specific activity can be very accurately measured by the rate of acid added, at a stable pH. The influence of inhibitors can be estimated by adding that specific inhibitor in the wastewater. The same principle can potentially be used for the other steps in anaerobic digestion.

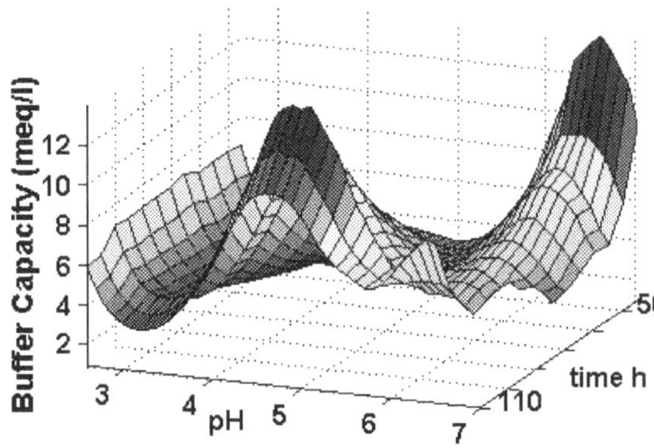

Figure 1 Buffers evolution from bicarbonate alkalinity to VFA during organic overload (from Zaher *et al.*, 2004). The bicarbonate concentration decreases over time, and organic acids increase.

Advances in Control

Control of anaerobic digestion aims not only to improve stability of the process to disturbances, but also to increase potential for both degradation rate and extent (Verstraete and Vandivivere, 1999).

During the last years, it has been practically demonstrated that anaerobic processes can be very efficiently controlled despite large changes in the influent concentrations. Very high organic loading rates (more than 100 $kg_{COD}.m^{-3}.d^{-1}$) have been achieved while keeping carbon removal above 75% (Steyer *et al.*, 1999). Very good control of intermediate products (*e.g.* volatile fatty acids and alkalinity) or end products (*e.g.*, effluent COD, total biogas produced or methane flowrate) was also possible over a long period of time. Many control approaches have been applied to AD, including PID (Alatiqi *et al.*, 1994), cascade PID (Liu *et al.*, 2004), fuzzy logic (Estaben *et al.*, 1997; Puñal *et al.*, 2003), artificial neural networks (Wilcox *et al.*, 1995; Holubar *et al.*, 2002), non parametric adaptive control (Hilgert *et al.*, 2000), adaptive control (Bastin and Dochain, 1990; Johnson *et al.*, 1995; Bernard *et al.*, 2001), disturbance accommodating control (Harmand *et al.*, 2000), interval based approaches (Alcaraz-González *et al.*, 2001), geometric control (Mendez-Acosta *et al.*, 2002), robust output feedback (Antonelli *et al.*, 2003; Mailleret *et al.*, 2003) each of them demonstrating several interesting characteristics for the control of AD processes.

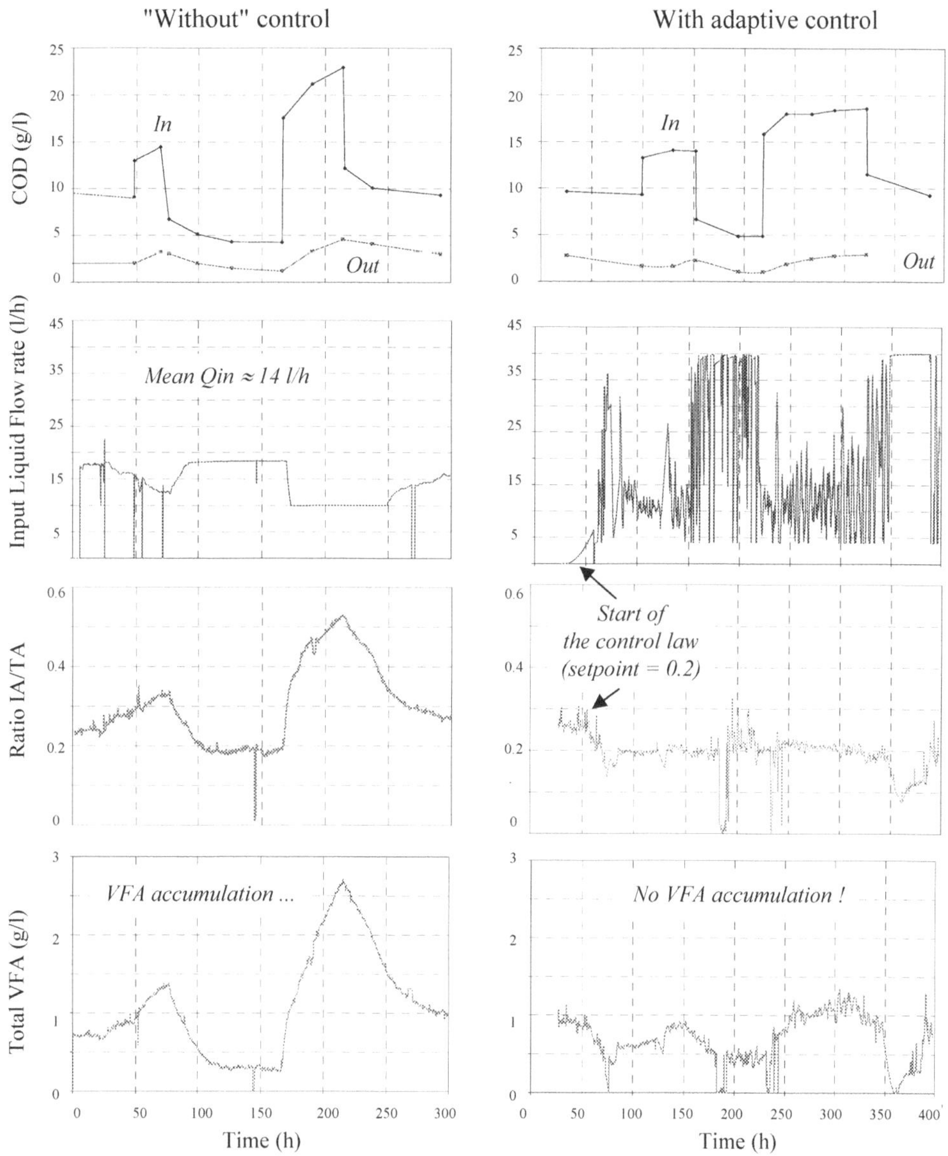

Figure 2 Experimental illustration of the benefits of on-line control. (with control, the objective is to regulate the ratio of the intermediate alkalinity over the total alkalinity at a setpoint of 0.2) (from Bernard *et al.*, 2001)

Most commonly, these control laws have been implemented through feed flow manipulation (See Figure 2). However, this degree of freedom is not always available, as it is generally determined by the upstream factory discharge. There is thus a lack of actuators to manipulate in anaerobic digestion, even though other options could include change in recycle flow, either directly from the reactor, or to a preacidification and CO_2 stripping mixed reactor.

Moreover, the main application has been in retained solids digesters, with mostly or completely soluble substrates (*i.e.,* application 'c' in the introduction) and control has generally not been applied in municipal sludge, or solids/manure digesters. This is a problem, as both systems could be improved by control. In municipal digesters, recycle of disturbances to the activated sludge section needs better management. In solids and manure digesters, large fractions of the available COD is lost as organic acids, as they generally operate with organic acid levels of above 1 g L^{-1}. In both cases, manipulated variables could include feed flow, recycle flow, gas recycle flow, and acid or base dosing volumes. Feed flow can be managed by diversion to a holding tank, or extended retention within clarifiers or thickeners. The new developed VFA sensors from the previous section are effective as inputs to digester controllers and there are very good, economically effective on-line ammonia and sulfide sensors available to detect inhibitory conditions.

However, these studies were mostly devoted to lab or pilot scale digesters. In addition, it is important to note that all these control laws were to meet the specific objective they have been designed for. As a consequence, it is not to be expected that, for example, a control law will manage a sensor fouling or a technical problem in the feeding circuit (*e.g.,* clogging of a pipe) if its goal is to control volatile fatty acids in the output of the reactor. In addition – and since there does not exist a "universal" control law that could manage all the disturbances occurring in a process – it is mandatory to couple control laws with an advanced diagnosis scheme. It is felt that it is the only possible way for achieving successful optimisation of AD processes at *industrial* scale.

In the past, these objectives were tackled using quantitative (Aubrun *et al.*, 2000) and qualitative (Genovesi *et al.*, 1999; Puñal *et al.*, 2002) model based diagnosis approaches, sometimes combining them together (Steyer *et al.*, 2002b) or with process history based methods (Steyer *et al.*, 1997). Nevertheless, it is our strong belief that a unified approach could be of great help for managing faults in sensors network while optimising the overall state of AD processes (Lardon *et al.*, 2004).

Advances in Benchmarking for Instrumentation and Control

Methods for evaluating and benchmarking control and operation strategies has improved, with release of the IWA anaerobic digestion model No.1 (ADM1; Batstone *et al.*, 2002). This is the result of several years co-operative research driven by international experts. The ADM1 has been applied to diverse systems, including anaerobic biosensors, sludge and manure digesters, high-rate digesters, and fermenters. The model code is available in most popular environmental simulators with standard scenarios (contact corresponding author for source code). It is also being used as a standardised basis to estimate parameters describing anaerobic digestion, and results so far indicate that variation is within an order of magnitude (Batstone *et al.*, 2003). There are now a number of excellent test cases available in the literature. A commonly used model benchmark for primary sludge digestion is Siegrist *et al.,* (2002). Although Siegrist did not use the ADM1 as such, their model structure is essentially similar, and the model parameters are directly transferable. This data set has been used to cross-verify a large number of implementations. The ADM1 as applied to this data set to simulate gas flow is shown in Figure 3.

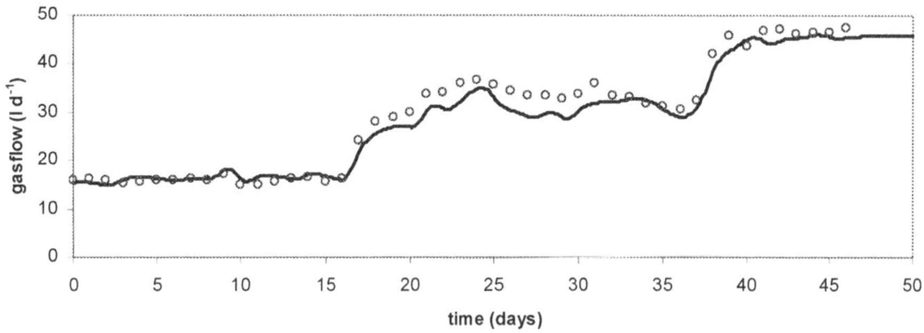

Figure 3 ADM1 applied to the data of Siegrist *et al.*, (2002) showing gasflow (o). Solid lines indicate simulations.

WWTP-wide benchmarking including municipal digesters is set to improve, as the European COST 624 WG1 group has proposed an expanded municipal wastewater treatment benchmark simulation model that includes an anaerobic digester, simulated by the ADM1 (Figure 4). The first simulation benchmark (Copp *et al.*, 2001) only focusing on the activated sludge plant, was very well received, and used to benchmark a wide variety of control strategies. The new benchmark will allow practitioners and control implementers to evaluate the influence of operation and control strategies in the anaerobic digester on the overall plant performance. It will also allow evaluation of complicated scenarios, such as management of reject water nutrients, or influent COD load. Indeed, operation of the primary clarifier will determine which fraction of the influent COD will be used for energy production (digestion of primary sludge), whereas the remainder in the effluent of the primary clarifier will be used for nutrient removal in the treatment plant (denitrification or bio-P processes). Another major advantage of this initiative is that the whole plant model, including ADM1 will be cross-verified via parallel implementations on different simulation platforms, to the general benefit of the users of these models.

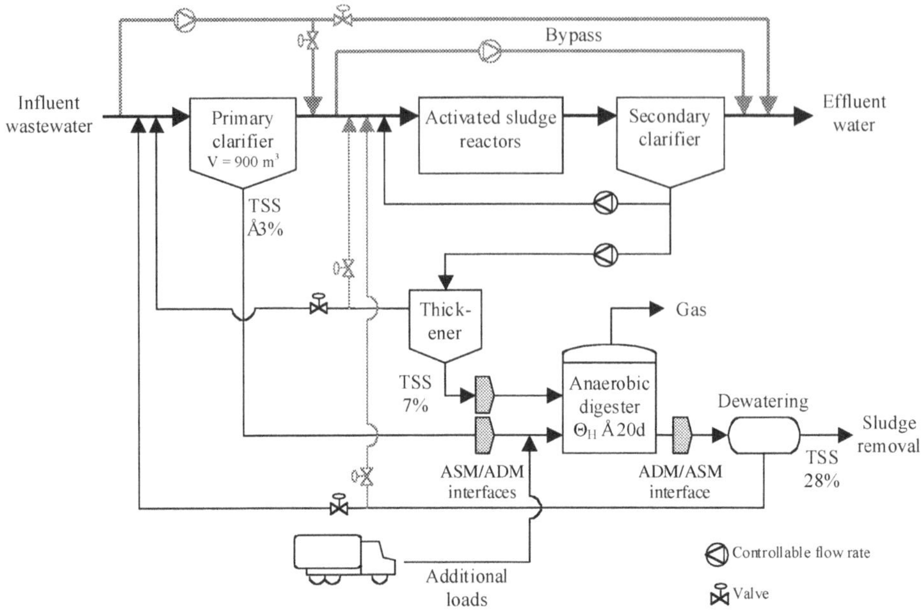

Figure 4 The .Cost 624 BSM2 plant as proposed by the Benchmarking working group. (From Jeppsson *et al.*, *submitted*)

Conclusions

Anaerobic digestion has classically been regarded as a difficult process to control and manage, because of slow, or indirect sensors, limitations in available inputs for manipulation, and a lack of understanding of the process. While manipulated variables are still limited by the nature of the process, the other two problems have been solved by the availability of fast VFA or soluble COD sensors, and release of the ADM1. High-rate digesters can now be very effectively controlled using model, or rule-based control, with input feed flow as manipulated input, and outlet COD as measured variable. Some of the sensors (particularly titration sensors) are simple, and require minimal maintenance. There is still a lack of control applications in solids/manure, and municipal digesters, but we are confident that this will improve in the near future, due to the improved sensors available, and application of the ADM1 for sensor and control benchmarking and analysis methods. Improved control of anaerobic processes also has the potential to widen the competitive scope of anaerobic digestion, and make application in low COD systems (e.g., municipal wastewater treatment) a realistic option.

Acknowledgements

Damien Batstone's research fellowship was funded by the EU 5th Framework Program under the project BIOWASTE contract no. QLK5-CT-2002-01138.

References

Alatiqi I.M., Dadkhah A.A., Akbar A.M. and Hamouda M.F. (1994) Comparison between dynamics and control performance of mesophilic and thermophilic anaerobicnext term sludge digesters, *The Che.l Eng. J. and Bioch. Eng. J.*, **55**(3):B55-B66

Alcaraz-González V., Harmand J., Rapaport A., Steyer J.P., González-Alvarez V., Pelayo-Ortiz C. (2001) Robust interval-based SISO regulation under maximum uncertainty conditions in an anaerobic digester, *IEEE CCA/ISIC International Conference on Control Applications CCA*, Mexico city, Mexico, September 2001, 6 pages on CDROM.

Antonelli R., Harmand J., Steyer J.P., Astolfi A. (2003) Set-point regulation of an anaerobic digestion process with bounded output feedback, *IEEE Trans. Contr. Syst. T.*, **11**(4):495-504.

Aubrun C., Harmand J., Garnier O., Steyer J.P. (2000) Fault detection filter design for an anaerobic digestion process, *Bioproc. Eng.*, **22**(5):413-420.

Bastin G., Dochain D. (1990) *On-Line Estimation and Adaptive Control of Bioreactors*, Elsevier, Amsterdam, 1990, 379 pages.

Batstone, D. J., Keller, J., Angelidaki, I., Kalyuzhnyi, S. V., Pavlostathis, S. G., Rozzi, A., Sanders, W. T. M., Siegrist, H., and Vavilin, V. A. (2002). *Anaerobic Digestion Model No. 1 (ADM1), IWA Task Group for Mathematical Modelling of Anaerobic Digestion Processes.* IWA Publishing, London.

Batstone, D.J., Pind, P.F., and Angelidaki, I. (2003) "Kinetics of thermophilic, anaerobic oxidation of straight and branched chain butyrate and valerate" *Biotech. Bioeng.* **84**:195-204.

Bernard O., Polit M., Hadj-Sadok Z., Pengov M., Dochain D., Estaben M., Labat P. (2001) Advanced monitoring and control of anaerobic wastewater treatment plants: software sensors and controllers for an anaerobic digester, *Wat. Sci. Tech.*, **43**(7):175–182.

Björnsson, L., Hörnsten, E.G., and Mattiasson, B. (2001) Utilization of a palladium-metal oxide semiconductor (Pd-MOS) sensor for on-line monitoring of dissolved hydrogen in anaerobic digestion. *Biotech., Bioeng.*, **73**(1):35-43

Copp, J.B. (2002) *The COST simulation benchmark Manual*, Office for publications of EC, Luxembourg, *also at:* http://www.ensic.inpl-nancy.fr/COSTWWTP/

Estaben M., Polit M., Steyer J.P. (1997) Fuzzy control for an anaerobic digester, *Control Eng. Practice*, **5**(9):1303-1310.

Feitkenhauer H., von Sachs J and Meyer U. (2002) On-line titration of volatile fatty acids for the process control of anaerobic digestion plants, *Wat. Res.*, **36**(1):212-218.

Genovesi A., Harmand J., Steyer J.P. (1999) A fuzzy logic based diagnosis system for the on-line supervision of an anaerobic digestor pilot-plant, *Biochem. Eng. J.*, **3**:171-183.

Harmand J., Manh A.G., Steyer J.P. (2000) Identification and disturbance accommodating control of a fluidized bed anaerobic reactor, *Bioproc. Eng.*, **23**(2):177-185.

Hilgert N., Harmand J., Steyer J.P., Vila J.P. (2000) Nonparametric identification and adaptive control of an anaerobic fluidized bed digester, *Control Eng. Practice*, **8**:367-376.

Holubar, P., Zani, L., Hager, M., Fröschl, W., Radak, Z. and Braun, R. (2002) Advanced controlling of anaerobic digestion by means of hierarchical neural networks, *Wat. Res.*, **36**(10): 2582-2588

Jeppsson, U., Rosen, C., Alex, J.,Copp, J., Gernaey, K.V., Pons, M.-N. and Vanrolleghem, P.A. Towards a benchmark simulation model for plant-wide control strategy performance evaluation of WWTPs *submitted* Watermatex 2004, Nov 3-5 2004, Bejing, China

Johnson K.A., Wheatley A.D., Fell C.J. (1995) An application of an adaptive control algorithm for the anaerobic treatment of an industrial effluent, *Trans. IChemE*, **73**(B):203-211.

Lardon L., Punal A. and Steyer J.P. (2004) On-line diagnosis and uncertainty management using evidence theory—experimental illustration to anaerobic digestion processes, *J. Proc. Control*, (in press).

Liu, J. and Mattiasson, B. (2002) Microbial BOD sensors for wastewater analysis. *Wat. Res.* **36**(15):3786-3802.

Liu, J., Olsson, G. and Mattiasson, B. (2004) Monitoring and Control of an anaerobic up-flow fixed bed reactor for high loading rate operation and rejection of disturbances *Biotech. Bioeng. in-press.*

Mailleret, L., Bernard, O. and Steyer, J.P. (2003) Robust regulation of anaerobic digestion processes. *Wat. Sci. Tech.* **48**(6):87-94.

Mendez-Acosta H.O., Femat R., Gonzalez-Alvarez V., Steyer J-P. (2002) Substrate regulation in an anaerobic digester based on geometric control, *IFAC Congreso Latinoamericano de Control Automatico CLCA 2002*, 4-6 December 2002, Guadalajara, Mexico, 6 pages on CDROM.

Moletta R., Escoffier Y., Ehlinger F., Coudert J.P. and Leyris J.P. (1994) On-line automatic control system for monitoring an anaerobic fluidized bed reactor: Response to organic overload, *Wat. Sci. Tech.*, **30**(12):11-20.

Mösche, M. and Meyer, U. (2003) Factors affecting constancy of acetate concentration and correct determination of methanogenic activity in pH-stat experiments *Wat. Sci. Tech.* **48**(6):111-118

Pauss A., Beauchemin C., Samson R., Guiot S. (1990) Continuous Measurement of Dissolved H_2 in an Anaerobic Reactor Using a Commercial Probe Hydrogen/Air Fuel Cell-Based, *Biotech. Bioeng.*, **35**:492-501.

Pind, P.F., Angelidaki, I., and Ahring, B.K. A new VFA sensor technique for anaerobic reactor systems, *Biotech., Bioeng.*, **82**(1):54-61

Puñal A., Roca E. and Lema J.M. (2002) An expert system for monitoring and diagnosis of anaerobic wastewater treatment plants, *Wat. Res.*, **36**(10):2656-2666.

Puñal A., Palazzotto L., Bouvier J.C., Conte T., Steyer J.P. and Delgenes J.P. (2003) Automatic control of VFA in anaerobic digestion using a fuzzy logic based approach, *Wat. Sci. Tech.*, **48**(6):103-110.

Rozzi, A., Castellazzi, L., and Speece, R.E. (2002) Acetoclastic methanogenic activity measurement by a titration bioassay *Biotech. Bioeng.*, **77**(1): 20-26

Siegrist, H., Vogt, D., Garcia-Heras, J., and Gujer, W. (2002). Mathematical model for meso and thermophilic anaerobic sewage sludge digestion. *Environ. Sci. Tech.* **36** (5): 1113 -1123.

Steyer J.P., Rolland D., Bouvier J.C., Moletta R. (1997) Hybrid fuzzy neural network for diagnosis - Application to the anaerobic treatment of wine distillery wastewater in a fluidized bed reactor, *Wat. Sci. Tech.*, **36**(6-7):209-217.

Steyer J.P., Buffière P., Rolland D., Moletta R. (1999) Advanced control of anaerobic digestion processes through disturbances monitoring. *Wat. Res.*, **9**:2059-2068.

Steyer, J.P., Bouvier, J.C., Conte, T., Gras, P., Harmand, J., and Delgenes, J.P. (2002a) On-line measurements of COD, TOC, VFA, total and partial alkalinity in anaerobic digestion processes using infra-red spectrometry. *Wat. Sci. Tech.* **45**(10):133–138

Steyer J.P., Harmand J., Delgenes J.P. (2002b) Handling uncertainty in diagnosis using a combined interval based and a fuzzy logic based approach - Application to wastewater treatment, *Int. Conf on Information Processing and Management of Uncertainty in Knowledge-Based Systems, IPMU2002*, Annecy, France, 1-5 July 2002, vol. II, pp. 899-904.

Van De Steene M., Van Vooren L., Ottoy J.-P. and Vanrolleghem P.A. (2002) Automatic buffer capacity model building for advanced interpretation of titration curves. *Environ. Sci. Technol.*, **36**(4): 715-723.

Vanrolleghem P.A. (1995) Sensors for Anaerobic Digestion : An Overview, *International Workshop on Monitoring and Control of Anaerobic Digestion Processes*, Narbonne, France, 7 pages, 6-7 December, 1995.

Verstraete W., Vandevivere P. (1999) New and broader applications of anaerobic digestion, *Crit. Rev. Environ. Sci. and Tech.*, **29**(2):151-165.

Wilcox, S.J., Hawkes, D.L., Hawkes. F.R., and Guwy, A.J. A neural network, based on bicarbonate monitoring, to control anaerobic digestion, *Wat. Res.* **29**(6):1465-1470

Zaher U., Bouvier J.C., Steyer J.P. and Vanrolleghem P.A. (2004) Titrimetric monitoring of anaerobic digestion: VFA, alkalinities and more, *submitted*, 10th IWA AD Congress, Montreal, Canada, August 29 - September 2, 2004.

Optimisation of WWTP Units Using CFD - A Tool Grown For Real Scale Application

K Essemiani*, S Vermande*, S. Marsal, L Phan*, J Meinhold***

*Anjou Recherche Veolia Water, Chemin de la Digue B.P 76, 78603 Maisons Laffitte, France.
* *Direction Technique Veolia Water, 1 Place Montgolfier, 94417 Saint-Maurice, France.

Abstract : Case studies for secondary settler as well as activated sludge basins using Computational Fluid Dynamics (CFD) as design, optimisation and diagnosis tool are presented. Validation of the models is demonstrated and their application for full scale installations is presented and discussed. It is illustrated that CFD has become a necessary tool dedicated to engineering teams to increase the performance and optimisation of different WWTP units, and to the operators for the diagnosis and rehabilitation stages.

Keywords Computational Fluid Dynamics (CFD), wastewater, activated sludge basins, secondary settling tanks.

Introduction

In recent years Computational Fluid Dynamics (CFD) has been recognised as a valuable R&D tool for hydrodynamic investigation for a variety of reactor systems. Enabling the assessment of necessary hydrodynamic parameters, CFD is used for design, retrofitting and for diagnosis of malfunctioning systems. Full scale applications within the water business are well documented for the drinking water sector (e.g. oxidation, chlorination and UV units [Levecq *et al*]). Further examples for various water treatment processes include: the dissolved air flotation [Ta *et al*] and combined sewers overflows [Lepene Kouyi *et al*].

This paper concentrates on the recent advances of CFD application for wastewater treatment units. Case studies (secondary settling tanks and activated sludge basins) are used to illustrate its viability for real scale applications, in which the efficiency is closely linked to hydraulic behaviour. Discussion is focussed on the high added value of the CFD tool to:

- Engineering companies, for the design and performance optimisation by testing different operating conditions and/or different geometrical parameters.
- Operators, for the retrofitting, troubleshooting and diagnosis of problems.

Optimisation & design for engineers – Secondary Settling Tanks

The solid/liquid separation of the treated wastewater from the biomass in an activated sludge processes is achieved by gravity sedimentation in Secondary Settling Tanks (SST). The sludge settling needs to be in a laminar flow regime to avoid re-suspension of the settled sludge. The optimisation of the SST is then controlled by the optimisation of their hydraulic functioning. The settling efficiency are directly related to :

© 2004 IWA Publishing. *2nd IWA Leading-Edge Conference on Water and Wastewater Treatment Technologies* edited by Mark Van Loosdrecht & Jonathan Clement. ISBN: 1 84339 508 8.

- Nature of the Influent to the SST (Sludge Volume Index SVI, Mixed Liquor Suspended Solids MLSS,..).
- Hydraulic load (Surface Overflow Rate (SOR), recirculation rate,…).
- Dimensions (surface, height,…) and the geometry of the SST equipment, affecting the hydraulic behaviour of the SST (inlet, recirculation, deflector, clifford,…).

Improvement of the settling can be achieved through a better understanding of the impact that these parameters have on the SST performance (sludge blanket) and of the hydraulic behaviour.

Validation of existing installations

The hydrodynamics of the SST has been subject to several studies [e.g. Krebs *et al*, De Clercq]. CFD has been used for simulating the sludge settling in a SST aiming at the optimisation via the prediction of the sludge blanket [Anjou Recherche]. For the case study presented, being performed in the frame of validating an existing installation, Table 1 resumes the operating conditions of the SST during normal weather and a storm event (high flowrate). The SST is equipped with a clifford and a deflector, as presented in Figure 1.

Table 1 Measured Operating conditions

	Normal weather	Storm event
SST diameter [m]/ Central axis depth [m]	27/7.2	27/7.2
Volume [m^3]/ projected surface [m^2]	2325/573	2325/573
Sludge Volume Index SVI [$mL.g^{-1}$]	260	250
Influent flowrate [$m^3.h^{-1}$]	238	1000
Recirculation flowrate [$m^3.h^{-1}$]	145	570
Mixed Liquor SS [$g.L^{-1}$]	2.30	2.50
Return Activated Sludge SS [$g.L^{-1}$]	3.75	2.00
Outlet Total Suspended Solids (TSS) [$mg.L^{-1}$]	3.00	10.00
Solid Loading Rate SLR [$kg.m^{-2}h^{-1}$]	0.96	4.36
Surface Overflow Rate SOR [$m.h^{-1}$]	0.16	0.75

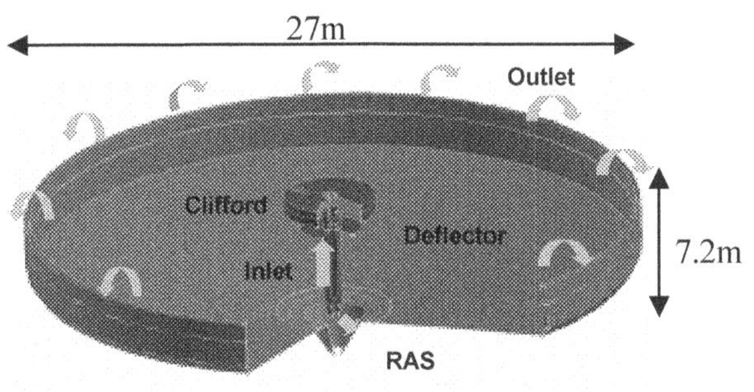

Figure 1 : Geometry of the studied SST

A prerequisite for the intensive use of any simulation tool is the validation step. Table 2 presents the comparison of the numerical and experimental values for the sludge blanket height, indicating a deviation within an acceptable range.

Table 2 Sludge blanket prediction

Case	Height of clarified water (m)		Deviation %
	Measured	Numerical	
Normal weather	3.25	3.20	2
Storm event	1.75	2.10	17

Figures 2 and 3 represent the TSS concentration profiles (numerical and experimental) at different positions and depths in the SST during both weather conditions. The numerical profiles are very close to the experimental ones, except in the bottom part where the rheological properties of the activated sludge have a major impact. Based on the deviation between predicted and measured values, the model can be regarded as sufficiently validated and hence be applied to certain support work for design engineers and operators in the different projects (design and diagnosis troubleshooting).

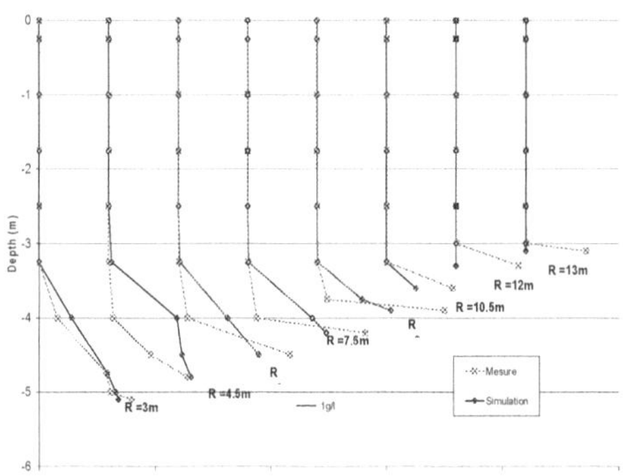

Figure 2 TSS profiles in the SST : normal weather

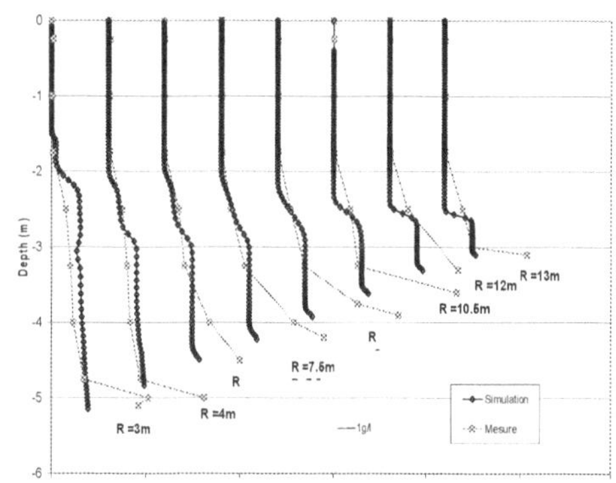

Figure 3 TSS profiles in the SST : storm event

Based on the validated model some more detailed scenario analysis can be performed. Figure 4 illustrates the velocity vectors field in the central vertical plane of the SST during the storm event. The influent induces a strong recirculation loop (n°1) inside the clifford area. The flow is then directed to the bottom of the SST (sludge recycle) inducing two further recirculation loops below and close to the deflector (n°2-3). Whether they represent an negative impact on the SST performance has to be further evaluated. CFD allows to access the TSS concentration at every point of the SST and thus the sludge blanket. Its form and TSS profile depend on the fluid movements. Figure 5 shows the sludge volume fraction during the storm event in the vertical central plane. A small "bump" can be detected in the area of the recirculation n°2. However, no negative impact is detected for the clear water outlet and for the sludge recycle. In fact the sludge is well concentrated in the STT bottom.

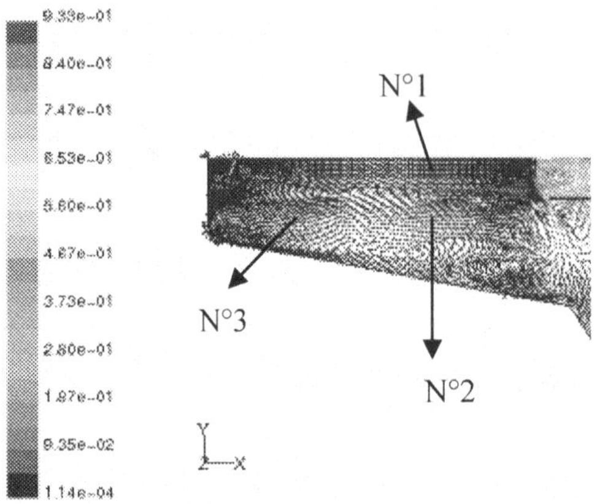

Figure 4 : Velocity vectors (m.s^{-1}) - storm event

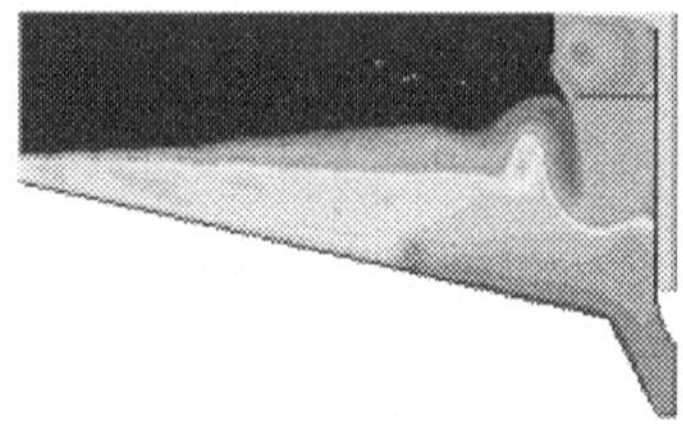

Figure 5 : Sludge vol. fraction : storm event

Application to the SST design approaches

An interesting application of CFD is the evaluation of different design approaches. Based on the validated model, a comparative study of 2 SST design approaches has been performed. The sludge blanket height is used as the criteria for the efficiency of the SST design. The configurations evaluated are illustrated in the figures 6 and 7 and consist of :

- Traditional SST design (deep clifford going in the sludge blanket),
- Design recommended by ATV norms [ATV], characterised by the presence of a clifford and a deflector, which has a larger diameter than the one of the clifford.

Table 3 presents the operating conditions, maintained constant for the both configurations.

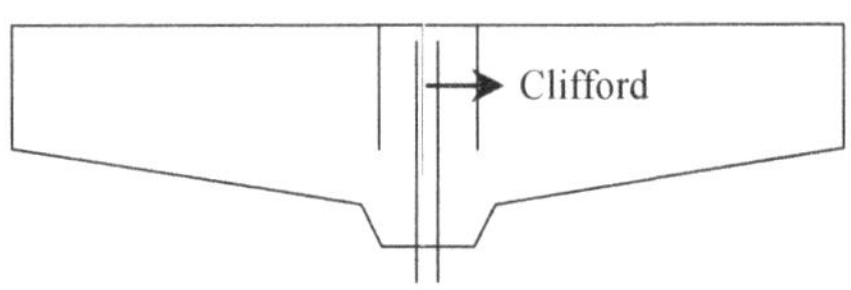

Figure 6 : Traditional SST

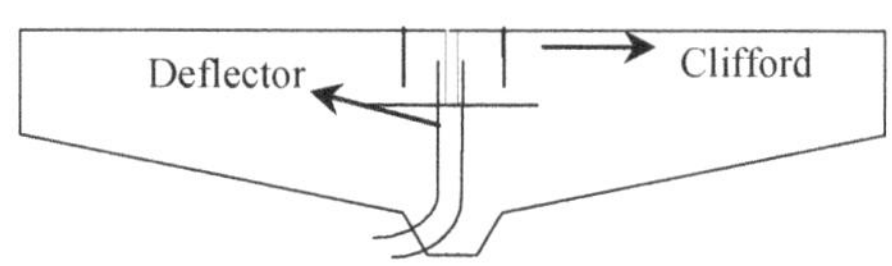

Figure 7 : ATV SST

Table 3 Constant operating conditions

Diameter (m)	27	Volume (m^3)	2325
Inlet pipe diameter (m)	0.7	Recirculation rate (%)	80
Mixed Liquor SS (g.L^{-1})	2.7	Solid Loading Rate(kg.h^{-1}.m^{-2})	1.05
Return Activated Sludge SS (g.L^{-1})	6	Sludge Volume Index (mL.g^{-1})	200

The geometrical elements of the two configurations (immersion and diameter of the clifford and the deflector) have been defined using the design criterions based on the surface overflow rate, which are specific for each SST design type. In order to account for normal and peak flow conditions a variety of surface overflow rates (SOR) have been tested for each configuration. Figures 8 and 9 represent the velocity field vectors for both configurations in the vertical central plane at a SOR of 0.8 m.h^{-1}, which represents the admissible velocity taken for design/dimensioning (peak flow):

- For the ATV configuration, a recirculation loop occurs in the Clifford area (n°1). The influent exits the deflector with an horizontal orientation. No vertical direction, that light induce a sludge recirculation, is detected.
- For the traditional configuration the recirculation loop is more important and it goes directly to the sludge recycle area (n°1).

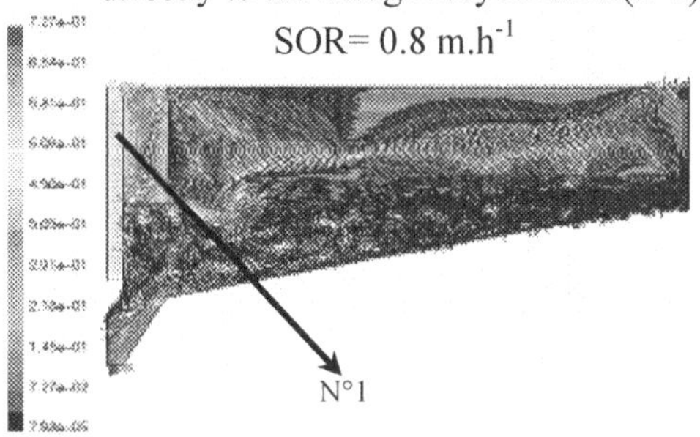

Figure 8 : Velocity vectors ($m.s^{-1}$) - ATV SST

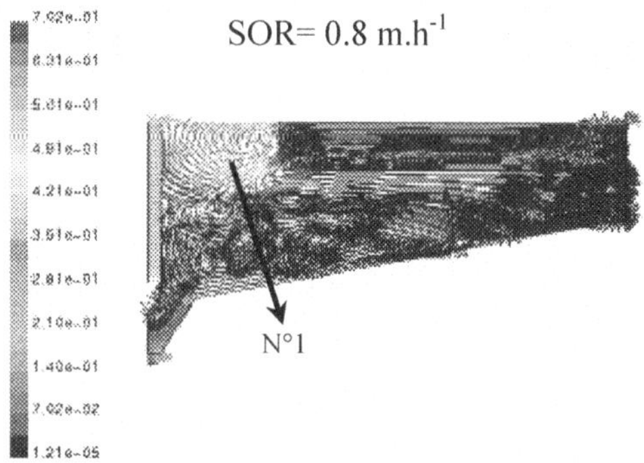

Figure 9 : Velocity vectors - traditional SST

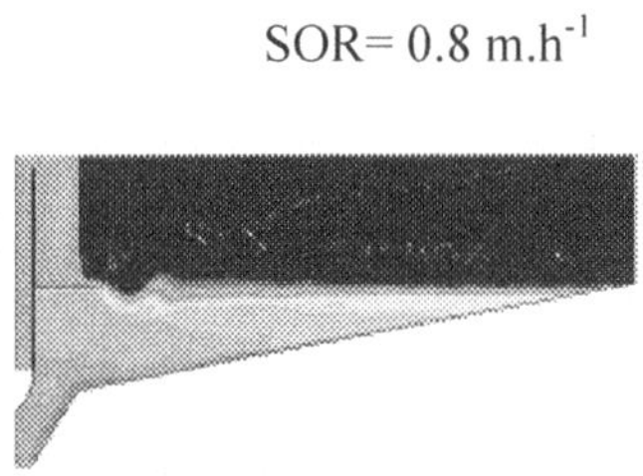

Figure 10 : Sludge vol. fraction - ATV SST

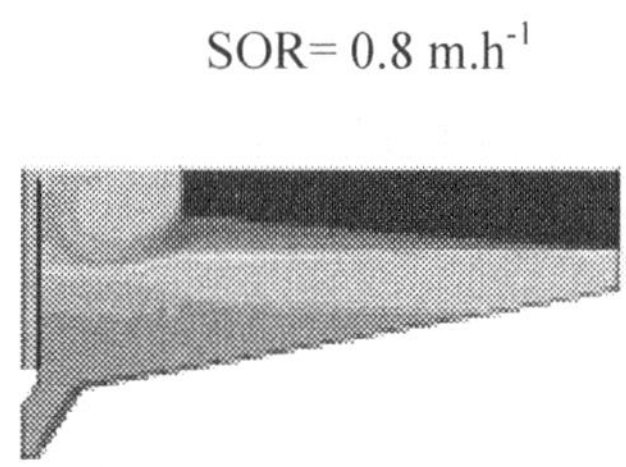

Figure 11 : Sludge vol. fraction – traditional SST

The influent orientation at the clifford exit has thus a very important impact on the sludge blanket height as illustrated in figures 10 and 11, which exhibits a higher sludge blanket for the traditional design.

Figure 12 summarises the results of all tested configurations, presenting variation of the sludge blanket height for every configuration as a function of the solid overflow rate. The Clifford plus deflector is redesigned according to the corresponding design approach or each velocity tested.

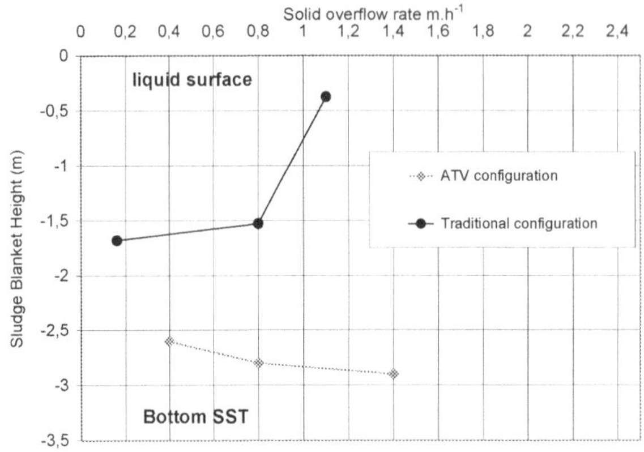

Figure12 : Comparison of the two configurations for various solid overflow rates (SOR)

At every solid overflow rate, the ATV configuration results in a lower and more constant sludge blanket height (varying from –3.4 to –2.7m). In contrast to this, the traditional configuration exhibits a strong increase of the sludge blanket height at SOR above 0.8 $m.h^{-1}$. The efficiency of the ATV design is due to the fact that the influent keeps a horizontal direction at the exit of the deflector area. This hydraulic behaviour results from the ATV design rules, based on a limit velocity in the clifford section and in the fluid comprised between the deflector and the clifford.

This comparative study illustrates well, the efficiency of CFD, allowing to compare different design approaches at much lower costs than experimental studies at a full scale.

Retrofitting and troubleshooting for operators - Activated Sludge Basins (ASB)

Agitation and aeration are closely linked with respect to the wastewater treatment performances. One essential parameter for correct operation of ASBs is the average horizontal velocity without aeration under clean water conditions. A value of 0.3-0.35 m.s-1 for this parameter is recommended (Da Silva, Gillot]) for ASB of up to 6 m depth, to ensure best conditions for mixing and aeration. However, the impact of aeration on the horizontal velocity still needs further hydrodynamics investigation (Gillot, 2003). Classical equipment consists of submerged impellers to ensure mixing, and fine bubble diffuser to supply oxygen. The choice of these are mainly based on supplier's data, as well as their position in the basin. However some studied cases exhibit erroneous design predictions as lower liquid velocity (Vermande, 2003) and/or poor oxygen transfer efficiency (OTE).

Two case studies are presented addressing the verification of the hydraulic functioning of different ASB types (figures 13, 14 and table 4). The systems were double checked via CFD and experimental measurements to evaluate the hydrodynamics and to validate the model applied for full scale installation. As a first phase CFD is used in clean water conditions without aeration to support start-up procedures and check equipment performances.

Measurements of the velocity profile under clean water conditions were performed in a vertical plane located far away from the impellers, in order to estimate the axial velocity in a well established flow area. Twenty measures were recorded in this section at different vertical and horizontal positions.

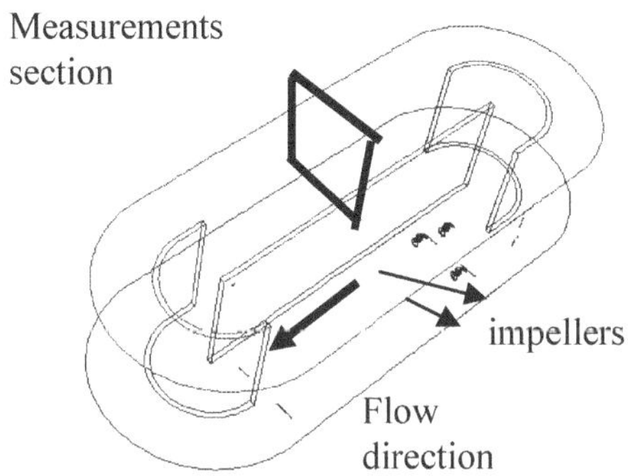

Figure 13 : Geometry of the ASB n°1

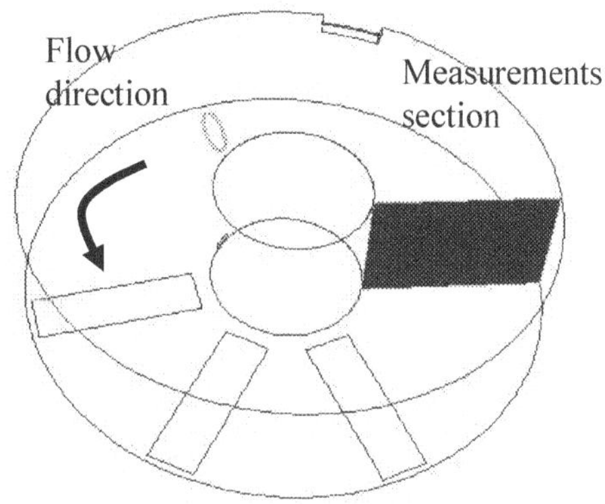

Figure 14 : Geometry of the ASB n°2

Table 4 : Characteristics of the studied oxidation ditches

		ASB n°1	ASB n°2
	Type	Channel Medium – high loaded	Annular low loaded, prolonged aeration
	Volume (m^3)	6200	1900
Impeller	Diameter (m)	1.8	1.8
	Number	3	1
	Rotation speed (rpm)	40	40
Diffuser	Type	Fine bubbles	Fine bubbles
	Air flowrate ($m^3.h^{-1}$)	3200	1624

The simulations are in agreement with the experimental measurements locally (velocity profiles) and globally (mean average) for both systems (figure 15 and table 5). The comparison of the horizontal velocity profiles are presented in the figure 15 for the ASB n°1 (ASBn°2 not shown). The profile is well predicted regardless the horizontal or vertical position. Table 5 presents the measured and the computed mean horizontal velocity, as well as the according deviation. The maximum difference is 12.5% for the ASB n°2, which is in the precision range of the experimental measurements (estimated at 15%). Hence, the model can well be used for further analysis and applications for other full scale installation.

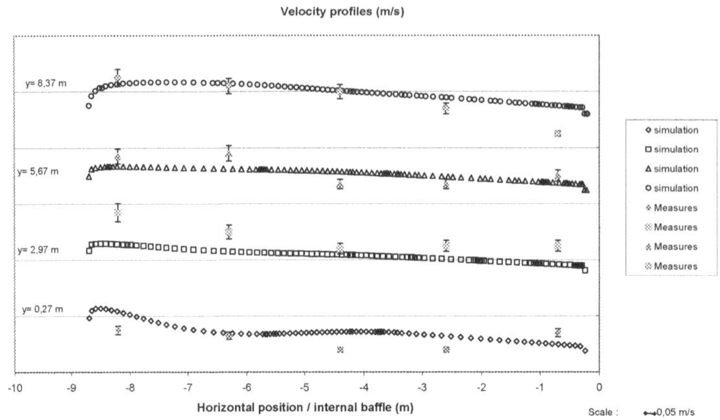

Figure 15 : Exp/CFD horizontal velocity profiles -ASB n°1

Table 5 : Comparison of the mean horizontal velocity

Case	Measured mean horizontal velocity ($m.s^{-1}$)	Computed mean horizontal velocity	Difference
ASB n°1	0.20 $m.s^{-1}$	0.18 $m.s^{-1}$	6%
ASB n°2	0.32 $m.s^{-1}$	0.36 $m.s^{-1}$	12.5%

The analysis of the hydrodynamics illustrates the expected results with maximum velocities located close to the impellers (figure 17). ASBn°1 exhibits distinct areas of low velocity right after the change of direction of the waterflow.

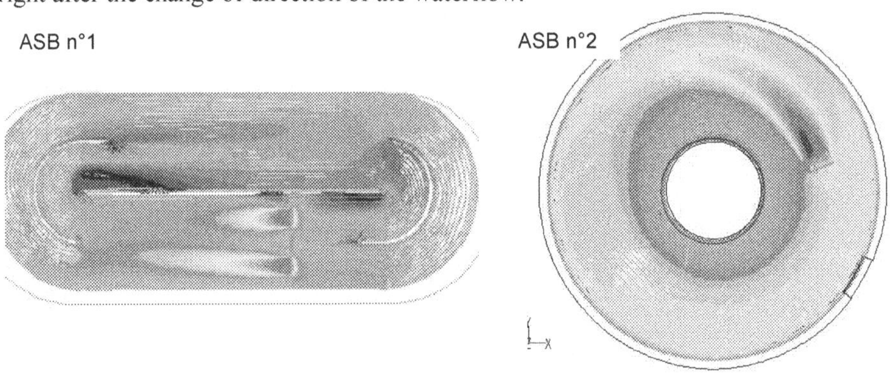

Figure 17: Velocity contours (m/s) at z = 2 m (red : higher velocity, blue : lower velocity)

The analysis of ASBn°2 verified its normal hydraulic functioning and ensured the operator with respect to the choice of the agitation unit of his systems. More important is the is low horizontal velocity predicted and also obtained experimentally for ASBn°1. Further analysis via CFD exhibited a poor agitator performance. The evaluation of different settings for the rotational speed and the type of agitator revealed that only a combination of the two can solve the problem. Increasing only the rotational speed or only the impeller's diameter did result in an increase of the horizontal velocity, which, however, still remained below the recommended range (0.30-0.35 m.s-1). The exchange of the impellers (2.5 m; 46 rpm) at the wwtp and subsequent experimental measurements verified the numerical analysis.

The work has shown the viability of the CFD model for first full scale diagnosis under clean water without aeration and is used for current and future design projects of ASB installation. It is evident that "clean water without aeration" investigations are only the first step in the analysis of ASB, as aeration and agitation do interact. Aeration will have an impact on agitation and consequently on the horizontal velocity. Agitation will help to avoid/ break up the so called spiral flows and hence affect oxygen transfer. Despite recent advances on modelling the aeration (oxygen transfer) and its impact on the hydraulics in different kind of reactors (bubble column, oxidation ditch pilot plant), no full scale validated model for ASB is available yet. With the validated and applied model for "clean water without aeration" as a sound basis, R&D is directed to the validation of a 2-phase model for ASB of full scale installations, in order to feed directly on-going design and diagnosis engineering projects.

Conclusion

Computational Fluid Dynamics (CFD) is a powerful and precise tool that allows the study of a very wide variety of applications in the water industry. It has become a necessary tool not only for R&D but also, dedicated to engineering teams to increase the performance and optimisation of different WWTP units, and to the operators for the diagnosis and rehabilitation stages.

Retrofitting of secondary settlers is often based on experience. Frequent questions are how to improve the SST efficiency in order to increase the acceptable overflow rate and/or reduce the sludge blanket height. If the modifications (e.g. inserting deflectors) do not achieve the intended improvement, it will become a costly exercise. The CFD approach enables the engineer to test different configurations for a wider range of retrofitting possibilities within less time and for lower costs. It also offers a good method to validate the adequate design approach. Evaluation of new secondary settler processes /units designed is another task that engineering companies will be able to perform with less efforts.

Analysis and troubleshooting of activated sludge basins is a further valuable area for CFD applications in wastewater treatment. Despite being currently restricted to models not including aeration, CFD evaluation can already add valuable insight and help for the choice and placement of adequate agitation units – one of the essential aspects in ASB operation and design. The tested model has been shown valid and valuable for two different treatment units and represents a sound tool for ongoing and future engineering design projects. Furthermore the model is to be taken as the basis for the 2 phase model (aeration included) development and validation for full scale ASB – the next compulsory step for full scale CFD use.

Main driving force for CFD applications is, and will also be in the future, related to cost benefits (energy consumption, volume) and/or increasing performance at a lower cost and in shorter time periods compared to experimental approach.

References

Abwasser Techniche Vereinigung ATV. (1991) Dimensioning of a single stage activated sludge plant upwards from 5000 total inhabitants and population equivalent. *Doc CEN/TC 165/WG 43, worksheet* A131, pp 1-67.

Anjou Recherche. (2001) Optimisation du fonctionnement hydraulique des clarificateurs secondaires par simulation numérique. *AESN report*, France.

Da Silva, G., (1994) Eléments d'optimisation du transfert d'oxygène par fines bulles et agitateur séparé en chenal d'oxydation. *PhD Thesis. CEMAGREF Antony – France.*

De Clercq, B. (2003) Computational Fluid Dynamics of settling tanks : Development of experiments and rheological, settling, and scraper submodels. *PhD thesis*, University of Ghent, Belgium.

Ekama, G.A., Barnard, J L., Günthert, F W., Krebs, P., McCorquodale, J A., Parker, DS., Wahlberg E J. (1997). Secondary settling tanks : Theory, modelling, design and operation. *IAWQ, Scientific and technical report N°6.*

Fluent User's Guide. (2003).

Gillot, S., Deronzier, G., Héduit, A. (1997) Oxygen transfer under process conditions in an oxidation ditch equipped with fine bubble diffusers and slow speed mixers. *WEFTEC, Chicago, USA.*

Gillot, S., Héduit, A. (2003) Predicting oxygen transfer in annular ditches equipped with fine bubble diffusers and mixers. WEFTEC, Los Angeles, USA.

Krebs, P., Armbruster, M. (2003) Improvement of clarifiers design and operation by numerical modelling. *IWA leading edge conference*, Noordwijk, 26-28 May.

Levecq, C., Féliers, C., Cervantès, P., De Traversay, C. (2003) Advantages of two complementary approaches for the evaluation of UV performance system. *2nd international congress on ultraviolet technologies*. Vienna, Austria, 9-11 July.

Vermande, S., Essemiani K., Meinhold , J., de Traversay Ch., Fonade Ch. (2003). Trouble shooting of agitation in an oxidation ditch: applicability of hydraulic modeling. WEFTEC, Los Angeles, USA. 2004.

Ta, C T., Brignal, W J. (1997) Application of single phase Computational Fluid dynamics techniques to dissolved air flotation tank studies. *In Proceeding of CIWEM*, London, 16-19 April.

Lepene Kouyi, G., Vasquez, J., Poulet, J B. (2003) 3D free surface measurement and numerical modelling of flows in storm overflows. *Flow and instrumentation, volume 14, issue 3.*

An optimised result of the sludge dewatering process during twenty-four hours per day

H.F. van der Roest*, E. Koornneef*,

* DHV Water BV, P.O. Box 484, 3800 AL Amersfoort, The Netherlands

Abstract The application of a self-optimising control-unit in combination with reliable measurement equipment results in optimised operation of a sludge-dewatering facility during twenty-four hours per day and seven days per week. Substantial savings have been achieved on sludge treatment costs (sludge deposits, transport, poly-electrolyte usage, make-up water, and energy). Furthermore the application of a self-optimising control-unit causes a stabile progress of the process, requiring less supervision of the dewatering facility as a secondary beneficial effect. The development and the testing of the control-unit has successfully concluded the last phase of the STOWA–project "Process control in the sludge line".

Keywords Sludge, Dewatering, Automation

Introduction

In the year 1999 a benchmark study among the water quality managers of the water authority has revealed that 12% – 45% of the total waste water treatment costs is needed for the treatment and deposition of the excess sludge. An important, but relatively complex, step in the sludge treatment process is the sludge dewatering. It is not simple to obtain an optimal result in sludge dewatering due to variations in the properties of the sludge. Especially in the weekends and during night-hours "safe" process conditions are chosen to keep the operation free of interference. A self-optimising control-system is capable to maintain "ideal" conditions in the sludge dewatering facility and to adapt these conditions when required, due to variations in the properties of the sludge. Under the authority of the STOWA the development of a self-optimising control-unit has been realised. In the year 1996 investigations have been started aiming at a complete inventory of all existing control equipment used in the sludge dewatering process together with the proven functionality. The inventory[1)] revealed the existence of a large number of measuring apparatus, but the non-existence of any enhanced process control units. Furthermore the experience with the existing equipment was very limited; any comparative tests hardly existed. Because of the foregoing, the second phase of the investigations was targeted at comparative testing of the several measurement apparatus. During the period of 1998 through 1999, nine different meters have been tested thoroughly at two locations with different types of sludge (aerobic stabilised and anaerobic). The investigations[2)] have revealed that a good measurement of the dry solids content of the dewatered sludge could be obtained by means of a multipoint infrared absorption measurement under different angles. In those days the available equipment for the measurement of the centrate or the filtrate properties could not produce a reliable value for the dry solids concentration in those streams. Some apparatus, however, could signal the moment and the direction in the event of a change of properties. Based on

© 2004 IWA Publishing. *2nd IWA Leading-Edge Conference on Water and Wastewater Treatment Technologies* edited by Mark Van Loosdrecht & Jonathan Clement. ISBN: 1 84339 508 8.

the results of the tests it was concluded to be feasible to develop a control-unit that could control the dewatering process adequately. Since a control-unit as described was not available on the market, a manufacturer of dewatering equipment (Alfa Laval) was selected to assist in the development and construction of such a unit. Prior to the design of the control-unit the boundary conditions have been formulated as a result of a brainstorm session with a delegation of the Dutch water quality managers. It seemed quite apparent that besides an adequate process-control with the control-unit, an on-line data registration of all relevant process data should take place as well. Based on these boundary conditions a control-unit has been designed for the optimisation of the sludge dewatering process. It has been taken into account to provide the unit with a very high level of flexibility, including sufficient possibilities to adapt the control-unit to the many different types of sludge and their dewatering facilities. Parallel with the Dutch investigations a similar program of investigations has been carried out by the American sister organisation of the STOWA: the Water and Environmental Research Foundation (WERF). At several different Sewage Treatment Plants (STP's) they have the application tested of some existing simple controlling concepts of peripheral equipment. It appeared not to be feasible to control the dewatering process optimally, because the controlling parameters of the dewatering equipment were not included in the controlling process. The investigations of the WERF were aiming predominantly at the development of a methodology to properly select and judge control equipment.

The control-unit

In figure 1 a process flow diagram (PFD) of the dewatering process shows the location of the control-unit schematically. The measurement of the dry solids concentration (DS) in the sludge to be dewatered and in the centrate is essential for a proper functioning of the control-unit. The dry solids concentration in the dewatered sludge is calculated, because a proper measurement of this parameter is still non-existent at this moment. The value should be verified on installation of the equipment. The accuracy of the measurement of the dry solids concentration in the centrate turned out to be substantially higher, when properly installed, than one would have expected looking at earlier tests. During the investigations the accuracy of the measurement could be further improved by means of semi-automatic cleaning of the measuring vessel, initiated by the control-unit at regular intervals. The operation of the control-unit is very user-friendly. By means of a PC the operator has the possibility to simply input (or change) several process conditions and adjustments for the sludge dewatering process. In this way input is provided for the sludge flow rate, the addition of polyelectrolyte, the dry-solids yield, the dry solids content of the dewatered sludge, and the set points of the dewatering equipment. Next, input of is provided of all relevant cost-factors, such as costs for PE, make-up water, sludge-transport, energy, and deposit of the sludge. With the several different set points and the on-line measurement of the concentration of the dewatered sludge, the dewatering facility is started up. After running a certain period on these set points, it is switched automatically to an optimisation program. In dependence of the selected program the control-unit optimises the dewatering process towards minimisation of costs or towards a maximised dry solids content of the dewatered sludge. If, for instance, the minimisation of the dewatering costs has been selected, the control-unit will change one or more process parameters, after which the result is compared with the previous adjustments. If the dewatering costs are decreased after the change and the boundary conditions are still met (i.e. clean centrate), then the next step is initiated. If the boundary conditions are not met anymore, or the dewatering costs are increased, then the change will be eliminated first, after which another parameter is changed, or a different step-size is tried. In this way the control-unit directs the process towards an optimum that is adapted continuously to the changing sludge properties within the limits of the boundary conditions.

Results of the investigations.

The control-unit has been installed at one of the two centrifuges of the STP Almere and has been tested under practical conditions during more than half a year. The other centrifuge has been used as a reference and was operate manually by an operator. The results of both centrifuges have been compared under different conditions and during longer periods. The control-unit has been operating predominantly in the cost-optimisation mode, aiming at a decrease of the PE addition and the stabilisation of the dry solids content of the dewatered sludge. The cost calculation for this treatment is based on tons of dry solids of final product and not on tons of raw material. In this case sludge with a higher percentage of dry solids will not give rise to extra savings for the water authority. In figure 2 a typical graph shows the dewatering results manually and automatically controlled with the control-unit over a period of one day. At the reference centrifuge (without control-unit) the PE-dosing is increased outside the working hours and simultaneously the pressure on the sludge is decreased to prevent any interference. On changing sludge properties there will be sufficient PE and the maximum pressure will not be exceeded, while keeping the quality of the centrate at level. However, if the properties of the sludge do not change this means an overdose (waste) of PE and a non-optimal dry solids content of the dewatered sludge. During daytime, when the operators are present, the PE-dosing is stepwise, manually, adjusted until the minimum dosing is achieved, while the pressure is increased. Since this adjustment takes place in relatively large steps, it takes a while before the process is stabilised and the next step can be taken. Before the end of the working day the PE-dosing and the sludge pressure are re-adjusted to the "safe" position. As can be seen from figure 2, with the control-unit a stabile and optimal dewatering result can be achieved during 24 hours per day. The result is constant high dry solids content and a low PE usage. In figure 3 the total costs of the dewatering of the selected example have been presented. With the control-unit the costs of dewatering are rather constant and average € 418.-- /ton dry solids, while without the control-unit the costs vary between € 425.-- /ton dry solids and € 440.-- /ton dry solids with an average of € 434.-- /ton dry solids. In this example the savings amount € 16.-- /ton dry solids, excluding the decreased supervision of the operator.

Costs and benefits

The costs of a control-unit are very much depending on the local conditions and specific demands. A rough estimate results in more than € 60,000.- per dewatering facility. In these costs are included the necessary hardware (dry solids meters, PC, controllers, piping), software, and the costs of labour for installation and calibration of the control-unit. In analogy the benefits are depending on the local situation as well. Furthermore, it is not easy to express all of the benefits in an exact amount of money. The savings can be achieved because of:

- Transport and treatment costs will be lower, due to a more constant and higher average dry solids content of the dewatered sludge;
- Labour costs will decrease now most of the work is done by the control-unit;
- Decreased usage of PE and make-up water;
- Increased efficiency of the facility, so less sludge has to be dewatered again.

During the period of the investigations the PE-usage at the STP Almere has decreased by 25% and a high and constant final dry solids content has been achieved at 0.5% more than the reference centrifuge. On top of that a significant and important improvement of the dry solids yield has been realised. Summarising, € 15.- /ton dry solids has been saved resulting in an annual saving of € 53,000.- Not included is the reduction of labour costs and sludge feed back to the aeration tanks.

Further developments

After the previous described STOWA research, the development and test of the control unit is continued. In the beginning of 2004 the system was installed at 8 different WWTP all over the world. As an example how the control unit works two sets of results are presented in graphs in figure 4 and 5. The graphs (trend curves) show the results that are available for the operators to evaluate the working of the system. The two graphs are from two different treatment plants, one in the Netherlands that dewaters aerobic stabilized sludge and one in Denmark that dewaters digested sludge. Both graphs give an impression how the control unit reacts on a change of a parameter. In figure 5 the result is shown after lowering the minimum polymer dose in two steps from 10,5 to 8,5 g PE/kg DS. In figure 6 the result is shown after a change in sludge concentration/quality.

Conclusions

It has been proven to be possible to continuously optimise the sludge dewatering process with a control-unit, achieving better dewatering results than before. An important condition for a proper operation of the control-unit is a reliable dry solids measurement in the sludge feed and the centrate or filtrate. The location of the meters is of the utmost importance, because the dry solids measurement in the centrate is functioning as a control and supervisory parameter for the dewatering process. The sensors used have been functioning excellently with a minimum of maintenance throughout the whole period of the investigations. The feasible savings depend very much on the local conditions. In principle the investment pays back in a relatively short period. With the successful concluding of the STOWA-project a first step has been made towards automatic sludge treatment. Because of the modular construction of the system it is feasible to be extend the control-unit in the nearby future. Possibilities point in the direction of automation of the sludge thickening process and even the withdrawal of surplus sludge from the aeration tanks.

References

[1] STOWA 97-13, Meet- en regelapparatuur bij slibontwatering: *Inventarisatie van apparatuur en mogelijkheden,* 1997

[2] STOWA 99-26, Meet- en regelapparatuur bij slibontwatering: *Vergelijkend praktijkonderzoek aan meetapparatuur,* 1999

[3] Water Environmental Research Foundation, project 98-REM-3, Thickening and dewatering processes: How to evaluate and implement an automation package, Alexandria, Virginia, USA , 2001.

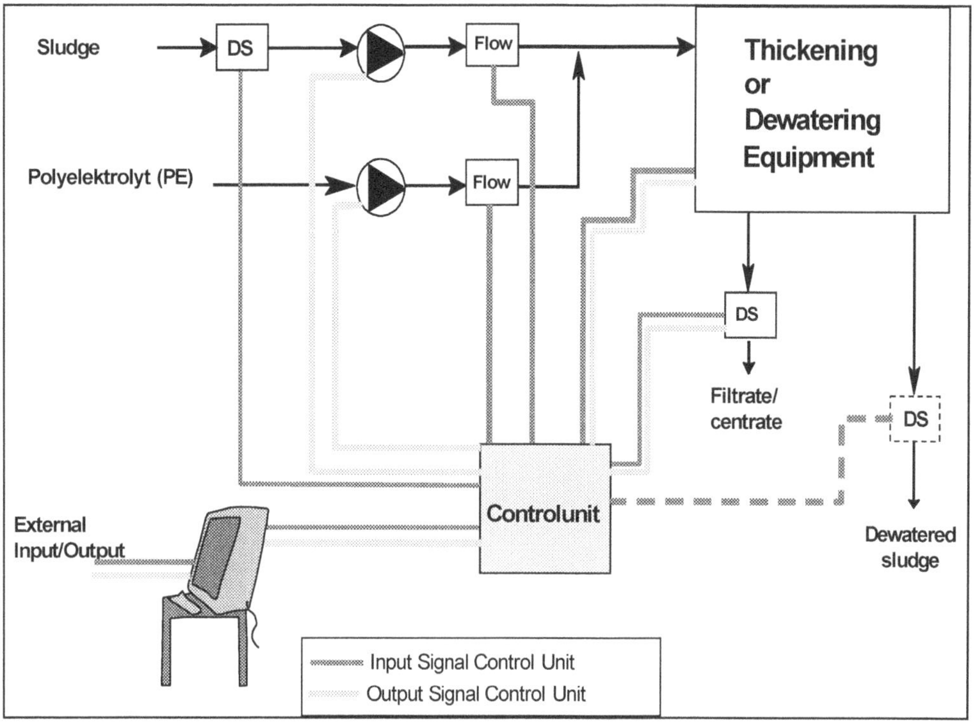

Figure 1 Location of the control unit in the dewatering installation

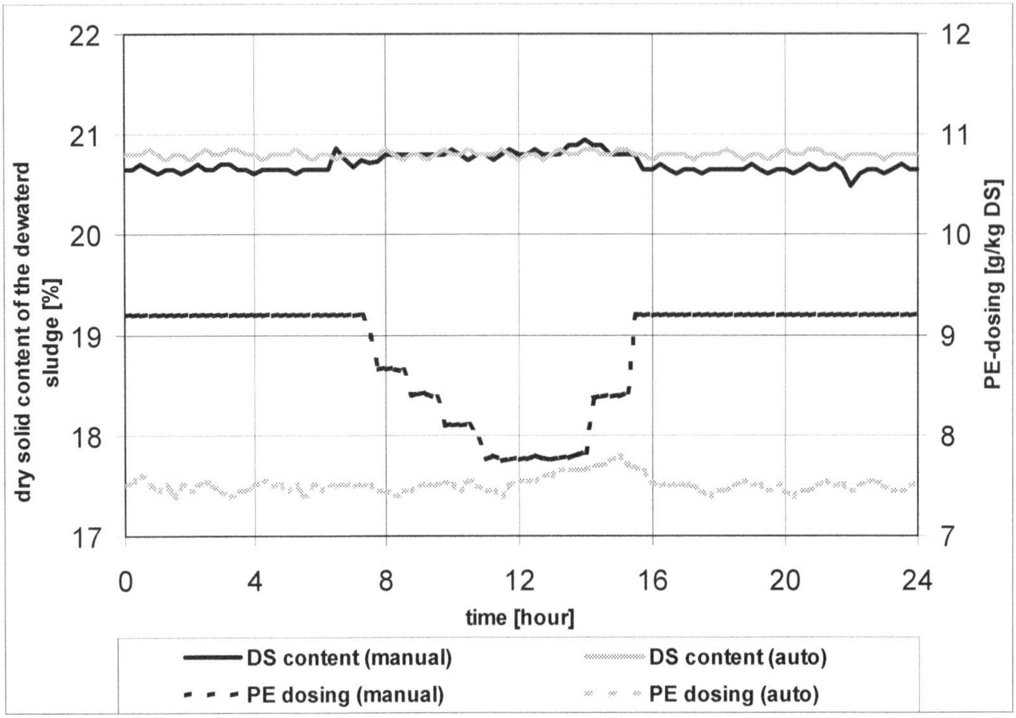

Figure 2 The results of a manually operated installation in comparison with an automatic controlled installation

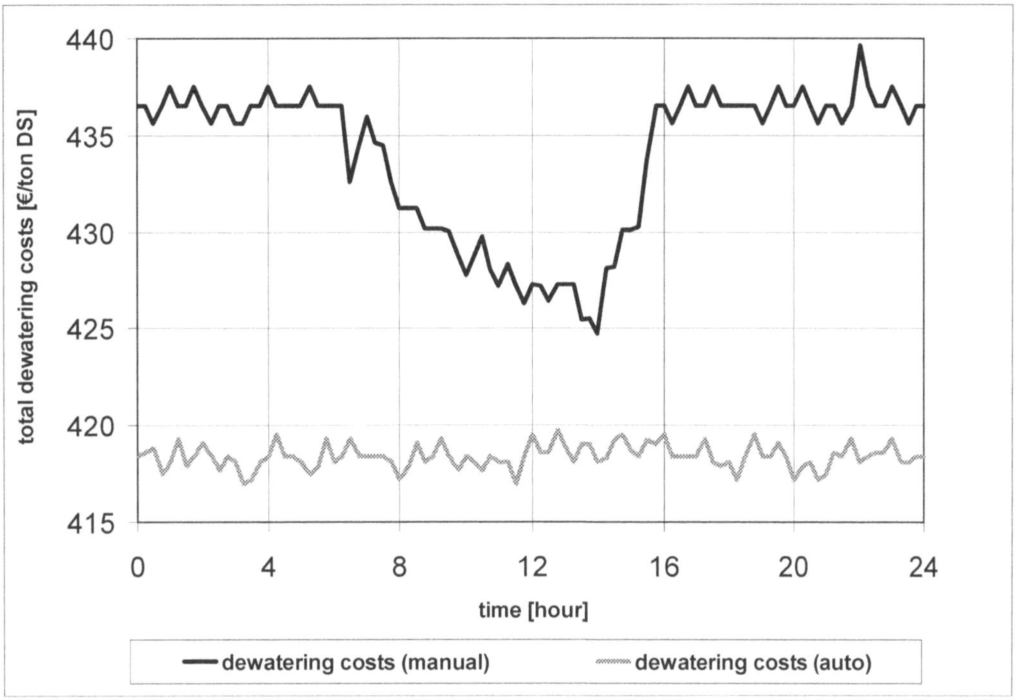

Figure 3 Difference in total dewatering costs of the manually operated installation in comparison with the automatic controlled installation

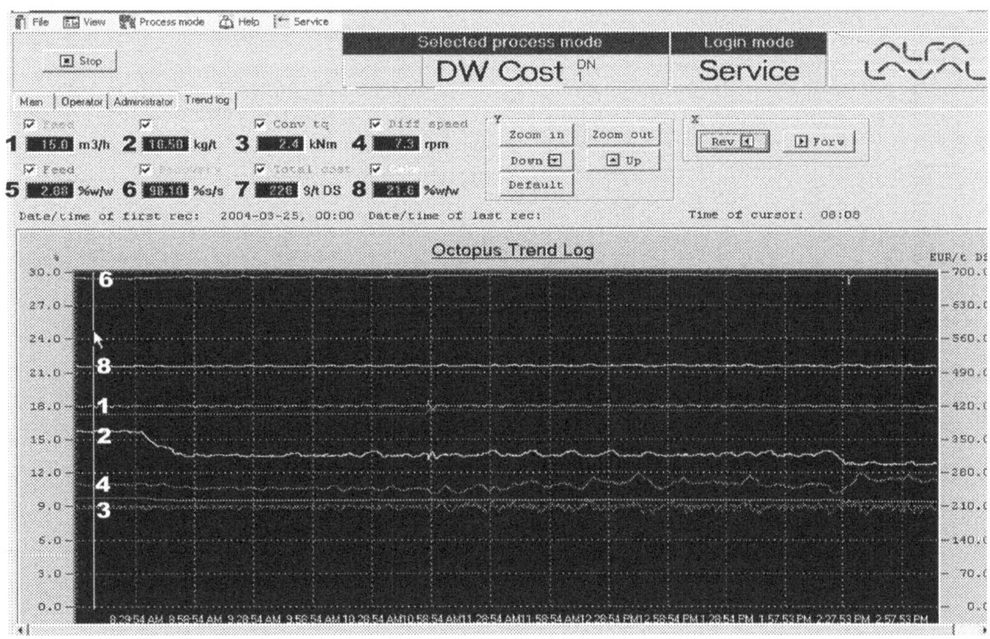

Figure 4 Example of polymer optimisation (line 2) from 10,5 to 8,5 g PE/kg DS. The system controls the other parameters to minimize the dewatering costs (line 7, between line 3 and 4) (WWTP with aerobic stabilized sludge, the Netherlands)

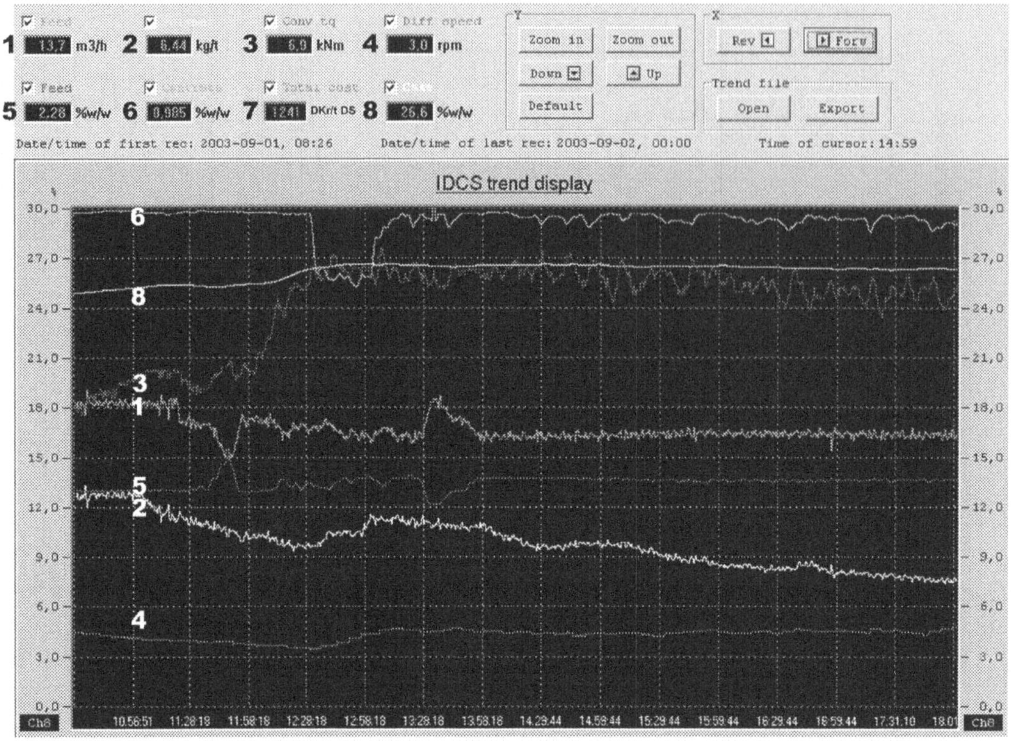

Figure 5 Optimisation of the dewatering after a change in digested sludge concentration and quantity (line 1 and 5) by increasing the torque (line 3) and decreasing the polymer (line 2) (WWTP Helsinborg, Danmark)

Removal of estrogenic trace contaminants from wastewater and landfill leachate with advanced treatment processes

T. Wintgens*, S. Lyko*, T. Melin*, A. I. Schäfer, S. Khan**, P. Sherman*** and A. Ried******

*RWTH Aachen University, Turmstrasse 46, 52056 Aachen, Germany. (E-mail: *wintgens@ivt.rwth-aachen.de*

**Civil, Mining & Environmental Engineering, University of Wollongong, NSW, 2522, Australia. (E-mail: a.schaefer@uow.edu.au, skhan@uow.edu.au)

***Environmental Protection Agency, QLD, Australia. (E-mail: Paul.Sherman@epa.qld.gov.au)

****WEDECO Umwelttechnologie GmbH, Boschstr. 6, 32051 Herford, Germany (E-mail: achim.ried@wedeco.net)

Abstract Several endocrine disrupting chemicals (EDCs), including nonylphenol and bisphenol A, have been found in wastewater treatment plant effluents and surface waters. As conventional processes in municipal wastewater treatment plants are not able to eliminate EDCs below a no-effect level, advanced treatment processes need to be considered to minimise the discharge of those compounds. Therefore, the application of membrane technology and advanced processes like activated carbon adsorption and ozonation is investigated here with the aim to evaluate the potential impact on increased endocrine disrupter removal. A number of example applications are given and removal of target compounds like 17ß-estradiol has been 97% in reverse osmosis treatment for effluent polishing and even 99% for xenoestrogenic BPA in a membrane bioreactor configuration for landfill leachate treatment.

Keywords endocrine disrupters, membrane technology, advanced treatment

Introduction

There are many different substances, which can disruptively interact with the hormone system of humans and animals. Natural and synthetic estrogens belong to the most effective endocrine disrupting chemicals (EDCs). Among the more important compounds are estradiol and ethinylestradiol, which is a synthetic hormone used in contraceptives. Potentially endocrine disrupting industrial chemicals are, for example, nonylphenol (NP), bisphenol A (BPA), and tributyl-tin. Xenoestrogens like NP and BPA are not as estrogenically active as estradiol, but can be found in much higher concentrations in environmental compartments due to their widespread production (Körner et al., 2000).

Several endocrine disrupting substances, including NP and bisphenol A as well as steroid hormones, have been found in wastewater treatment plant effluents and surface waters (Ahel, 1987, Spengler et al., 2001). As conventional processes in municipal wastewater treatment plants are not able to eliminate EDCs below a no-effect level (Stumpf et al., 1996), advanced treatment processes need to be considered to minimise the discharge of those compounds.

© 2004 IWA Publishing. *2nd IWA Leading-Edge Conference on Water and Wastewater Treatment Technologies* edited by Mark Van Loosdrecht & Jonathan Clement. ISBN: 1 84339 508 8.

Therefore, the application of membrane technology and advanced processes like activated carbon adsorption and ozonation is investigated here with the aim to evaluate the potential impact on increased endocrine disrupter removal. Special attention will be paid to hybrid processes like membrane bioreactors and dense membrane processes like nanofiltration and reverse osmosis as well as chemical oxidation process with ozone.

Different relevant applications will be considered like effluent treatment for wastewater reuse as well as the purification of highly contaminated landfill leachates, which can be considered as a highly challenging and relevant example of a polluted aqueous matrix (Paxeus, 2000). The aim of this paper is to investigate the removal efficiency of different advanced treatment processes. The three different case studies presented cover a range of advanced treatment processes which are regarded as key components in state-of-the-art water treatment and particularly promising with respect to the removal of organic trace contaminants.

Analytical methods

Samples have been taken from various pilot to full-scale treatment plants described below in different studies and have been analysed in terms of estrogenic organic micropollutants.

Analysis of Hormones

In the first case study on effluent treatment presented, carried out in an advanced water recycling demonstration plant in Queensland/Australia, 1L samples were extracted on preconditioned 1g C18 cartridges under vacuum after the addition of estradiol-D4 as surrogate standard. The cartridges were dried under nitrogen and the extracts were eluted with acetonitrile. The eluents were then dried under vacuum. The estrogens were determined by GC-NICI-MS after a two-step derivatization with pentafluorobenzyl bromide followed by *N*-trimethylsilylimidazole. This method was applied for the analysis of 17β-estradiol (E2), estrone (E1) and ethinylestradiol (EE2) in aqueous matrices (Braga et al., 2001).

Analysis of Xenoestrogenes

All samples from the landfill leachate treatment and the second effluent treatment study were immediately stabilised by addition of NaN_3. Within several hours, the samples had been carried to the laboratory, where they were kept refrigerated in brown glass bottles. The samples were prepared and analysed within 72h. Sample processing and analysis for measurement of nonylphenol and bisphenol A was performed according to Meesters et al. (2002) and basically consists of an in-situ derivatisation step and liquid-liquid-extraction prior to the GC-MS measurement. The chemical structures of the target analytes are shown in Figure 1.

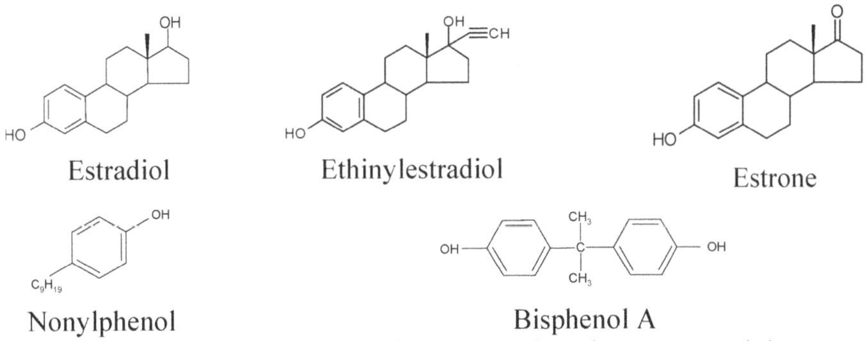

Figure 1 Molecular structures of the estrogenic and xenoestrogenic hormones

Removal of estrogenic steroid hormones in effluent treatment

Microfiltration, Nanofiltration and Reverse Osmosis Treatment

Advanced Water Recycling Demonstration Plant in Queensland, Australia

The Queensland Government, as a component of the Queensland Water Recycling Strategy, commissioned the construction of the Advanced Water Recycling Demonstration Plant (AWRDP) (Gibson and Apostolidis, 2001). The plant consists of eight modules, each housing a different water treatment technology. For these studies, the influent to the AWRDP was taken from tertiary treated effluent from the nearby Brendale Wastewater Treatment Plant (Pine Rivers Shire Council).

The modules of the AWRDP processed around 60L/min and included lime clarification (LC), dissolved air flotation (DAF), dual media filtration (DMF), ozonation (O_3), biological activated carbon filtration (BAC), microfiltration (MF), combined reverse osmosis/ nanofiltration (RO/NF) and UV-disinfection. Chlorine can be added if the use of the water requires the presence of a disinfection residual. Individual modules may be brought on-line or bypassed very quickly and simply. The AWRDP is described in detail in Khan et al. (2003). Figure 2 shows the flow sheet of the membrane units with microfiltration as well as reverse osmosis and nanofiltration modules.

As the measurement of background samples of the effluent did not reveal measurable concentrations of the targeted hormones at all sampling points spiking tests were performed to ascertain the measurement of retention in different process configurations. By spiking measurable concentrations of the three steroid hormones compounds, it was possible to gain significantly more information on the behaviour of these compounds during the advanced treatment processes. The study incorporated both the microfiltration (USF Memcor CMF hollow fibre modules) and RO/NF modules (Trisep spiral wound modules with X-20 RO membranes and XN-45 nanofiltration membranes). The target compounds were spiked into the feed tank of the microfiltration module. Samples for analysis were taken from the microfiltration module (feed and permeate) and the RO/NF module (RO permeate, NF permeate, RO/NF retentate).

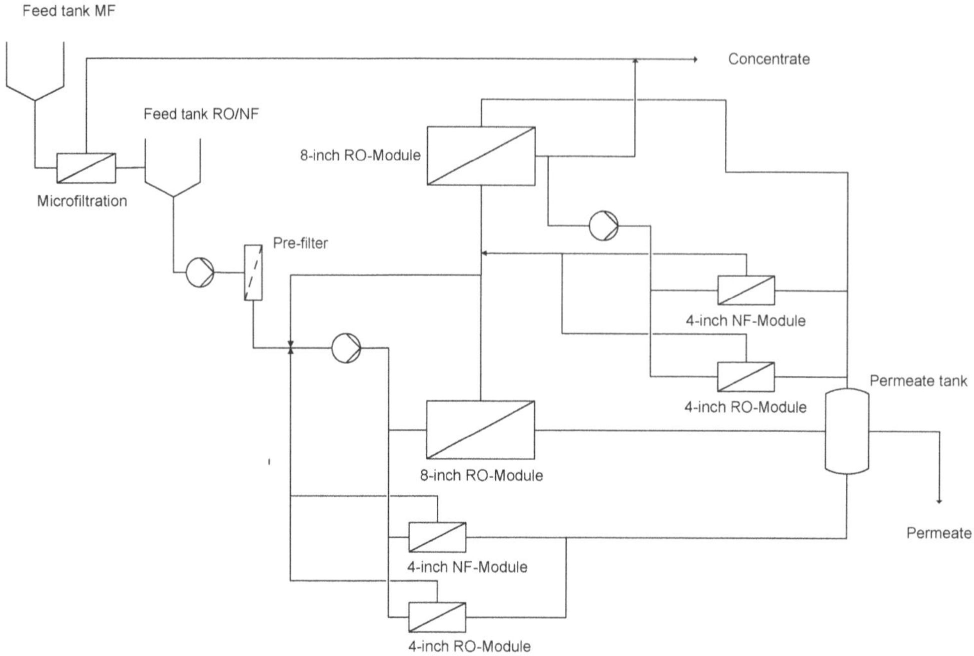

Figure 2 Flowsheet of the membrane units at the AWRDP in Queensland/Australia

Results and Discussion

The hormone retention by the membrane processes nanofiltration and reverse osmosis is shown in Figure 3 (mean of five samples). The permeate concentrations are very low, but not zero, confirming results from stirred cell experiments (Schäfer et al., 2003) where NF concentrations are slightly higher than RO permeates. Of interest, again, is the high concentration of all hormones in the RO/NF retentates.

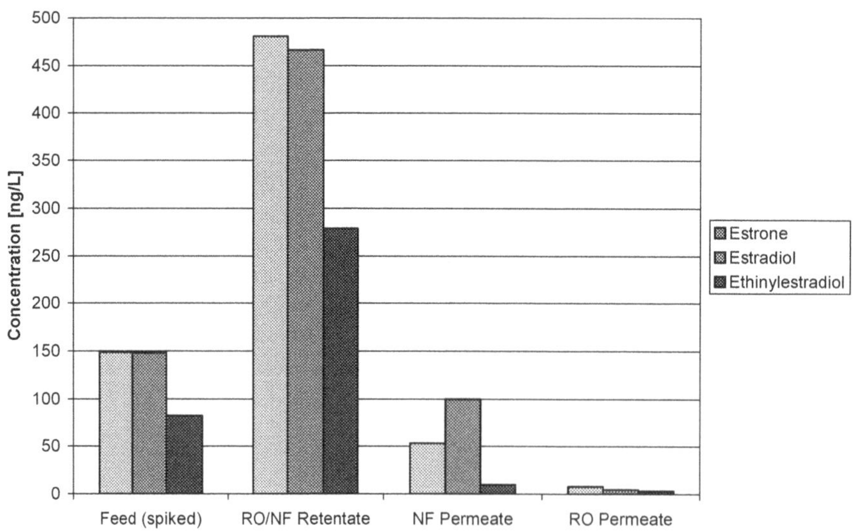

Figure 3 Hormone concentrations in NF and RO feed, permeate and concentrate

The hormone concentrations determined in spiking experiments including the microfiltration unit are given in Figure 4 (mean of three samples).

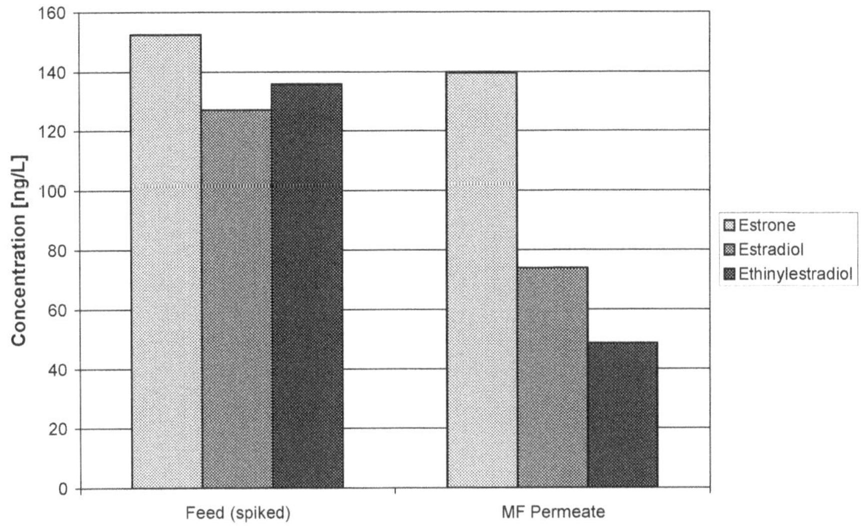

Figure 4 Hormone concentrations in microfiltration feed and permeate

The results indicate that even the porous MF membranes display some retention of these small organic molecules, which can only be explained by adsorption to particulate matter which is retained by the MF membranes or by non-steady state adsorption effects on the membranes itself occurring during spiking experiments. This confirmed the results by Schäfer et al. (2002) which indicated an interaction of the hormones with particulates and subsequent retention by porous membrane processes.

Ozone treatment

Effluent treatment in Braunschweig, Germany

In drinking water treatment it has already been shown that ozonation is very effective in oxidising micropollutants (Ternes et al., 2002). For wastewater treatment this technology is not yet optimised and deserves further studies. The objective of the presented tests was to determine whether ozonation may be capable to remove trace levels of the industrial chemicals NP and BPA in treated wastewater. Therefore a pilot plant for ozonation received effluent from a German municipal wastewater treatment plant (WWTP). The WWTP with a mean flow rate of 60,000 m^3 d^{-1} consists of mechanical pre-treatment, followed by simultaneous nitrification and denitrification, biological phosphate removal and secondary clarification.

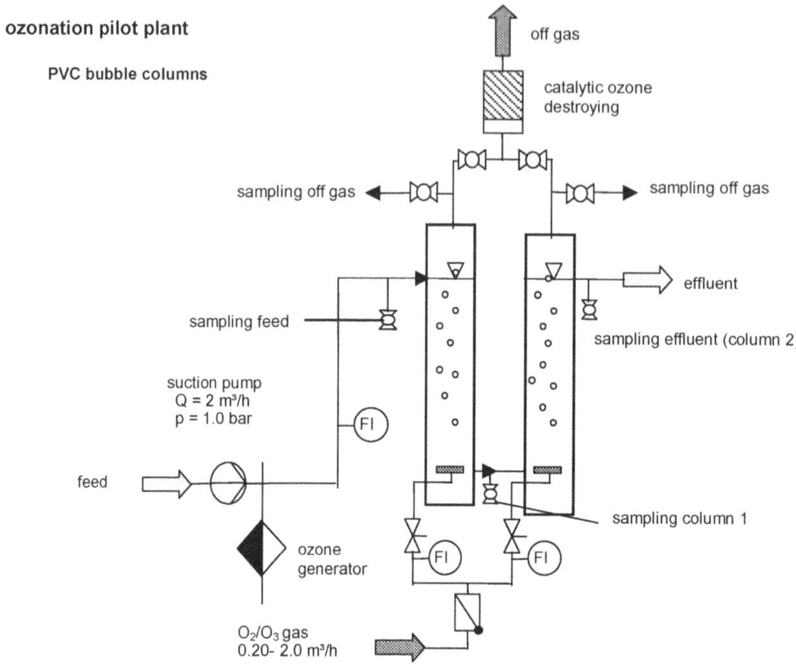

Figure 5 Pilot plant at a German municipal WWTP

Results and Discussion

The WWTP effluent (2 m^3 h^{-1}) was fed by a suction pump into the continuously operated pilot plant for ozonation operated by WEDECO. The pilot plant consists of an ozone generator (100 g h^{-1}), 2 diffusers and 2 PVC bubble columns (Figure 5). The dimensions of the 2 bubble columns were: height 5 m, diameter 0.2 m, volume 150 L. Ozone is generated by electrical discharge from oxygen. The contact time in these ozone columns was 9 min. The effective ozone doses resulted from the difference between the ozone gas concentration in the inlet and outlet of the bubble columns. Former studies detected only 2-5% of the ozone in the gas outlet (Ternes et al., 2003).

Samples of effluent of the WWTP showed concentrations on a non-detectable level of NP and BPA. Therefore the effluent (pH = 7.55, T = 12.9°C) was spiked with the selected chemicals. After spiking with a stock solution (10 g/L of NP and BPA in acteone) the theoretical concentration of NP and BPA should be 2.0 mg/L. Two trials were carried out with the same conditions. For every trial two different ozone doses (10 and 15 g O_3/m^3) were used. The settings of the experiments are given in table 1.

Table 1 Test settings of the ozone generator during the trials

Parameter	Dimension	O_3 dose: 10 g O_3/m^3	O_3 dose: 15 g O_3/m^3
inverter power	%	54	95
O_3 concentration.	g O_3/m^3 gas	40	76.5
O_3 capacity	m^3 gas/h	0.4	0.4

Table 2 Results of the ozonation of NP- and BPA-spiked WWTP effluent

Samples	O_3 dose [g O_3/m³]	NP [µg/L]	BPA [µg/L]
trial 1			
effluent WWTP	-	n.d.	n.d.
spiked effluent	8	900.7	1022.0
column 1	8	216.3	134.1
column 2	8	62.5	2.1
spiked effluent	15	671.7	1197.0
column 1	15	106.9	26.0
column 2	15	17.6	-
trial 2			
effluent WWTP	-	n.d.	2.9
spiked effluent	10	715.2	910.0
column 1	10	286.1	301.0
column 2	10	24.2	29.5
spiked effluent	15	569.8	759.4
column 1	15	175.4	98.8
column 2	15	18.3	23.3

The results given in table 2 indicate that it is possible to decompose NP and BPA using ozonation. Noticeable are the high losses of the chemicals after spiking (28 – 45 %). This is caused by the adsorption of these compounds. After column 1, operating at counter-current flow the bigger part of NP resp. BPA was removed with slightly better removing rates for BPA. The second column, operating at co-current flow reduced the concentrations to a level near the detection limit. The removal rates of column 2 are marginally better than those of column 1. The effect of increased ozone doses was not as pronounced as expected. In fact, the remaining concentrations were reduced but the enhancement of removal rates is negligible. The ozonation products formed are still unknown. But it has to be noted that in the present study only the primary target degradation was investigated, thus further research is necessary to identify and confirm the structures of metabolites formed by ozonation.

Xenoestrogenes in landfill leachate treatment

Membrane bioreactor and activated carbon treatment

Landfill leachate treatment plant in Alsdorf Warden, Germany

Within the presented study a full-scale landfill leachate treatment plant with a membrane bioreactor configuration and a parallel reverse osmosis installation was investigated. Samples from different sample points within the process were analysed with regard to estrogenically active substance bisphenol A.

The membrane bioreactor configuration consists of three activated sludge tanks and an external ultrafiltration unit (Figure 6). This process train treats an average feed flow of 7m³/h landfill leachate. The membrane bioreactor configuration works with denitrification and a two-stage nitrification. The bioreactors have a gross volume of 180m³ each and operate under 3.5bar total pressure with internal re-circulations. The total suspended solids concentration within the bioreactors is kept at about 25g/L by discontinuous excess sludge removal. The ultrafiltration stage is equipped with tubular modules in six parallel racks and operated in cross-flow mode. The cross-flow velocity is kept at about 5m/s and the trans-membrane pressure difference is around 6bar. A total membrane area of 120m² is installed.

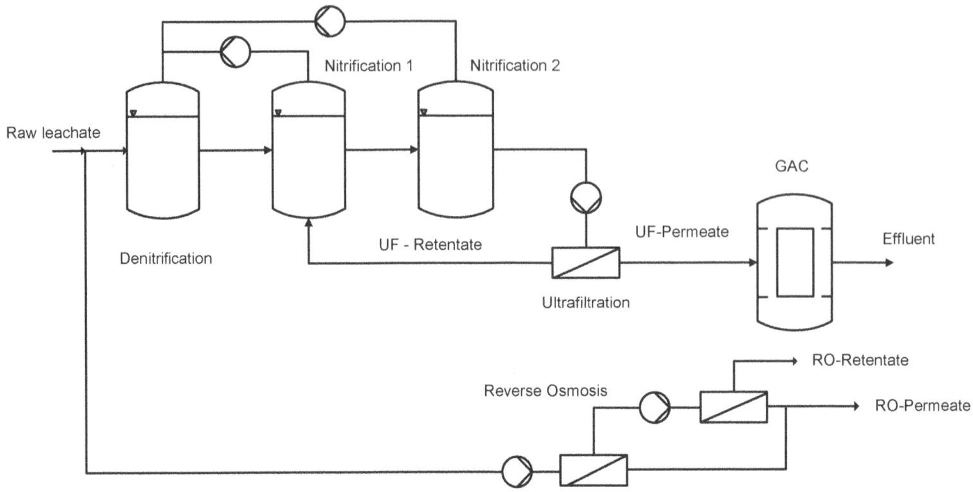

Figure 6 Membrane bioreactor configuration at a landfill leachate treatment plant

Results and Discussion

Bisphenol A is taken an indicator for the behaviour of xenoestrogenic compounds in the plant. Considerably high concentrations, ranging up to milligrams of BPA per litre, were measured in the raw landfill leachate. There is a 99% BPA-removal in the MBR plus additional polishing the granular activated carbon systems. Combining the hydraulic data of the plant with the average BPA concentrations measured at the different sampling points a mass balance can be derived (Figure 7).

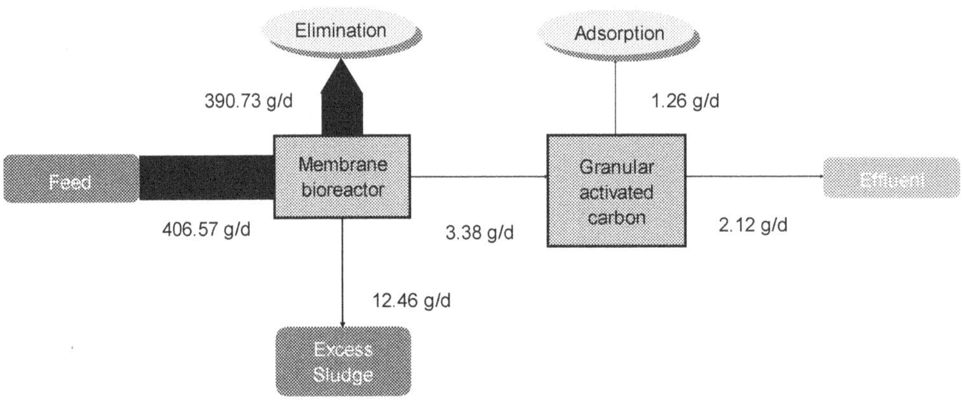

Figure 7 Bisphenol A loads in landfill leachate treatment plant

Conclusion

The aim of this paper was to make a broad quantitative assessment and comparison of different treatment processes on the basis of three major case studies. Except microfiltration all considered technologies were able to demonstrate a significant and consistent removal of the investigated estrogenic trace contaminants. The studied cases showed that advanced treatment processes like reverse osmosis, nanofiltration (to a lesser extent than RO) and even membrane bioreactors with ultrafiltration membranes (mainly attributed to biological removal) as well as ozonation are appropriate to minimise the discharge of endocrine disrupting compounds through different wastewater matrices. In case of the investigated landfill leachate application activated carbon adsorption displayed low removal capacity for BPA, but this might be due to the fact that the GAC units are designed for very limited effluent upgrade in terms of bulk organic load (about 50% COD removal).

Acknowledgements

The authors would like to acknowledge The Queensland State Government and the Queensland Water Recycling Strategy for access to the AWRDP as well as Jan Zaricky from the Queensland EPA for the support of this study. The PhD candidates Olga Braga (UNSW) and Long Nghiem (UoW) are thanked for the support with hormone analysis. A/Prof George Smythe at the Biomedical Mass Spectrometry Unit, UNSW is thanked for his support with hormone analysis method development.

The authors wish to thank as well the German Research Association (DFG) for the funding of the Graduate College 546 AGEESA at the RWTH Aachen University, the plant manager Mr Reichbauer of the Alsdorf-Warden plant (AWA Entsorgung GmbH) for the support and access to the plants as well as Prof. Schröder and R. Meesters from the laboratory of the Sanitary Engineering Department at RWTH Aachen for analytical support within the project.

References

Ahel M. (1987) Biogeochemical behaviour of alkylphenol polyethoxylates in the aquatic environment. Dissertation University of Zagreb.

Braga O., Schäfer A.I., Smythe, G. (2001) Analysis of endocrine disrupters and pharmaceuticals in the aquatic environment. In: Recent Advances in Water Recycling Technology, Workshop Proceedings, 26 November 2001, Brisbane, QLD, Australia.

Gibson H. E. and Apostolidis N. (2001) Demonstration, the solution to successful community acceptance of water recycling *Water Sci. Technol.*, **43**(10), 259-266.

Khan S. J., Schäfer A. I., Wintgens T., Sherman P. and Zaricky J. (2003) Removal of hormones and pharmaceuticals in the "Advanced Water Recycling Demonstration Plant" in Queensland, Australia. In Proceedings of the *4th International Water Association Specialized Conference on Assessment and Control of Hazardous Substances in Water – Ecohazard*. International Water Association, Aachen, Germany.

Körner W., Bolz U., Süßmuth W., Hiller G., Schuller W., Hanf V., Hagenmaier H. (2000) Input/Output balance of estrogenic active compounds in a major municipal sewage plant in Germany. *Chemosphere*, **40**, 1131-1142.

Meesters R.J.W. and Schröder H. F. (2002) Simultaneous Determination of 4-Nonylphenol and Bisphenol A in Sewage Sludge. *Anal. Chem.*, **74**, 3566-3574.

Paxeus N. (2000) Organic compounds in municipal landfill leachates. *Wat. Sci. Tech.*, **42**(7-8), 323-333.

Schäfer A. I., Mastrup M. and Lund Jensen R. (2002) Particle interactions and removal of trace contaminants from water and wastewaters. *Desalination*, **147**, 243-250.

Schäfer A. I., Nghiem L. D. and Waite T. D. (2003) Removal of the natural hormone estrone from aqueous solutions using nanofiltration and reverse osmosis *Environ. Sci. Technol.*, **37**(1), 182-188.

Spengler P., Körner W. and Metzger J. W. (2001) Substances with estrogenic activity in effluents of sewage treatment plants in southwestern Germany. 1. chemical analysis. *Environ. Tox. Chem.* **20**(10), 2133-2141.

Stumpf M., Ternes T.A. and Haberer K. (1996) Nachweis von natürlichen und synthetischen Östrogenen in Kläranlagen und Fließgewässern. *Vom Wasser*, **87**, 251-261.

Ternes T.A., Meisenheimer M., McDowell D., Sacher F., Brauch H.-J., Haist-Gulde B., Breuss G., Wilme U., and Zulei-Seibert N. (2002) Removal of pharmaceuticals during drinking water treatment. *Environ. Sci. Technol.* **36**, 3855-3863.

Ternes T. A., Stuber J., Herrmann N., McDowell D., Ried A., Kampmann M. and Teiser B. (2003) Ozonation: a tool for removal of pharmaceuticals, contrast media and musk fragrances from wastewater? *Water Res.*, **37**(8), 1976-1982.

Comparison of the Elimination of Endocrine Disrupters in Conventional Wastewater Treatment Plants and Membrane Bioreactors

P. Ivashechkin*, P. F. X. Corvini, M. Fahrbach***, J. Hollender***, M. Konietzko*, R. Meesters*, H. Fr. Schröder* , M. Dohmann***

*Institute of Environmental Engineering (ISA), Aachen Technical University, Templergraben 55, Aachen 52056 Germany. (E-mail: ivashechkin@isa.rwth-aachen.de)
**Institute of Biology V - Environmental Chemistry, Aachen Technical University, Templergraben 55, Aachen 52056 Germany. (E-mail: philippe.corvini@bio5.rwth-aachen.de)
***Institute of Hygiene and Environmental Health, Aachen Technical University, Templergraben 55, Aachen 52056 Germany (E-mail: michael.fahrbach@rwth-aachen.de)

Abstract Conventional wastewater treatment does not provide satisfactory elimination of some endocrine disrupting compounds (EDC). Membrane bioreactors are reported to possess higher removal rates. In order to verify it a conventional pilot-scale treatment plant (CTP) and a membrane bioreactor (MBR) were operated in parallel with the same BOD_5 sludge loading. Nonylphenol (NP), bisphenol A (BPA) and 17α-ethinylestradiol (EE2) were used as model substances. Two sludge retention times (SRT), 12 and 25 days, were independently tested. NP and BPA were almost completely eliminated in both pilot plants. EE2 turned out to be more persistent: <60% removal at the SRT=12 d and <80% removal at the SRT=25 d. Positive influence of the higher SRT was evident, but difference between MBR and CTP performance was found only for BPA elimination. No considerable retention of the model substances by the membranes or residing on the membranes biofilm occurred. The output with excess sludge was under 5% of the influent load for each EDC. Calculated NP volatilisation was <35%. On this basis, observed elimination of EDCs was considered to be biodegradation. Since EDC removal rates in MBR were not significantly higher than in CTP, in order to enhance EDC elimination other techniques like nano-filtration, ozone treatment or adsorption on activated carbon must be applied.
Keywords wastewater, endocrine disrupters, membrane bioreactor

Introduction

It is known that conventional wastewater treatment techniques reduce EDC concentration of endocrine disrupting chemicals (EDCs) to some extent, but frequently not to levels lower than effective concentrations for fish (Jobling *et al.*, 2002; Johnson and Sumpter, 2001). With the introduction of membrane technology into wastewater treatment there rose a hope of enhancing EDC elimination.

Micro-filtration membranes (0.4 μm) themselves are not able to retain EDC molecules, but filtration cake built on the membrane surface during operation could cause elimination of endocrine disrupters (Schäfer et al., 2003). Since in membrane bioreactors (MBR) suspended solids are completely held back, EDCs adsorbed on sludge are not likely to contaminate the effluent as it may be the case in conventional wastewater treatment plant (CTP).

© 2004 IWA Publishing. *2nd IWA Leading-Edge Conference on Water and Wastewater Treatment Technologies* edited by Mark Van Loosdrecht & Jonathan Clement. ISBN: 1 84339 508 8.

Furthermore, as almost all the bacteria are retained by membranes, development of EDC-biodegrading "specialists" in MBRs is more probable than in CTPs.

It has been reported that EDC elimination rates in MBR are higher than in CTP (Holbrook et al., 2002; Zühlke et al., 2003). In some studies no significant difference between these techniques was found (Kreuzinger et al., 2003; Kroiß et al., 2003). It is difficult to compare MBR and CTP performance, because operation conditions (sludge loading, temperature, wastewater composition) are seldom exactly the same. Besides, due to analytical problems with the determination of EDCs in the solid phase there are only few publications where EDC concentrations in excess sludge were measured (Andersen et al., 2003).

In this study xeno-estrogens nonylphenol (NP), bisphenol A (BPA) and synthetic steroid 17α-ethinylestradiol (EE2) were chosen as model EDCs to compare efficiency of conventional aeration tank and membrane bioreactor concerning EDC elimination. To insure significance of the study, experiments were performed in pilot-scale plants with real wastewater. The CTP and the MBR were operated in parallel with the same BOD_5 sludge loading (SL) to enable a direct comparison. Two sludge retention times (SRT), 25 and 12 days were independently tested in order to model two mostly widespread types of wastewater treatment. Simultaneous sludge stabilising (SRT=25 d) is usually applied at smaller treatment plants. SRT of 12 days requires a smaller aeration tank, but additional anaerobic sludge digestion is necessary. It is applied at medium and large plants.

Influent and effluent concentrations of the model substances were measured. NP and BPA were determined in sludge as well. Complete balances including output of the EDCs with excess sludge and suspended solids in the effluent of secondary sedimentation tank as well as NP volatilisation were performed. In order to evaluate possible retention of the EDCs by the membranes, samples of permeate and MBR sludge supernatant were simultaneously taken.

Materials and Methods

Experiment Set-up

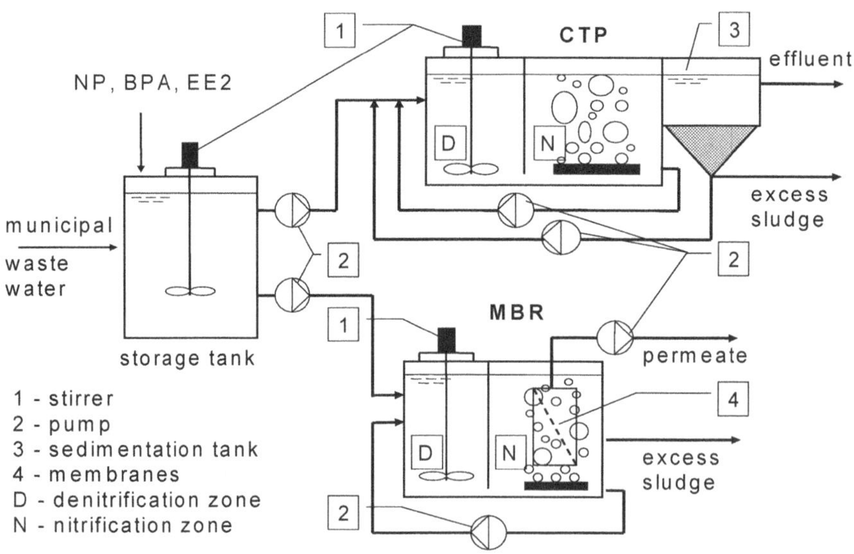

Figure 1 Experiment set-up.

The pilot plants are shown in the Figure 1. Municipal wastewater taken from the output of primary sedimentation tanks of the main wastewater treatment plant of the city Aachen was stored in a 6 m^3 container under stirring. Two electronically adjustable progressing cavity pumps provided continuous influent for the plants.

CTP consisted of an aeration tank (V_{MAX}=1350 L) with an upstream denitrification zone and a secondary sedimentation tank (V=150 L). Two progressing cavity pumps were recirculating NO_3 containing sewage (2*Q_{CTP}) and activated sludge from the sedimentation tank (1*Q_{CTP}), where Q_{CTP} is CTP influent flow rate. Excess sludge was removed manually once a day.

Membrane bioreactor (V=220 L) possessed an upstream denitrification zone and a recirculation pump as well (3*Q_{MBR}), where Q_{MBR} is MBR influent flow rate. Twelve micro-filtration (0,4 μm pores) flat sheet membrane panels 210X300 mm were installed in the nitrification zone. The aeration element was positioned under the panels to generate a cross-flow and provide membrane cleaning. Permeate was drained from the panels by a peristaltic pump (Δp=-0.2 bar): 3 min "on" and 1 min "off" for a better cleaning of the membranes. Excess sludge was removed daily.

All the tanks were of stainless steel and the hoses (polyvinyl chloride) were made as short as possible to minimise adsorption effects.

Experiment Performance

Activated sludge for the CTP was taken from the aeration tank of the wastewater treatment plant of the city Aachen. Activated sludge for the MBR was taken from another pilot MBR that had been fed with Aachen wastewater for two years.

Experiment performance can be divided in three periods: run-up, stable operation, sampling. The aim of the run-up period was to acquire a constant dry matter content (DM) in the aeration tank, BOD_5 sludge loading and good elimination of COD and NH_4^+. Nitrate and total nitrogen elimination were evaluated as well. After that bisphenol A (donation by Dow GmbH, Stade, Germany; technical purity), nonylphenol isomer mixture (Fluka Chemie AG, Buchs, Switzerland, purity >98%) and 17α-ethinylestradiol (Fluka Chemie AG, Buchs, Switzerland, purity >85%) were dissolved in methanol and added to the wastewater in the storage container resulting in the following concentrations: 500 μg/L for NP and BPA;125-210 μg/L for EE2. The pilot plants were fed with this mixture during 1 SRT to provide bacteria adaptation and saturation of possible adsorption sites.

Temperature, O_2 content and pH were measured daily. DM of CTP and MBR sludge, as well as of the influent and the CTP effluent; COD, BOD_5, N, NO_3^{2-} and NH_4^+ of the influent and the effluents were determined twice a week (DEV, 2002). Besides, total phosphorus in the influents, effluents and excess sludges was determined and balances were drawn to verify accuracy of mass and volume flow measurements.

During the third period of five days, three 24 h composite samples (1st, 3rd and 5th day) were taken from the blank wastewater and from the storage tank. EDC concentrations in the effluents of the pilot plants, MBR sludge supernatant, CTP sludge and MBR sludge were measured daily (five 24 h composite samples). The samples were stabilised with sodium azide (0,01%). The liquid ones were stored in glass flasks at +4°C, the sludges were freeze-dried and stored at +4°C as well.

Analytical Methods

NP and BPA concentrations in the activated sludge were determined simultaneously by Gas Chromatography/Mass Spectrometry (GC/MS) with detection limit of 1 µg/gDM for both substances (Meesters and Schröder, 2002). Liquid samples were filtered (0.2 µm filter, Schleicher and Schuell, Germany), NP and BPA were simultaneously extracted with cyclohexane by means of liquid/liquid extraction. Further steps were the same as for the sludge analytic. Detection limit for both substances was 1 µg/L.

EE2 concentration in the liquid phase was quantified by means of High Performance Liquid Chromatography (HPLC). 5 µL of the filtered (0.2 µm filter, Schleicher and Schuell/Germany) sample was injected into the HPLC (1100 Hewlett Packard) equipped with a diode array detector and a fluorescence detector. Separation was achieved on a C18 column (150X2.1 mm, 3 µm, Hypersil, Germany) with a guard column C18 (2X2 mm, 3 µm, Phenomenex, Germany) at 35 °C and flow rate of 0.2 mL/min. Acetonitrile and water were used for gradient elution. Detection limit was 5 µg/L.

Calculation of EDC Elimination Rates

In the liquid phase NP, BPA and EE2 were determined (C_i is measured concentration in the storage tank, µg/L; C_{E0} is effluent concentration, µg/L). NP and BPA concentrations were measured in sludge as well (C_S, µg/g). EE2 concentration in sludge was calculated using partitioning coefficient K_D=0.27 l/g (Siegrist et al., 2003): $C_S=K_D*C_{E0}$.

In order to take into account the amount of EDCs carried out of the CTP sedimentation tank with suspended solids, effluent EDC concentrations in the CTP were calculated as $C_E=C_{E0}+0.01*C_S$, where C_E is total CTP effluent concentration of a model substance, µg/L; 0.01 is dry matter content in the effluent, g/L. For MBR $C_E=C_{E0}$.

Total EDC elimination rates (%) were calculated as $100\%*(1-C_E/C_i)$. Output of EDCs with excess sludge (%) was calculated as $100\%*Cs*Q_{ES}/(Ci*Q)$, where Q_{ES} is excess sludge output, g/d; Q is influent flow rate, L/d. Biodegradation of the model substances (%) was acquired by subtracting EDC output with sludge from total EDC elimination, because it was assumed that all the adsorption sites were saturated with EDCs by the moment of sampling.

Results and Discussion

Operation Stability

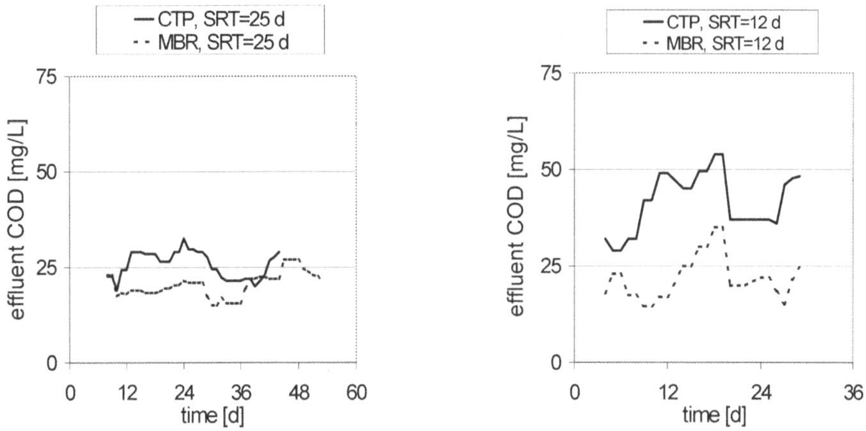

Fig. 2. COD in the effluents of the pilot plants. (a) SRT=25 days, influent COD=492 mg/L. (b) SRT=12 days, influent COD=498 mg/L.

Already 1 week after the beginning of the experiment both pilot plants fulfilled German minimal requirements on BOD_5, COD and NH_4^+ elimination (Wasserhaushaltsgesetz): effluent BOD_5<8 mg/L, COD< 67 mg/L, NH_4^+<0.9 mg/L. Comparison of COD elimination in the pilot plants is shown in the Figure 2. MBR performance was a little better than that of the CTP because the membranes held back all suspended solids. In the effluent of the CTP DM content was 0.01 g/L.

Nitrate was insufficiently eliminated in both plants (15-25 mg/L in the effluent), but good elimination of COD and NH_4^+ in the CTP and MBR provides a possibility to transfer the obtained results concerning the elimination of EDCs to large-scale plants. Performed phosphor balance proved that wastewater and sludge flows were measured correctly.

Other operation parameters are put together in the Table 1. Due to technical problems by the SRT=25 days MBR had to be operated 2 weeks longer than CTP. This is the reason why the mean measured temperatures differ. By the SRT=12 days the pilot plants were operated simultaneously but 1°C difference was still present.

Table 1. Average measured operation parameters of the pilot plants, where V is volume of activated sludge tank; DM is dry matter content of activated sludge; Q is daily influent; BOD_5 is biological oxygen demand of the influent; Q_{ES} is amount of daily removed excess sludge; Q_{DM} is daily discharge of suspended solids with the effluent of CTP; pH, t are pH and temperature in the activated sludge tanks (mean of the last two weeks); O_2 is oxygen content in denitrification tank (D) and nitrification tank (N); SL is BOD_5 sludge loading.

SRT [d]		V [L]	DM [g/L]	Q [L/d]	BOD_5 [mg/L]	Q_{ES} [g/d]	Q_{DM} [g/d]	pH	t [°C]	O_2 [mg/L]		SL [kgBOD_5/ kgDM*d]
										D	N	
25	CTP	1350	3.6	864	255	194	8.6	7.5	23	0	3.0	0.045
	MBR	220	11.8	490	254	104	0	7.4	19	0	3.2	0.048
12	CTP	1000	2.5	1224	260	208	12.2	7.3	13	0	3.1	0.127
	MBR	220	5.2	576	260	95	0	7.2	12	0	3.1	0.131

EDC Recovery Rates in the Influent

The concentrations of NP, BPA and EE2 in the municipal wastewater were lower than the detection limits. After the addition of the model substances to the storage tank and 24 h homogenisation influent samples were taken, filtered and analysed. The measured concentrations (Tab. 2) were much lower than the added. This can be explained by the adsorption of the EDCs on the suspended solids in the influent (DM=0.1-0.2 g/L) and possible biological processes in the storage tank. For the calculation of elimination rates the measured, not the added, concentrations were used in order to avoid overestimation of the pilot plants performance.

Table 2. Added amount of the model substances and recovered concentrations in the influent, where C_O is added concentration; C_i is mean measured concentration in the filtered influent (number of samples n=3).

SRT [d]	25 (CTP)			25 (MBR)			12 (CTP, MBR)		
EDCs	NP	BPA	EE2	NP	BPA	EE2	NP	BPA	EE2
C_O [µg/L]	502	503	125	502	501	126	512	512	210
C_i [µg/L]	257	251	90	191	233	88	353	331	191

EDC Recovery Rates in the Effluent

In order to evaluate recoveries in the effluent, blank samples of CTP effluent and permeate were stabilised with sodium azide (0.01%), spiked with 50 µg/L of each model substance, shaken and analysed (three times). Obtained recoveries in the effluent and the permeate did not differ significantly and were 89-97% for NP, 84-94% for BPA and 81-98% for EE2.

EDC Elimination in the Pilot Plants

In general elimination of EDCs from wastewater can take place due to biodegradation, adsorption on sludge and volatilisation. The Henry's law constants (H) for BPA and EE2 at 25°C are very low: $4.3*10^{-6}$ $Pa*m^3*mol^{-1}$ (Cousins et al., 2002) and $3.7*10^{-7}$ $Pa*m^3*mol^{-1}$ accordingly. The latter is calculated from the vapour pressure of $P=6*10^{-9}$ Pa (Schweinfurth et al., 1996), solubility of $C=4.8$ g/m^3 (Tabak et al. 1981) and M=296 g/mol: H=P*M/C. These substances are considered to be low-volatile.

Henry's law constant for NP (25°C) is much higher: 11.02 $Pa*m^3*mol^{-1}$, air/water partitioning coefficient (25°C) is $4.65*10^{-3}$ m^3/m^3 (Hansen et al., 2002). By dissolved NP concentration of 250 µg/L, aeration of 2 m^3/h per 1 m^3 bioreactor volume and t=25°C, NP volatilisation can be approximated as 2.32 mg/h per 1 m^3 bioreactor volume or up to 35% of the daily NP load (C_i=250 µg/L). As during this experiment NP volatilisation could not be differentiated from biodegradation, difference between total NP elimination and output with sludge will be referred to as "biodegradation".

Most of the measured EDC concentrations in the effluents and the excess sludges were above the detection limits which allowed plausible calculation of total elimination rates (Fig. 4). Subdivision of the total elimination in biodegradation and output with excess sludge is shown in the Table 3.

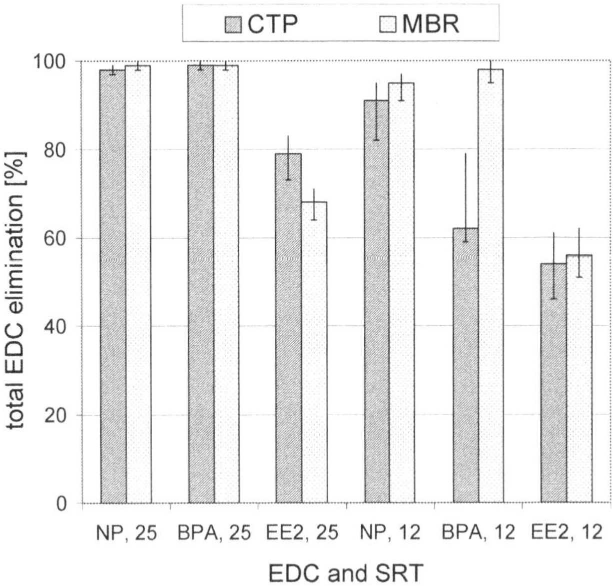

Figure 3 Total EDC elimination rates in the CTP and MBR by various SRT

BPA was almost completely biodegraded in both pilot plants by the SRT=25 days. This corresponds to the results obtained by Kreuzinger et al. (2003). By the lower SRT elimination rates decreased. BPA elimination in the CTP by the SRT=12 days was significantly lower than in the MBR by the same SRT. Better biodegradation in the MBR must have been the reason, because the output with sludge was insignificant, output with the solid phase in the effluent of the CTP was <0.1% and no BPA retention by the membranes occurred (tab. 4). Better biodegradation could have taken place due to a higher concentration of micro-organisms able to metabolise BPA in the MBR than in the CTP, but further studies are required to confirm it.

Table 3. EDC elimination rates.

SRT [d]	EDCs	Elimination in CTP [%]			Elimination in MBR [%]		
		Bio-degradation	Excess sludge	Total	Bio-degradation	Excess sludge	Total
25	NP	96	1.1	98	97	1.5	99
	BPA	99	0.1	99	99	0.1	99
	EE2	78	1.3	79	66	1.8	68
12	NP	86	4.6	91	91	4.1	95
	BPA	61	0.9	62	98	0.1	98
	EE2	52	2.1	54	54	2.0	56

For NP no considerable difference between the CTP and MBR performance was observed, although NP seems to be held back by the membranes to some extent (Tab. 4). NP adsorbed on the suspended solids in the effluent of the CTP (only 0.3% of the measured influent NP daily load) did not influence the elimination rate significantly. NP output with sludge was lower than 5% which means that even for this very hydrophobe compound biodegradation was the main eliminating mechanism.

Table 4. Mean EDC concentrations (number of samples n=5) in the supernatant of the MBR sludge and in the permeate

SRT		25 days			12 days	
EDCs	NP	BPA	EE2	NP	BPA	EE2
C in supernatant [μg/L]	3.8	2.7	26.4	24.8	6.1	80.4
C in permeate [μg/L]	2.1	3.0	28.3	17.4	5.9	84.1

Elimination of EE2 was less intensive than that of NP and BPA which corresponds to the literature (Paumann and Vetter, 2003; Kreuzinger et al., 2003). Output with sludge was around 2% of the daily influent load, retention by membranes did not take place. By the higher SRT EE2 was eliminated better than by the SRT=12 d. This can be ascribed to the difference between biological processes: by the higher SRT micro-organisms receive less organic matter and are forced to adjust themselves to degradation of persistent compounds. No advantage of MBR for EE2 elimination was noticed. By the SRT=25 days EE2 was better eliminated in the CTP than in the MBR. The higher mean temperature in the CTP by the SRT=25 days that could lead to the 10% difference in degradation rate. By the SRT=12 days elimination rates in both plants were equal.

Conclusions

Contrariwise to the widespread assumptions, MBR did not show significantly better EDC elimination rates. No considerable retention of the model substances by the membranes took place. Elimination of the studied EDCs in both pilot plants must have occurred mostly due to biodegradation. Up to 35% NP elimination could take place because of volatilisation. The advantage of the higher SRT was significant. Differences in the BPA behaviour in the MBR and the CTP must be further investigated.

The amount of EDCs carried out with excess sludge is fairly low, but as the results of this study show, by the NP effluent concentration of 3.8 μg/L concentration in sludge can be over 13000 μg/kgDM. In case of agricultural application of sewage sludge local NP concentrations can cause negative effects. The remaining concentrations in the treated wastewater are not to underestimate either, because EDCs can act additive (Kortenkamp *et al.*, 2003). For the enhancement of EDCs elimination SRT must be increased and/or further treatment steps like chemical oxidation, nano-filtration and activated carbon filtration of the effluent must be applied. Contaminated sludge must be incinerated.

In MBR higher SRT can be reached just by the increasing of activated sludge content. For the same procedure in a CTP costly extension of aeration and sedimentation tanks would be necessary, because the concentration of activated sludge in CTP is limited by the sedimentation processes. Furthermore, free from suspended matter permeate does not need to be pre-treated (in contrast to CTP effluent) before ozonation or nano-filtration. Thus, MBR is more flexible than CTP and still might have some advantages concerning the elimination of EDCs.

Acknowledgement

We thank technical personnel of the Institute of Environmental Engineering for support by the experiment set-up and Environmental Laboratory of the Institute of Environmental Engineering as well as Ms M. Meindorf (Institute of Hygiene and Environmental Health) for assistance in routine analytic. The study was financed by the DFG (German Research Foundation) as a part of postgraduate program AGEESA.

References

Andersen H., Siegrist H., Halling-Sorensen B. and Ternes T. (2003). Fate of estrogens in a municipal sewage treatment plant. *Environ. Sci. Technol.*, **37**(18), 4021-6.

Cousins I. T., Staples C. A., Klecka G. M. and Mackay D. (2002). A multimedia assessment of the environmental fate of bisphenol A. *Human and Ecological Risk Assessment* **8**(5), 1107-1135.

DEV (2002). Deutsche Einheitsverfahren zur Wasser-, Abwasser- und Schlammuntersuchung. Wiley-WCH Verlag, Berlin 2002.

Hansen B.G., Munn S.J., Pakalin S., Heidorn C., Allanou R., Pellegrini G., Anfossi V., Berthault F., Naughton A., Praderio L., Loonen H. (2002). European Union Risk Assessment Report. 4-Nonylphenol (Branched) and Nonylphenol. CAS No. 84852-15-3 and 25154-52-3. EINECS No. 284-325-5 and 246-67-0 - EUR 20387/EN.

Holbrook R.D., Novak J. T., Grizzard T. J., and Love N. G. (2002). Estrogen receptor agonist fate during wastewater and biosolids treatment processes: a mass balance analysis. *Environ. Sci. Technol.*, **36**(21): 4533-4539.

Jobling S., Beresford N., Nolan M., Rodgers-Gray T., Brighty G. C., Sumpter J. P. and Tyler C. R. (2002). Altered sexual maturation and gamete production in wild roach (Rutilus rutilus) living in rivers that receive treated sewage effluents. *Biol. Reprod.* **66**(2): 272-281.

Johnson A. C. and Sumpter J. P. (2001). Removal of endocrine-disrupting chemicals in activated sludge treatment works. *Environ. Sci. Technol.*, **35**: 4697-4703.

Kortenkamp A., Rajapakse N. and Silva E. (2003). Something from nothing? – combination of multi-component mixtures of estrogenic chemicals. Oral presentation at SETAC Europe meeting in York, UK , 31.03-1.04.2003.

Kreuzinger N., Clara M., Strenn B. and Kroiss H. (2003). Relevance of the sludge retention time (SRT) as design criteria for waste water treatment plants for the removal of endocrine disrupters and pharmaceuticals from waste water. In: Schröder, H. Fr. [Edt.]: Ecohazard 2003, proceedings of the 4th IWA specialized conference on assessment and control of hazardous substances in water. 14-17.09.2003, Aachen, Germany.

Kroiß H., Strenn B., Clara M., Kreuzinger N. (2003). Verhalten von bestimmten Arzneimittelrückständen, Industrie- und Umweltchemikalien in Membranbioreaktoren (VALIUM). Forschungsbericht, Dezember 2003. TU Wien, Institut für Wassergüte und Abfallwirtschaft.

Meesters R. J. W. and Schröder H. F. (2002). Simultaneous determination of 4-nonylphenol and bisphenol A in sewage sludge. *Anal. Chem.* 74, 3566-3574

Paumann R. and Vetter S. (2003). Hormonwirksame Stoffe in Österreichs Gewässern – ein Risiko? ARCEM Endbericht, Modul IV, Risikomanagement. Wien. (www.arcem.at)

Schäfer A., Nghiem L., Gallenkemper M., Wintgens T. and Melin T. (2003). Rückhalt hormonell wirksamer Spurenschadstoffe durch Membranverfahren. In Membrantechnik in der Wasseraufbereitung und Abwasserbehandlung. 5. Aachener Tagung Siedlungswasserwirtschaft und Verfahrenstechnik. 2003. Aachen, Germany: Institut für Verfahrenstechnik und Institut für Siedlungswasserwirtschaft der RWTH Aachen.

Schweinfurth, H., Länge, R. and Gunzel, P. (1996). Environmental fate and ecological effects of steroidal estrogens. IBC Conference Proceedings of "Oestrogenic chemicals in the Environment", London, 9-10 May 1996.

Siegrist H., Joss. A., Adler A., McArdell-Bürgisser C., Göbel A., Keller E. and Ternes T. (2003). Microverunreinigungen – Abwasserentsorgung vor neuen Anforderungen? *EAWAG news* 57d.

Tabak H. H., Bloomhuff R. N. and Bunch R. L. (1981). Steroid hormones as water pollutants. II. Studies on the persistence and stability of natural urinary and synthetic ovulation-inhibiting hormones in untreated and treated wastewaters. *Developments in Industrial Microbiology,* 22, 497-514.

Zühlke S., Dünnbier U., Gnirss R., Lesjean B., Buisson H., Recherche A. and Water V. (2003). Langzeituntersuchungen zur Entfernung organischer Spurenstoffe mit zwei Membranbelebungsanlagen im Vergleich zu einem konventionellem Klärwerk. In Membrantechnik in der Wasseraufbereitung und Abwasserbehandlung. 5. Aachener Tagung Siedlungswasserwirtschaft und Verfahrenstechnik. 2003. Aachen, Germany: Institut für Verfahrenstechnik und Institut für Siedlungswasserwirtschaft der RWTH Aachen.

Electrochemical regeneration of a carbon based adsorbent as a process for the removal of organic compounds from wastewaters

N.W.Brown*, E.P.L. Roberts*, A.A. Garforth*, R.A.W. Dryfe**

*Department. of Chemical Engineering, UMIST, PO Box 88, Manchester, M60 1QD
**Department of Chemistry, UMIST, PO Box 88, Manchester, M60 1QD

Abstract This paper describes a process for the adsorption of organics onto a non-porous adsorbent material and its subsequent electrochemical regeneration using a three stage operation. The use of a non-porous adsorbent results in quick adsorption and regeneration at the expense of low adsorptive capacity. The generation of chlorinated hydrocarbons from the use of sodium chloride electrolytes within the electrochemical cell is not found to be a problem as the adsorbent preferentially adsorbes chlorinated phenols. 100% regeneration can be achieved by 10 minutes electrochemical treatment using a current density of 22 mA cm^{-2} and passing a charge of 25 C g^{-1}. The energy consumption of this process is found to be in the range of 6.7 to 24.8 kWh kg COD^{-1}. Current efficiencies greater than 100% have been found, possibly due to the creation of additional adsorption sites and the removal of COD through the generation of solids.

Keywords Adsorption, Electrochemical regeneration, Nyex, Phenol .

Introduction

Industries are under increasing pressure to reduce the levels of toxic organics in their effluent, which has led to significant academic and industrial interest. The most cost effective processes for the removal of organics are usually biological, however for toxic and non-biodegradable organics these processes are not viable and other physical and/or chemical methods are used. These include chemical and electrochemical oxidation, coagulation and flocculation, membrane and adsorption processes. For the removal of low concentrations of these compounds adsorption processes are attractive as very low discharge consents can be achieved.

The conventional approach is to select or develop highly porous adsorbents with a high capacity (for example activated carbons) that are excellent adsorbents, but require complex and costly regeneration. The most significant adsorbent in wastewater treatment is activated carbon which can achieve high adsorptive capacities. However once it has been exhausted it must be disposed of (by landfill or incineration) or regenerated. Regeneration is widely used as it often represents the most commercially viable and environmentally acceptable option (San Miguel, et al., 2001). Industrially thermal regeneration is most often used. However this is a high energy, high cost process which results in material losses of around 5-10% and, except for large users, off-site regeneration (Narbaitz and Cen, 1994). Therefore research into the development of alternative adsorption processes has been undertaken.

© 2004 IWA Publishing. *2nd IWA Leading-Edge Conference on Water and Wastewater Treatment Technologies* edited by Mark Van Loosdrecht & Jonathan Clement. ISBN: 1 84339 508 8.

Two approaches have been widely investigated to overcome the problems associated with the use of activated carbon and its thermal regenerated. One approach investigates low cost adsorbents that can be used on a once through basis (Sivaraj, et al., 2001, Namasivayam, 2003), although this merely transfers the pollutant from the liquid to the solid phase. The second approach investigates different techniques to regenerate activated carbon. Methods investigated include chemical or solvent (Leng and Pinto, 1996, Streat and Horner, 2000), microbial (Hutchinson and Robinson, 1990), ultrasonic (Zhang, 2003), wet air oxidation (Mundale, et al., 1991) and electrochemical.

The first reference to the electrochemical regeneration of activated carbon adsorbents was Owen and Barry (1972), who achieved regeneration efficiencies of 61% and concluded that the process merited further investigation. Since that time a number of researchers (Sveshnikova, et al., 1987, Narbaitz and Cen, 1994, Zhang, 2002) have investigated the regeneration of activated carbon, with Canizares et al (2004) recently describing a combined adsorption/electrochemical regeneration device for the treatment of aqueous phenolic wastes. Their system operates in batch mode with a period of activated carbon adsorption, followed by electrochemical regeneration, during which the adsorbed organics are removed from the system and converted into carbon dioxide and electrocoagulated solids.

This paper introduces a different approach to adsorption processes using a non-porous adsorbent material which can be rapidly and easily regenerated. The adsorbent material, Nyex 100, being used in this work has a high electrical conductivity allowing effective electrochemical regeneration. The use of non-porous adsorbents significantly increases the rate of adsorption as it eliminates intra-particle diffusion (often the rate limiting step). This would also facilitate electrochemical regeneration as the electrochemical effects are believed to be restricted to the external surface of the carbon (Narbaitz and Cen, 1994).

Previous work by the authors has shown that the adsorption process is very quick with up to 88% of the adsorptive capacity being utilised within 2 minutes, with equilibrium in less than 30 minutes (Brown, et al., 2004b). Low discharge consents can be achieved with, for example, Atrazine being removal to below 1 $\mu g\ l^{-1}$ (Brown, et al., 2004a).

Materials

Nyex 100 was supplied by Nykin Developments in the form of a wet powder with an average dry solids content of 50%. Nyex 100 is a low cost, non-porous carbon powder, that has been treated by a proprietary process. It has a carbon content of over 95% with a mean particle diameter of 124 microns and a range of 10-600 microns. The BET surface area of the particles was determined by nitrogen adsorption and was found to be 2.75 $m^2\ g^{-1}$. This is very low compared with an activated carbon with surface areas of up to 2,000 $m^2\ g^{-1}$. Mercury porosimetry indicated that there were virtually no internal pores in the material explaining the low surface area obtained. All other chemicals were supplied as analytical grade by Aldrich.

Methods

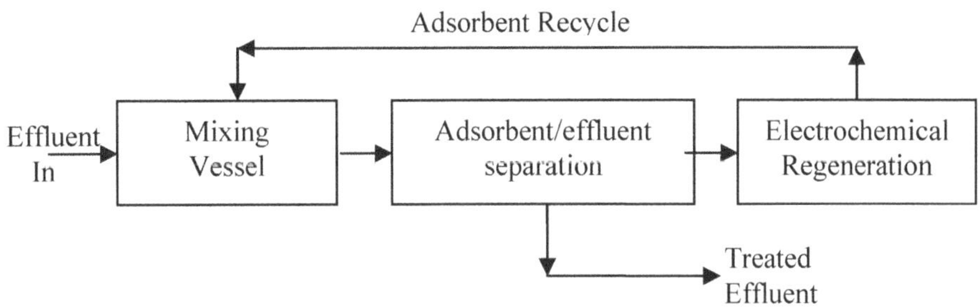

Figure 1 – Schematic of proposed adsorption/electrochemical regeneration process for the treatment of wastewater

The proposed process is shown in Figure 1. It comprises a three stage process of adsorption, solid/liquid separation and electrochemical regeneration. In the laboratory experiments reported in this work, adsorption occurs by mixing the adsorbent and effluent in a 250ml volumetric flask for 30 minutes, with vacuum filtration being used to separate the adsorbent from the liquid phase. Electrochemical regeneration is achieved in a simple electrochemical cell (Figure 2).

Adsorption Studies

Adsorption isotherms were prepared by adding various known weights of adsorbent to 100 ml of 100 mg l^{-1} phenol/chlorinated phenol solutions in a 250 ml flask and mixing for 30 minutes (a time shown to achieve equilibrium). After adsorption the solution was vacuum filtered (Whatman GF/C) and analysed. Since the regenerated Nyex 100 was to be reused straight from the anodic compartment of the electrochemical cell without any further treatment, it was confirmed that the presence of sodium chloride and the acid conditions (pH 3.5) did not effect phenol adsorption.

Electrochemical Regeneration

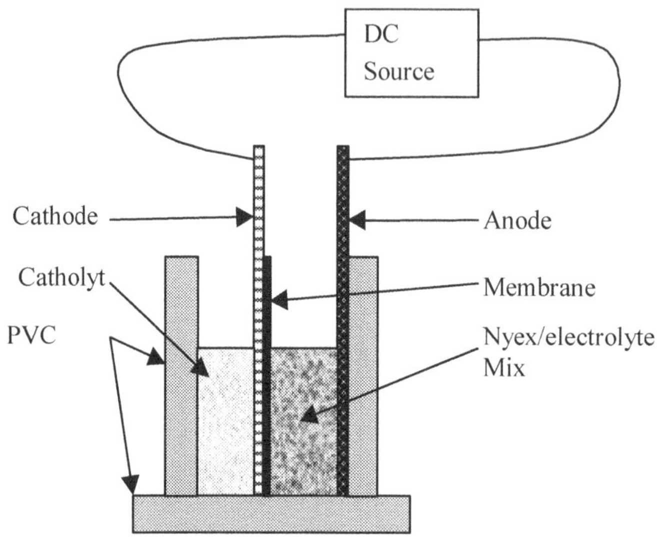

Figure 2 – Schematic of electrochemical cell

This cell was divided into anodic and cathodic compartments by a microporous polypropylene Daramic 350 membrane. The cathode was perforated 316 stainless steel and the anode was a mixed metal oxide coated titanium (supplied by Electrode Products Technology Ltd). The active area of the electrodes was 22.5 cm^2, with anode and cathode compartments being 10 mm wide. The electrolyte in the cathode compartment was a 2% w/w NaCl solution (except for the trial using sodium sulphate as electrolyte when 2% sodium sulphate solution was used). DC power was supplied by a Thurlby Thandar Instruments power source (PL320QMD). The procedure to determine the regeneration efficiency was:

(i) Initial adsorption – 25g of wet Nyex 100 was added to 100 ml of 100 mg l^{-1} phenol solution in a 250ml flask and mixed for 30 minutes. After adsorption the flask contents were vacuum filtered (Whatman GF/C) and the phenol concentration of the filtrate determined.

(ii) Electrochemical regeneration – The loaded, filtered (but still wet) adsorbent was pressed into the anode compartment. If sodium salts were to be added they were mixed with the adsorbent prior to filling the anode compartment (Sodium salts were added on a wt/wt basis of the wet adsorbent, with the adsorbent mix in the anode compartment being approximately 50% dry solids). A DC current in the range 100 – 1,000 mA was applied for regeneration times in the range 2 - 10 minutes. There was no flow in the cell and the only mixing was due to the gas bubbles produced at the electrodes.

(iii) Re-Adsorption - The full contents of the anode compartment were transferred to a 250 ml and adsorption was undertaken using identical conditions to the initial step.

For the adsorption/electrochemical regeneration over a number of cycles, steps (ii) and (iii) were repeated.

The adsorptive capacity, q, defined as the mass of adsorbate per gram of adsorbent (mg g^{-1}), can be calculated from the initial (C_o) and final (C_f) phenol concentrations according to equation 1.

$$q = [(Co - Cf)V] / m_d \qquad (1)$$

where V is the volume of solution used and m_d is the dry weight of Nyex 100.

From the initial and regenerated adsorptive capacities the regeneration efficiency, *RE,* was calculated. This was defined as:

$$RE = q_r/q_i \times 100 \qquad (2)$$

Where q_r is the adsorptive capacity of the regenerated material and q_i is the initial adsorptive capacity of that same sample of fresh Nyex 100 under identical adsorption conditions.

The current efficiency was calculated from:

$$CE = [(COD_i - COD_f)\ 8It]\ FV \qquad (3)$$

where COD_i and COD_f are the chemical oxygen demands (g O_2 l^{-1}) of the initial solution and filtrate respectively, F is the Faraday constant (96,487 C mol^{-1}), V is the volume of solution being treated (l), I is the current (A) and t is the regeneration time (secs). In this case the COD removed during the adsorption cycle is compared to the electrochemical regenerating conditions during the previous regeneration. (Note that the first adsorption cycle always uses fresh adsorbent).

Analysis

Phenol and chlorinated phenolic compounds were determined using GC using a Varian CP 3800 instrument fitted with a Varian VF-1ms column and a FID detector.

Results and Discussion

In this process one of the most important considerations is the recycling of the adsorbent particles back into the process. Hence the effect of varying the charge passed (0 – 25 C g^{-1}) over a number of cycles has been investigated (Figure 3). The current was kept constant at 0.5 amps (current density 22 mA cm^{-2}) and the regeneration time varied from 0 – 10 minutes. 100% regeneration of the Nyex required a charge of around 25 C g^{-1}, although a charge of 12.5 C g-1 achieved nearly 90% regeneration. As would be expected from Figure 3, the current efficiency is higher during the initial period of regeneration (Figure 4). Two interesting points that are highlighted in Figure 4 are that current efficiencies over 100% can be achieved and that the current efficiency falls over a number of cycles. These suggest that anodic oxidation is not the only process that is occurring. Other process that are believed to be causing these results include:

(I) Adsorption onto some of the sites that were not occupied during the first adsorption cycle. As Figure 3 shows, adsorption onto loaded material can still occur resulting in an apparent increase in the current efficiency.

(II) Electrochemical regeneration could be modifying the surface of the adsorbent giving an increase in the adsorptive capacity. This was shown by electrochemically treating fresh adsorbent prior to loading. This gave an increase in the adsorptive capacity of 21.5%.

(III) Solid particles could be formed resulting in the removal of COD. Various workers, for example Gattrell and Kirk (1993), investigating the electrochemical oxidation of phenol at electrodes have noted that polymerised phenolic compounds can be produced (often fouling the electrode and causing a reduction in the electrode performance). Canizares (2004) found that electrochemical regeneration of activated carbon using a steel anode and a stainless steel cathode resulted in the generation of electrocoagulated solids, possibly due to the generation of ferric ions from the anode.

All of these are believed to be factors influencing adsorption after electrochemical treatment and it will be necessary to run the process for a significantly greater number of cycles to determine how far the reduction in current efficiency falls.

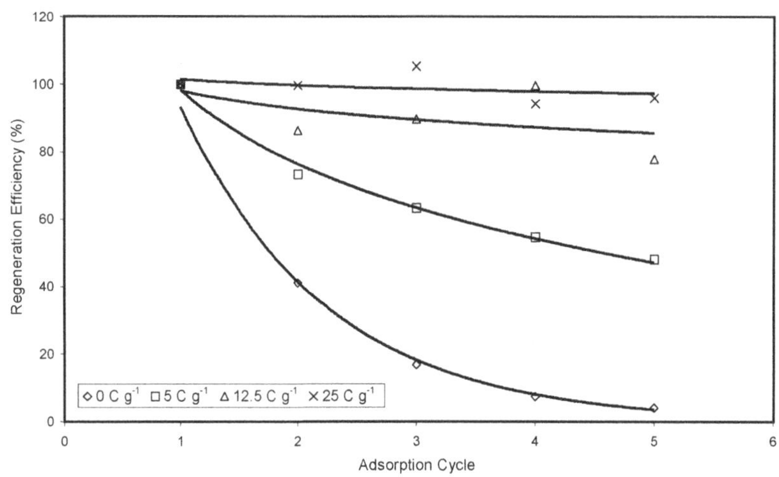

Figure 3 – The regeneration efficiency at different regeneration charges over a number of adsorption/regeneration cycles. Cycle 1 is fresh adsorbent. Note even when no charge is passes (0 C g^{-1}) there is still additional adsorption giving an apparent regeneration efficiency. Regeneration current is 0.5 A.

In this case there appears to be no fouling of the electrode particles with polymerised phenolics as there was little change in the cell voltage over the cycles. However as the particles have a high surface area (compared to a flat electrode) the coating coverage may be so small that a large number of cycles is required for noticeable effect. Alternatively free solid particles may be being produced and may build up within the bed of particles. Whether this is a cause for concern will require further work.

The above work was undertaken without the addition of sodium chloride to the adsorbent mix in the anodic compartment. However previous work (Narbaitz and Cen, 1994, Zhang, 2002, Brown, et al., 2004b) had suggested that the addition of electrolyte to the cake within the anode compartment resulted in an increase in the regeneration efficiency. In the current work regenerating adsorbent loaded with phenol (Figure 5) the presence of 0.1% sodium chloride electrolyte gave an increase in regeneration efficiency. However there was no benefit in increasing this level, although the data is limited. This work also confirms their findings that sodium chloride results in higher regeneration efficiencies than sodium sulphate. This is beneficial as the sodium chloride is a cheaper raw material, causes a greater reduction in cell potential and is easier to dissolve.

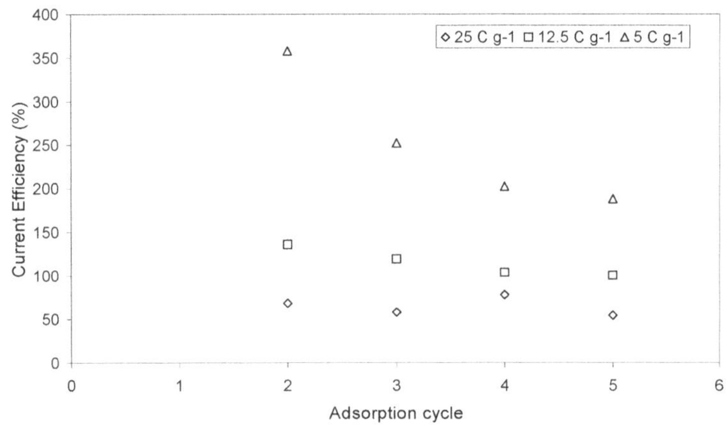

Figure 4 – The current efficiency over a number of adsorption/regeneration cycles for different charge passed. Cycle 1 is fresh adsorbent and COD removed during the adsorption cycle is compared with the charge passed during the preceding regeneration cycle. Regeneration current is 0.5A

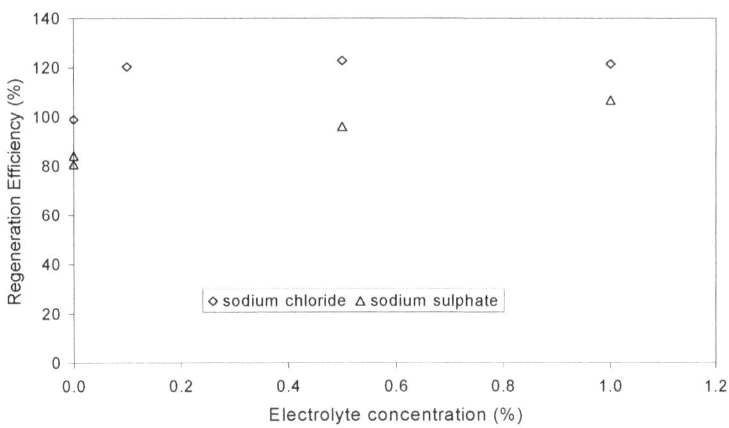

Figure 5 - The effect of the electrolyte composition and concentration on the regeneration efficiency using freshly loaded Nyex 100. Regeneration conditions are 0.5 A for 5 minutes giving a charge passed of 12.5 C g^{-1}.

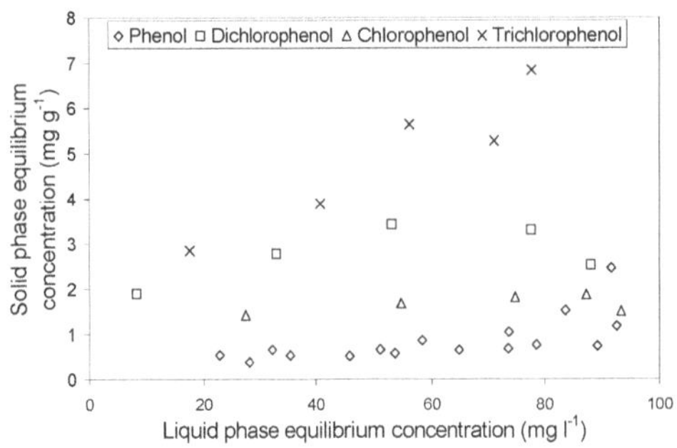

Figure 6 Adsorption Isotherms for phenol and chlorinated phenols onto Nyex 100.

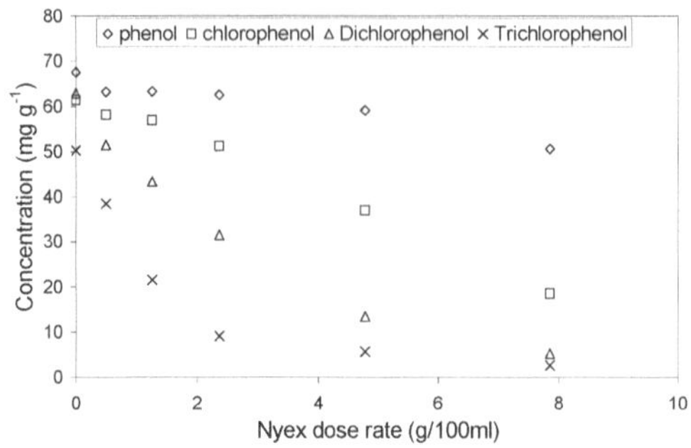

Figure 7 Adsorption of phenol/chlorinated phenols at different Nyex additions.

However the presence of sodium chloride during the electrochemical treatment of wastewater containing phenol has been shown to result in the production of chlorinated compounds (Comninellis and Nerini, 1995, Korbahti, et al., 2002). This is a potentially significant problem as electrochemical breakdown products may be more toxic than the original phenol. Hence the presence of chlorinated phenol compounds was investigated. However on analysing the filtrates for mono-, di- and tri-chlorinated phenols, only trace levels of mono-chlorophenols were found in a couple of the samples after regeneration. This was observed even when the charge passed was less than that required for 100% regeneration. In order to investigate why so few chlorinated compounds were found adsorption isotherms were prepared for these compounds. These isotherms (Figure 6) show that increased phenol chlorination resulted in increased adsorptive capacity. The preferential adsorption of the chlorinated compounds is shown in Figure 7 where a solution containing approximately equal quantities of phenol and mono-, di- and tri-chlorinated phenols was mixed with various quantities of Nyex. This demonstrates that there is preferential adsorption of tri-chlorinated phenol, followed by the less chlorinated phenols and then phenol. Hence any chlorinated compounds produced appear to remain preferentially bound to the adsorbent surface.

A trial was undertaken to see if washing the regenerated adsorbent (Charge passed 12.5 C g^{-1}) in deionised water would release the chlorinated compounds. No chlorinated compounds were detected.

The effect of current density on regeneration was investigated. This shows that whilst there is no major effect, there is a gradual decrease in regeneration efficiency with increasing current density under conditions of constant charge passed. This is similar to previous data reported by the authors (Brown, et al., 2004b).

The energy consumption required to remove a kg of COD was calculated from the energy used for regeneration and the adsorptive capacity of the adsorbent in the subsequent adsorptive cycle. The cell potential was found to be approximately 4 volts for all trials when a current of 0.5 A was passed and no sodium salts were added. This calculation shows that energy consumption is from 6.7 to 24.8 kWh kg COD^{-1}. This compares with an energy figure of 84.03 kWh kg COD^{-1} for the 80% regeneration of activated carbon loaded with phenol (Canizares, et al., 2004), 12.9 to 73.4 kWh kg COD^{-1} for the removal of COD from domestic wastewater at Ti/Pt anode (Vlyssides, et al., 2002) and 30 – 50 kWh kg COD^{-1} at highly doped SnO2 anodes (Stucki, et al., 1991).

Conclusions

The feasibility of this combined approach of phenol adsorption onto a non-porous adsorbent and electrochemical regeneration has been demonstrated by the results presented in this paper over a number of cycles. Electrochemical regeneration of the adsorbent has successfully recovered 100% of the adsorptive capacity using a current of 0.5 A for 10 minutes giving a charge passed of 25 C g^{-1}. This gives a current efficiency of over 100% as additional adsorption is occurring and COD is being removed as solid particles that are believed to be generated. Energy consumption of 6.7 – 24.8 kWh kg COD^{-1} has been calculated which is lower than competitive electrochemical processes. However a reduction in the current efficiency over succeeding cycles needs further investigation. Whilst chlorinated phenolic compounds are probably being produced during regeneration, they appear to remain preferentially adsorbed on the Nyex during subsequent adsorption stages. This makes it less likely that more toxic chlorinated compounds will be released into the environment. Such a process is likely to be viable for treating low concentrations of organic pollutants with on-site regeneration of the adsorbent.

Acknowledgements

The authors would like to acknowledge the financial and material support received from the Engineering and Physical Sciences Research Council (EPSRC), Severn Trent Water Ltd., Nykin Developments and Electrode Products Technology Ltd.

References

Brown N. W., Roberts, E. P. L., Chasiotis, A., Cherdron, T. and Sanjhrajka, N., (2004a) Atrazine removal using adsorption and electrochemical regeneration, Submitted to Water Research.

Brown N. W., Roberts, E. P. L., Garforth, A. A. and Dryfe, R. A. W., (2004b) Electrochemical Regeneration of a carbon based adsorbent loaded with crystal violet dye., Submitted to Electrochimica Acta.

Canizares P., Lobato, J., Garcia-Gomez, J. and Rodrigo, M. A., (2004) Combined adsorption and electrochemical processes for the treatment of acetic aqueous phenol wastes., Journal of Applied Electrochemistry. 34, 111-117.

Comninellis C. and Nerini, A., (1995) Anodic oxidation of phenol in the presence of NaCl for wastewater treatment, Journal of Applied Electrochemistry. 25, (1),23-28.

Gattrell M. and Kirk, D. W., (1993) A Study of Electrode Passivation During Aqueous Phenol Electrolysis, Journal of the Electrochemical Society. 140, (4),903-911.

Hutchinson D. H. and Robinson, C. W., (1990) A Microbial Regeneration Process for Antigranulocytes Activated Carbon .2. Regeneration Studies, Water Research. 24, (10),1217-1223.

Korbahti B. K., Salih, B. and Tanyolac, A., (2002) Electrochemical conversion of phenolic wastewater on carbon electrodes in the presence of NaCl, Journal of Chemical Technology and Biotechnology. 77, (1),70-76.

Leng C. C. and Pinto, N. G., (1996) An investigation of the mechanisms of chemical regeneration of activated carbon, Industrial & Engineering Chemistry Research. 35, (6),2024-2031.

Mundale V. D., Joglekar, H. S., Kalam, A. and Joshi, J. B., (1991) Regeneration of Spent Activated Carbon by Wet Air Oxidation, Canadian Journal of Chemical Engineering. 69, (5),1149-1159.

Namasivayam C., and Kavitha, D., (2003) Adsorptive removal of 2-chlorophenol by low cost coir pith carbon., Journal of Hazardous Materials. B98, 257-274.

Narbaitz R. M. and Cen, J. Q., (1994) Electrochemical Regeneration of Granular Activated Carbon, Water Research. 28, (8),1771-1778.

Owen P. H., and Barry, J.P., (1972), Electrochemical carbon regeneration., PB 239156, Environics Inc., Huntington Beach, California.

San Miguel G., Lambert, S. D. and Graham, N. J. D., (2001) The regeneration of field-spent granular-activated carbons, Water Research. 35, (11),2740-2748.

Sivaraj R., Namasivayam, C. and Kadirvelu, K., (2001) Orange peel as an adsorbent in the removal of Acid violet 17 (acid dye) from aqueous solutions, Waste Management. 21, (1),105-110.

Streat M. and Horner, D. J., (2000) Adsorption of highly soluble herbicides from water using activated carbon and hypercrosslinked polymers, Process Safety and Environmental Protection. 78, (B5),363-382.

Stucki S., Kotz, R., Carcer, B. and Suter, W., (1991) Electrochemical Waste-Water Treatment Using High Overvoltage Anodes .2. Anode Performance and Applications, Journal of Applied Electrochemistry. 21, (2),99-104.

Sveshnikova D. A., Ramazanov, A. K., Aliev, Z. M., Shakhnazarov, T. A., Kamalutdinova, I. A. and Babaev, M. E., (1987) Electrochemical regeneration of activated carbons, Soviet Journal of Water Chemistry and Technology. 9, (5),115-117.

Vlyssides A., Karlis, P. K., Rori, N. and Zorpas, A. A., (2002) Electrochemical treatment in relation to pH of domestic wastewater using Ti/Pt electrodes., Journal of Hazardous Materials. B95, 215-226.

Zhang H. P., (2002) Regeneration of exhausted activated carbon by electrochemical method, Chemical Engineering Journal. 85, (1),81-85.

Zhang H. P., Wang, S. and Liu, Z., (2003) Ultrasonic regeneration of granular activated carbon., Environmental Engineering Science. 20, (1),57-64.

Ozone and UV processes for additional wastewater treatment to remove pharmaceuticals and EDCs

A.Ried, J. Mielcke, M. Kampmann

WEDECO Umwelttechnologie GmbH, Boschstr. 6, 32051 Herford, Germany
Tel.: +49 5221 930-144 Fax: +49 5221 33834 E-mail: achim.ried@wedeco.net
Thomas A. Ternes
Bundesanstalt für Gewässerkunde, Koblenz, Germany E-mail: ternes@bafg.de
Bernhard Teiser
Abwasserverband Braunschweig, Braunschweig, Germany

Abstract Recycling of polluted water is limited in principle by possible contamination by persistent substances, germs, odors and/or colors that cannot be degraded in municipal sewage works using established clarification technology.

In areas with scarce ground water resources, though, it may be rational to treat waste from municipal sewage works to allow the water to be used for example for irrigation purposes. Similarly if the waste is discharged into bathing water, standards applicable to germ reduction in the treated waste water must be met.

Substances that are not or hardly biodegradable are referred to as persistent substances. These may be of various anthropogenic origins. The main compounds/areas of origin discussed in this connection include industrial chemicals (e.g. nonyl phenol, organotin compounds, PCB, phthalates), pesticides (e.g. DDT, Gammexane), pharmaceuticals and cosmetics.

Public debate is currently focusing on what are called endocrine substances. Pilot trials, using the combined treatment technique ozone/UV, were carried out at an municipal waste water treatment plant. The results show a positive treatment effect by ozonation. The comparison of the treatment techniques "membrane, Ozone/UV and AOP`s demonstrates advantages for Ozone/UV in regard to cost and operation expenses.

Keywords Endocrine disrupters, pharmaceuticals, waste water treatment, Ozone, UV, AOP

Introduction

Public debate is currently focusing on what are called endocrine substances. The term *endocrine* denotes substances that affect the hormone system [Seibert,1996]. A summary of these compounds is given in Figure 1. Negative effects on the hormone system of fish in surface water have been reported [Schlumpf,1996]. Effects such as reduced fertility of fish are to be observed in particular in the area of influence of sewage plant outlets. Possible effects on humans are still the subject of controversy. Aspects under discussion include general reduction of fertility, impaired development of sexual organs in young males and greater incidence of tumors [DVGW information,1998].

There are various pathways for the compounds under discussion. Our water cycle merits special attention. Substances that enter the water cycle via waste water will, if clarification is inadequate, ultimately make their way into the potable water – our main foodstuff. A number

© 2004 IWA Publishing. *2nd IWA Leading-Edge Conference on Water and Wastewater Treatment Technologies* edited by Mark Van Loosdrecht & Jonathan Clement. ISBN: 1 84339 508 8.

of anthropogenic substances enter the water cycle via municipal and industrial waste water and landfill leachate which cannot be degraded by the natural clarification mechanisms. Established clarification techniques in the form of sewage works (biology, filtration, flocculation/precipitation…) permit a large degree of clarification to specified discharge values. The persistent substances referred to above are not regularly/systematically dealt with, however.

It is also known that these substances are not or are only insufficiently degraded in conventional sewage plants [Ternes, 1999a,b]. They are thus continually entering the surface water, ground water and finally, potable water.

In the case of endocrine substances in particular, no limits currently apply. Nor are there any guidelines on possible treatment techniques. A three-year EU study (*POSEIDON*) has been launched to establish analytical methods, examine pathways and develop possible clarification techniques.

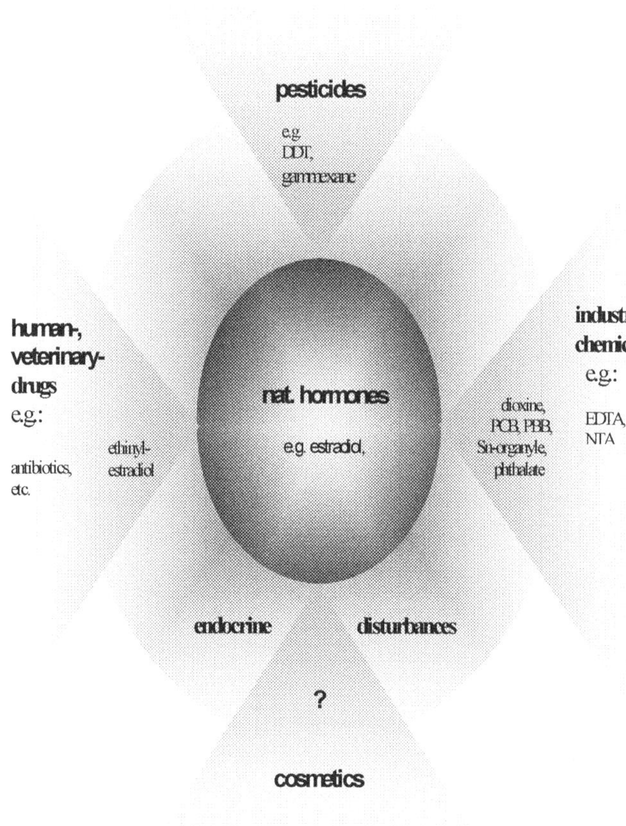

Figure 1 Which anthropogenic substances affect the hormone balance?

On the other hand, the requirements for disinfection of water for recycling have long been established by international guidelines.

EU Bathing Waters Directive (76/160/EC)	Total coliforms:	500 CFU/100ml (E-coli: 100 CFU/100ml)
WHO guidelines (1989)	Fecal coliforms:	1000 CFU/100ml
Californian Title 22 (1993)	Total coliforms:	2.2 MPN

Use of UV and/or Ozone Techniques for Wastewater Disinfection

It is desirable for waste water polluted by germs hazardous to health to undergo disinfection before that water is returned to the natural water cycle.

UV Technique

Ultraviolet radiation with a wavelength between 240 and 290 nm is able to penetrate pathogenic microorganisms (bacteria and viruses). In particular, the nucleic acid bases contained in the DNA absorb the high-energy UV radiation. This causes chemical changes as a result of which the microorganism is deprived of the ability to divide and reproduce. Consequently, the germ count in the water drops significantly. At the same time, reactions with other components of water are virtually absent.

The use of UV radiation to disinfect waste water is an inexpensive, reliable process that has been in use industrially for a long time.

According to international standards the dose necessary to disinfect waste water is between 300 and 400 J/m^2.

Ozone Technique

Ozone (O_3) is a very reactive form of elementary oxygen. Due to its high oxidation potential, ozone reacts quickly with organic components of water and also for example with the cell wall of microorganisms. Under this chemical attack, the germs concerned die. Ozone thus has a disinfectant effect.

The use of ozone for disinfection purposes is worthwhile especially where UV technique comes up against limiting factors due to unfavorable water properties (low UV transmission, high proportions of undissolved substances) and in cases where, besides disinfection, other treatment aims (such as deodorization and decolorizing) are to be attained [Graham,1999,Notornicola,1999].

Combination of Ozone and UV Technique for More Efficient Clarification

Waste Water Treatment – Principle Effects

In the case of discharge from municipal sewage works which is heavily polluted by an increased proportion of industrial discharges or unsatisfactory final clarification, the use of ozone can enable more efficient disinfection to be achieved. Such discharges are typically characterized by higher turbidity (levels of sludge and solids) or inherent coloration, leading to reduced transmission in the UV-c range. Such waste can be successfully treated with a combination of ozone and UV.

The ozone/UV process combination is one of the so-called *Advanced Oxidation Processes (AOP)*, which can further degrade toxic and hardly degradable substances such as AOX, HCX, PAK, dioxins, pesticides, etc.

Table 1 Principle effects of ozone and/or UV treatment

Treatment step	Ozone	ozone/UV	UV
Effect			
	Oxidation Disinfection	continuing oxidation Disinfection (radical reactions)	disinfection
Parameters	Persistent substances Colors, odors Bacteria, viruses, parasites	persistent substances bacteria viruses, parasites	bacteria viruses

The applicability of the ozone/UV process to polluted waste water, leachate and ground water treatment has been documented by extensive results from pilot studies and the operation of commercial-sized plants (leachate, ground water) [Ried,2000]. Degradability data are available in particular for the following substances / compounds: AOX, PAK, organotin compounds, pesticides, phenols, chlorinated aromatics and phthalates. It is also known that hardly biodegradable substances more readily undergo biodegradation following oxidative treatment (ozone/UV). Consequently, such after-treatment of waste water in sewage works discharges could help to make the residues remaining after ozone/UV treatment more accessible to the natural clarification mechanisms (receiving water, soil filtration).

An "activated" natural clarification process of this kind contributes to keeping our potable water resources virtually free from such compounds. The use of ozone/UV in potable water conditioning can provide additional security.

Pilot Trials MSP Braunschweig

The pilot tests were carried out at the municipal sewage treatment plant (MSP) in Braunschweig (city in Germany, near Berlin). This treatment plant is designed for 385.000 population equivalences and comprises different treatment steps with mechanical pre-treatment and biological treatment for carbon, nitrogen and phosporus reduction. Under normal conditions the flow rate is approximately 60.000 m^3/d (15,8 mgd).

MSP Braunschweig is a potential end user of the EU-project (POSEIDON). WEDECO as an additional potential industrial end user carried out pilot tests to investigate the treatment effect of ozone/UV- technique on the effluent of the MSP (Braunschweig).

This pilot plant comprises an ozone generator (100g/h), 2 diffusor / bubble columns and an UV-reactor and was installed at the facilities in Braunschweig. The pilot trials were done in September/ October 2001 and June/July 2002.

The chemical and physical parameters (conductivity, TOC, COD, BOD_5, AOX, filtrable solids, transmittance, SAC 254 nm and SAC 436 nm) of raw water are listed in table. 1.

Table 2 Analytical Results of Raw Water

test run 1		raw-water
pH-value		---
Conductivity	mS/cm	7,5
TOC	mg/l	23,0
COD	mg/l	30,0
BOD_5	mg/l	2,8
AOX	µg/l	100
Filterable solids	mg/l	4,5
Transmittance	% (254 nm)	59,2
SAC* 254 nm	m^{-1}	22,8
SAC* 436 nm	m^{-1}	1,8

* SAC = spectral absorption coefficient

The values of the raw water parameters are typical for this type of effluents in Germany.

TOC and COD stayed nearly constant during the ozone treatment, maybe a slight reduction of COD could be detected. The low BOD with 2,8 mg/L increased to approx. 5 mg/L. After a treatment with ozone a certain increase of the BOD is observed usually. The average of the total reduction of AOX was 50%.

Due to the ozone treatment a significant influence is noticed for the parameters transmittance, SAC 254 nm and SAC 436 nm. Both absorption coefficients decreased and consistent the UV transmittance was improved clearly from 59,2 % to 84,1 %. So the ozone treatment step leads to better conditions (higher transmittance) for the UV-disinfection step.

The analyzed organic trace pollutants (more than 50 compounds) belong to different classes such as antibiotics, beta-blockers, antiphlogistics, antiepileptics, estrogens, musk fragrances and iodinated contrast media (ICM), phytosteroids, flavonides.

All are frequently present in municipal waste water. The concentration of the single substances varied between 0,01 µg/L and 10 µg/L in the effluent of the municipal waste water treatment plant.

The applied ozone doses during the pilot trials varied between 5 mg/L and 15 mg/L. The results showed a very positive treatment effect by ozonation for most of the compounds with removal efficiencies of frequently higher than 90 %.

One highlight of the ozonation was the effective oxidation/degradation of the three major endocrine disrupters (17α-ethinylestradiol, 17β-estradiol and estrone), which probably lose most of their estrogenic potency. Thus, ozonation might drastically reduce estrogenic effects on fish caused by discharging treated municipal waste water into rivers and streams.

Furthermore, it can be predicted that the potential for the formation of resistant bacterial strains is lowered significantly because antibiotics were no longer detected in the ozonated waste water.

Only for iodinated contrast media (ICM) ozonation was inappropriate (removal 30-90%) and AOPs were conducted which gave only slightly higher oxidation quantities.

The pilot study displayed a high potential for the application of Ozone and Ozone based AOP's to remove trace pollutants (e.g. estrogens) from waste water.

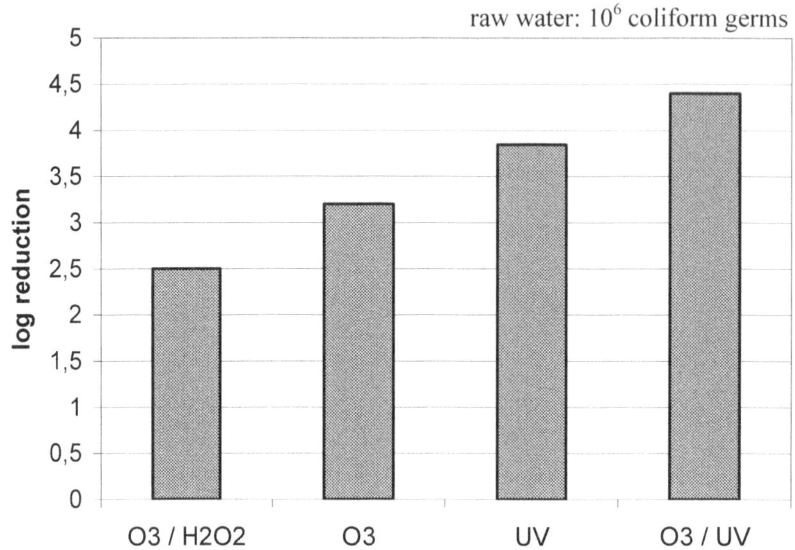

Figure 2 Comparison of disinfection results for E-coli (O_3 dosage: 10 mg/L, H_2O_2 dosage: 10 mg/L, UV dosage: 400 J/m²)

During ozonation and AOPs a log-reduction of up to 4 was achieved for measured microorganism such as clostridium perfringens, streptococci and E-coli. Thus in addition to the removal of persistent organic pollutants, the water is disinfected to a high degree [Ternes 2003].

What about Cost?

Beside the principle treatment effects, in practice the cost is a major factor for the selection of the most suitable process. The following calculations give an overview for the cost related to the use of the above processes. For comparison the pilot data from the case study at MSTP *Braunschweig* is used:

Effluent MSTP:	**Supposition for calculation:**
COD level [mg/l]: 30 – 50	Energy [€/kWh]: 0.07
SS [mg/l]: < 10	LOX [€/m³]: 0.10
UV(T) [%]: 50 – 60	H_2O_2 [€/kg]: 0.40

Operation conditions: Ozone: 10-15 g/m3; H2O2: 10 g/m3; UV(low pressure): 20 W/m3; UV(medium pressure): 2 KW/m3.

The data displayed derives from our assessments but might not represent every single case Nevertheless this data indicates a general tendency, as described in Fig. 3.

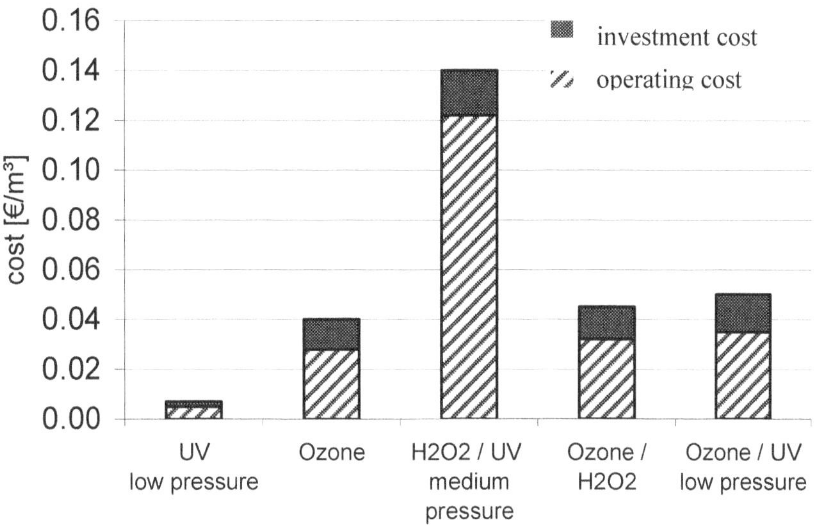

Figure 3 Comparison of investment and operating cost for UV, Ozone and AOP

The cost for a membrane treatment step will be in the range of 0.40 to 0.50 € (Engelhardt and Rothe, 2001). During membrane treatment the persistent substances are extracted from the effluent and concentrated. For that reason the waste from membrane treatment require further treatment.

AOP is an on-site process during which the relevant substances are oxidised and removed during the treatment. However, the possible risk of by-product formation needs to be considered.

For the economical use of AOP it is not possible to mineralise all persistent substances. The goal of this treatment is to reduce the toxicity of the substances and to increase their biodegradability. This latter effect can then be used either to enhance subsequent biofiltration or to improve removal through a natural biological means and resources.

In comparison to the use of membranes the cost for an AOP process is significantly lower.

What Is The Best Available Technique For Additional Wastewater Treatment?

The following matrix should be helpful to choose the best technique for a particular sample of waste water in relation to the treatment goal. The matrix is estimated at a pre-treated waste water (e.g. filtration, flocculation, biology), characterized by parameters as follows:

COD	50 – 150 mg/l
SS	10 – 50 mg/l
UV (T%)	50 – 70 %

Table 3 Comparison of treatment goal, available technique and operation cost

Treatment goal	**Available technique**	**Dosage range**	Operation cost **€ / m³**
Disinfection	UV (low pressure)	300 – 400 J/m²	0,0045 – 0.01
COD reduction	Ozone	50 – 150 g/m³	0.11 – 0.35
Color reduction	Ozone	10 – 50 g/m³	0.025 – 0.15
Reduction of persistent substances e.g. endocrine disruptors, antibiotics	Ozone	10 – 15 g/m³	0.025 – 0.035
Reduction of parasites e.g. threadworm	Ozone	10 – 15 g/m³	0.025 – 0.035

Conclusions

The main conclusions from these comparisons are:

- UV is an economic technique for disinfection.
- Ozone treatment achieves additional benefits, e.g. reduction of persistent substances, colour removal and COD reduction.
- In case a combined oxidation process (AOP) is needed, the combinations Ozone/H_2O_2 and Ozone/UV (low pressure) are more cost effective than the UV/H_2O_2 (medium pressure).
- For scale-up and process optimisation, pilot trials are essential to provide data for the best plant design.
- The goal of the oxidation process is to reduce the toxicity of the substances and increase their biodagradability
- The optimized combination of Ozone and UV achieves very economical and ecological treatment results regarding disinfection and removal of organic trace pollutants
- The comparison of the treatment techniques "membrane, Ozone / UV and AOPs" demonstrates advantages for ozone / UV in regard to cost and operating expenses.

Acknowledgements

The authors would like to thank the employees of the municipal sewage treatment plant, *Braunschweig* for permission to present data collected at the plant and to acknowledge the contributions made by colleagues involved in this study from the *POSEIDON* project (EU – EVK1-CT-2000-00047) and the *ESWE Institute*, *Wiesbaden/ BfG Koblenz.*

References

[i] DVGW Information "Relevanz von Arzneimitteln, hormonell wirksamen Substanzen und Kosmetika in Gewässern (Relevance of medicines, endocrine substances and cosmetics in waters)", Wasser-Information Nr. 54 (edition 6/98)1-6 (1998).

[i] Graham, N.;Parasheva,P (2001) "Recent Studies of Ozone Disinfection of secondary Municipal Effluents" Proceedings (London, UK ,15th ozone World Congress, 2001), Vol. 1, pp. 276-291.

[i] Notornicola, M. and Lopez A. "Ozone Disinfection for Municipal Waste Water Reuse in Agricultur", Proceedings (Dearborn, USA 1999), Vol. 1 pp. 65-79.

[i] Ried, A.; Mielcke, J.; Stapel, H. "Application of Ozone and its Combination with UV Technique for the advanced Treatment of Waste Water for Reuse", (Tunis WATRAMA conference, 2000).

- Schlumpf,M. and Lichtensteiner, W. (1996), Hormonaktive Xenobiotika – Störungen von Fortpflanzungs- und Entwicklungsprozessen; Z. Umweltchem. Ökotoxologie 8 (6) 321-333.
- Seibert,H. (1996), Störungen der Entwicklung und Funktion des männlichen Reproduktions-systems; UWSF-Z. Umweltchem. Ökotoxologie 8 (4) 275-284.
- Ternes, T.A.; Kreckel, P. and Müller J. "Behavior and Occurence of Estrogens in Municipal Sewage Treatment Plants (II. Aerobic batch experiments with activated sludge.)",Sci. Total Environ, (1999a), 225 pp. 91 – 99.
- Ternes, T.A.; Stumpf, M.; Müller, J; Haberer, K.; Wilken, R-D.; Servos, M. , "Behavior and Occurence of Estrogens in Municipal Sewage Treatment Plants (I. Investigations in Germany and Canada)" ,Sci. Total Environ, (1999b), 225 pp. 81 – 90.
- Ternes T., Stüber J., Herrmann N., McDowell D., Ried A., Kampmann M., Teiser B. (2003), Ozonation: A tool for removal of pharmaceuticals, contrast media and musk fragrances from wastewater; Water Research 37/8: 1976-1982.

Detection of estrogenic effect formation potential in chlorinated drinking water

S. Itoh*, Y. Yoshimura*, T. Okada* and Y. Tsujimura*

*Department of Urban Management, Kyoto University, Kyoto 606-8501, Japan
(E-mail: *itoh@urban.env.kyoto-u.ac.jp*)

Abstract The estrogenic effect of chlorinated water increased even without residual chlorine. The components that can be called "estrogenic effect formation potential" and "estrogenic effect intermediates" are defined in chlorinated humic acid solution. This understanding is very similar to that of the process of trihalomethanes (THMs) formation. An appropriate procedure to detect the estrogenic effect formation potential, which represents the maximum value of estrogenic effect, was demonstrated. On the other hand, organic matters of which estrogenic effect increase or decrease by chlorination are present in natural water. The important point is that the estrogenic effect of chlorinated drinking water is the sum of the increased activity and the decreased activity. An appropriate procedure that concentrates the increased estrogenic activity and the decreased activity in chlorinated drinking water simultaneously was established. The estrogenic effect formation potential of chlorinated Lake Biwa water was measured according to the procedure established in this study.

Keywords Endocrine disrupting chemicals; estrogenic effect; MVLN assay; chlorination; disinfection by-products

Introduction

The potential risk of endocrine disrupting chemicals (EDCs) in the aquatic environment has been an important issue also in waterworks. Hundreds of compounds are now listed as suspected EDCs. On the other hand, the Endocrine Disruptor Screening and Testing Advisory Committee (EDSTAC, 1998) has recommended in the final report that six kinds of mixtures in addition to single compounds be subjected to the assays to evaluate the potential for endocrine disruption. Disinfection by-products in drinking water are included in those mixtures listed.

© 2004 IWA Publishing. *2nd IWA Leading-Edge Conference on Water and Wastewater Treatment Technologies* edited by Mark Van Loosdrecht & Jonathan Clement. ISBN: 1 84339 508 8.

As a matter of fact, there are some researches and reviews (Waller *et al.*, 1998; International Programme on Chemical Safety, 1998; Zavaleta *et al.*, 1999) suggesting certain adverse reproductive effects and development outcomes that may be associated with trihalomethanes (THMs) and chlorination by-products. We consider that it is important to measure estrogenic effect of natural water and by-products by chlorination in addition to typical possible EDCs.

The previous study (Itoh, 2004) revealed that the estrogenic effect of Lake Biwa water decreased by coagulation and activated carbon treatment, however, it increased by chlorination. This phenomenon is very similar to the formation of THMs in drinking water treatment process in that natural organic matters (NOM) are major precursors of both estrogenic effect and THMs. It was also pointed out that the components that can be called "estrogenic effect formation potential" and "estrogenic effect intermediates" are defined in chlorinated water. This understanding is very similar to that of the process of THMs formation. One of the purposes of this study is to demonstrate the procedure to detect the estrogenic effect formation potential that means the maximum estrogenic effect after being treated with chlorine.

The previous study (Itoh, 2004) also found that organic matters of which estrogenic effect increases or decreases by chlorination are present in natural water. The increase or the decrease by chlorination can be detected depending on a sample preparation procedure. The other purpose of this study is to establish an appropriate concentration procedure of natural water and its chlorinated one to detect the estrogenic effect including the estrogenic effect formation potential.

Materials and methods

Chlorination of humic acid

The effect of hydrolysis of chlorinated water was examined to investigate the characteristics of estrogenic effect of chlorinated water. Commercial humic acid (Wako Pure Chemical Industries, Ltd.) as a model substrate in water was used. TOC of the humic acid solution was 1080 mg/L. Chlorination was performed by addition of the desired amount of sodium hypochlorite solution (Wako Pure Chemical Industries, Ltd.). First, the humic acid solution in a phosphate buffer of 100 mM at an initial pH of 6.0 was treated with 1100 mg-Cl_2/L of chlorine. The chlorination proceeded in a dark room for 48 hours, after which residual chlorine was not detected. Dechlorination was not carried out so as not to change the estrogenic effect in chlorinated waters even when the chlorination was performed with a higher concentration of chlorine. It was confirmed that chlorine had no influence on the MVLN assay up to a concentration of 30 mg-Cl_2/L in the culture media. After the treatment, the pH of the solution was readjusted. The solution was allowed to stand at 20℃ over a pH range of 7-10, after which estrogenic effect was measured.

Concentration of Lake Biwa water

Lake Biwa is the largest lake in Japan and also the major water source for 14 million people in Kansai area in Japan. Organic matters in Lake Biwa were concentrated by adsorption and desorption method using a cartridge packed with 2 mL of OASIS HLB (Nihon Waters K.K.) resin. 1 L of Lake Biwa water filtered with a 0.45 m membrane filter of which DOC was 2.2 mg/L was concentrated using an OASIS HLB cartridge. After pH of a sample water was adjusted to 2, it was fed to the cartridge in upper flow at a flow rate of 20 mL/min. After drying the cartridge with nitrogen gas, the adsorbed substances were desorbed by 3 mL of 0.1 M NaOH at 0.5 mL/min. The recovery of TOC was 23% in this step. After feeding pure water and drying the cartridge with nitrogen gas, the remaining substances were desorbed by 3 mL of dichloromethane at 0.5 mL/min. After dichloromethane was evaporated with nitrogen gas, the residue was dissolved with ethyl alcohol. Since it was confirmed that ethyl alcohol had no influence on the MVLN assay up to a concentration of 1 % in the culture media, the ratio of an added sample to the media was less than 1 %.

Filtered Lake Biwa water was chlorinated at an initial concentration of 1.0 mg/L. Residual chlorine was not detected after 24 hours and TOX was 110 μg/L. Chlorinated water was concentrated by the same procedure for raw water.

The MVLN assay

EDSTAC has recommended the MVLN assay as an estrogen receptor transcriptional activation assay in the screening. This assay utilizes an MCF-7 derivative that has been stably transfected with the Vit-Luc reporter gene (Pons *et al.*, 1990). Thus, the MVLN cell line expresses the endogenous estrogen receptor of MCF-7 and the same time, contains an exogenous estrogen responsive reporter (luciferase). Therefore, the estrogen specific transcription activity of a test chemical is directly related to the activity of luciferase measured in the lysate of treated MVLN cells.

In this study, estrogenic effect was measured by the MVLN assay. The MVLN cell line was obtained from its source (Pons *et al.*, 1990). For routine maintenance, cells were grown in Dulbeco's modification of Eagle's medium (DME, Gibco BRL Life Technologies, Inc.) supplemented with 10% fetal bovine serum (FBS, Gibco BRL Life

Technologies, Inc.) at 37℃ in an atmosphere of 5% CO_2/95% air under saturating humidity.

MVLN cells used to the assay have to be withdrawn from estrogen 6 days prior to passage. Cells in 10% FBS media are rinsed three times with PBS and fed 10% dextran coated charcoal (DCC) FBS (Hyclone) media. These cells should be rinsed and fed 10% DCC FBS media every other day until seeded into a well plate. The MVLN cells withdrawn from estrogen are seeded into a 24 well plate at a density of approximately 5×10^4 cells/well, fed 1 mL/well of 10% DCC FBS media. Two days after seeding, cells are fed treatment media containing a test chemical and solution. The cells should have new treatment media applied the next day. Then, the cells are lysed by adding 0.5 mL of lysis buffer (TOYOINK MFG. Co., Ltd.) the next day. Cell lysate was harvested, and luciferase activity was detected with a luminometer (Lumat LB9507, Berthold). Concentration of protein in cell lysate was measured using sodium bicinchoninate (Smith *et al.*, 1985), and luminescence units per milligram of protein (relative light units, RLU) were calculated.

Relative luciferase activity was calculated as follows using 17β-estradiol (E_2) as a standard chemical (Itoh *et al.*, 2000).

$$\text{Relative luciferase activity (\%)} = \frac{L_t - L_C}{L_E - L_C} \times 100 \tag{1}$$

where L_E is RLU of E_2 at 1.0×10^{-9}M; L_c is RLU of control cell lysate (in 10% DCC FBS media); and L_t is RLU of a test chemical and solution.

Analytical procedures

TOX and TOC were measured by a TOX-10 analyzer (Mitsubishi Chemical Corporation) and a TOC-5000A analyzer (Shimadzu), respectively. E_2, estrone (E_1) and bisphenol A (BPA) were measured by an enzyme-linked immunosorbent assay (Takeda Chemical Industries, Ltd.). Concentration of chlorine in mg-Cl_2/L was determined by the DPD ferrous titrimetric method.

Results and discussion

Change in the estrogenic effect formed by chlorination

The change in the estrogenic effect after chlorination of humic acid was examined. After finishing the chlorination of humic acid as described in **Chlorination of humic acid**, the pH of the chlorinated solution was adjusted to 7.0 or 10.0, and change in the estrogenic effect was measured by the MVLN assay. The final concentration of chlorinated solution in the culture media was 25 mg/L as TOC.

Figure 1 shows the result of the assay. It was found that the estrogenic effect of the chlorinated solution increased even without residual chlorine at pH 7. It is well known that the concentration of THMs increases in the distribution systems. The obtained result suggests that some part of the estrogenic effect in drinking water also increases over time after chlorination. It increased faster at a pH of 10.0 than at a pH of 7.0. This probably occurs because the hydrolysis rate increases with increasing pH. In addition, the increased estrogenic effect decreased at pH 10. This result suggests that the estrogenic effect of chlorinated water increases by hydrolysis, however, it begins to decrease when hydrolysis proceeds furthermore.

In the previous study (Itoh, 2004), the effect of pH and water temperature on the estrogenic effect formed by chlorination was examined. It increased under the condition of pH 7 and 80℃ for one hour, however, the increased estrogenic effect decreased under the alkaline condition and 80℃. This result would support the above discussion. Since drinking water is maintained at the neutral pH and the normal temperature in the actual water supply system, it seems that the direction of change of the estrogenic effect would be the increase in the distribution systems.

Figure 2 shows the effect of added chlorine concentration on the estrogenic effect. The final concentration of chlorinated solution in the culture media was 60 mg/L as TOC. Figure 3 shows residual concentration of chlorine in the experiment shown in Fig. 2. Residual chlorine does not influence the MVLN assay up to a concentration of 30 mg/L in the media. Since 56 μL of sample water is added to 1 mL of the media, 540 mg/L of residual chlorine in chlorinated humic acid does not influence the MVLN assay. Thus, it was confirmed that residual chlorine in which Cl_2/TOC was 1 and 2 did not influence the assay results shown in Fig. 2. On the other hand, chlorinated solutions in which Cl_2/TOC was 4 and 10 had cytotoxic effect. Figure 2 shows that relative luciferase activity achieved the maximal value at 24 hours in the case Cl_2/TOC was 1. This activity was 2.2 times as strong as that of humic acid itself. The increased estrogenic effect, however, decreased at 65 hours.

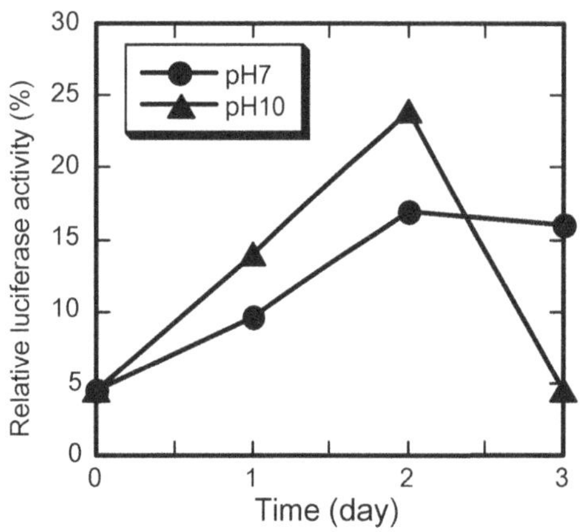

Figure 1 Change in the estrogenic effect of chlorinated humic acid.

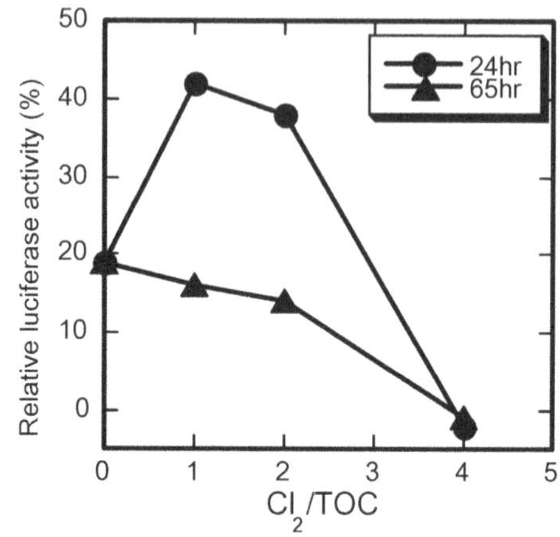

Figure 2 Effect of added chlorine concentration on the estrogenic effect of chlorinated humic acid.

This phenomenon is illustrated in Fig. 4. Judging from Fig.1, estrogenic substances formed just after chlorination would be a part of chlorination by-products. The components that can be called "estrogenic effect formation potential" and "estrogenic effect intermediates" are defined in chlorinated humic acid solution, and they change into estrogenic substances with the duration of time. "Trihalomethane formation potential" and "trihalomethane intermediates" in the formation process of THMs have been defined, which is very similar to the understanding illustrated in Fig. 4.

Based on the results shown in Figs 1 and 2, one of the appropriate conditions to detect the estrogenic effect formation potential would be as follows. The ratio of added chlorine to TOC is 1. The chlorination is proceeded for 24 hours at pH 7 at 20℃. After the chlorination, the pH is adjusted to 10 and the solution is allowed to stand for two days.

The effect of water temperature was also examined in the previous study (Itoh, 2004), however, high water temperature would be not practical to measure the estrogenic effect of chlorinated natural water. This is the reason why the effect of water temperature was not examined in the present study.

Sample preparation procedure to detect estrogenic effect of drinking water

The results shown in Figs 1 and 2 were obtained using commercial humic acid. In addition to NOM such as humic substances, individual compounds such as E_2 and alkyl phenols are present in natural water.

The previous study (Itoh, 2004) points out that organic matters of which estrogenic effect increases or decreases by chlorination are present in natural water. First, humic substances have some estrogenic effect (Itoh *et al.*, 2000). This estrogenic effect of NOM increases by chlorination. In contrast, estrogenic effect of most micro-pollutants would decrease by chlorination (Itoh, 2004; Takeda *et al.*, 2003). Most individual compounds could be decomposed by chlorine and those estrogenic effect decrease drastically, although some studies (Hu, *et al.*, 2002; Lenz, *et al.*, 2003) have reported that the estrogenic effect of BPA and 4-nonylphenol (4-NP) increases under the certain chlorination condition. The important point is that the estrogenic effect of chlorinated drinking water is the sum of the increased activity and the decreased activity by chlorination.

It has been demonstrated that the increase or the decrease by chlorination can be detected depending on a sample preparation procedure. The preparation procedure using XAD7HP resin and NaOH as a desorbing solvent can be recommended to detect the estrogenic effect of NOM in natural water. Itoh et al. (2003a) have examined recovery efficiencies of solid phase extraction procedures to concentrate estrogenic individual compounds. As a result, OASIS HLB resin and dichloromethane were selected as an adsorbent and a desorbing solvent, respectively. The preparation procedure using OASIS HLB resin and dichloromethane has to be used to detect the estrogenic effect of individual compounds in natural water.

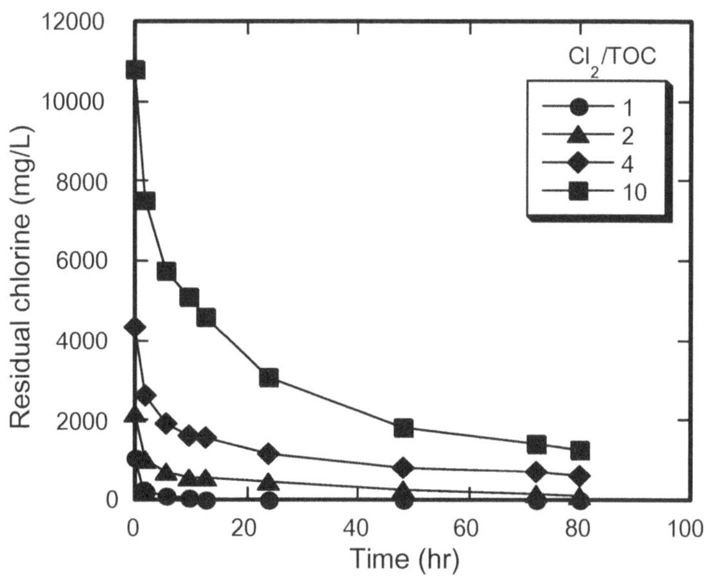

Figure 3 Residual concentration of chlorine in humic acid solution.

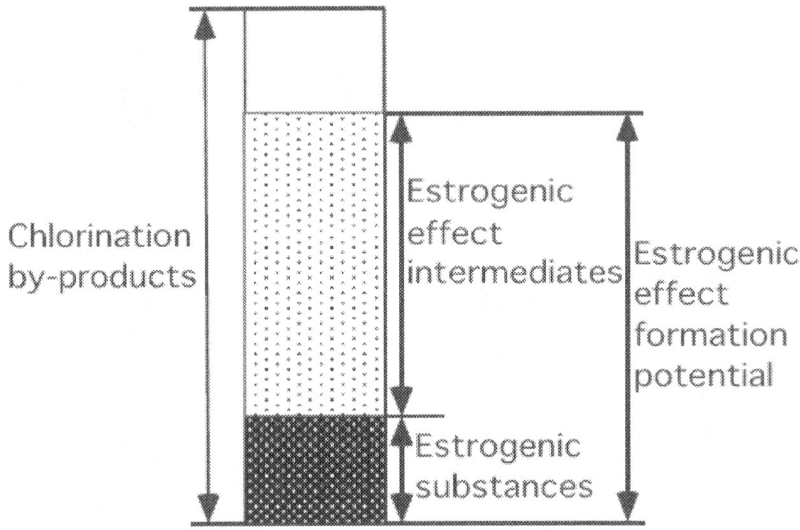

Figure 4 Components of estrogenic effect in drinking water.

It is convenient to measure the estrogenic effect of drinking water if an appropriate procedure that concentrates the increased activity and the decreased activity in chlorinated water simultaneously could be established. Itoh *et al.* (2003b) have already confirmed by a three-dimensional fluorescence spectrophotometer that NOM extracted using XAD7HP resin and NaOH could be extracted using OASIS HLB resin and NaOH. Based on these findings, the effectiveness of a preparation procedure shown in Fig. 5 is examined in this study. It is supposed that humic substances and their chlorination by-products are recovered by the extraction using NaOH. Subsequently, individual compounds and their by-products are supposed to be recovered by the extraction using dichloromethane.

Figure 6 shows the estrogenic effect of chlorinated humic acid and its recovery using OASIS HLB resin and NaOH. Chlorination of humic acid solution (TOC=1100 mg/L) was performed with 1500 mg/L of chlorine at pH 7 for 72 hours. This chlorinated humic acid was diluted to 20 mg-TOC/L, after which the diluted sample was concentrated by the procedure shown in Fig. 5. The recovery of the estrogenic effect was approximately 50%, while the recovery of TOC was 23%. On the other hand, Figure 7 shows the estrogenic effect of a desorbed solution by dichloromethane after NaOH was fed to an OASIS HLB cartridge. The estrogenic effect could not be detected until 720 mg/L of TOC was added to the media. Figures 6 and 7 show that the estrogenic effect formed by chlorination is extracted mainly by NaOH.

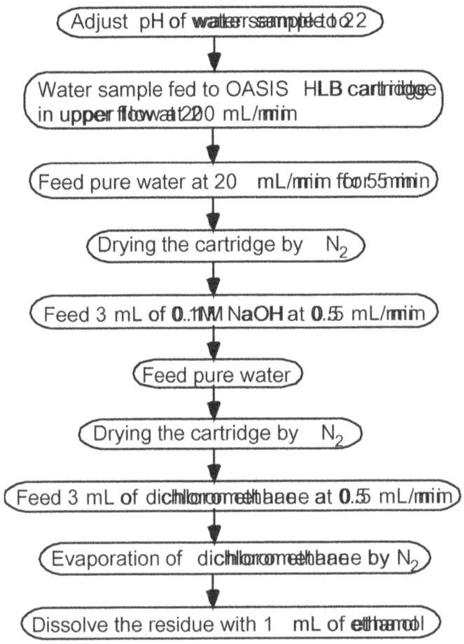

Figure 5 Procedure of concentrating the estrogenic effect of natural water and its chlorinated one by solid phase extraction using an OASIS HLB cartridge.

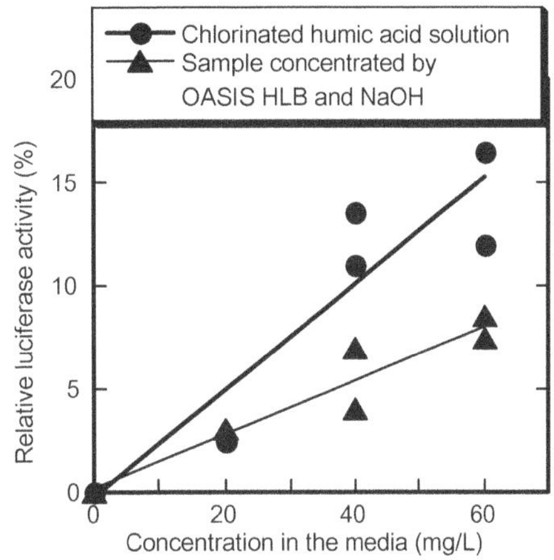

Figure 6 Recovery of the estrogenic effect of chlorinated humic acid using OASIS HLB resin and NaOH.

Figure 8 shows recovery efficiencies of E_2, E_1 and BPA by the procedure shown in Fig. 5. It is obvious that most of the substances can be concentrated by this procedure, which also means that feeding NaOH does not influence on the recoveries of these compounds. In addition, it was also found that most of the estrogenic effect of these compounds was recovered. When these compounds are chlorinated, it was confirmed that recovered concentrations of these compounds correspond to the recovered estrogenic effect (data are not shown).

Judging from these results, a sample preparation procedure shown in Fig. 5 can be proposed in order to measure the estrogenic effect of natural water and its chlorinated one.

The change in the estrogenic effect of natural water after chlorine injection was measured according to the preparation procedure shown in Fig. 5. Figure 9 shows the result of chlorinated Lake Biwa water. Relative luciferase activity of raw water was 8% and it shows that the estrogenic effect of raw water almost disappeared by chlorination. It clearly shows that the estrogenic effect was formed over time after chlorination even without residual chlorine. It increased faster at pH 10 than at pH 7 and began to decrease rapidly at pH 10. According to this result, an appropriate procedure to detect the estrogenic effect formation potential, which means the maximum estrogenic effect, is as follows. A concentration of added chlorine is 1.0 mg/L in this experiment. The chlorination is proceeded for 24 hours at pH 7 at 20℃. After the chlorination, the pH is adjusted to 10 and the sample water is allowed to stand for one day. In the experiment using humic acid (see Fig. 1), two days was needed to reach the maximum value after the adjustment of pH to 10. On the other hand, in Fig. 9, only one day was needed to reach the maximum value. It is obvious that there is a large difference in TOC concentration between humic acid and natural water, however, it is not clear why there exists a difference in the reaction rates.

NOM is well known as one of the major precursors of THMs. In addition, it has been reported that the concentration of THMs in drinking water does not decrease sufficiently

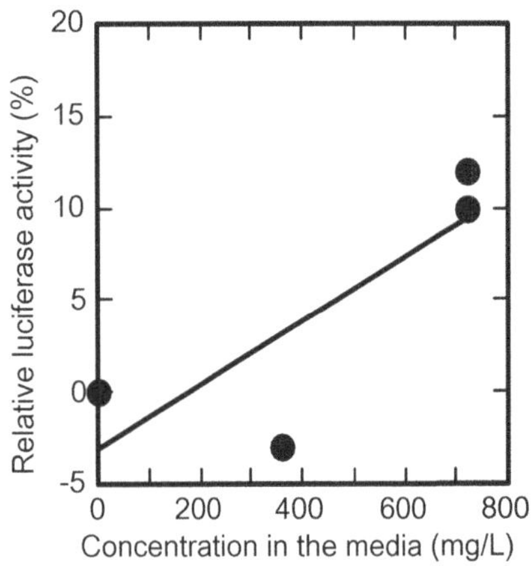

Figure 7 Recovery of the estrogenic effect of chlorinated humic acid by dichloromethane.

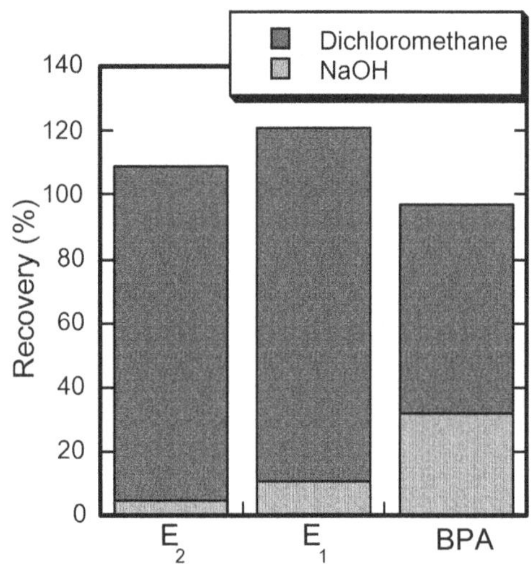

Figure 8 Recoveries of typical EDCs by the concentration procedure shown in Fig. 5. Initial concentrations: E2, 0.0027μg/L; E1, 0.001 μg/L; BPA, 0.5 μg/L.

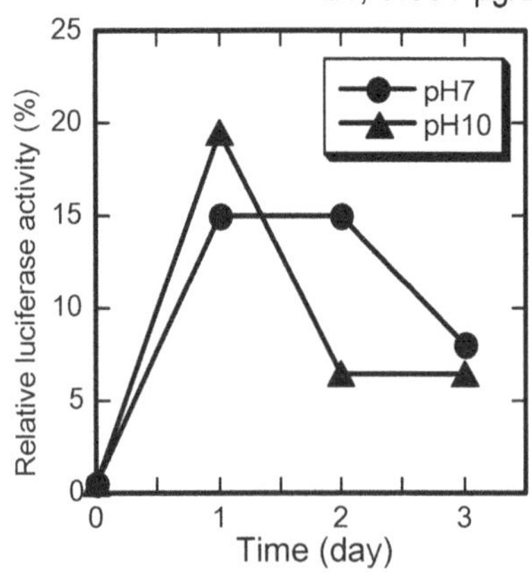

Figure 9 Change in the estrogenic effect of chlorinated Lake Biwa water.

even after a sewage treatment system is established. Recently, it has been found that hydrophilic acids that are produced by biological treatment in a wastewater treatment plant and are discharged into a reservoir would be one of the important precursors of THMs. In the future research, it has to be investigated which components of NOM contribute to the increased estrogenic effect of chlorinated drinking water.

Conclusions

The estrogenic effect of chlorinated water increased even without residual chlorine. The components that can be called "estrogenic effect formation potential" and "estrogenic effect intermediates" are defined in chlorinated humic acid solution. This understanding is very similar to that of the process of THMs formation. An appropriate procedure to detect the estrogenic effect formation potential, which represents the maximum value of estrogenic effect, was investigated. The effects of chlorine dosage, contact time and pH on the estrogenic effect of chlorinated humic acid were examined. As a result, an appropriate procedure was demonstrated as follows. The ratio of added chlorine to TOC is 1. The chlorination is proceeded for 24 hours at pH 7 at 20℃. After the chlorination, the pH is adjusted to 10 and the solution is allowed to stand for two days.

Organic matters of which estrogenic effect increase or decrease by chlorination are present in natural water. The important point is that the estrogenic effect of chlorinated drinking water is the sum of the increased activity and the decreased activity by chlorination. An appropriate procedure that concentrates the increased estrogenic activity and the decreased activity in chlorinated drinking water simultaneously was established. It was supposed that humic substances and their chlorination by-products are recovered by OASIS HLB resin and NaOH as an adsorbent and a desorbing solvent, respectively. Subsequently, individual compounds and their by-products are supposed to be recovered by the continuous extraction using dichloromethane.

Based on the obtained results, the estrogenic effect formation potential, which means the maximum estrogenic effect, of chlorinated Lake Biwa water was measured.

References

Endocrine Disruptor Screening and Testing Advisory Committee (EDSTAC) (1998) *Final Report*.

Hu J., Aizawa T. and Ookubo S. (2002) Products of aqueous chlorination of bisphenol A and their estrogenic activity, *Environ. Sci. Technol.* **36**(9), 1980-1987.

International Programme on Chemical Safety (1998) *Environmental Health Criteria, Disinfectants and Disinfection By-products*.

Itoh S., Ueda H., Nagasaka T. and Sumitomo H. (2000) Evaluating variation of estrogenic effect by drinking water chlorination with the MVLN assay. *Wat. Sci. Technol.* **42**(7-8), 61-69.

Itoh S., Nakanishi G., and Hayasaka T. (2003a) Change of estrogenic effect of natural water by chlorination and sample preparation procedure for a bioassay, *J. Japan Water Works Association*, **72**(4), 10-20 (in Japanese).

Itoh S., Hayasaka T. and Okada T. (2003b) Concentration of humic substances in Lake Biwa and its chlorinated water by fluorescence analysis. *J. Water and Waste*, **45**(6), 24-28 (in Japanese).

Itoh S. (2004) Research needs for EDC in drinking water – The similarity of the production of trihalomethanes and estrogenic effect -. *Wat. Sci. Technol.* (in press).

Lenz K., Beck V. and Fuerhacker M. (2003) Behavior of bisphenol A (BPA), 4-nonylphenol (4-NP) and 4-nonylphenol ethoxylates (4-NP1EO, 4-NP2EO) in oxidative water treatment processes, *Proceedings of The 4th IWA Specialized Conference on Assessment and Control of Hazardous Substances in Water –ECOHAZARD 2003-*, Aachen, Germany, 14-17 Sep., 5/1-5/7.

Pons M., Gagne D., Nicolas J.C., Mehtai M. (1990) A new cellular model of response to estrogens: A bioluminescent test to characterize (anti)estrogen molecules, *BioTechniques*, **9**(4), 450-459.

Smith P.K., Krohn R.I., Hermanson G.T., Mallia F.H., Gartner F.H., Provenzano M.D., Fujimoto E.K., Goeke N.M., Olson B.J., and Klenk D.C. (1985) Measurement of protein using bicinchoninic acid, *Anal. Biochem.*, 150, 76-85.

Takeda M., Kamata M., Ohno K., Kamei T., Magara Y. and Terao Y. (2003) Effects of chlorination on estrogenic activity induced by bisphenol A and chlorination by-products. *J. Japan Society on Wat. Environ.* **26**(11), 65-71 (in Japanese).

Waller K., Swan S.H., DeLorenze G., and Hopkins B. (1998) Trihalomethanes in drinking water and spontaneous abortion, *Epidemiology*, 9, 134-140.

Zavaleta J. O., Hauchman F. S., and Cox M. W. (1999) Epidemiology and toxicology of disinfection by-products. In: *Formation and Control of Disinfection By-products in Drinking Water*, Singer, P. C. (ed.), American Water Works Association, Denver, CO., pp.95-117.

Sludge Minimization in a Full-Scale Biological Denitrification Plant

S. Nishimura* and H. Yasui*

*Kurita Water Industries Ltd., 4-7 Nishi-Shinjuku 3-chome, Shinjuku-ku, Tokyo, 160-8383, Japan. e-mail: sousuke.nishimura@kurita.co.jp

Abstract The design criteria and full-scale plant experience on ozone-enhanced sludge minimization in a biological denitrification process are reported. More than 90% reduction in the excess sludge production was achieved by returning ozonated sludge to a denitrification tank, where denitrification and biological sludge degradation took place simultaneously. Methanol addition could also be reduced, indicating that the ozonated sludge acted as an electron donor. The maximum COD loading rate in this particular case was limited to be 4 kg/m^3/d to make appropriate sludge separation in the clarifier. During four years of operation, the reduced mass of the sludge was estimated to be over 4,000 tons of cake with a moisture content of 85%, and at the same time, the facility reduced nearly 1,500 tons of nitrogen discharge to an enclosed sea area. The total number of full-scale plants having similar process configurations reached 33 in Japan, both in industrial and municipal fields.

Keywords activated sludge treatment; denitrification; excess sludge reduction; ozone; full-scale plant

Introduction

Sludge minimization in wastewater treatment process is one of the major environmental concerns, since it can cause leachate, soil, and groundwater contaminations in landfill sites, air pollution when incinerated, as well as high costs to deal with. It has been also pointed out that methane release from dumped organic sludge can cause global warming.

For a biological treatment such as activated sludge process, a unique approach to efficient sludge reduction has been proposed by Yasui *et al.* (1997, 2003), with their data from both experimental and full-scale study. This approach combines two functions, i.e., water purification and sludge digestion into a common biological reaction tank. The conceptual flow scheme is shown in Figure 1.

© 2004 IWA Publishing. *2nd IWA Leading-Edge Conference on Water and Wastewater Treatment Technologies* edited by Mark Van Loosdrecht & Jonathan Clement. ISBN: 1 84339 508 8.

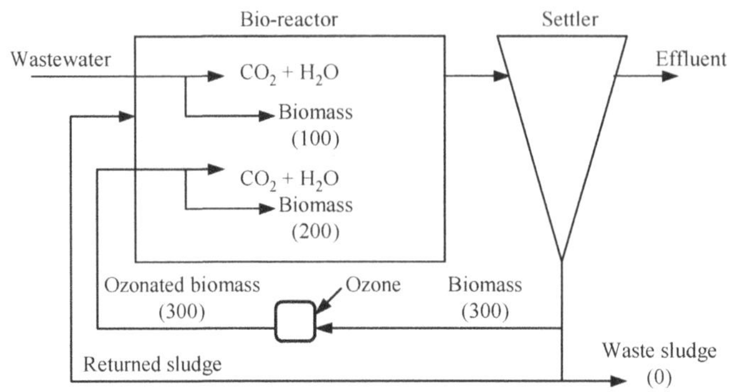

Figure 1 Loop for sludge reduction: Figures in parenthesis show typical mass balance of the sludge. Re-circulating ozonated biomass, 300% as much as that generated from wastewater, makes 100% reduction of the waste sludge.

Since 1995, Kurita Water Industries Ltd. has commercialised this technology under the trade name of "Bioleader^{TM}", and has constructed 33 full-scale treatment plants in Japan. The application fields and the number of plants are listed in Table 1.

Table 1 Number of full-scale installation in Japan: Motor vehicle industry seems to be most conscious on environmentally friendly production. Ditto for chemical and semiconductor industry. Sludge disposal cost for chemical industry can be higher due to its potential risk of chemical contaminant, thus sludge reduction process might have higher economical attraction for them.

Type of wastewater	Number	Type of wastewater	Number
Food	3	Municipal	7
Beverage	2	Motor vehicle	8
Chemical	6	Pharmaceutical	1
Semiconductor	5	Steel	1
		Total	33

Through the experience from the full-scale plants, the authors have learnt that controlling sludge concentration in the bio-reactor is quite important as it governs sludge separation in the clarifier (Vesilind, 1968), and this could not be achieved by simply increasing ozone dose, but careful designing of other affecting factors was needed.

This paper aims to emphasize that design criteria for biological treatment is as important as ozonation in Bioleader^{TM} process, by introducing a case study for denitrification process, for which such criteria were not known before.

Materials and methods

The concept of Bioleader^{TM} process is rather simple, however, to attest the practical effectiveness is not as easy as its principle. The degradation rate of the ozonated sludge is not sufficiently high to be measured with desktop analytical methods such as BOD or respirometer; its values are even as low as endogenous respiration rates. This fact causes difficulty in designing plant configurations, because it could be misleading to assume the ozonated sludge as a readily biodegradable BOD or COD loading.

Therefore, the first step for designing was to conduct a long-term experiment using a pilot scale plant, so to assess overall performance after slow degradation of ozonated sludge reached steady state. Based on the pilot plant results, a full-scale plant was constructed for a nitrate containing wastewater from stainless steel works.

Pilot scale test

Experimental set-up

A schematic diagram of the experimental set-up is shown in Figure 2. A nitrate wastewater was continuously fed to the two parallel series of the denitrification process. The wastewater was almost inorganic, thus methanol was added as an electron donor for nitrate reduction. Ozonation was conducted using a transparent column with a 100mm diameter, equipped with a re-circulation line for mixing, with its end, sludge was sprayed form the top, so to prevent foams come out to the off-gas from the top. The off-gas was led to an activated carbon scrubber to be released to the atmosphere.

Operational conditions

The test conditions are shown in Table 2. Two different reactor volumes were compared to see the effect of volumetric COD loading rate. Methanol dose was controlled according to the ORP level in the denitrification tank. The methanol dose rate was slightly raised in runs 2A and 2B, because nitrate leakage was found several times in the previous runs, 1A and 1B. Temperatures in the reactors were maintained in the range between 25-35 degrees C. Ozonation was conducted at a rate that made SRT_{BL}s shown in Table 2. SRT_{BL} refers to total VSS (kg) in the system divided by ozonated VSS (kg/d), which indicates average sludge degradation time in the bio-reactor. No sludge withdrawal was done during the test period.

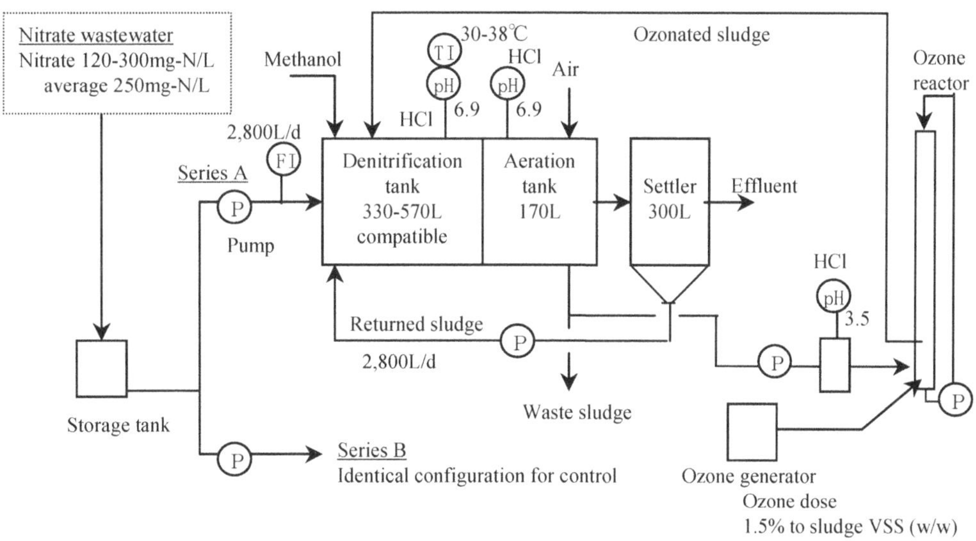

Figure 2 Schematic diagram of experimental set-up

The loading rate in Run-2A was programmed to be fluctuating so as to simulate the actual change in wastewater inflow. Ozonation rate was also controlled to be proportional to the nitrogen loading, at an average SRT_{BL} of 6.7 days. In the other runs, the loading rate was kept constant.

Table 2 Experimental conditions: SRT_{BL} refers to total VSS (kg) in the system divided by ozonated VSS (kg/d), which indicates average sludge degradation time in the bio-reactor. Higher ozonation rate (kg-VSS/d) results in shorter SRT_{BL}.

parameters	Run-1A	Run-1B	Run-2A	Run-2B	units
Volume of denitrification tank	330	330	570	570	liter
Nitrogen loading rate to DN tank volume	2.1 constant	2.1 constant	max. 1.2 fluctuating	1.2 constant	$kg/m^3/d$
Methanol / N	2.4	2.4	2.9	2.9	kg/kg
COD loading rate to DN+AT tank volume	5.0 constant	5.0 constant	max. 4.0 fluctuating	4.0 constant	$kg/m^3/d$
SRT_{BL}	3.6	3.0	6.7	3.6	day
Sludge discharge	0.0	0.0	0.0	0.0	kgSS/d

Denitrification rate measurement

Specific denitrification rates (kg-N/kg-VSS/d) of sludge samples were measured using apparatus shown in Figure 3. Sludge sample, nitrate reagent, and organic matter such as methanol or ozonated sludge were stirred under anoxic condition, and specific nitrate reduction rate was analysed.

The denitrification rate on methanol varied in a range of 0.1 to 0.4 kg-N/kg-VSS/d. On the contrary, the rate on ozonated sludge did not exceed 0.03 kg-N/kg-VSS/d. This fact indicated that anoxic degradation of the ozonated sludge was much slower than that of methanol. The measurement also showed that denitrification rate and hence the sludge degradation was even retarded when nitrate concentration was lower than 5 mg-N/L as shown in Figure 4 (Tokuhara *et al.* 2002).

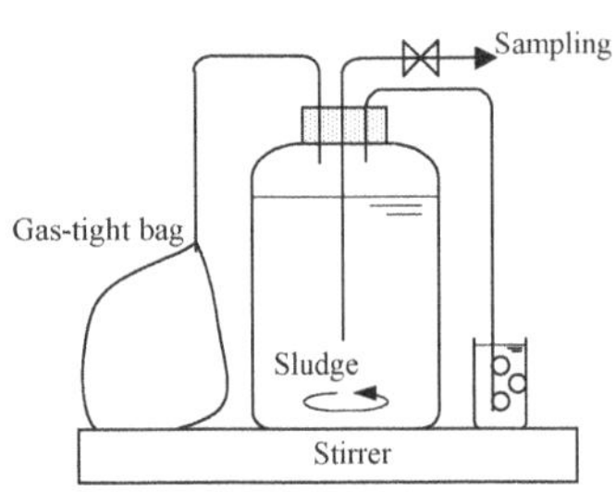

Figure 3 Apparatus for denitrification rate measurement

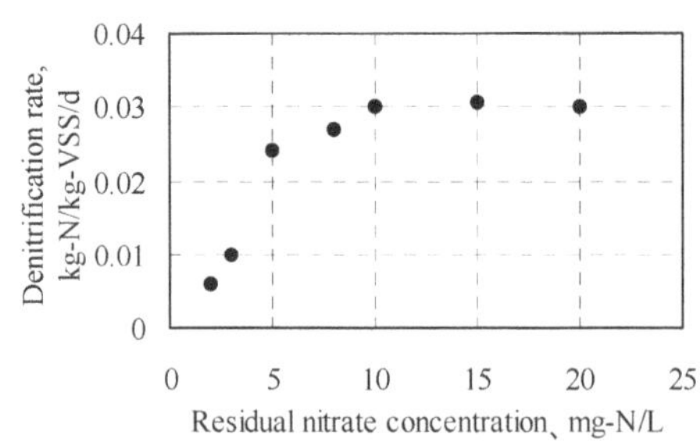

Figure 4 Anoxic degradation rate of ozonated sludge (Tokuhara, *et al.*, 2002)

Prediction of clarifier performance

Because of the land limitation in the full-scale case, the upflow linear velocity (LV) in the clarifier was required to be more than 14.4 m/d. One dimensional flux theory was used to predict the performance of the clarifier. Empirical relationship of Vesilind was used to calculate the settling velocity at different concentrations. The settling parameters of V_0 and k in Vesilind equation were calculated by conducting settling tests at various plants.

Empirical relationship between zone settling velocity and SS concentration (Vesilind, 1968)

$$V = V_0 \exp(-kX) \quad \textbf{Eq. 1}$$

where, V: zone settling velocity of the sludge bed, m/d
V_0: converged V value towards X=zero
X: SS concentration, g/L
k: coefficient dependent on specific sludge settling property, L/g

As shown in Figure 5, the maximum settling velocity was estimated to be 200 m/d. The k value was found to correlate to DSVI and the relationship used in the present study is as shown in Equation 2. Equations 2 and 3 were used to work out a relationship between the MLSS concentration and permissive LV in the clarifier. Figure 6 shows the result of such analysis for different DSVI values.

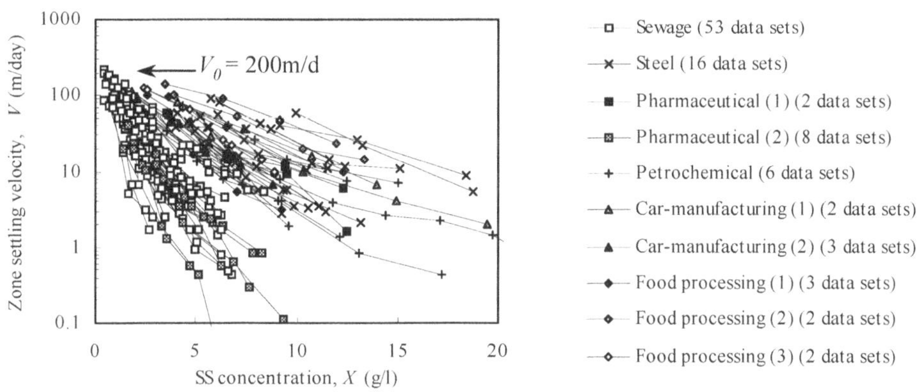

Figure 5 Zone settling velocities found in various types of sludge in Japan (Komatsu, et al. 2002)

Empirical relationship between Vesilind's coefficient k and SVI (Komatsu, *et al.*, 2002)

$$k = 0.098 + 0.0024 \cdot DSVI \quad \textbf{Eq. 2}$$

where, $DSVI$: sludge volume index, mL/g-SS, diluted with a clarifier effluent if SV_{30} exceeded 30%

Definition of sludge settling flux F

$$F = X(V + q_R) \quad \textbf{Eq. 3}$$

where, F: sludge settling flux in clarifier, kg-SS/m^2/d
q_R: down flow velocity associated with withdrawal from the bottom, m/d

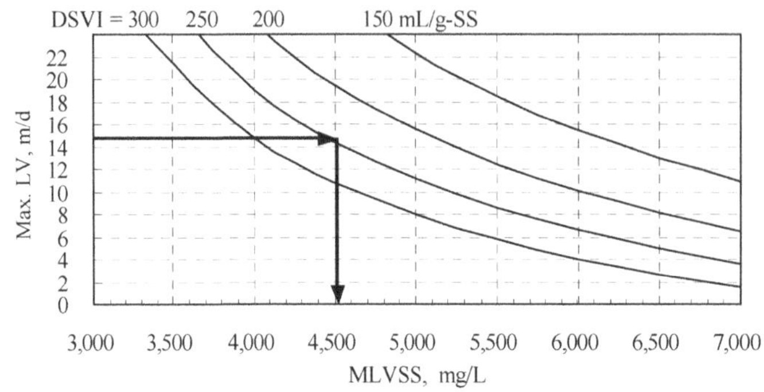

Figure 6 Relationship between maximum LV and sludge concentration, q_R=11.1m/d

Figure 6 showed that MLVSS in the clarifier influent should not exceed 4,500mg/L, when DSVI is 250 mL/g-SS. This value was used to evaluate the pilot test results.

Results

Pilot scale test

Sludge concentration

MLVSS concentrations in Run-1 and Run-2 are shown in Figure 7. The VSS/SS ratio in the sludge was almost constant at 0.88. In Run-1A, with a COD loading rate of 5.0 kg/m^3/d, MLVSS exceeded the maximum allowable value of 4,500 mg/L as determined in Figure 6. In Run-1B, a shorter SRT_{BL}, namely, higher ozonation rate was tested, however, this was not successful most likely because of a short degradation time, in which the sludge degradation was limited.

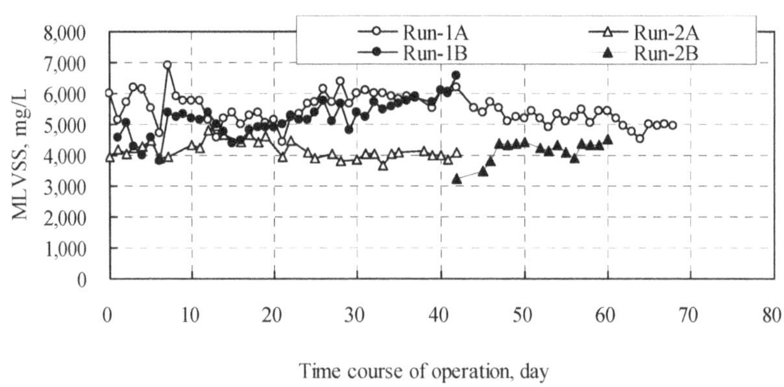

Figure 7 Sludge concentrations in pilot-scale test: Sludge samples were taken from post-aeration tank for the analysis. Run-1was with higher COD loading rate than Run-2

So to control the MLVSS at 4,500 mg/L or lower, it was needed to reduce COD loading rate down to 4.0 kg/m^3/d, as shown in Run2A and 2B in Figure 7.

Full-scale plant

Plant configuration

The maximum design flow rate of the wastewater was 9,360 m^3/d, with its average nitrate concentration of 260 mg-N/L. A bio-reactor including a denitrification and post-aeration tank with its total volume of 2,450 m^3 was constructed, followed by two parallel clarifiers and high-rate sand filters. Ozone was dosed into the split line of the return sludge at 3 kg-O_3/h, using three stainless-steel-made completely mixed tanks in series.

The plant was well equipped with automatic control devices, which included methanol and phosphorous dose control using total nitrogen analyser, pH in bio-reactors, and emergency coagulant dose using sludge height detector. As shown in Figure 8, the automatic methanol control realized a methanol/nitrogen ratio of as low as 2.4. This value was substantially lower than typical ratio of more than 3, and thus made maximum use of ozonated sludge as substitute for methanol as an organic source.

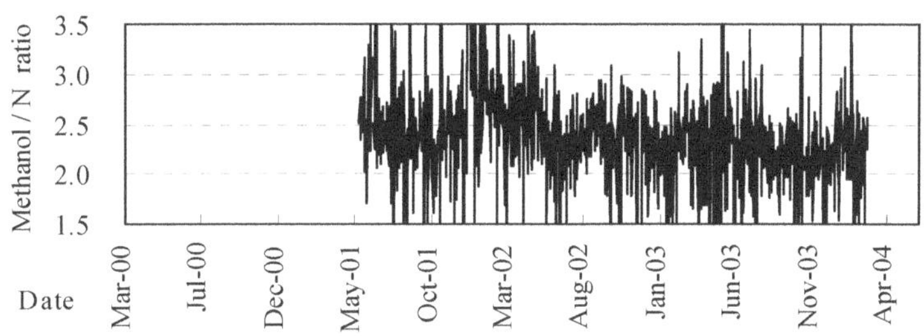

Figure 8 Methanol / nitrogen ratio

The final effluent nitrogen concentration was also controlled automatically by adjusting the flow rate of treated and untreated stream, as well as another nitrogen containing stream from a different section in the works. This realized optimal operational cost (Fujiyasu, et al., 2000).

Effluent quality

The nitrate concentrations in the treated wastewater has been less than 10 mg-N/L, and the total nitrogen in the final effluent to the sea after mixed with other waste streams has been well controlled to be less than 60 mg-N/L to meet a local regulation, as shown in Figure 9.

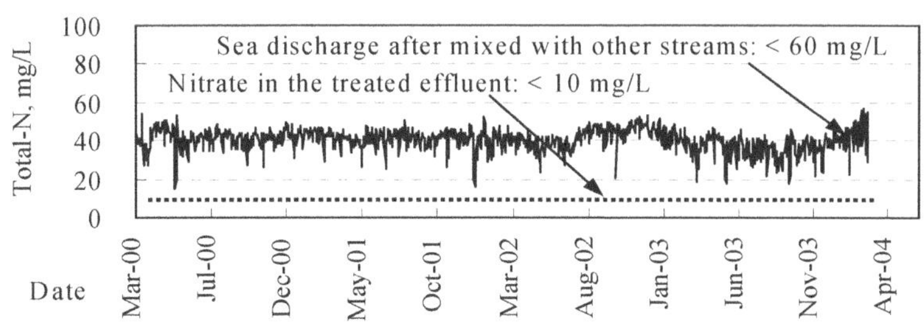

Figure 9 Total nitrogen concentration in the discharged water to the sea: Nitrate in the treated effluent was roughly checked using an on-site colorimetric test kit, and total nitrogen in the sea discharge was measured and recorded using a TN meter with an automatic sampler.

Sludge concentration and settling property

The MLVSS concentration in the denitrification tank has been maintained at desired values of around 4,500 mg/L, as shown in Figure 10.

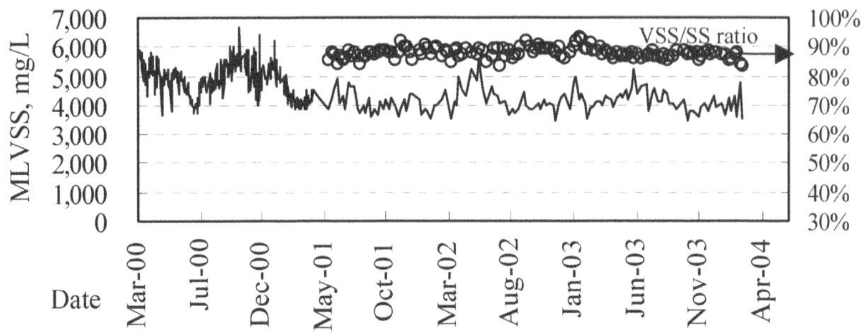

Figure 10 MLVSS concentration in denitrification tank

The ratio of VSS/SS has been almost constant in the range between 85 to 90%, indicating negligible mineral accumulation in the system without sludge discharge. DSVI value has been less than 250 mL/g-SS, almost through the year, although periodical deterioration was found in mid summer and winter season, possibly attributed to temperature shocks. This deterioration resulted in temporary sludge withdrawal. However, the total mass of the withdrawn sludge was less than 10% to the sludge generation, as shown in Figures 11.

Overall reduction in sludge and nitrogen loading to environment

In the past four years of full-scale operation, the total reduction in sludge and nitrogen discharge has reached over 4,000 ton-cake and nearly 1,500 ton-N, as shown in Figures 11. The overall reduction efficiency of the sludge was over 90%. The fact could make a clear evidence of the effectiveness of the environmental management, committed by the works.

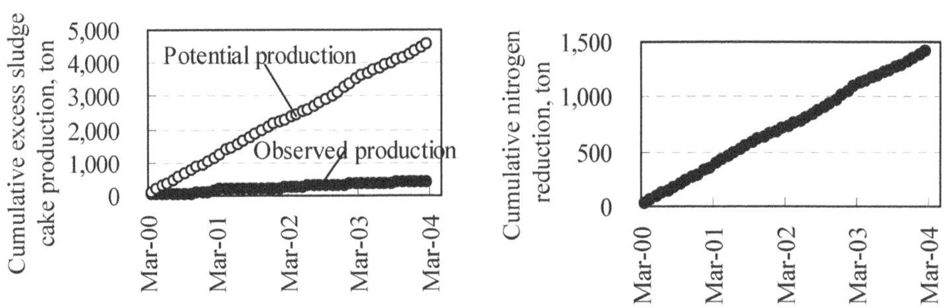

Figure 11 Cumulative reduction in sludge (left) and nitrogen (right): Potential production of the sludge was calculated based on nitrogen loading record, using the following figures: methanol addition = 2.9 kg/kg-N, sludge yield = 0.15, and moisture content of the cake = 85%.

Conclusion

A full-scale denitrification plant showed 90% reduction in its sludge production, by returning ozonated sludge to the denitrification tank. The ozonated sludge was utilized as an electron donor to the denitrifying reaction, resulting in reduced requirements of methanol addition. The COD loading rates should be designed so as to make appropriate MLSS concentrations, which affect the sludge separation in the clarifier. This fact impressed that the proper designing of biological process is by no means less important than ozonation part. In the past four years of the full-scale operation, the overall reduction efficiency of the sludge was over 90%, and total amount of the reduced sludge cake was estimated to be over 4,000 ton, assuming its water content to be 85%. In the same period, nearly 1,500 tons of nitrogen was removed from the waste stream to an enclosed sea area. The total number of the full-scale plants having similar process configurations reached 33 in Japan, with application both in industrial and municipal activated sludge process.

References

Fujiyasu, H., Tokuhara, M., Kitagawa, M., Okano, H. (2000) Method and apparatus for biological treatment of nitrate nitrogen, *Published Patent Application Gazette*, JP-2000-325985-A, Japan Patent Office.

Komatsu, K., Tokutomi, T., Nishimura, S., Tanaka, M., Yasui, H. (2002) The relationship between zone-settling velocity of activated sludge and SVI, *Environmental Engineering Research*, **39**, 21-29, *in Japanese*.

Tokuhara, M., Nishimura, S., Fujii, H. (2002) Method for treating nitrogen containing wastewater, *Published Patent Application Gazette*, JP-2002-192189-A, Japan Patent Office.

Vesilind, P.A., (1968) Design of prototype thickeners from batch settling tests. *Wat. Sewage Wks.*, **115**, July.

Yasui, H., Nakamura, K., Sakuma, S., Iwasaki, M., and Sakai, Y. (1997) A full-scale operation of a novel activated sludge process without excess sludge production, *Wat. Sci. Tech.* **34**(3/4), 395–404.

Yasui, H., Goel, R., Fukase, T., Matsuhashi, R., Wakayama, M., Sakai, Y., Noike, T. (2003) Bioleader^{TM}, a novel activated sludge process minimizing excess sludge production, *Proceedings, Global Conference on Leading Edge Water and Wastewater Treatment Technologies, IWA*

Hydrogen-Based Membrane Biofilm Reactor for Wastewater Treatment

Bruce E. Rittmann, Robert Nerenberg

Northwestern University, Department of Civil and Environmental Engineering, 2145 Sheridan Road, Evanston, IL 60208-3109 U.S.A.; b-rittmann@northwestern.edu, Nerenberg.1@nd.edu

Beverley Stinson, Dimitrios Katehis, Echo Leong, and James Anderson

Metcalf & Eddy of New York, Inc., 60 East 42nd Street, New York, NY 10165 U.S.A.; Beverley.Stinson@m-e.com, Dimitrios.Katehis@m-e.com, Echo.Leong@m-e.com, James.Anderson@m-e.com

Abstract The H_2-based membrane biofilm reactor (MBfR) delivers H_2 gas to a biofilm that naturally accumulates on the outer surface of bubbleless membranes. Although the MBfR is proven for the reduction of nitrate and perchlorate in drinking-water and groundwater settings, its most extensive application may be for advanced nitrogen removal in wastewater treatment, where existing approaches either fail to achieve the goals of advanced-N removal or have severe problems of cost, reliability, and safety. By utilizing H_2 gas as the electron donor to drive denitrification, the MBfR completely eliminates an added organic electron donor, which overcomes major problems: a large increase in excess biomass generation, over- or under-dosing of donor, safety concerns, and relying on specialized methanotrophs. In addition, the MBfR is simple to operate and can be used for tertiary denitrification or placed within a pre-denitrification process. Preliminary results reported here show that a biofilm of autotrophic denitrifiers accumulates rapidly in the wastewater setting, the MBfR can drive NO_3^- concentrations below 1 mgN/L, and the H_2 pressure controls the NO_3^- flux.

Keywords Biofilm, Denitrification, Hydrogen, Membrane, Nutrients, Wastewater

Introduction

One of the emerging challenges for wastewater treatment is achieving very low effluent concentrations of total nitrogen (TN) and total phosphorus (TP). Increasingly severe problems with eutrophication and hypoxia in lakes, reservoirs, estuaries, and the near-shore ocean are forcing environmental regulators to impose more stringent effluent requirements on TN and TP. For example, an effluent standard for TN could be 1 mgN/L when the discharge is to a sensitive water body; it is possible that a receiving-water standard of 0.12 mgN/L could be applied if the wastewater were the dominant water input.

Existing wastewater-treatment technology is capable of taking effluent TN down to the range of 10 – 15 mg//L, but it is neither reliable nor cost-effective for achieving ≤1 mgN/L. The key for taking TN down to the 1-mg/L level is stable denitrification that drives NO_3^--N to a few tenths of a mg/L. Stable nitrification can drive NH_4^+-N to a few tenths of a mg/L, and filtration can bring organic N to almost zero (Rittmann and McCarty 2001). If soluble organic nitrogen can be held to a few tenths of a mg/L, the total N can be in the region of 1 mg/L: e.g., 0.2 mg/L NH_4^+-N, 0.3 mg/L NO_3^--N, and 0.5 mg/L soluble organic N.

© 2004 IWA Publishing. *2nd IWA Leading-Edge Conference on Water and Wastewater Treatment Technologies* edited by Mark Van Loosdrecht & Jonathan Clement. ISBN: 1 84339 508 8.

Pre-denitrification is an excellent approach to utilize the influent BOD to fuel denitrification, but realistic constraints on the mixed-liquor recycle rate limit it to about 75% N removal, which leaves around 10 mg/L TN in the effluent when the influent is around 40 mgN/L (Rittmann and McCarty 2001). Furthermore, a high influent TKN:BOD ratio can foil the pre-denitrification strategy as a means for total N removal. Return of digester supernatants is a common situation leading to a high influent TKN:BOD ratio.

Tertiary denitrification using an organic electron donor, such as methanol or acetate, could, in principle, drive effluent NO_3^- to a few tenths of a mgN/L. However, the dosing of the organic donor cannot be controlled well enough to ensure full NO_3^- removal without massive donor overdosing that increases effluent BOD and wastes money. In addition, tertiary denitrification using an organic donor significantly increases excess sludge production and often involves special chemical handling. For example, methanol (CH_3OH) is popular for its relatively low cost, but methanol is a dangerous chemical that is toxic to humans, is regulated, has very difficult handling properties, and is oxidized only by specialized methanotrophs.

A revolutionary new approach that should overcome most of the limitations of traditional denitrification exploits the H_2-based membrane biofilm reactor (MBfR), which has been extensively studied for treating drinking water or groundwater contaminated with nitrate, perchlorate, or other oxidized contaminants (Rittmann et al. 2004). The MBfR offer great potential to augment or replace traditional denitrification so that very low TN – in the region of 1 mg/L – can become a reality.

This paper first reviews the MBfR approach and previous experience oriented toward drinking water and groundwater treatment. It then introduces the several ways in which the MBfR can be adapted for treating wastewater to achieve advanced TN removal. Finally, the paper summarizes preliminary results in which the MBfR was applied in a wastewater setting.

Review of the MBfR and Previous Experience

The technological goal of the MBfR is to deliver H_2 gas as an electron donor to autotrophic bacteria that reduce NO_3^- or other oxidized contaminants. H_2 is the ideal electron donor for fueling denitrification and other reductions because of the following inherent advantages over organic electron donors:

- H_2 is a low-cost source of electrons
- H_2 supports autotrophic bacteria, which totally eliminates the need for an organic C source
- H_2 produces far less excess biomass
- H_2 cannot have a significant residual in the water and cannot increase effluent BOD
- H_2 is non-toxic to humans
- H_2 can be purchased in bulk or generated on-site
- Bacteria use H_2 to reduce most oxidized contaminants

Despite its inherent advantages, H_2 has not been used an electron donor in the past, because no efficient and safe delivery system was available. Its low water solubility (only 1.2 mg/L in equilibrium with 1 atm) means that it cannot be supplied in a water stream. Its flammability and low water solubility mean that H_2 cannot be sparged.

The MBfR overcomes the limitations on H_2 delivery. H_2 gas is supplied to the interior of special hollow-fiber membranes that are "bubbleless," or have no continuous pores. We have been using a composite bubbleless membrane produced by Mitsubishi-Rayon, but other membranes may be suitable. H_2 diffuses through the membrane wall. Molecules of H_2 reaching the outside of the membrane are oxidized by H_2-oxidizing, autotrophic bacteria that form a naturally occurring biofilm. The biofilm bacteria use most of the electrons to reduce the oxidized contaminants to harmless products: e.g., NO_3^- is reduced to N_2 gas, and perchlorate (ClO_4^-) is reduced to Cl^- ion. Supplying H_2 directly to the biofilm results in rapid and nearly 100% utilization of H_2, both of which enhance process efficiency and safety. In addition, the supply rate of H_2 to the biofilm is self-regulated by the loading of oxidized contaminant to the biofilm. Thus, H_2 delivery with the MBfR is efficient, safe, and simple.

Denitrification

The original research on the MBfR addressed denitrification in the drinking-water setting (Lee and Rittmann 2000, 2002, 2003). For denitrification, NO_3^- is the electron acceptor, and H_2 oxidation supplies the electrons:

Acceptor reaction:	$NO_3^- + 6H^+ + 5e^- \rightarrow 0.5N_2 + 3H_2O$
Donor reaction:	$2.5H_2 \rightarrow 5H^+ + 5e^-$
Overall:	$NO_3^- + 2.5H_2 + H^+ \rightarrow 0.5N_2 + 3H_2O$

In drinking-water treatment, the goal is to remove NO_3^- to below the standard, which typically is 10 mgN/L. Therefore, achieving partial NO_3^- removal to well below the standard is the goal, as long as partial removal is reliable and no other water-quality problems are introduced. Other problems could include accumulation of nitrite, release of too much biomass, production of biodegradable dissolved organic carbon (BDOC), or a large pH increase.

The extensive research with a bench-scale MBfR (Lee and Rittmann 2000, 2002, 2003) showed that partial or full NO_3^- removal was possible and easily controllable. Adjusting the H_2 pressure to the interior of the membranes simply and reliably controlled the effluent concentration of NO_3^--N. Within the H_2-pressure range of about 0.15 to 0.6 atmospheres, the effluent NO_3^- concentration could be controlled systematically from less than 0.1 mgN/L for the higher H_2 pressures and low to moderate NO_3^- surface loading to 10 mgN/L for the lower H_2 pressure and a high NO_3^- surface loading. Increasing the H_2 pressure inside the membranes increased the H_2-delivery capacity, making it possible to drive the NO_3^- concentration to very low level, treat a higher surface loading of NO_3^-, or a combination of both. In most cases, the NO_2^- concentration was less than 1 mgN/L, and increasing the H_2 pressure made it possible to drive effluent NO_2^- to less than 0.1 mgN/L.

The denitrification research also showed that the biofilm that accumulated on the outside of the membranes in the bench-scale MBfR was dense and strong. The specific detachment rate was very low (< 0.02/d), and the effluent biomass concentration was correspondingly low (c. 1 mg/L). The autotrophic biofilm produced some soluble microbial products, giving a typical effluent BDOC of 0.5 mg/L, which can be eliminated by downstream biofiltration (Rittmann and McCarty, 2001). The net acid consumption of denitrification required attention to pH buffering, but this situation is true for all denitrification processes.

Perchlorate Reduction and Pilot Testing

Being a major component of rocket fuel, perchlorate is an emerging oxidized contaminant in areas affected by military bases and rocket manufacturing and testing. Perchlorate affects thyroid function and is an endocrine disruptor. Although no drinking-water standard is yet in place, the State of California has an action level of 4 μg/L, and the U.S.E.P.A. anticipates that its health-based standard ultimately will be in the range of 1 – 4 μg/L. Perchlorate can be bacterially respired in a stepwise 8-electron reaction that produces Cl^- ion.

Acceptor reaction: $ClO_4^- + 8H^+ + 8e^- \rightarrow Cl^- + 4H_2O$

When 4 moles of H_2 provide the 8 electrons, the overall reaction is

Overall reaction: $ClO_4^- + 4H_2 \rightarrow Cl^- + 4H_2O$

Bench-scale experiments (Nerenberg et al. 2002, 2004; Nerenberg and Rittmann 2002; Rittmann et al. 2004) proved that an MBfR active in denitrification reliably reduces ClO_4^- to below the action level of 4 μg/L, that the H_2 pressure to the membrane is the sensitive control on the capacity of reduce ClO_4^-, and that prolonged feeding of ClO_4^- enriches the biofilm in perchlorate-reducing bacteria, although they are present in natural denitrifying populations. The bench-scale work also showed that oxygen and nitrate are good electron acceptors to support perchlorate-reducing bacteria, although their concentrations in the MBfR must be very low to preclude inhibition of perchlorate reduction.

Field-scale pilot testing was carried out at La Puente, California (Rittmann et al. 2004; Nerenberg et al. 2004, Adham et al. 2003). The pilot system consisted of two columns each having ~7,000 hollow-fiber membranes and received a flow rate around 2 L/min. The La Puente groundwater contained approximately 60 μg/L of ClO_4^- and 5.6 mgN/L of NO_3^-. After a start-up period in which practical operating problems were overcome, the pilot-scale system achieved excellent ClO_4^- removal, typically at or below the 4-μg/L action level. Nitrate also was removed to about 0.2 mgN/L, and O_2 was completely removed. One of the most important contributions of the pilot study was quantifying the H_2 use rate, which could not be measured with the small gas flows in the bench-scale studies. The measured H_2 use rate was very close to 100% of the theoretical use rate based on the consumption rate of the three acceptors entering the MBfR: NO_3^-, O_2, and ClO_4^-. The 100% H_2 use means that the MBfR wastes no electron donor, which is essential for good economy, safety, and effluent quality.

Other Oxidized Contaminants

One potential advantage of using H_2 gas with the MBfR is that most oxidized contaminants should be reduced by microbial catalysis. A screening study (Nerenberg and Rittmann 2004) tested eight new oxidized contaminants in short term tests involving two bench-scale MBfRs that had biofilm grown with NO_3^- or O_2 as the primary electron acceptor. Approximately 1 mg/L of one new contaminant was fed for two hours (enough to give hydraulic steady state), and the removal of the applied contaminant was then measured. Each contaminant showed significant reduction, at least 29%. The oxidized contaminants and their % removals in the nitrate MBfR are summarized here: arsenate ($H_2AsO_4^-$) >50%, bromate (BrO_3^-) >95%, chlorate (ClO_3^-) 29%, chlorite (ClO_2^-) 67%, chromate (CrO_4^{2-}) >75%, dichloromethane (CH_2Cl_2) 45%, selenate (SeO_4^{2-}) 74%, and selenate (SeO_3^{2-}) 57%. Based on the results with perchlorate (Nerenberg et al. 2002), the removal percentages are likely to increase with continuous feeding of the oxidized contaminants; thus, the removals obtained in the screening study probably are minima. More in-depth research is on going.

Extending the MBfR to the Wastewater Setting

Adapting the MBfR to wastewater treatment should achieve two major goals.

- Eliminate any organic electron donor, which will to minimize excess sludge production, minimize chemical costs, eliminate the need to use specialized methanotrophs, and eliminate the possibility of donor over-dosing.
- Provide a simple system that is easily integrated into existing wastewater-treatment systems.

The MBfR can be integrated into existing or new activated-sludge designs in two distinct ways.

- Using the MBfR for tertiary denitrification, or post-treatment to remove NO_3^- remaining after conventional treatment, such as pre-denitrification.
- Placing the MBfR units directly in a pre-denitrification system to enhance its performance without constructing a tertiary-treatment process.

Tertiary Denitrification

The goal of tertiary denitrification with the MBfR is to reduce the effluent NO_3^- concentration to an advanced-treatment standard (e.g., $\leq$ 1 mgN/L) when a conventional pre-denitrification process brings the NO_3^- concentration down to 10 – 15 mgN/L. Figure 1 illustrates how the tertiary system would be employed in a typical post-treatment scenario. This application is similar to the drinking-water settings that have been investigated for denitrification, although two differences are evident. First, the effluent criterion for NO_3^- is lower for wastewater treatment: $\leq$ 1 mgN/L versus well below the drinking-water standard of 10 mgN/L. Second, the influent to the MBfR is likely to contain a significant concentration of suspended solids, which are absent or negligible in drinking-water treatment. The physical configuration of the MBfR must be adjusted to accommodate influent solids without fouling the MBfR. One advantage of using the H_2-based MBfR in tertiary denitrification is that the autotrophic biomass yield is low, which means that the solids concentration in the effluent will be low (Lee and Rittmann 2002). This minimized solids load to downstream filters, if they are employed. If no downstream filters are in place, the effluent suspended solids, COD, and N due to released biomass will be low.

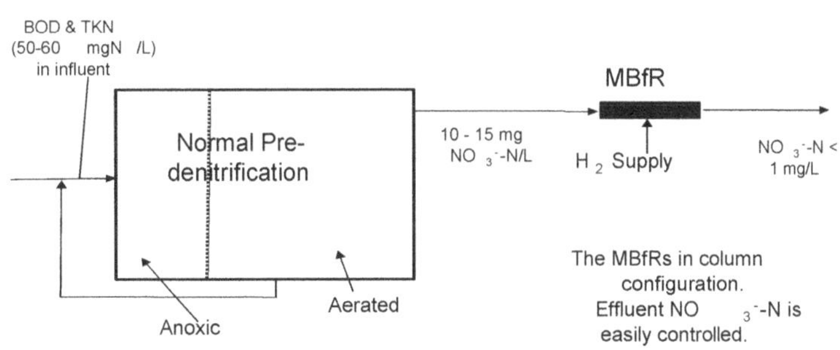

Figure 1 Schematic of how the MBfR can be used for tertiary denitrification to bring the effluent NO_3^- concentration from the typical level of pre-denitrification to an advanced-treatment level.

Integration into Pre-denitrification

Integrating the MBfR into pre-denitrification obviates the need to construct any tertiary-treatment process. The benefits for capital costs and space are obvious. Figure 2 illustrates how MBfR units could be integrated into a multiple-pass, pre-denitrification system to augment the capacity for NO_3^- removal. Integration creates a hybrid biofilm/suspended-growth system in which the biofilm is dominated by H_2-oxidizing autotrophs, while the suspended bacteria are BOD-oxidizing heterotrophs and autotrophic nitrifiers. Achieving integration involves a number of technical challenges: preventing fouling of the membranes from the suspended biomass and influent solids, good mass transfer to the membranes, biofilm control on the membranes, controlling the oxygen concentration to allow good nitrification but not create a large H_2 demand to reduce O_2, and locating the MBfR units in the best place to gain very low effluent NO_3^- concentration while using the influent BOD as much as possible for denitrification.

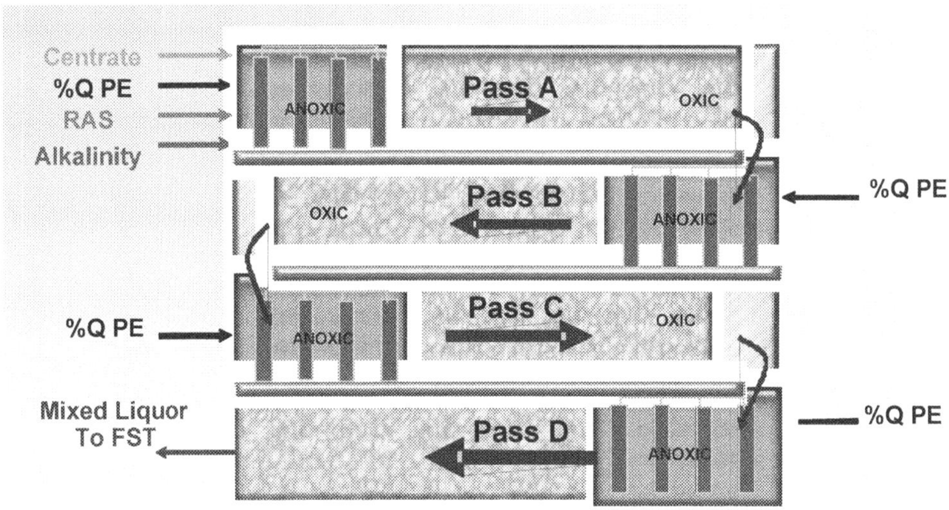

Figure 2 Illustration of how MBfR units could be integrated into a pre-denitrification system to augment the removal of NO_3^-.

Experience in the Wastewater Setting

Preliminary studies were carried out as preparation for pilot studies on wastewater denitrification with the MBfR. The preliminary studies utilized a novel "open matrix" MBfR having 206 cm^2 of membrane area in a volume of 300 cm^3, which gives a specific surface area of 0.69 cm^{-1} = 69 m^{-1}. The concept of the "open matrix" is to allow mixed liquor to move between the membrane fibers without being filtered out or fouling the membrane surface. The influent contained the effluent from the first-stage of the multiple-stage pre-denitrification plant in New York City. The influent to the MBfR had a NO_3^- concentration of 10 - 20 mgN/L. The flow rate was 2.3 L/d, giving an empty-bed hydraulic retention time of 3 h. No inoculum was provided before feeding the wastewater to the MBfR.

Denitrification is the MBfR started immediately and achieved a high level of denitrification within a few days. Effluent NO_3^- was driven to well below 1 mgN/L, and H_2 pressure gave sensitive control of the denitrification capacity. For example, when the H_2 pressure was only 2 psi (0.14 atm) and influent NO_3^- was 13 mgN/L, the effluent NO_3^- was

0.85 mgN/L, giving a NO_3^- flux of 1.4 gN/m^2-d. Increasing the H_2 pressure to 5 psi (0.34 atm) when the influent NO_3^- was 16.4 mgN/L gave effluent NO_3^- of only 0.4 mgN/L, with a NO_3^- flux of 1.8 gN/m^2-d.

Despite the open-matrix configuration, extended operation led to excess biofilm or suspended-solids accumulation, which caused some membrane fibers to clump together. Clumping reduced the biofilm surface area and the mass-transport rate to the biofilm. For example, a H_2 pressure of 5 psi (0.34 atm) gave a nominal NO_3^- flux of 1 gN/m^2-d and had effluent NO_3^- and NO_2^- concentrations of 3.6 and 1.6 mgN/L, respectively, after clumping. The deficiency of the open-matrix design in the 300-mL reactor was too-low turbulence and mixing around the membranes, and this will be a primary design goal for forthcoming pilot studies.

Conclusions

The H_2-based MBfR has been proven for the reduction of nitrate and perchlorate in drinking water and groundwater settings, and it shows promise for a range of other oxidized contaminants in water. Perhaps its most extensive application will be for advanced nitrogen removal in wastewater treatment, where existing approaches fail to achieve the goals of advanced-N removal; have severe problems of cost, reliability, and safety; or both. By utilizing H_2 gas as the electron donor to drive denitrification, the MBfR completely eliminates an added organic electron donor, which overcomes major problems: a large increase in excess biomass generation, over- or under-dosing of donor, safety concerns, and relying on specialized methanotrophs. In addition, the MBfR is simple to operate, and it can be used for tertiary denitrification or integrated into a pre-denitrification process. Preliminary results show that autotrophic denitrifiers accumulate rapidly in the wastewater setting, the MBfR can drive NO_3^- concentrations below 1 mgN/L, and the H_2 pressure controls the NO_3^- flux. Some special features will be needed to deal with suspended solids and excess biofilm accumulation in the wastewater setting.

References

Adham S., Gillogly T., Lehman G., Rittmann B. and Nerenberg R. (2003). *Application of Bioreactor Systems to Low-Concentration Perchlorate-Contaminated Water*, Amer. Water Works Research Foundation, Denver, CO.

Lee K.-C. and Rittmann B.E. (2000). A novel hollow-fiber membrane biofilm reactor for autohydrogenotrophic denitrification of drinking water. *Wat. Sci. Tecl.*, **41**(4-5), 219 – 226.

Lee K.-C. and Rittmann B.E. (2002). Applying a novel autohydrogenotrophic hollow-fiber membrane biofilm reactor for denitrification of drinking water. *Wat. Res.*, **36**, 2040-2052.

Lee K.-C. and Rittmann B.E. (2003). Effects of pH and precipitation on autohydrogenotrophic denitrification using the hollow-fiber membrane biofilm reactor. *Wat. Res.*, **37**, 1551-1556.

Nerenberg R. and Rittmann B.E. (2002). Perchlorate as a secondary substrate in a denitrifying hollow-fiber membrane biofilm reactor. *Wat. Sci. Tech.*, **2**(2), 259-265.

Nerenberg R. and Rittmann B.E. (2004). Reduction of oxidized contaminants with a hydrogen-based, hollow-fiber membrane biofilm reactor. *Wat. Sci. Tech.*, in press.

Nerenberg R., Rittmann B.E., and Najm I. (2002). Perchlorate reduction in a hydrogen-based membrane biofilm reactor. *J. Amer. Wat. Works Assn.*, **94**(11), 103-114.

Nerenberg R., Rittmann B.E., Gillogly T.E., Lehman G.E., and Adham S.S. (2004). Perchlorate reduction using the hollow-fiber membrane-biofilm reactor: bench and pilot-scale studies. *Proc. 2003 Battelle Symposium on In Situ and On Site Bioremediation,* Orlando, FL, June 2003. Paper C-08 on CD-ROM.

Rittmann B.E. and McCarty P.L. (2001). *Environmental Biotechnology: Principles and Applications,* McGraw-Hill Book Co., New York.

Rittmann B.E., Nerenberg R., Lee K.-C., Najm I., Gillogly T.E., Lehman G.E., and Adham. S.S. (2004). The hydrogen-based hollow-fiber membrane biofilm reactor (HFMBfR) for reducing oxidized contaminants. *Wat. Sci. Tech.: Water Supply* **4**(1): 127-133.

Biofilm Treatment of MTBE

M. A. Sedran*, M. T. Suidan*, A. D. Venosa***

* University of Cincinnati, Civil and Environmental Engineering, 701 Engineering Research Center, P.O. Box 210071, Cincinnati, OH 45221-0071
** U. S. Environmental Protection Agency National Risk Management Research Laboratory, 26 W. Martin Luther King Drive, Cincinnati, OH 45268

Abstract Methyl tert-butyl ether (MTBE) is a gasoline additive used to improve air quality by reducing emissions. Unfortunately, MTBE has leaked from underground storage tanks into the groundwater and threatens drinking water supplies. This study investigates the degradation of MTBE and BTEX by a fluidized bed reactor (FBR). The culture in the aerobic FBR with granular activated carbon (GAC) as a biological attachment medium degraded MTBE and BTEX. The culture in the FBR degraded 99.7% of the influent MTBE with and without BTEX present. Aeration resulted in minimal volatilization. At an empty bed contact time (EBCT) of one-hour, the MTBE effluent concentration averaged 0.0215 mg/L and each BTEX compound was at an effluent concentration below 0.003 mg/L. The effluent concentrations of MTBE, TBA, and BTEX increased at decreased EBCTs.

Keywords Biofilms, MTBE, BTEX, biodegradation.

Introduction

MTBE (methyl tert-butyl ether) is an octane-enhancing gasoline additive used in areas with poor air quality to reduce air emissions. Although MTBE has improved air quality, its use is now under question due to widespread groundwater contamination through leakage from underground storage tanks. MTBE groundwater contamination is a concern because of the potential adverse health effects from MTBE entering drinking water sources. The U.S. E.P.A. classified MTBE as a possible human carcinogen and released an advisory recommends drinking water levels of MTBE in the range of 0.020 to 0.040 mg/L (U.S. E.P.A., 1997).

BTEX (benzene, toluene, ethylbenzene, xylene) is a combination of compounds in gasoline that will also leak from USTs. BTEX groundwater contamination can also cause damaging health effects to exposed individuals. Benzene is a known carcinogen and has a drinking water maximum contaminant level (MCL) of 0.005 mg/L. Toluene, ethylbenzene, and the total xylenes have MCLs of 0.7, 1.0, and 10.0 mg/L, respectively. Fuel oxygenates like MTBE are far more water soluble than gasoline hydrocarbons like BTEX and have poor adsorption characteristics (Sulfita and Mormile, 1993). Therefore, MTBE travels further and faster in the groundwater than other gasoline components that may enter into groundwater systems. MTBE and BTEX can be present together near the source of the plumes.

© 2004 IWA Publishing. *2nd IWA Leading-Edge Conference on Water and Wastewater Treatment Technologies* edited by Mark Van Loosdrecht & Jonathan Clement. ISBN: 1 84339 508 8.

Biodegradation of MTBE

Several studies (Jenson and Arvin, 1990; Fayolle et al., 1998) indicated that MTBE is resistant to biodegradation, probably because of the high energy required to cleave its ether bond. Salanitro et al. (1994) were the first to report aerobic biodegradation of MTBE by a mixed bacterial culture, with the formation of tert-butyl-alcohol (TBA) as a degradation by-product. Later studies (Fortin et al., 1999; Wilson et al., 2000; Bradley et al., 2001; Hunkeler et al., 2001;) reported successful aerobic biodegradation of MTBE as the sole carbon source by mixed cultures. Others (Steffan et al., 1997; Hanson et al., 1999) isolated pure cultures capable of degrading MTBE.

Previous studies have shown the presence of another carbon source to have a negative effect on the degradation of MTBE by pure cultures (Mo et al., 1997; Hardison et al., 1997 Deeb et al., 2001; Liu et al., 2001). Mixed cultures have been observed to be resistant to biodegradation of MTBE in the presence of other carbon sources (Hunkeler et al., 2001; Schroeder et al., 2000). In contrast, other studies with mixed cultures have shown the degradation of MTBE to improve in the presence of other substrate (Garnier et al., 1999; Kharoune et al., 2001).

Biofilm degradation of MTBE has been investigated. Studies have also shown biofilm treatment of MTBE in air streams (Eweis et al., 1998; Fortin et al., 1999) with biofilters. MTBE removal from the liquid phase has also been seen with fixed biofilms in a biotrickling filter (Converse et al., 1998) and in activated carbon packed-bed columns (Chang et al., 2000).

Fluidized Bed Reactors

Biological fluidized bed reactors (FBRs) utilize a medium such as granular activated carbon (GAC), anthracite, or sand to provide a surface of attachment for biological activity, thus creating a biofilm to degrade contaminants. In a FBR, contaminated water enters the bottom of column containing the attachment medium and the upward flow of the water is sufficient to suspend or fluidize the medium. The fluidizing mechanism of the FBR is beneficial as it reduces the thickness of the hydrodynamic boundary layer, which decreases the resistance to mass transport in the reactor to provide more opportunities for successful biodegradation (Suidan et al., 1988). However, fluidization may be a problem as it may shear off biomass. Another disadvantage to FBRs may be a long start-up period required to grow the biomass on the medium.

GAC surface roughness and high surface area are favorable for microbial attachment, therefore making it an especially effective medium for FBRs. Although GAC has an internal structure that is not completely solid, Fox et al. (1990) reported that microbial growth occurs on the surface of the GAC particle and not in the internal areas of the particle. However, the growth on the GAC particles is still greater than the growth on sand or anthracite because of the deep crevices in the GAC surface that allow protection for the biomass from the shear that can occur from the fluidization in FBRs.

Reports have shown that biomass retention is critical to MTBE degradation because of the low biological yield of MTBE (Morrison et al., 2002; Salanitro et al., 1994; Wilson et al., 2000). The use of FBRs may prove effective for the biological removal of MTBE because of the biological attachment medium, GAC, which will help retain biomass.

Research Objectives

This study investigates the biodegradation of MTBE and BTEX using an aerobic fluidized bed reactor (FBR) with granular activated carbon (GAC) as a biological attachment medium, thus creating a biofilm. The FBR was operated at decreased empty bed contact times (EBCTs) and the effluent concentrations of MTBE, TBA, and BTEX were monitored.

Materials and Methods

The FBR, with a total volume of 7.9 L, was designed with a main column, an influent header space, an effluent header space, and a recycle line, each constructed with Plexiglass and PVC tubing (Corro-Flo Engineering, Cincinnati, OH). The main column was surrounded by a temperature control system using a flowing propylene glycol/water mixture coolant to maintain a reactor temperature of 20°C (68°F). The influent header contained marbles to uniformly distribute the fluidizing liquid and prevent GAC from entering the recycle line. Initially, 500 g of 16 x 20 U.S. Mesh F400 GAC (Calgon Corporation, Pittsburgh, PA) was placed in the reactor, with a measured dry volume of 1.025 L. The reactor was inoculated with 10 mL of an MTBE only degrading mixed culture on the fifth day of operation (Pruden et al., 2001). Aeration for the reactor was provided by two aeration spargers (Alltech Associates, Inc., Deerfield, IL) at a rate of 7.5 mL/min controlled by a variable flowmeter (Cole-Parmer, Chicago, IL) in order to maintain a dissolved oxygen (DO) level above 3 mg/L.

Initially, the FBR had a one hour empty bed contact time (EBCT) and a total flowrate of 24.6 L/day. The recycle flowrate was set to achieve 150% bed expansion. Three feed solutions were fed into the recycle loop, including a nutrient solution, a buffer solution, and deionized water. The concentrated nutrient solution was acidified and added to provide salts and vitamins needed for biological growth. The sodium carbonate buffer solution was added to balance the acidic nutrients and to maintain a constant pH between 7.4 and 8.1. Initially, the only influent contaminant was MTBE, which was fed at a concentration of 8 mg/L. After 225 days of operation, 2 mg/L of each BTEX compound was added while maintaining the MTBE influent concentration of 8 mg/L. The MTBE and the MTBE and BTEX mixture was fed to the buffer feed line using a Model 11 high precision syringe infusion pump (Harvard Apparatus, Inc, South Natick, MA) equipped with a fixed needle gastight syringe (Hamilton Co., Reno, NV). Once stable performance was achieved, the flowrate was incrementally increased to analyze the performance of the reactor with decreased EBCTs, which are desirable in order to minimize the cost of full-scale operations. The reactor was operated at EBCTs of 1.08 hours, 0.89 hours, 0.63 hours, 0.53 hours, 0.26 hours, and 0.13 hours.

For all experiments, the following chemicals were used: MTBE (Mw = 88.15, 99%, Aldrich, Milwaukee, WI), TBA (Mw = 74.12, 99%, Aldrich, Milwaukee, WI), benzene (Mw = 78.11, 99%, Fisher Scientific, Pittsburgh, PA), toluene (Mw = 92.14, 99%, Fisher Scientific, Pittsburgh, PA), ethyl-benzene (Mw = 106.17, 99%, Fisher Scientific, Pittsburgh, PA), and p-xylene (Mw = 106.17, 99%, Fisher Scientific, Pittsburgh, PA).

FBR performance was monitored for effluent concentrations of MTBE, TBA, and BTEX, pH, dissolved oxygen, and dissolved organic carbon. Liquid and gaseous concentrations of MTBE, TBA and BTEX were determined using a Hewlett Packard 5890 Series II gas chromatograph (GC) (Hewlett Packard, Palo Alto, CA) equipped with a flame ionization detector (FID) and 60/80 Carbopack B/5% Carbowax 20 M glass column (Sulpelco, Bellefonte, PA). Liquid concentrations of MTBE, TBA, and BTEX when present at low concentrations were measured using a Tekmar Dohrmann (Cincinnati, OH) 3100 Sample concentrator with a Hewlett Packard 6890 Series Plus GC. Analysis of dissolved oxygen was monitored with a Corning Checkmate II dissolved oxygen sensor (Corning, New York) and COD was monitored using Hach (Loveland, CO) ultra-low range digestion vials. DOC was measured on a Shimadzu TOC-5000 Analyzer (Shimadzu, Kyoto, Japan) using the difference between the total carbon and inorganic carbon or using non-purgeable organic carbon analysis when the effluent concentrations of MTBE and BTEX were below 0.1 mg/L.

Results and Discussion

An aerobic fluidized bed reactor (FBR) used GAC as the biological attachment medium to degrade MTBE and BTEX. The operation and additional analysis of the reactor provide further insights to FBRs degrading MTBE than previous studies. The culture in the FBR degraded 99.7% of the influent MTBE at an EBCT of one-hour and a temperature of 20°C after a 30-day start-up period. Aeration resulted in minimal volatilization. The gaseous concentration of MTBE averaged 0.0021 mg/L. Table 1.1 summarizes the removal efficiency of the culture in the FBR at an EBCT of one-hour and an influent of 8 mg/L MTBE. DO averaged 5.1 mg/L, and the pH ranged from 7.52 to 8.09. Similar FBR effluent concentrations of MTBE have been reported when operating at temperatures of 25°C, but effluent concentrations increased at lower temperatures (Vainberg et al., 2002). Another FBR required iso-pentane as a cometabolite to degrade MTBE, and a start-up period of up to 200 days was observed before MTBE degradation occurred (Stringfellow et al., 2000 & 2002).

Table 1.1 FBR mass removal of MTBE at an EBCT of one-hour feeding MTBE only.

	Mass flow, mg/day	**Percent of total MTBE influent mass, %**
MTBE in influent	179.3	–
MTBE in effluent	0.43	0.24
MTBE in exhaust air	0.18	0.10
Overall MTBE removal from liquid	178.7	**99.7**

The short start-up period of 30 days was probably observed in this reactor because the reactor was seeded with 10 mL of an MTBE degrading culture (Pruden et al., 2001). The seed culture was from a porous pot reactor being fed with 150 mg/L of MTBE. The VSS concentration of the seed culture was 3500 mg/L, so introducing 10 mL of this sample added 35 mg of VSS to the FBR. It was visually clear that the seeding resulted in a biomass increase in the reactor. After the seeding, the tubing in the recycle line and the marble pack at the bottom of the reactor both began to appear orange-brown in color, indicating that biological growth was occurring. Biological growth was probably also occurring on the black GAC but could not be visually identified.

After 225 days of operation, 2 mg/L of each BTEX compound was added to the feed. Table 1.2 summarizes the MTBE removal at an EBCT of one-hour with MTBE and BTEX in the influent. BTEX addition did not affect the degradation of MTBE. There was no spike in the MTBE effluent concentration after BTEX addition, and the reactor continued to remove 99.7% of the MTBE from the liquid.

Table 1.2. FBR mass removal of MTBE at an EBCT of one-hour feeding MTBE and BTEX.

	Mass flow, mg/day	**Percent of total MTBE influent mass, %**
MTBE in influent	202.9	–
MTBE in effluent	0.48	0.24
MTBE in exhaust air	0.32	0.16
Overall MTBE removal from liquid	202.4	**99.7**

The effluent concentrations of MTBE, TBA, and BTEX increased when the EBCT was decreased. Figure 1.1 displays the FBR performance during EBCTs of 0.89, 0.63, 0.54, and 0.27 hours. Figure 1.2 displays the FBR performance at EBCTs of 0.27 and 0.13 hours and during a double in the influent MTBE and BTEX feeds at an EBCT of 0.27 hours. The MTBE effluent increased slightly from 0.02 mg/L to 0.03 mg/L when the EBCT was decreased from one-hour to 0.89 hours. This stabilized for nine days before the conversion to an EBCT of 0.79 hours.

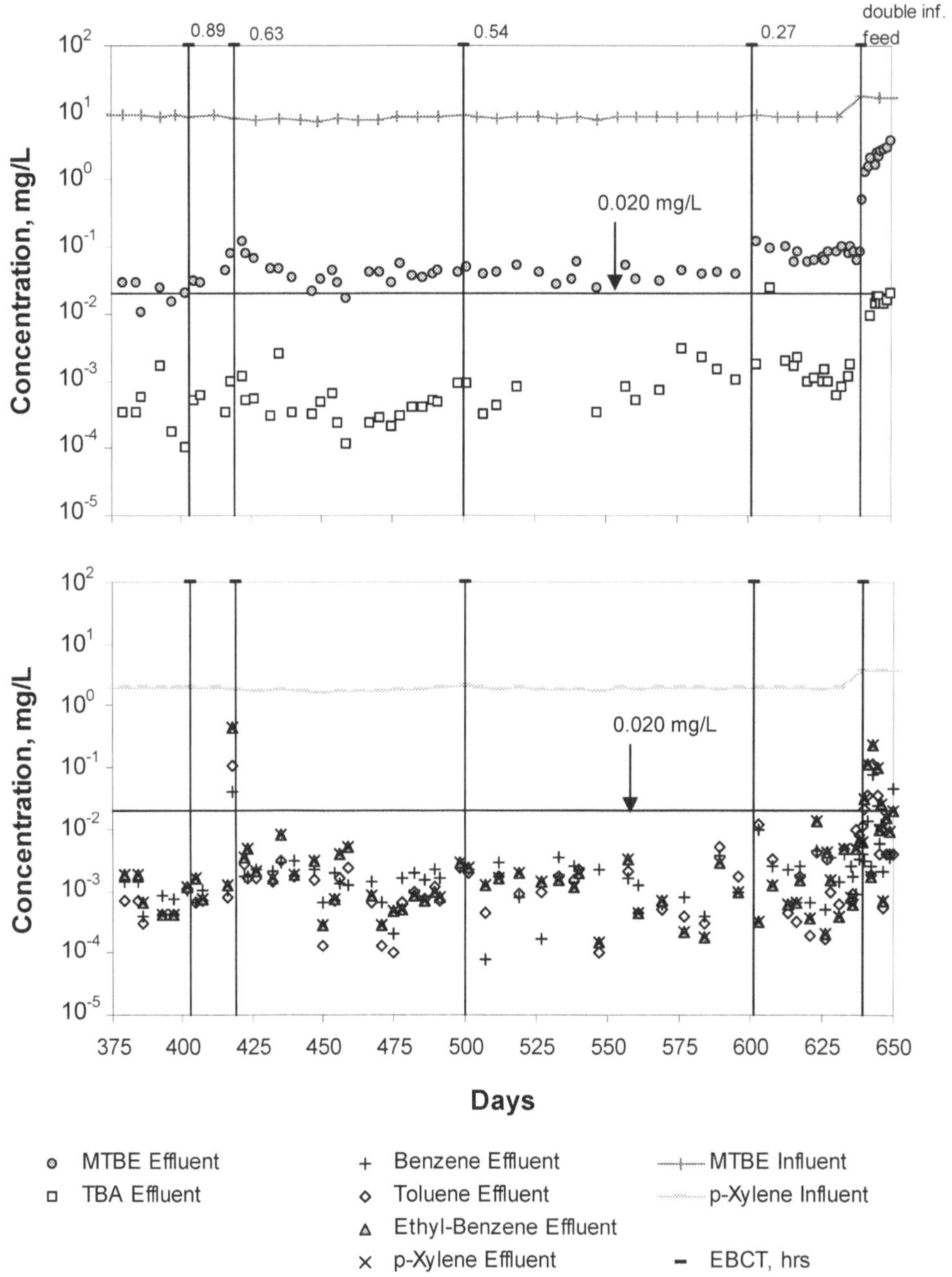

Figure 1.1 FBR performance (log scale) showing influent MTBE effluent MTBE, TBA, (top) and influent p-xylene and effluent BTEX (bottom).at an EBCTs of 0.89, 0.63, 0.54, and 0.27 hours.

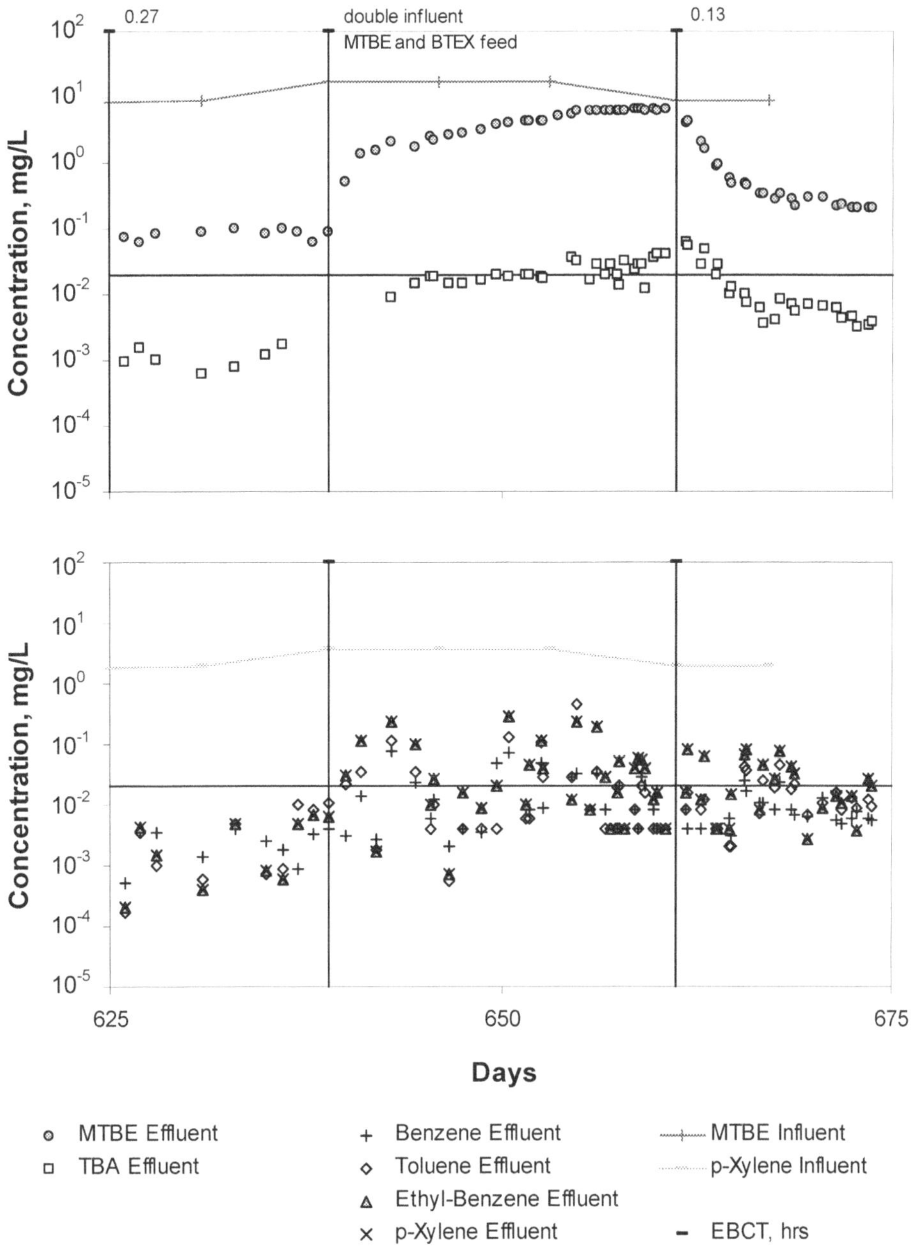

Figure 1.2. FBR performance (log scale) showing influent MTBE effluent MTBE, TBA, (top) and influent p-xylene and effluent BTEX (bottom).at EBCTs of 0.27 and 0.13 hours and during a double of the influent MTBE and BTEX feeds at an EBCT of 0.27 hours.

The EBCT was reduced to 0.63 hours after seven days of operation at an EBCT of 0.79 hours. At this time, the effluent concentrations of MTBE and BTEX increased and then decreased. This initial increase in effluent concentration could be due to an acclimation period required at the EBCT of 0.63 hours. Also, it is possible that the FBR was not operated at an EBCT of 0.79 hours long enough to stabilize. The effluent concentration from the EBCT of 0.79 hours was limited to one sampling event, and the effluent at that time was 0.044 mg/L. The reactor probably required more acclimation time before decreasing the EBCT to 0.63 hours. There was a one month period in which the MTBE effluent concentration continuously decreased before once again reaching stable performance.

The effluent MTBE increased to an average of 0.04 mg/L at an EBCT of 0.53 hours and to 0.08 mg/L at an EBCT of 0.27 hours. After stable performance was achieved at an EBCT of 0.27 hours, the MTBE and BTEX influent concentrations were doubled. The MTBE liquid effluent concentration increased to 6.3 mg/L. The MTBE gaseous effluent concentration increased to MTBE of 0.3 mg/L. This reactor transformation rate was not observed in the batch experiments at high concentrations, indicating a possible limitation with the FBR at high concentrations of an MTBE and BTEX sustained spike. When the flow rate was doubled and the influent concentrations of MTBE and BTEX were decreased to the original levels, the reactor stabilized at a lower concentration. The FBR MTBE effluent concentration averaged 0.21 mg/L at an EBCT of 0.13 hours.

Conclusions

This study showed that a culture in an aerobic FBR with granular activated carbon (GAC) as a biological attachment medium successfully degraded MTBE and BTEX. The culture in the FBR degraded 99.7% of the influent MTBE with and without BTEX present. Aeration resulted in minimal volatilization. At an empty bed contact time (EBCT) of one-hour, the MTBE effluent concentration averaged 0.0215 mg/L and each BTEX compound was at an effluent concentration below 0.003 mg/L. The effluent concentrations of MTBE, TBA, and BTEX increased at decreased EBCTs.

Biomass retention is critical to MTBE degradation. Studies with complete biomass retention systems (Wilson et al., 2000, Morrison et al., 2002) were capable of achieving lower effluent MTBE concentrations than the FBR. While the FBR is capable of retaining biomass, complete biomass retention systems can achieve lower MTBE standards.

References

Bradley, P.M., Landmeyer, J.E., Chapelle, F.H. (2001) Widespread Potential for Microbial MTBE Degradation in Surface-Water Sediments, *Env. Sci. Tech.* **35,** 658-662.

Chang, D.P.Y. (2000) Experience with Laboratory and Field-Scale Ex Situ Biodegradation of MTBE, *USEPA/API Workshop on Biodegrad. of MTBE*, Cincinnati, OH.

Converse, Brett et al. (1998) Cost Estimate Basis of Liquid Phase Biological MTBE Removal, June 16, 1998 UC Toxic Substances Research and Teaching Program Workshop, UC Davis, CA.

Deeb, R.A., Hu, H.Y., Hanson, J.R., Scow, K.M., Alvarez-Cohen, L. (2001) Substrate Interactions in BTEX and MTBE Mixtures by an MTBE-Degrading Isolate, *Env. Sci. Tech.* **35**, 212-317.

Eweis, J.B. (1998) Biodegradation of MTBE in a Pilot-Scale Biofilter, *Proc. from the First Int. Conf. on Rem. of Cholorinated and Recalcitrant Compounds in Monterey, CA*, Battelle Press.

Fayolle, F., Hernandez, G., Le Roux, F., Vandecasteele, J.P. (1998) Isolation of Two Aerobic Bacterial Strains that Degrade Efficiently Ethyl tert-Butyl Ether, *Biotech. Letters*, **20,** 283.

Fortin, N.Y., Deshusses, M.A. (1999) Treatment of Methyl tert-Butyl Ether Vapors in Biotrickling Filters: Analysis of Rate Limiting Step and Behaviour under Transient Conditions, *Env. Sci. Tech.* **33**, 2987-2991.

Fox, P., Suidan, M.T., Brady, J.T. (1990) A Comparison of Media Types in Acetate Fed Expanded-Bed Anaerobic Reactors, *Wat. Res.*, **24** (7)827-835.

Garnier, P., Magana, M., Revah, S. (1999) Cometabolic Biodegradation of Methyl t-Butyl Ether by a Soil Consortium, *Proc. Fifth Int. Sym. In Situ and On-Site Biorem.*, Battelle Press, **5**(3)31-35.

Hanson, J.R., Ackerman, C.E., Scow, K.M. (1999) Biodegradation of Methyl tert-Butyl Ether by a Bacterial Pure Culture, *Appl. Env. Microbio.* **65**, 4788-4792.

Hardison, L.K., Curry, S.S., Ciuffetti, L.M., Hyman, M.R. (1997) Metabolism of Diethyl Ether and Cometabolsim of Methyl tert-Butyl Ether by a Filamentous Fungus, a Graphium sp., *Appl. Env. Microbio.* **63**, 3059.

Hunkeler, D., Butler, B.J., Aravena, R., Barker, J.F. (2001) Monitoring Biodegradation of Methyl tert-Butyl Ether (MTBE) Using Compound-Specific Carbon Isotope Analysis. *Env. Sci. Tech.* **35**, 676-681.

Jenson, H.M., Arvin, E. (1990) Solubility and Degradability of the Gasoline Additive MTBE, and Gasoline Compounds in Water, In *Contaminated soil*, Arendt F, Hinsenveld M, Van den Brink WJ eds. Kluwer, Dorderecht, 445-448.

Kharoune, M., Pauss, A., Lebeault, J. M. (2001) Aerobic Biodegradation of an Oxygenates Mixture: ETBE, MTBE and TAME in an Upflow Fixed-Bed Reactor, *Wat. Res.* **35,** 1665-1674.

Liu, C.Y., Speitel Jr., G.E., Georgiu, G. (2001) Kinetics of Methyl t-Butyl Ether Cometabolism at Low Concentrations by Pure Cultures of Butane-Degrading Bacteria, *Appl. Env. Microbio.* **67,** 2197-2201.

Mo., K., Lora, C.O., Wanken, A.E., Javanmardian, M., Yang, X., Kulpa, C.F. (1997) Biodegradation of Methyl t-butyl ether by Pure Bacterial Cultures, *Appl. Microbio. Biotech.* **47,** 69-72.

Morrison, J.R., Suidan, M.T., Venosa, A.D. (2002) Use of Membrane Bioreactor for Biodegradation of MTBE in Contaminated Water, *J. of Env. Eng.* **128**(9) 836-841.

Purden, A.

Pruden, A., Wilson, G.J., Suidan, M.T., Venosa, A.D. (2001) Biodegradation of methyl tert-butyl ether under various substrate conditions. *Env. Sci. Tech.* **35**, 4235-4251.

Salanitro, J.P., Diaz, L.A., Williams, M. P., Wisniewski, H.L. (1994) Isolation of a Bacterial Culture that Degrades Methyl t-Butyl Ether, *Appl. Env. Microbio.* **60,** 2593-2596.

Schroeder, E.D., Eweis, J.B., Chang, D.P.Y., Veir, J.K. (2000) Biodegradation of Recalcitrant Components of Organic Mixtures, *Water, Air, and Soil Pollution*, **123,** 133-146.

Steffan, R.J., McClay, K., Shannon, M., Rothmel, R., Pitre, M., Condee, C.W. (1997) Biodegradation of Gasoline Oxygenates Methyl tert-Butyl Ether, Ethyl tert-Butyl Ether, and tert-Amyl Methyl Ether by Propane-Oxidizing Bacteria, *Appl. Env. Microbio.* **63**, 4216-4222.

Stringfellow, W.T., Hines, R.D., Cockrum, D.K., Kilkenny, S.T. (2000) Factors Influencing Biolgical Treatment of MTBE in Fixed Film Reactors, *Proc. Sec. Int. Conf. on Rem. Chlor and Recal. Compounds. in Monterey, CA*, Battelle Press, 175-181.

Stringfellow, W.T., Oh, K.C. (2002) Initiation of MTBE Biotreatment in Fluidized-Bed Bioreactors, *J. of Env. Eng.* **128**(9) 852-861.

Suflita, J.M., Mormile, M.R. (1993) Anaerobic biodegradation of known and potential gasoline oxygenates in the terrestrial subsurface, *Env. Sci. Tech.* **27**(5) 976-978.

Suidan, M.T., Pfeffer, J.T., Nakhla, G.F. (1988) Anaerobic Expanded-Bed GAC Reactor for the Treatment of Biologically Inhibiting Wastes Generated During Coal and Petroleum Distillation, In *Anerob. Dig. 1988: Proc. of the Fifth Int. Symp. Anaerob. Dig.*, Pergamon Press, New York.

U.S. E.P.A. (1997) Drinking Water Advisory: Consumer Acceptability Advice and Health Effects Analysis on Methyl Tertiary-Butyl Ether (MtBE), *EPA-822-F-97-009* Washington, D.C.

Vainberg, S., Tonga, A.P., Sutton, P.M., Steffan R.J. (2002) Treatment of MTBE-Contaminated Water in Fluidized Bed Bioreactor, *J. of Env. Eng.* **128**(9) 842-851.

Wilson, G.J., Richter, A.P., Suidan, M.T., Venosa, A.D. (2000) Aerobic Biodegradation of Gasoline Oxygenates MTBE and TBA, *IWA, Int. Meeting Conf. Proc.*, Paris, France.

Landmark Advances in Sludge Conditioning and Dewatering

S.K. Dentel *

*Dept. Civil & Environmental Engineering, Univ. of Delaware, Newark, DE USA 19716
dentel@udel.edu

Abstract There have been many studies of the nature of sludge, its response to mechanical stress, the workings of sludge conditioners, and the connections between sludge origins, conditioning, and the desired final state. Nonetheless, sludge conditioning and dewatering have continued to be practiced as art rather than science. This presentation reviews recent developments that finally may change this picture.

Keywords Sludge, biosolids, dewatering, conditioning, flocculants, solid-liquid separation.

Overview and some caveats

Sludge conditioning and dewatering have long been considered as art rather than science. It has even been pessimistically stated that, in sludge dewatering, "the water always wins." However, recent developments—both scientific and technological in nature—are establishing new landmarks that advance the state of the art, to the extent that we soon may be able to describe these processes as science rather than art.

This short paper can only provide an overview of these advances, and some are so recent that it is difficult to provide archival references in all cases. On the other hand, most of these advances have been evolutionary in nature and do stem from insightful work over the past decades which cannot be cited at length here.

It must also be stated that the growing need for efficient sludge processing technologies has led to intensified competition involving corporate, consulting, and academic sectors, and that, as a consequence, an impartial review may omit significant citations in spite of the best intentions. A final caveat is that the term "biosolids" is not used in this paper: though its function toward the public perception of sludge is recognized, any technical discussion of solid-liquid separation is hindered by terminology that defines the starting material as a solid.

New ways of describing sludge properties

Sludges as gels

An important development in the understanding of sludges is in discerning how a relatively small solids fraction can provide the extensive water structuring that makes dewatering so difficult. Recent research has led away from the earlier pictures of sludge structure, which envisioned spherical "sludge particles" as defining the material's physical and chemical behaviour. A new paradigm is to focus instead on the large-scale (but molecularly based) hydrophilic matrix necessary for water structuring. The chemical and physical

© 2004 IWA Publishing. *2nd IWA Leading-Edge Conference on Water and Wastewater Treatment Technologies* edited by Mark Van Loosdrecht & Jonathan Clement. ISBN: 1 84339 508 8.

characterization of this matrix appears to offer a better model of sludge behaviour as well as a conceptual basis for how sludge properties can be changed (Dentel, 2001).

Poxon gave this physical description of sludges based on her scanning electron micrograph studies: "a space-filling structure (matrix) that is gel-like in its ability to reduce the mobility, and hence the separability, of sludge water." The analogy to a gel was given further validity by LeGrand et al. (1998), based on their experimental comparisons of sludges and synthetic gels.

Three types of gel characteristics are pertinent in describing behaviours of relevance to sludges: rheological, colligative, and macromolecular. These are summarized briefly with respect to recent advances.

Rheology of Sludges and Gels

Until recently, rheological characterisations of sludges relied on static experiments with concentric cylinder viscometers and on fairly simplistic models, such as the Bingham description (see Dentel et al., 1997). Such methods cannot account for such behaviours as thixotropy during mixing, network formation during conditioning, or increasing visco-elasticity as dewatering transpires. More robust models are now being used for this purpose, and they allow the gel-like properties of sludges to be identified and quantified. More advanced rheological measurements are required, however, to inform these models.

The newer rheological approaches include

- Network strength measurements by either constant or ramped shear rate (Yen et al., 2002; Örmeci et al., 2004);
- Oscillatory tests (such as frequency or amplitude sweep tests) to characterize the viscoelastic response of intermediate-state sludges (also known as dynamic mechanical analyses) (Ayol et al., 2004);
- Creep tests, which apply to more concentrated, "paste-type" sludges (e.g. Baudez and Coussot, 2001); and
- Immobilization cell rheometry, which allows dynamic mechanical analyses while filtration is occurring (Ayol et al., 2004).

These types of analyses are new to the field of sludge characterization, and some are new in the larger field of rheology. Ultimately, they will provide a realistic physical model of sludge behaviour during conditioning and dewatering. For example, oscillatory tests provide characterising parameters known as the storage modulus (G') and loss modulus (G''); a comparison of these indicates whether the sludge behaviour is gel-like ($G'>G''$) or dispersion-like ($G'<G''$).

Colligative Properties of Sludges and Gels

Another means of characterizing gels is through their colligative properties. When the "solid" phase and aqueous phase of a suspension are intimately involved, much of the water is within the interfacial region subject to electrical charge interactions. Aside from the imbalance of charge near the charged surfaces, this also effects a significant elevation in total ion concentration in the region. If additional water is to be withdrawn—through dewatering, freezing, or drying—this becomes opposed by significant osmotic pressure differences between the water being withdrawn and the water remaining in the gel. In a sense, dewatering can then be likened to desalination, which works against a similar barrier.

This analogy is an extreme contrast to the "sludge particle" concept of the past. Nonetheless, Mikkelson (1999) and Keiding et al. (2001) provided preliminary but quantitative simulations of filtration, freezing, and drying processes based only on these principles. The striking success of this approach represents an important advance in the understanding of sludge properties. Particularly as processes are developed and used to attain increasing solids concentrations, the colligative properties are sure to become more significant. On the other hand, this conceptualisation by itself cannot describe all important aspects of sludge behaviour, such as the importance of physical shear or the irreversibility of dewatering processes.

Macromolecular Properties of Sludges and Gels

If a sludge truly shares some of the traits of familiar household gels (such as yogurt, gelatine, or toothpaste) but also include particulate material embedded its structure, what are the chemical commonalities that make this so? Based on the above description of gel properties, the water must be contained within a matrix of hydrophilic "solids." The "solids" then include not only particulate material, but also colloidal, subcolloidal, and macromolecular fractions, with the latter of more importance due to the need for abundant interfacial area at the nanoscale. Inorganic constituents may serve this purpose, as with silica gels or aluminium hydroxide gels (explaining the poor dewaterability of alum sludges). However, in wastewater-derived sludges, macromolecular organic substances apparently impart the gel-like properties.

Goodwin and co-workers were the first to argue that these sludge constituents are responsible for the significant water retention and poor dewaterability of many sludges (Pavoni et al., 1972; Forster, 1982; Goodwin and Forster, 1985). Others have elaborated upon this concept in identifying extracellular protein and polysaccharide as particular contributors (e.g. Murthy, 1998). Novak and Bivins (2000) have shown the correlation for higher molecular weight fractions of these organics. Recent findings even suggest that

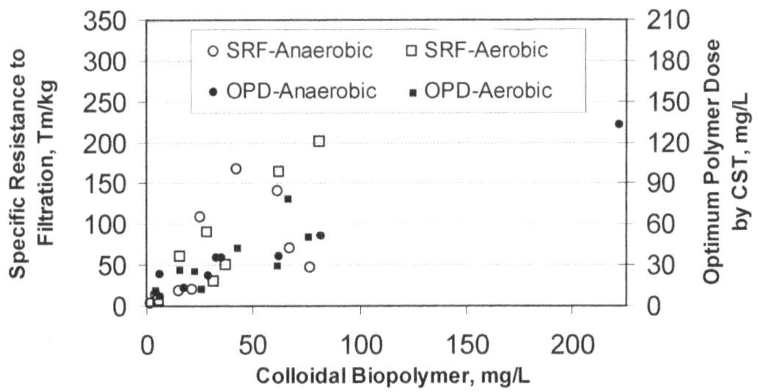

Figure 1 Relationship of specific resistance and optimum polymer doses (based on CST minima) reported by Park et al. (2004). Data readings are approximate.

measured levels of these biopolymers exhibit a linear relationship to both dewaterability and polymer demand (Figure 1, with dewaterability indicated by the specific resistance to filtration of the unconditioned sludge).

The correlations in Figure 1 are logical because such macromolecules should be capable of providing the large-scale hydrophilic matrix necessary for the extensive water structuring that takes place in a sludge. However, these results do not establish that colloidal biopolymer is the *only* factor contributing to conditioning or dewatering requirements.

The likening of sludge to a gel offers more than theoretical advances. As discussed in the following sections, the rheological, colligative, and macromolecular properties can be defined and manipulated to provide practical benefits.

New ways of controlling sludge properties

Utilising the gel model for sludge structure and chemistry offers two advantages. Firstly, it may indicate new ways to improve *existing* methods of modifying sludge properties—particularly, for the purpose of sludge conditioning—and secondly, it may suggest new ways of improving dewaterability and other properties.

Understanding and Optimizing Existing Conditioning Methods

Based on her scanning electron micrographs of sludge conditioned by ferric chloride, Poxon (1996) concluded that the ferric ions collapsed the gel structure into conditioned aggregates. Results of LeGrand et al. (1998), who conditioned a synthetic gel with La^{+3} (as an analog to Fe^{+3}), were consistent with such a collapse mechanism. The metal ions appear to diffuse uniformly throughout a gel, collapsing it into a dense and poorly filterable network. Essentially, the inorganic cations' ability to collapse a gel may be related to both the ability to penetrate into the gel structure (via the hydrated radius) and the reactivity with the gel constituents.

Due to their larger size, cationic polymers may function by destabilizing exterior portions of a gel matrix. The heterogeneous structure that results is more likely to have voids that allow for porous media flow. This distinction relative to metal salt conditioning is supported by the findings of Poxon (1996) and LeGrand (1998). This points to possible conditioning strategies through the combination of inorganic and organic conditioning chemicals. Some studies have indicated benefits from such a combination (Chitikela and Dentel, 1998; Murthy et al., 2000).

If the role played by polymeric flocculants is not to bridge rigid particles, but rather to provide a backbone themselves for floc rigidity, then polymers with a less flexible structure may impart certain advantages. This possibility has in fact been confirmed: the addition of a cross-linking agent during polymer synthesis gives a more "structured" conformation that leads to greater floc strength, albeit at a greater dosage requirement (Dentel et al., 2000; Dentel, 2001). Many newer flocculant products utilize this finding, particularly for processes (such as high-shear mechanical dewatering) that require maximum floc strength. It would be difficult to explain this phenomenon with a "sludge particle" model.

New Ways of Conditioning Biosolids

The gel model also supports two new ways of attacking the problem of sludge conditioning to produce high dewaterable sludges can then be developed: the use of trivalent aluminium during secondary treatment, which complexes these biopolymers as they are formed, and the use of enzymes which degrade them prior to conditioning.

Cation effects and additions

It has long been known that the trivalent metals Al(III) and Fe(III) interact selectively with certain sludge constituents (e.g. Rudolfs and Setter, 1931). Gossett et al. (1978) and Dentel and Gossett (1987, 1988) established that these cations preferentially complex with proteinaceous materials, relative to polymeric polysaccharides. In fact, all of these reports showed the digestion of sludges coagulated with these chemicals to be less efficient, which was the main concern of the above studies. Decreases in digestibility have not been significant when the metal is precipitated as a hydroxide prior to addition, and the use of Fe(II) (ferrous) salts have not caused significant effects.

None of these studies examined the effects of Al(III) or Fe(III) salts on dewaterability of the sludges following digestion. Of course, ferric chloride has been used for many years as a sludge conditioner, although organic polyelectrolytes replaced this additive in many applications. The ferric doses were generally quite high, leading to increased sludge volumes and disposal costs.

New investigations have examined whether the strategic use of these cations in smaller amounts could offer synergistic benefits when combined with polymer conditioning. The hypothesis evolved from observations connecting the level of divalent cations to sludge dewaterability. Logically, if cations of higher valency bind proteins selectively, and extracellular proteins are responsible for the gel-like properties of sludges, then this binding could improve dewaterability.

Historically, the understanding of multivalent cation effects on dewaterability dates to the DLVO model of flocculation. However, works specifically with pure bacterial cultures (Endo et al., 1976; Busch and Stumm, 1968) and with sludge samples (Pavoni et al., 1972; Novak and Randall, 1986; Bruus at al., 1992) suggest a specific binding takes place. Higgins and Novak (1997) were able to reject the DLVO explanation (i.e. non-specific double-layer compression) by contrasting the effects of high Ca^{2+} concentrations with those of high Na^{+} concentrations, the latter actually leading to poor dewaterability.

The landmark insight here is that Fe(III) or Al(III) salts be added to the biological process during floc formation, rather than afterward. In this manner, the cation has more immediate access to the extracellular biopolymer, giving more efficient chemical use and conditioning (Kelly et al., 2000; Murthy et al., 2000). This realization was initially with respect to autothermal aerobically digested sludges but may be applicable more broadly. And, though Al(III) salts have previously been found less effective as sludge conditioners than is Fe(III), an advantage in its addition to anaerobic digestion is that it cannot be chemically reduced and thereby solubilised as can the ferric iron.

It should be noted that these concepts may be limited in their applicability, or require more specific understanding of the chemical nature of the "biopolymers" involved. For example, Alavanja (1999) was unable to demonstrate a cation bridging mechanism with respect to Ca^{+2}, Na^{+}, and NH_4^{+} additions to anaerobically digested sludge. There is also some disagreement in the literature cited here as to the relative importance of proteins vs. polysaccharides, and Novak and co-workers have recently suggested that there is more than one type of protein involved, in order to explain different results with different sludge types.

Enzyme additions

If microbially generated macromolecules are responsible for water retention and structuring in sludges, another promising strategy is the use of enzymes that will degrade these constituents. This notion was first explored by Thomas et al. (1993) with success at both lab and plant scale. There was no information on whether the enzyme use changed the polymer dose required to treat the anaerobically digested sludge following various enzyme additions, but the dewatered cake solids were up by 2-4% in full scale tests. The enzyme product used was a formulation including carbohydrase, lipase and protease activities.

Zhou et al. (2002) treated thermophilically digested sludge with three different types of proteases and found statistically significant improvements in dewaterability for all three, as indicated by CST measurements. It does not appear that residual protein was measured. The experiment was intended to test the role of proteins in dewaterability rather than as a potential treatment process.

Ayol (2004) added a commercial mix of enzymes to pre-treat anaerobically digested sludges prior to polymer conditioning. The mix included enzymes to hydrolyse proteins, polysaccharides, and lipids. Dewaterability was improved, based on CST, solid contents of final product (using a bench-scale filter press), filtrate turbidity, and suspended solid measurements. These results are exemplified in Figure 2 for one of the two biosolids used.

Although field emission scanning electron microscopy revealed no obvious differences in sludge structure, analyses showed lower protein and polysaccharide concentrations after enzyme treatment, also shown in Figure 2.

Ayol and Dentel (2004) evaluated the practical feasibility of using this product for sludge pre-treatment and found it to be favorable. For example, a 10 mg/L enzyme dose prior to a 250 mg/L polymer usage increased the dewatered biosolids TS by approximately 50% (e.g. from 20% solids up to 30%), which would decrease the mass and volume of dewatered material by about 33%. The cost for this enzyme addition is estimated at 9% of the polymer cost. Further work in this direction is clearly warranted.

Parenthetically, it should be noted that the results in Figure 2 appear to conflict with the hypothesis of Novak et al. (2004) that proteinaceous extracellular polymer is responsible for poor dewaterability in anaerobically digested sludges. For these data, improvements in dewaterability seem to correspond more closely to the decrease in polysaccharides than to the decrease in protein, even though this is an anaerobically digested sludge. Similar results were found for a second sludge from a different plant.

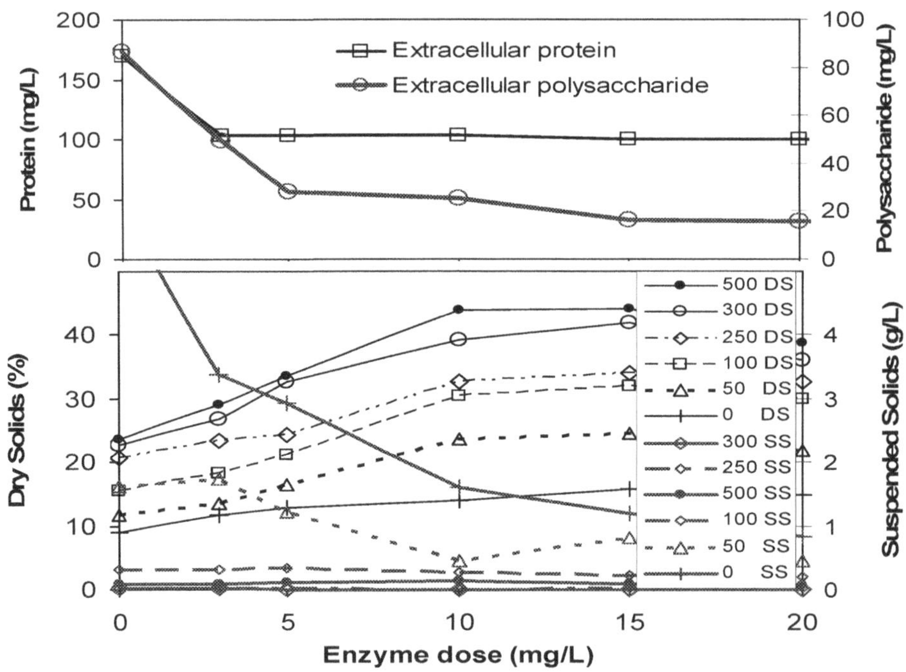

Figure 2 Results of pre-treating an anaerobically digested sludge with enzyme mix. Upper graph shows extracellular polymeric substances after pre-treatment, with protein concentrations as equivalent concentrations of bovine serum albumen, and polysaccharide concentrations as equivalent glucose concentrations. Lower graph shows cake solids and filtrate suspended solids concentrations obtained with a belt press simulator at the polymer doses (in mg/L) shown in the legend.

Advances in solid-liquid separation

Chemical conditioning is used to modify sludge or suspension properties so that the subsequent dewatering process will be efficient. The properties modified are invariably those of the colloidal or solid phase rather than those of the solution. This principle can be extended to the efficient development of physical conditioning processes. Furthermore, mechanical dewatering can be improved if physical conditioning takes place simultaneously to dewatering. This is a unifying feature of several promising dewatering processes.

Principle of "in-situ" physical conditioning

A number of methods have been proposed and attempted in order to pre-treat or pre-condition biosolids so that they will exhibit improved dewatering properties. These include conventional heating, microwave heating, freezing, electrical arc application, and sonic treatment, as well as various chemical treatments with alkali, acids, oxidizing or reducing agents. Yet few of these methods are presently in use. One reason is simply that the biosolids start at ca. 2–5% total solids (or 95-98% moisture). The physical or chemical treatment is thus applied mostly to water, an inefficiency that leads to excessive energy or chemical costs. Furthermore, the water then released in dewatering (centrate or filtrate) is likely to contain increased concentrations of solubilised solids, which will be recycled to primary or secondary treatment. This leads to greater loadings throughout the plant.

Recent developments in dewatering technology suggest a different strategy. Biosolids may be conventionally dewatered in a mechanical dewatering device, allowed to approach the maximum solids concentration attainable. At this state, the biosolids are within a reactor (filter press, frame press, or centrifuge) in an ideal state to impose the "pre-" treatment: the material is fully enclosed and a large proportion of the moisture has been removed. Consequently, the physical or chemical treatment is more efficiently applied, reducing the energetic and economic costs. A lesser amount of water suffers from degraded quality if the treatment releases any solubilised organics. The treatment is in conjunction with dewatering, so it can be applied until the dewatering reaches a desired extent, and no further.

Perhaps the most promising use of this concept (applying both heat and vacuum) has been shown to produce wastewater solids at well above 90% TS. This is a recessed chamber filter press such as the J-Vap® with heat and vacuum application. The heat increases the fluid viscosity and the rate of evaporative loss, while the vacuum also speeds evaporation. In fact, the vacuum can provide conditions exceeding the effective boiling point of water, which obviously will enhance dewatering. Details on this technology have been presented elsewhere (Abu-Orf et al., 2003), including results on a variety of sludge types. It is also plausible that the boiling effect may lead to considerable pathogen destruction. This particular advance disproves the claim that "the water always wins."

In a sense, high-solids centrifuges accomplish physical conditioning while dewatering. The filtration and expression processes exemplified on the earlier process of vacuum filtration (or with the traditional specific resistance test) are accompanied by considerable shear which modified the floc properties. These effects can be duplicated in the laboratory by pre-shearing and then subjecting the sludge to a specific resistance or CST test. Full-scale processes that combine heat or shear with dewatering such as the Centridry® and Lysate centrifuges (respectively) can also be viewed as accomplishing physical conditioning by this means: at a point in the process when the concentrated solids are already contained, they may subjected to further treatment, providing a drier material in subsequent treatment. The Lysate process is primarily intended to enhance digestion by shearing during centrifugal thickening, but may be expected to also provide some improvements in dewatered solids due to the disruption of floc structure.

The anionic nature of the sludge gel can also be used to advantage in-situ, and recent developments in "electrodewatering" by belt press incorporate electro-osmosis during dewatering. More on these developments is available elsewhere (Abu-Orf et al., 2004).

Related questions and future possibilities

What does the future portend for sludge conditioning and dewatering? As summarized here, advances are clearly being made in both areas. And yet, there is unlikely to be a "one size fits all" solution that emerges, because sludge properties, energy and transportation costs, and public acceptance of various management options will vary geographically. Perhaps more vexing is that the attainment of greater solids concentrations by dewatering has led to unexpected increases in odours and in pathogen growth. Concentrated research efforts will identify the causes of these problems, but these advances are still in the future. Nonetheless, our improved understanding of the phenomena and processes involved should help in this direction.

Acknowledgements

The author thanks Dr. Mohammad Abu-Orf for helpful suggestions toward the development of this paper. Ms Azize Ayol is acknowledged for contributions in the rheological section, and for the data in Figure 2. Mr. Chul Park is thanked for clarifications on data in Figure 1.

References

Abu-Orf, M., Muller, C., Park, C., and Novak, J.T. (2004) State of the art innovative technologies to reduce water content of dewatered municipal residuals. Jour. Resid. Sci. Technol. 1(2).

Abu-Orf, M.M., Trautmann, J., Laquidara, M., Junnier, R., Dentel, S.K., and Mah, J. (2003) Demonstration of the J-Vap technology for dewatering and drying municipal residuals. Proc. WEF/AWWA/CWEA Joint Residuals Biosolids Mgmt Conf., Baltimore, February 21.

Alavanja, C.M. (1999) The effects of Ca^{2+}, Na^{+}, and NH^{4+} on the dewaterability of anaerobically digested sludge. Master's thesis, Cornell Univ., Ithaca N.Y.

Ayol, A. (2004) Enzymatic treatment effects on dewaterability of anaerobically digested biosolids-I: Performance evaluations. Process Biochem. In submission.

Ayol, A., Dentel, S.K., and Filibeli, A. (2004) Rheological study of synthetic and real sludges: Determination of the dewatering kinetics using immobilization cell. In submission.

Ayol, A., Dentel, S.K., and Filibeli, A. (2004) Rheological study of synthetic and real sludges: Determination of the dewatering kinetics using immobilization cell. In submission.

Ayol, A. and Dentel, S.K. (2004) Enzymatic treatment effects on dewaterability of anaerobically digested biosolids-II: Laboratory characterizations of drainability and filterability. In submission.

Baudez, J.C. and Coussot, P. (2001) Rheology of aging, concentrated, polymeric suspensions: Application to pasty sewage sludges. *Jour. Rheology* **45**(5), 1123-1140.

Bruus, J.H., Nielsen, P.H., and Keiding, K. (1992) On the stability of activated sludge flocs with implications to dewatering. Water Research **26**(12), 1597-1604.

Busch, P.L. and Stumm, W. (1968) Chemical interactions in the aggregation of bacteria bioflocculation in waste treatment. Environ. Sci. . Technol. **2**(1) 49-53.

Chitikela, S., and Dentel, S.K. (1998). Dual-chemical conditioning and dewatering of anaerobically digested biosolids: laboratory evaluations. Water Envir. Res., 70(5), 1062-1069

Dentel, S.K., "Conditioning." Ch. 16 in *Sludge into Biosolids*, ed. L. Spinosa and P.A. Vesilind. London: IWA (2001).

Dentel, S. K. and Gossett, J. M. (1987) Coagulation of organic suspensions with aluminum salts. Jour. Water Poll. Control Fed. **59**, 101-108.

Dentel, S.K., Gucciardi, B.M., Griskowitz, N.J., Chang, L., Raudenbush, D.L., and Arican, B. (2000), Chemistry, function, and fate of acrylamide-based polymers. In Chemical Water and Wastewater Treatment VI , H.H. Hahn, H. Odegaard, and E. Hoffmann, Eds.; Springer Verlag, Berlin, 35-44.

Endo, T., Nakamura, K., and Takahashi, H. (1976) Pronase-susceptible floc forming bacteria: Relationship between flocculation and calcium ion. Agricultural and Biological Chemistry **40**(11) 2289-2295.

Goodwin J.A.S. and Forster, C.F. (1985) A further examination into the composition of activated sludge surfaces in relation to their settlement characteristics. Water Research 19, 527-533.

Higgins, M.J. and Novak, J.T. (1997) The effect of cations on the settling and dewatering of activated sludges: laboratory results. Water Environment Research **69**(2), 215-224.

Kelly, Harlan G., Hart Frese, Allan Gibb and Jianpeng Zhou (2000) "Dewatering Improvements of Thermophilically Digested Biosolids through Addition of Metal Salts at the Salmon Arm Water Pollution Control Centre" Canadian Society for Civil Engineering, 6th Environmental Engineering Specialty Conference, June 7-10, 2000, London, Ont.

Murthy, S.N. (1998) Bioflocculation: implications for activated sludge properties and wastewater treatment. Ph.D. dissertation, Dept Civil Eng., VPI&SU, Blacksburg VA, USA.

Murthy, S.N., Novak, J.T., and Holbrook, R. D. (2000) Optimizing dewatering of biosolids from autothermal thermophilic aerobic digesters (ATAD) using inorganic conditioners. Water Environment Research 72(6), 714-721.

Novak, J.T. and Bivins, J.L., "Changes in dewatering properties between the thermophilic and mesophilic stages in TPAD systems," presented, WEFTEC 2000, Anaheim CA, Oct. 18, 2000.

Novak, J.T. and Randall, C.W. (1986) Sludge flocculation at an industrial waste facility. Proc. 18th Mid-Atlantic Indust. Waste Conf. Technomic Publ. Co. Lancaster Pa., 322-330.

Novak, J.T., Park, C., and Abu-Orf, M. (2004) Conditioning and dewatering of digested waste activated sludges. Jour. Resid. Sci. Technol. 1(1), 45-51.

Örmeci, B., Cho, K., and Abu-Orf, M. (2004) Development of a laboratory protocol to measure network strength of sludges. Journal of Residuals Science and Technology, 1(1), 35-44.

Pavoni, J.L., Tenney, and M.W., Echelberger Jr. W.F. (1972) Bacterial exocellular polymers and biological flocculation. J. Wat. Pollut. Control Fed. **44**, 414-431.

Poxon, T. (1996) Structure and dewaterability in anaerobically digested sludge. In Proc. Water Envt. Fed. 69nd Ann. Conf. (WEFTEC'96), Dallas, 5-9 October, 167-178.

Thomas, L., Jungschaffer, G., and Sprossler, B. (1993) Improved sludge dewatering by enzymatic treatment. Water Sci. Technol. **28**(1), 189-192.

Yen P., Chen L.C., Chien C.Y., Rome-Ming W., Lee D.J. (2002) Network strength and dewaterability of flocculated activated sludge, Water Research, 36, pp. 539-550.

An innovative application for an established technology to improve sludge dewatering

M. Boehler*, D. Pinnow, W. Krauss**, H. Brauchli*** and H. Siegrist***

*EAWAG, Swiss Federal Institute for Environmental Science and Technology, Duebendorf, Switzerland
** Bucher-Guyer AG, Foodtech, Niederweningen, Switzerland
*** Abwasserverband Glarnerland (AVG), Bilten, Switzerland

Abstract This paper presents the results of full-scale waste water treatment plant (WWTP) sludge dewatering trials using a Bucher de-juicing press model HP2500 (Boehler et al., 2003). Besides digested sludge from the WWTP at Glarnerland sludge from five other WWTP's was also dewatered. The investigations show that with hydraulic system tested dry solids (DS) contents of 32-41% (with a throughput of approximately 5 m^3 h^{-1}) could be reached. With extended cycle / batch time DS contents of up to 43.5% could be achieved. Dewatering trials with raw sludge (not putrefied) from two different WWTP's resulted in DS content of 35% and 41%. The dosage of coagulant aids, particle size distribution and volatile suspended solids (VSS) have a significant influence on the dewatering results. An economic efficiency calculation is presented with direct comparison to a centrifuge used at the WWTP. Return of investment is expected to be achieved within 3 years.

Keywords sewage sludge, sludge dewatering, sludge disposal, drainage technology

Introduction

Today in Europe biological waste water treatment is primarily carried out by the activated sludge process. Because of the physical-chemical processes involved in activated sludge waste water treatment, the sludge tends to accumulate heavy metals and low levels of organic compounds which are not readily biodegraded (e.g. pesticides, household chemicals, trace pollutants, etc.) and are present in the waste water. As a result of a detailed ecotoxicological risk evaluation (FAL, 2001) agricultural use of stabilised sewage sludge will be banned in Switzerland from 2005. Because disposal in landfills is already prohibited in Switzerland, sewage sludge has to be dewatered, dried, incinerated and the ash disposed of in landfills. In future 100% of the sewage sludge (200,000 mt DS/year) will be incinerated (BUWAL, 2001). Energy consumption calculations has shown there could be advantages in burning raw sludge, this has brought new discussions about burning non digested sludge. Sludge digestion is labor and cost-intensive and uses more than 60% of the total manpower of a waste water treatment plant (WWTP; Gujer, 1999).

A first step in the disposal of the resulting sludge (excess sludge + primary sludge) is a reduction of volume. An important partial step after digesting is the mechanical separation of sludge liquor from the solid. The efficiency of drainage has a significant impact on the cost of subsequent sludge treatment and disposal. If incineration is the end of the treatment, the dewatered sludge must have minimum content of dried solid matter (>35% DS) in order to ensure an energy neutral incineration. The efficiency of the sludge dewatering equipment has

© 2004 IWA Publishing. *2nd IWA Leading-Edge Conference on Water and Wastewater Treatment Technologies* edited by Mark Van Loosdrecht & Jonathan Clement. ISBN: 1 84339 508 8.

a significant impact on the overall cost of waste water treatment. The company Bucher Foodtech A.G. developed and patented (1964) an innovative hydraulic press system for de-juicing fruits and vegetables, this press system was used for full scale trials to dewater digested sludge and raw sludge from the WWTP Glarnerland (AVG) and other WWTP's.

Mechanical Sludge Dewatering

Overview

Sewage sludge drainage is a common unit operation of solid-liquid separation. Drainage can take place by filtration or centrifugation. In order to operate efficiently, mechanical dewatering requires conditioning of the sewage sludge, flocculation increases the speed of dewatering. Physical limits are set for the mechanical dewatering of sewage sludge and are determined by the size of the water fraction of a sludge suspension. The free water (or bulk water) fraction is not bound to sludge particles and can be separated by mechanical dewatering. The capillary interstitial water fraction, trapped in the crevices and interstitial spaces of the flock and organisms, the water bonded on surfaces and in cells, and the water of hydration, is bound water which mostly remains in the sludge filter cake after dewatering (Kopp, 2001; Vesilind, 2003). The established mechanical de-watering technologies like belt presses (20-28%), decanting centrifuges (20-35%) and filter presses (28-*45%; * including aggregates, e.g. chalk) achieve DS contents indicated in the brackets (Jung et al., 1995).

Materials and methods

Hydraulic press system HP2500 – configuration and function

The Bucher HP2500 hydraulic de-juicing press used in these trials consists of a cylinder (total volume 2.5 m^3) equipped with a flexible drainage system and a piston (Figure 2.1). The main component of the system is the drainage system, which comprised 130 flexible drainage elements each element consists of a drainage core over which is fitted a filter sock (in the model HP2500 each drainage element is 1200mm long x 50mm∅). The filter sock is a strong, hard wearing filter cloth (similar to recessed-plate filter cloth) this holds back the solids of the sludge. Two filter cloths were initially trialled (Type M1 and Type V1), both exhibited very similar throughput. The drainage cores are provided with drainage channels (similar in appearance to the splines on a shaft) and lead the filtrate into collection chambers in the cylinder and piston and then to the collection tank.

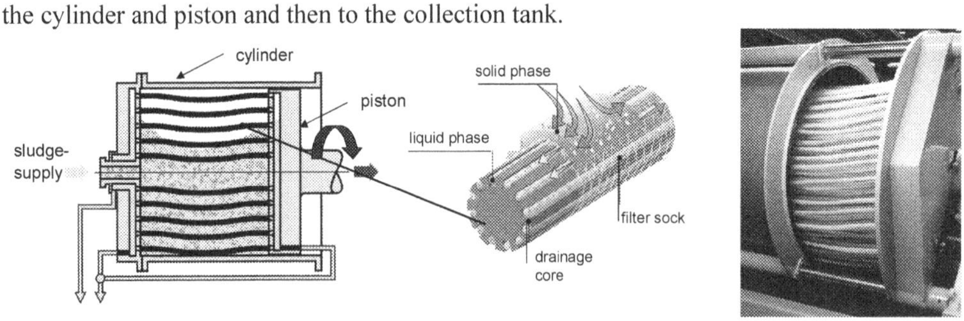

Figure 2.1: Hydraulic press system and functional principle of drainage elements, on the right side a opened cylinder with drainage elements

During a complete filtration batch / cycle, sludge is filled into the cylinder in several steps with a pump. Between each filling step pressure is gradually increased (up to 10 bar) as the piston reduces the basket volume and presses the liquid through the drainage elements into the filtrate collecting tank. A solid filter cake is developed with each filling step (e.g. 0.4 m^3 sludge is added at each step) throughout the first cycles. During the filling step the piston is drawn back, the drainage elements stretched and the filter cake broken into pieces. At the same time air flows back through the drainage elements to expand, clear and clean the filter socks. Due to the continuous rotation and compression of the cylinder new drainage paths are developed in the filter cake. In a single batch a total of 8 to 12 m_3 (depending on initial DS content of the sludge) sludge is processed in 30 to 40 compression steps. In order to achieve optimal DS contents, the last compression steps (about 10) occur without sludge being added. The principle procedure is shown in Figure 2.2. The Bucher HP hydraulic press system is fully automated. Operating status, yield and performance values as well as alarms are continuously and automatically logged by an internal data acquisition system.

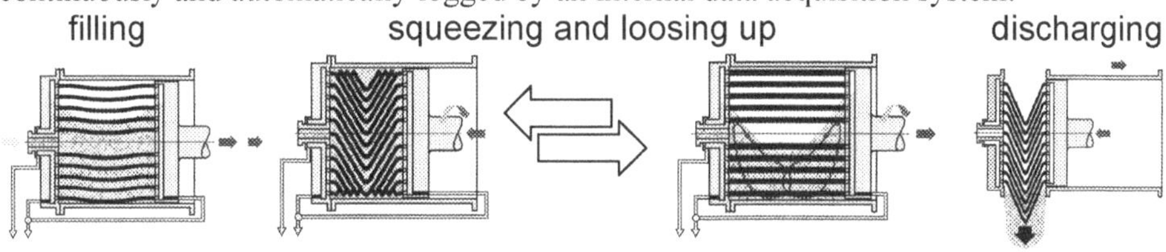

Figure 2.2: Principle procedure of sludge dewatering with the HP hydraulic press system

Full scale experimental set up at the WWTP - digested and raw-sludge used in trials

Three machines were used during the sludge drainage trials at the WWTP Glarnerland (Figure 2.3), an older belt press (ROEDIGER), a current generation decanter centrifuge (NOXON, typ DC 12 FC HS) and the Bucher type HP2500 press. Dewatering trials on sludges from other sites were delivered to the WWTP Glarnerland by road tanker, the trials on these sludges were made only on the Bucher type HP2500 press.

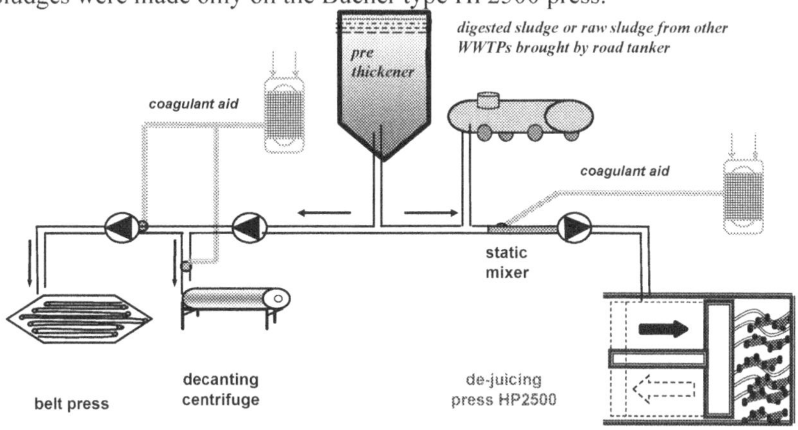

Figure 2.3: Experimental set up for full scale trials at the WWTP Glarnerland

Digested sludge was taken from the sludge pre-thickener of the WWTP Glarnerland, also digested sludge from another five WWTP's were dewatered. The trials with raw sludge (not digested) were carried out with sludge from the WWTP's Glarnerland and Hombrechtikon.

The raw sludge used in the trials consists of a fraction of primary sludge and a fraction of excess biomass. The initial DS content of the sludge was in the range of 2 and 8% w/w.

Analytical methods – coagulant aid

Sludge samples were analysed for total solids (DS; 103°C, 24h) and ash content (volatile suspended solids (VSS); 550°C, 2h). Examinations of sludge particle size distribution were made with a MASTERSIZER X (Malvern Instruments Ltd., U.K.). The sludge was conditioned using an organic polymer (CP31; FLONEX AG, Sissach, CH) as a coagulant aid. The coagulant aid was added as a 0.5%-solution (dilution with water) by a dosing pump (up to 8.5 m^3hr^{-1}) into the sludge line (∅ 100 mm, about 25 m long). To ensure good dispersion the polyelectrolyte was dosed against the direction of sludge flow and immediately before a static mixer and the sludge filling pump (rotary piston pump; $50m^3hr^{-1}$). Coagulation was complete within the sludge tube.

Results and Discussions

Dewatering efficiency and throughput of HP2500

About 170 drainage trials were carried out at the WWTP Glarnerland with 1100 m^3 of digested and raw sludge. Figure 3.1 displays the results of the trials with digested sludge from 5 different plants, the differences in drainage behaviour with the different digested sludges is of interest.

With the Bucher HP2500 hydraulic press system a DS content in the range of 32 – 41% (circle) could be achieved. A throughput of approximately 150 kg dry solid per hour (about 5 m^3 with a DS content of approx. 3% of liquid sludge per hour) could be treated. Note that due to variations in DS of the various sludges the throughput is given as kg DS per hour to enable comparison. Up to 43.5% DS content could be achieved by extending the batch time but this gives a significant reduction in throughput.

The worse drainage results were with digested sludge from the WWTP Werdhoelzli/Zuerich (DS content within 25-26%). The drainage result for this mud slurry may be caused by a very high VSS of 53%. A recessed-plate filter press was operated in parallel on this sludge and showed the same poor dewatering.

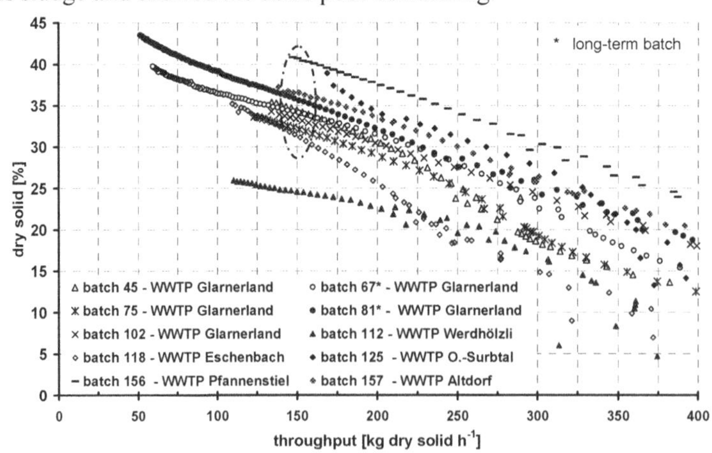

Figure 3.1: Dewatering trials with sludge from different WWTP's in the investigation period

Figure 3.2 shows de-waterability of raw sludge with the Bucher HP2500 hydraulic press system. The drainage of raw sludge from the Glarnerland WWTP resulted in an average DS content of 35% and a DS content of 41% from the WWTP Hombrechtikon.

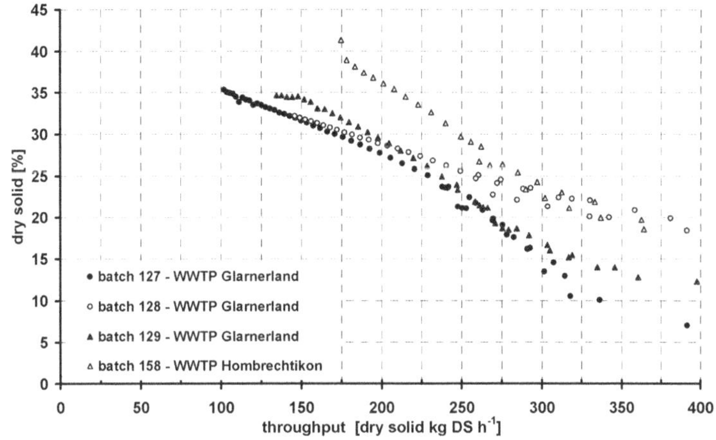

Figure 3.2: Dewatering trials with raw sludge from two different WWTP's

Impact of pressure

The pressure of the Bucher HP2500 may be varied to accommodate variations in the drainability of the feed sludge and to prevent clogging of the filter socks. The pressing pressure at the beginning of the batch is low and increased continuously after a predetermined DS concentration. Trials were made with different end compression phases to determine the influence of pressure. Limiting the pressure to 5 bar at the end of the batch reduces the DS by about 1.5% to 3% compared with operating to 10 bar (Figure 3.3).

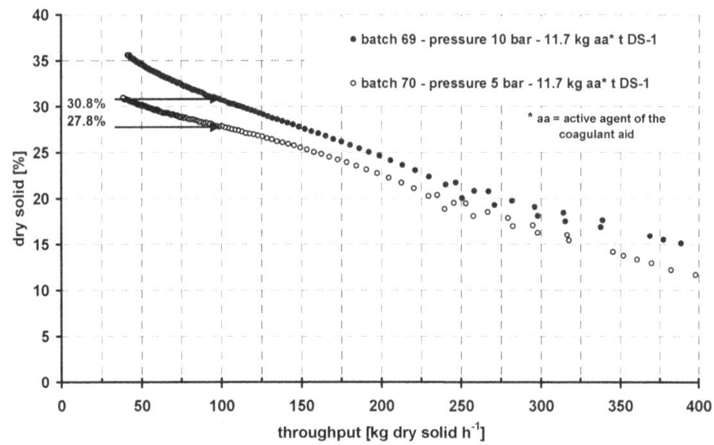

Figure 3.3: Trials with different compression phases at end of batch

Impact of coagulant aid

The addition of cationic organic polymers as a coagulant aid leads to a significant improvement in the de-waterability and thus has a large influence on the drainage performance of mechanical sludge dewatering, this is known as "Filter Cake Permeability". During flake formation two reactions occur simultaneously;

a) Coagulation by compensation of the electrostatic forces b) Flocculation.

With the coagulation the electrostatic repelling forces are destabilized and the separated particles can move closer together, so the free water fraction can better run from the sludge matrix. During the flocculation bridges are formed between the sludge particles by adsorbed polymers, resulting in the formation of macro flakes. The dosage of coagulant aid is also necessary in the Bucher HP2500 hydraulic press system (≥12 kg coagulant aid per metric tonne DS). Increasing the coagulant dosage resulted better drainage results (Figure 3.4).

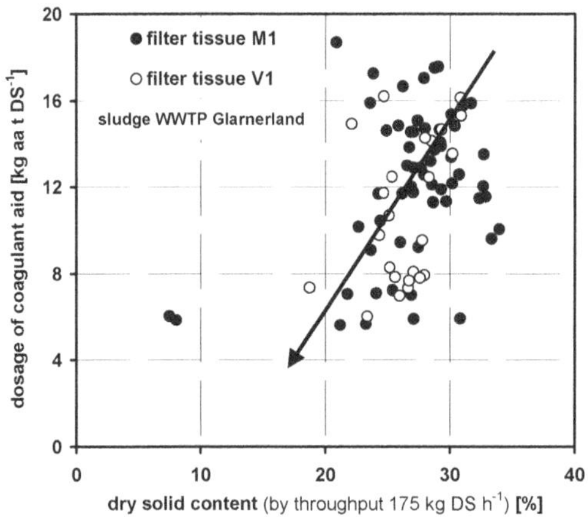

Figure 3.4: DS content achieved with different dosage of coagulant aid

Figure 3.5 shows four de-watering trials with digested sludge of uniform quality. The 4 dewatering operations show the significant influence coagulant aid dosage has on achievable DS content and throughput. With increasing dosing quantity the achievable DS content rises.

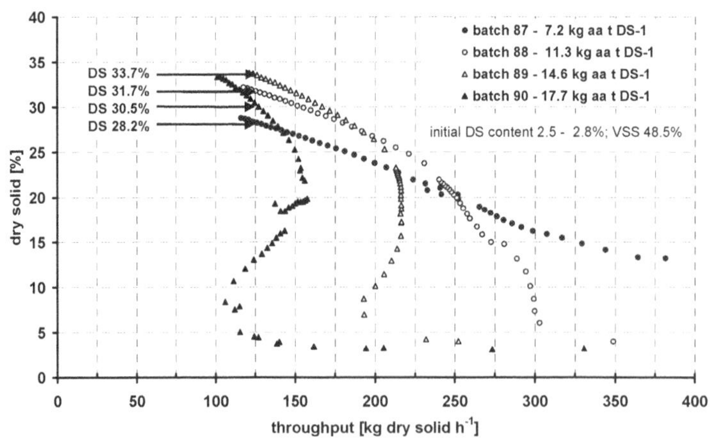

Figure 3.5: Impact of coagulant on de-waterability of digested sludge

Of note is the reduced dewatering speed at the beginning of the trials if dosage of coagulant is increasing. This is probably due to surplus flocculant increasing the viscosity of the filtrate and/or mycels of flocculant or a gel formed on the filter sock. With further processing this effect is reversible.

Impact of volatile suspended solid and particle size distribution of sludge

It is well-known from the literature that volatile solids content has a significant influence on drainage. During this investigation it was confirmed that a higher DS contents can be achieved with a lower less volatile solids content. The arrow marks the trend in Figure 3.6.

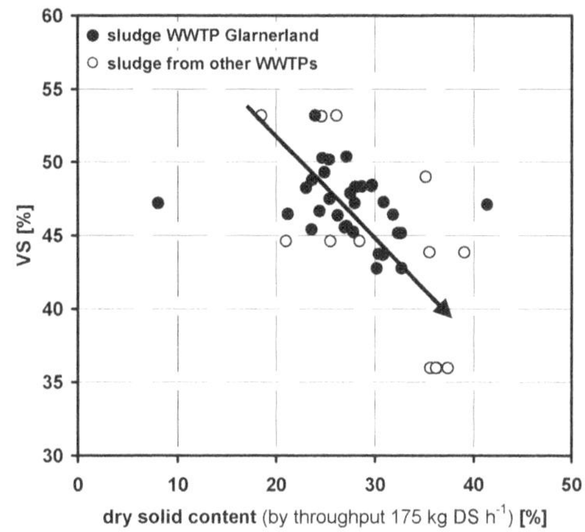

Figure 3.6: DS content achieved with different volatile suspended solids

Average particle size of sludge from WWTP Glarnerland used in these trials is significantly finer and has a smaller size variation compared to sludge samples from other WWTP's. A smaller average particle size is expected to correlate with increased water detention, due to higher specific surface area. Further, a smaller size distribution causes a relatively high amount of capillary water (Kopp, 2002; Vesilind, 2003). It can therefore be stated that the sludge from the WWTP Glarnerland tested in our trials has unfavourable dewatering characteristics (confirmed also by the performance of the conventional dewatering devices at this WWTP). Due to the homogeneity of the particle distribution found in the sludge tested, the performance of the dewatering process described here cannot be correlated directly with particle distribution (see Figure 3.7).

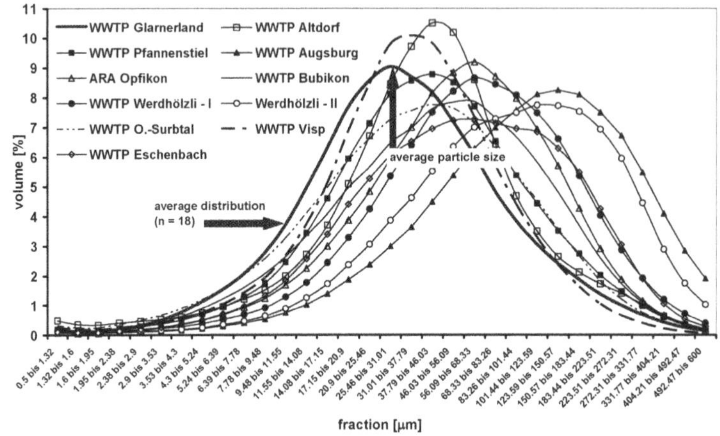

Figure 3.7: Particle size distribution of different digested sludge

Costs of sludge de-watering with Bucher HP-system

The economic efficiency of using a Bucher HP Press has been calculated in direct comparison with the new decanter/centrifuge at the WWTP Glarnerland. Investment cost, maintenance costs, operating expenses (personal costs, sludge conditioning etc.) and costs for the disposal (transport + incineration) were considered. To guarantee a throughput of 8~10m^3/hr for the WWTP Glarnerland a Bucher HP5000 fruit press is required. Based on an average increase of 3% DS using the Bucher HP-System compared with the installed decanter/centrifuge and a real disposal cost of cost of €145/t wet sludge the payback for period of the additional capital cost of the Bucher HP-System press over the capital cost of the decanter/centrifuge is 3 years. If the disposal costs increase (for example, due to lack of incineration capacity) the economic efficiency of the press improves (figure 3.8) rapidly.

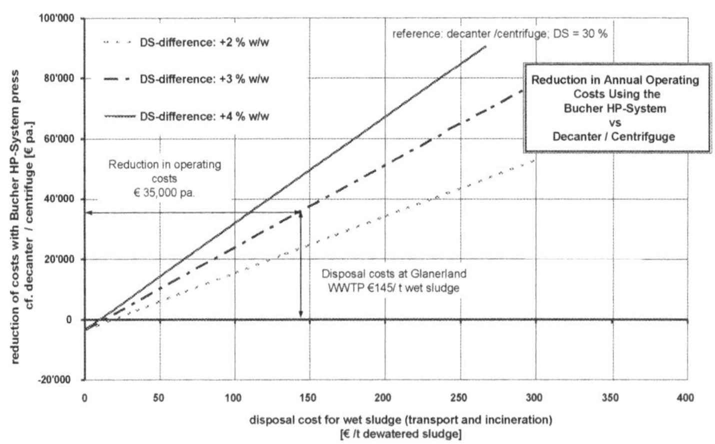

Figure 3.8: Economic efficiency calculation between Bucher HP-System Press and decanter/ centrifuge

Conclusions

The investigations using the Bucher HP2500 hydraulic press system at the WWTP Glarnerland shows that using a technology established in one field but used for a totally different application can open a new domain. In this case well established technology from the fruit processing industry can fulfil the needs of ecological processing.

In comparison to other established mechanical dewatering technologies the innovative application of the Bucher HP2500 hydraulic press system achieved higher DS content on waste water sludges and can be automatically operated without supervision under the practical conditions of a WWTP. The economics show that this alternative technology can compete with conventional systems.

Further optimisation (drainage system, hydraulics, size....) will increase the efficiency and the economics of the system in relation to other technologies. Here further test series have been successfully accomplished by the water management federation Wupperverband (Germany; Kolisch et al., 2004).

The planned commercial adoption of this technology by a number of WWTP's will give the opportunity to show whether this hydraulic system will, in the long term become established in the field of sewage sludge treatment.

Acknowledgements

The research was financially supported by the Swiss Commission for Technology and Innovation (CTI). The authors would like to thank the operating crew of the WWTP (AVG) for their excellent technical assistance in collecting experience and data.

References

Boehler M., Pinnow D., Krauss W., Brauchli H. and Siegrist H. (2003) Neue Presstechnologie zur verbesserten Klaerschlammentwaesserung, EAWAG final report, Duebendorf, Switzerland, Juni

BUWAL (2001) Klaerschlamm in der Landwirtschaft, press report, Bern, Switzerland, September.

FAL (2001) Risikoanalyse zur Abfallduengerverwertung in der Landwirtschaft, final report, Zuerich-Reckenholz, Switzerland, July

Gujer W. (1999) Siedlungswasserwirtschaft, Springer-Verlag Berlin Heidelberg New York

Jung J., Nellenschulte T., et al., Entwicklung in der Klaerschlammentwaesserung, in awt-Abwassertechnik 2/95, 56 – 60.

Kolisch G., Boehler M., Arancibia C., Pinnow D. and Krauss W. (2004) A new approach to improve sludge dewatering using a semi-continuous hydraulic press system, submitted IWA-conference, Marrakech 2004.

Kopp, J. B. (2001) Wasseranteile in Klaerschlammsuspensionen – Messmethode und Praxisrelevanz, dissertation, Institut fuer Siedlungswasserwirtschaft, Technische Universitaet Braunschweig, Germany.

Vesilind P.A, (2003) Future research in sludge dewatering, Proceedings IWA Specialised Conference BIOSOLIDS 2003 Waste water sludge as a resource, Trondheim, Norway, 23-25. Juni, 2003

Kemicond – Acid Oxidative Sludge Conditioning Process

Peter Cornel*, Christian Schaum*, Alexander Voigt*, Göran Karlsson, Michael Recktenwald****

* TU Darmstadt, Institut *WAR*, Petersenstrasse 13, 64287 Darmstadt, Germany
(E-mail: c.schaum@iwar.tu-darmstadt.de)
** Kemira Kemi AB, Industrigatan 70, 251 09 Helsingborg, Sweden

Abstract The Kemicond process for sludge conditioning consists of a chemical treatment by sulphuric acid and hydrogen peroxide followed by a two stage dewatering unit. It is shown that the chemical treatment can facilitate the dewatering, resulting in much higher solid concentrations (DS) compared to untreated sludge. Furthermore the bacterial counts indicate that the germs are reduced to an extend similar to pasteurisation. With some modifications the process offers the possibility of phosphorus recovery and a reuse of precipitants.

As the process works with digested sludge containing iron phosphates, orientating lab scale results of sludge derived from treatment plants using aluminium coagulant for phosphate removal show very little effect. Further work is required to understand the mechanisms and to adopt the process to different sludge conditions.

Keywords sewage sludge, sludge conditioning, sludge dewatering, disinfection

Introduction

Wastewater treatment produces residuals such as primary sludge and activated sludge. In many European countries anaerobic digestion of the sludge is applied. The amount and structure of the digested sludge depends on several parameters, such as

- waste water treatment process and process conditions,
- type and amount of precipitant used (e.g. iron or aluminium salts for P-precipitation)
- the stabilization process itself and
- the dewatering process.

As dewatering is generally facilitated by anaerobic digestion, the precipitants used in the treatment process as well as in the sludge conditioning will lead to a wide range of dewaterability. Studies by Kemira Kemi AB in Sweden indicate that the dewatering of ferruginous digested sludge can be improved by the conditioning with sulphuric acid followed by oxidation e.g. with hydrogen peroxide. Based on these experiments and the experience of the Krepro process (Cornel, 2002) the so-called Kemira conditioning process (Kemicond process) was developed. The goals of the process are the improvement of sewage sludge dewatering and disinfection. The process offers the potential for the recovery of precipitants and/or phosphate by slight process modifications.

© 2004 IWA Publishing. *2nd IWA Leading-Edge Conference on Water and Wastewater Treatment Technologies* edited by Mark Van Loosdrecht & Jonathan Clement. ISBN: 1 84339 508 8.

Process description

The Kemicond process is a five step process (Figure 1), consisting of

1. acidification
2. oxidation
3. neutralization and polymer addition
4. pre-dewatering in a screening drum and
5. main dewatering in a screw press

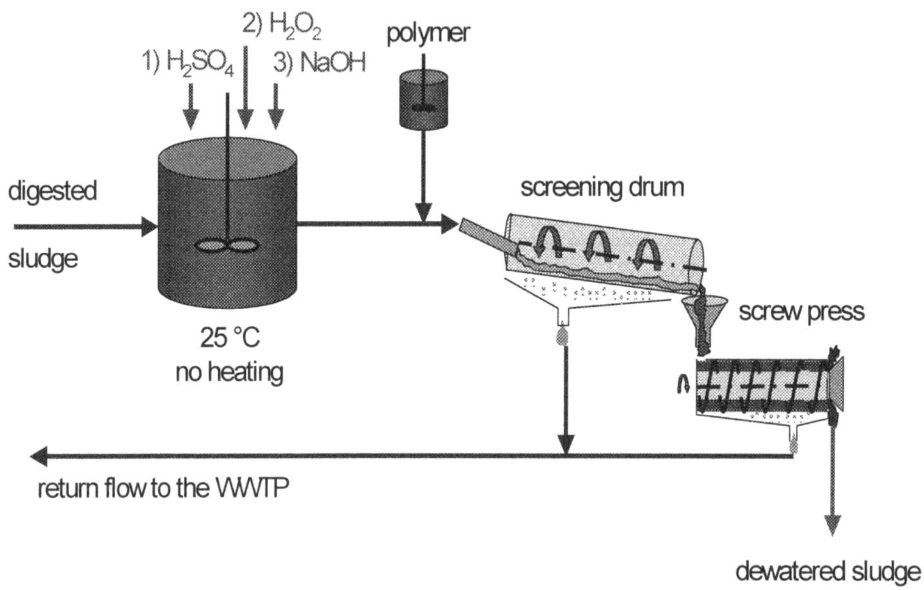

Figure 1 Flow diagram of the Kemicond process (batch operation)

Ad 1. Acidification of the sludge

By the addition of sulphuric acid (H_2SO_4) to the digested sludge at room temperature (approximately 25 °C), the pH-value is lowered from around neutral to a value between 3 and 4. This causes the dissolution of some ferrous hydroxides, phosphates and other salts as indicated by increasing concentrations of Fe^{2+} and phosphate in solution. The dissolved COD increases slightly, showing that part of the solid organic matter is oxidized and re-dissolved. At this pH-range heavy metals remain mainly in the solid phase. The acidification is accompanied by the release of large amounts of carbon dioxide.

Ad 2. Oxidation

By the addition of hydrogen peroxide (H_2O_2) to the acidified system, the dissolved ferrous ions (Fe^{2+}) are believed to be oxidised and transferred into ferric (Fe^{3+}). Fe^{3+} on its part can react with the dissolved phosphate, forming iron phosphate ($FePO_4$), which precipitates at the given pH-range between 3 to 4. In addition to that, the structure of the sludge seems to change in a way that it is favorable for the dewatering. In a batch process the required time for the acidification and the oxidation is about one hour.

Ad 3. Neutralization and Conditioning

Before dewatering, the pH-value has to be increased to a value between 5 and 6, preferably by caustic soda. At this pH-value, the flocculation with the added polymer works best.

Ad 4. Pre-dewatering in a rotating screening drum

The considerable diluted reaction mixture with a dry solids (DS) content of around 2.5 % is dewatered in a rotating screening drum where the reaction mixture releases 80 % of its water. The thickened sludge keeps a dry solid content between 12 and 15 %. The dosage of the polymer in step 3 plays a major roll for dewatering in this step. It has to be chosen according to the sludge properties and the pH-value. If the pH-value is too low, the polymer can be deactivated.

Ad 5. Main dewatering step in a screw press

The screw squeezes out the remaining free water from the pre-dewatered sludge under strong mechanical pressure. The sludge can achieve dry substance contents between 30 and 50 %. The sludge reacts slightly acidic at the adjusted pH-value between 5 and 6. It can be neutralized by lime or magnesium if necessary.The odour of the sludge is reduced considerably due to the strong oxidative potential caused by the peroxide. Sulphides and other odorous compounds are oxidized to a large extend. The dewatering step is almost odour-free. Biological activities and microorganisms are reduced by oxidation, as shown later.

Miscellaneous

The gases produced during and after the acidification contain a considerable amount of hydrogen sulphide (H_2S) and have to be treated.

Process background

Despite all uncertainties about how the process steps cause the changes in structure and dewatering behaviour of the sludge, it is assumed that a part of the Fe^{2+}-ions, released in the first step, are oxidized in the second step according to the following idealized and simplified chemical equations:

Oxidation: $Fe^{2+} \rightarrow Fe^{3+} + e^-$ (1)

Reduction: $H_2O_2 + 2\,H^+ + 2\,e^- \rightarrow 2\,H_2O$ (2)

Redox-Equation: $2\,Fe^{2+} + H_2O_2 + 2\,H^+ \rightarrow 2\,Fe^{3+} + 2\,H_2O$ (3)

Precipitation: $Fe^{3+} + H_2PO_4^- \rightarrow FePO_4 + 2\,H^+$ (4)

Thus the equation could be summarized as:

$$2\,Fe^{2+} + H_2O_2 + 2\,H_2PO_4^- \rightarrow 2\,FePO_4 + 2\,H_2O + 2\,H^+ \quad (5)$$

If the molar ratio of ferric (Fe^{3+}) and phosphate is at least 1:1, the phosphate is precipitated quantitatively as ferric phosphate. In case the molar ratio is > 1 the excess of iron will precipitate as hydroxide when the pH-value is raised. The excess of iron can be reused as a coagulant if the liquid phase is separated at low pH-value where iron remains predominantly in solution, and fed into the water treatment process.

Results of the Experimental work

1. Waste water treatment Plant Käppala, Sweden

The waste water treatment plant (WWTP) of Käppala on the island of Lidingö treats the sewage from eleven municipalities in the Stockholm (Sweden) region counting up to 500,000 population equivalent (PE) in total. The waste water treatment plant produces 38,000 Mg of dewatered sludge per year. This corresponds to 7,000 Mg/a (1 Mg = 1,000 kg) dry solids.

The basic parameters for the experiments such as the added amounts of sulphuric acid, hydrogen peroxide, caustic soda and polymer were determined by preceding laboratory tests on a small scale. The pilot scale tests were performed over a period of ten months, beginning in spring 2003.

Sludge properties of the WWTP Käppala

The digestion step of the WWTP Käppala consists of two digestion chambers. The first one treats exclusively primary sludge, the second one treats only biological sludge. After this parallel digestion, the two sludges are mixed and pumped to the dewatering step by belt filter presses. The dry solids content of the mixed digested sludge is on average 2.5 %. The temperature during the dewatering step is 20°C.

Table 1 Analysis on a spot sample of the dried digested feed sludge, WWTP Käppala

Compound	Analysis	Unit
Al	15	g/kg DS
Fe	69	g/kg DS
P	41	g/kg DS
N_{tot}	42	g/kg DS
Loss of ignition at 550°C	64	%

Chemical consumption

The conditioning starts with the loading of the pilot batch reactor and the acidification. The necessary pH-value between 3 and 4 needs acid additions between 250 and 350 kg sulphuric acid per Mg dry solids. The added amount of acid varies due to the changing composition of the sludge and also due to dissolved compounds like carbon dioxide. A good digestion produces high amounts of dissolved carbon dioxide which, as a consequence, demands an extra portion of acid because of the buffering properties of the dissolved gas. As ferrous hydroxide and phosphate salts are the easiest soluble compounds, their present concentration in the sludge is the other decisive parameter for the consumption of acid.

Just the decrease of the pH-value is not sufficient for a successful dehydration. In order to get an dewatering effect as high as possible, the addition of an oxidant like hydrogen peroxide is necessary. Using other oxidants is conceivable, but hydrogen peroxide does not give any side products. The best dosage of hydrogen peroxide ranges from 25 to 50 kg per Mg dry solids.

Table 2 Chemicals consumption (pilot plant WWTP Käppala)

Added chemical	Amount	Unit
Sulphuric acid (calc. as 100 % H_2SO_4)	250-350	kg/Mg DS
Hydrogen peroxide (calc. as 100 % H_2O_2)	26-50	kg/Mg DS
Caustic soda (calc. as 100 % NaOH)	7-37	kg/Mg DS
Polymer	4-5	kg/Mg DS

Correlation between the consumption of sulphuric acid and the dewatering

Laboratory tests on filterability (ATV (1996), Nellenschulte (1996)) were carried out by a suction filter test. The amount of filtrate per unit of time indicates the filterability. Figure 2 shows the changes in the filterability of the sludge with different chemical dosages.

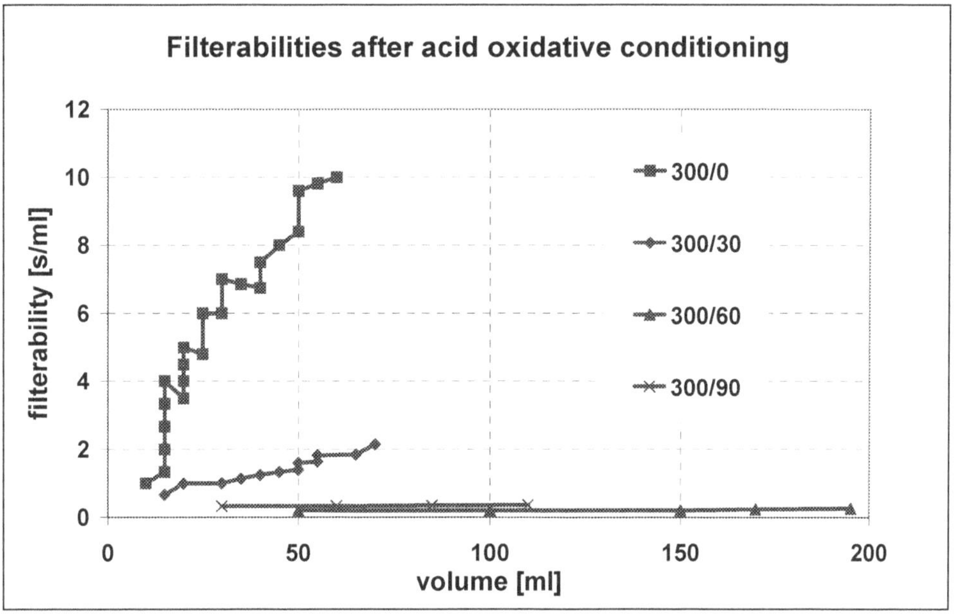

Figure 2 Lab tests on the filterabilities of conditioned sludge from Käppala with acid dosages of 300 (equivalent to a pH-value of ≈ 3)) and H_2O_2 dosages of 30, 60 and 90 kg per Mg DS

As indicated by the steep slop, the sulphuric acid solely, results in a relatively bad filterability. The addition of hydrogen peroxide improves the dewaterability decisively.

Correlation between the pH-value and the addition of peroxide for the dissolution of phosphorus

Figure 3 correlates the dissolution of phosphorus with the pH-value. At a pH-value below 3.5, more than 70 % of the phosphorus is dissolved after the addition of sulphuric acid. By adding peroxide the dissolved phosphorus decreases, as phosphorus precipitates as iron phosphate, and the pH-value decreases as well in accordance with equation 5. The two arrows point up the shift in pH and dissolved P for two selected experiments. The concentration profile of dissolved iron runs almost parallel to the phosphate concentrations (not depicted).

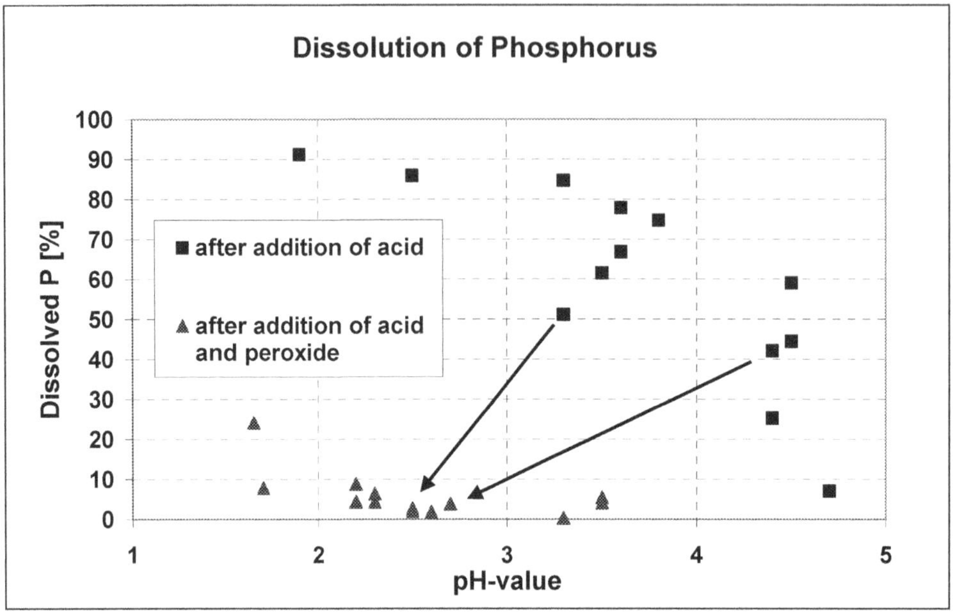

Figure 3 Lab tests on the dissolution of phosphorus depending on the pH-value and the addition of peroxide at Käppala

Disinfection

The acid conditioning together with the application of hydrogen peroxide creates a strongly oxidative potential in the presence of ferric ions, so that the biological activities in the sludge are almost entirely stopped. According to the analysis of the National Swedish Veterinary Institute (*Statens Veterinärmedicinska Anstalt*) the disinfecting effect of this process can be compared to a conventional pasteurisation as the results shown in Table 3 symbolize.

Table 3 Hygienisation tests on digested sludge, conditioned sludge and heat-treated sludge at 70°C (WWTP Käppala) cfu = cell forming units

Test	Digested sludge	Kemicond	Pasteurisation 70°C	Unit
Coliforms 37°C	63 000	<10	<10	cfu / ml
Thermotolerant coliforms	8 800	<10	<10	cfu / ml
Escherichia coli	7 100	<10	<10	cfu / ml
Chlostridium perfringens	112 000	55 000	198 000	cfu / ml
Salmonella Enteriditis	positive	negative	negative	
Salmonella Virchow	positive	negative	negative	

Analysis of the sludge liquor return flow to the influent of the WWTP Käppala

The return flow from the screw press is about 0.5 % of the total feed flow to the WWTP. Table 4 compares the total amounts of returned constituents per Mg dewatered sludge between the Kemicond process and the conventional sludge dewatering with a belt filter press. The much higher sulphate-concentration is a result of the addition of sulphuric acid. The dissolved Fe-salts might act as precipitant in the waste water treatment process.

Table 4 Comparison of contents of the return flow from screw press and belt filter press

	Screw press; pH-value = 4.7			**Belt filter press**	Unit
	Cylinder reject	Screw reject	**Total**		
SS	4	57	**61**	**62**	kg/Mg DS
Fe	14	2.4	**16.4**	**0.18**	kg/Mg DS
PO_4-P	0.4	0.4	**0.8**	**1.5**	kg/Mg DS
NH_4 -N	20-28	2-3	**22-31**	**10-27**	kg/Mg DS
SO_4-S	66	7	**73**	**2**	kg/Mg DS
COD	23	5.8	**28.8**	**21**	kg/Mg DS

Economics

Under the conditions of the WWTP Käppala it can be shown that the additional costs for the chemical treatment are far less than the savings due to the increase of the dry solid content in the dewatered sludge from around 20 % to more than 40 %.

2. Laboratory results at THE TU Darmstadt

The experiments at the Technische Universität Darmstadt (TUD) were carried out with mixed digested sludge (primary + excess sludge) from a typical WWTP in Germany. The size of the WWTP is 30,000 PE. Aluminum salts are used for the precipitation of phosphate. Table 5 compares the sludge qualities of the WWTP Käppala and a WWTP in Germany. The main differences are the concentrations of aluminum and iron, as expected. The degree of digestion is remarkably higher as indicated by the significant lower value for ignition loss.

Table 5 Analysis on a spot sample of the dried digested feed sludges

Compound	Analysis sludge Germany	Analysis sludge Käppala	Unit
Al	39	15	g/kg DS
Fe	9	69	g/kg DS
P	40	41	g/kg DS
Loss of ignition at 550°C	44	64	%

Figure 5 shows the filterabilities of untreated, acidified and both acidified and oxidized sludge. Compared to the results of Käppala in Figure 2, the filterabilities of the untreated and just acidified samples are much better. Consequently the filterability is much less influenced by the addition of peroxide.

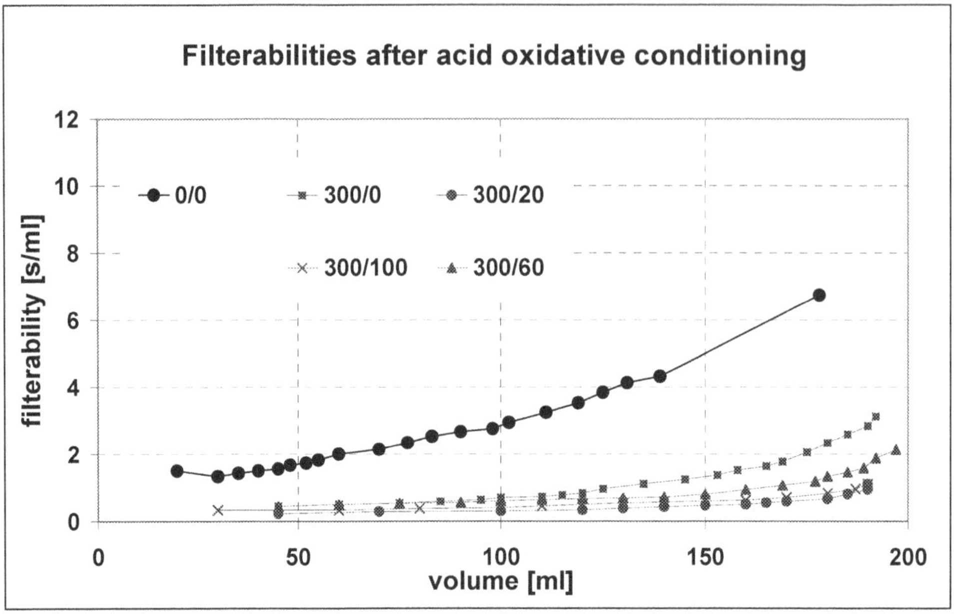

Figure 5 Lab tests on the filterabilities of conditioned sludge with acid dosages of 300kg H_2SO_4 (equivalent to a pH-value of ≈ 3) and H_2O_2 dosages of 20, 60 and 100 kg per Mg DS

The different reactions of the aluminium phosphates become obvious by the results shown in Figure 6. The plot shows the percentage of dissolved phosphorus after the addition of acid solely and after the succeeding oxidation step by peroxide. Compared to the results in Figure 3 the dissolution at pH 3 to 4 is as low as < 35 % compared to 60 – 80 % in Käppala. Furthermore the dissolved phosphorus remains almost unchanged by the hydrogen peroxide treatment.

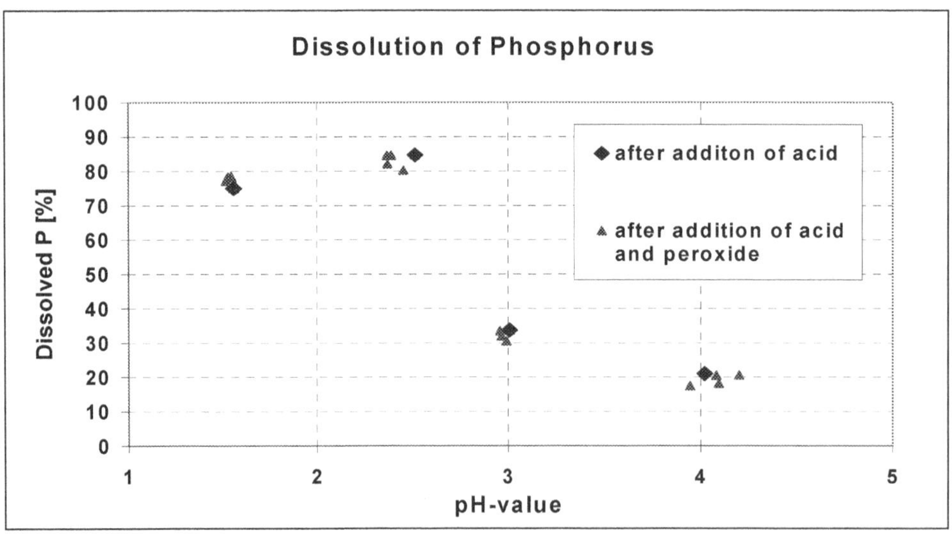

Figure 6 Lab tests on the dissolution of phosphorus depending on the pH-value and the addition of peroxide

Conclusions and future work

The Kemicond process can improve the dewaterability of digested sewage sludge that contains iron phosphate. The resulting higher solid content reduces the sludge quantity significantly. An additional benefit is the destruction of germs and the reduction of odour. Together this leads to improvements and savings in sludge handling and disposal.

Future investigations will focus on the impact of feed sludge composition on the process, with a special view on the applied coagulants and the mechanisms of dissolution, precipitation and other phase interactions during the treatment steps.

Modifications of the process can lead to the recovery of phosphorus and coagulants from the waste streams.

The economic aspects of different experimental set-ups and process alternatives have to be evaluated.

References

ATV (1996) ATV-Handbuch Klärschlamm, 4. Auflage, Ernst & Sohn Verlag, Berlin

Cornel, P. (2002) Prinzipielle Möglichkeiten der Rückgewinnung von Phosphor aus Klärschlamm und Klärschlammasche, aus: WAR Schriftenreihe 147 „Rückgewinnung von Phosphor aus Klärschlamm und Klärschlammasche", Institut *WAR*, TU Darmstadt

Nellenschulte, T. (1996) Modell zur Charakterisierung des Entwässerungsergebnisses von Klärschlämmen, Schriftenreihe TU Braunschweig Nr. 59, Dissertation, Braunschweig

Sludge hygienisation – techniques and experiences

B. Paulsrud*, S. Nybruket and H. Ødegaard*****

*) Aqateam – Norwegian Water Technology Centre
P.O. Box 6875, Rodeløkka, 0504 OSLO, Norway
bjarne.paulsrud@aqateam.no
**) Norwegian Water and Wastewater Works Association (NORVAR)
Vangsveien 143, 2317 HAMAR, Norway
steinar.nybruket@norvar.no
***) Norwegian University of Science and Technology (NTNU)
Dep. of Hydraulic and Environmental Engineering
S. P. Andersens vei 5, 7491 TRONDHEIM, Norway
hallvard.odegaard@bygg.ntnu.no

Abstract: Land application of sewage sludge (biosolids) for sustainable management of organic resources is of increasing interest worldwide, but it is crucial that the sludge is treated and handled properly in order to reduce the risk for negative effects to human, animal and plant health. Several countries have already regulations with criteria for sludge hygienisation prior to land application, and it is expected that a revised EU Directive on sewage sludge will include such criteria. When evaluating sludge hygienisation techniqes it is important to consider the types and limit values of pathogens or indicator organisms included in the regulation, as this can heavily influence the design and operation of actual hygienisation process.

The Norwegian regulation on treatment and use of organic waste products (including sewage sludge) reqires hygienisation and stabilisation of all organic waste to be applied on agricultural land or on green areas, and inactivation of viable helminth ova is one of the hygienisation criteria. The paper presents nine different processes for sludge hygienisation employed at Norwegian sludge treatment plants and briefly comments more than ten years of full scale experience with some of these processes (thermophilic aerobic digestion, thermophilic aerobic pre-treatment + mesophilic anaerobic digestion (dual digestion), pre-pasteurisation + mesophilic anaerobic digestion, mesophilic anaerobic digestion + thermal drying, composting, lime treatment, long-term storage of dewatered sludge, thermal hydrolysis + mesophilic anaerobic digestion, thermophilic anaerobic digestion).

Keywords: Biosolids, composting, hygienisation, lime treatment, long term storage, pre-pasteurisation, sewage sludge, thermal drying, thermal hydrolysis, thermophilic aerobic digestion, thermophilic aerobic pre-treatment, thermophilic anaerobic digestion.

Introduction

Beneficial use of sewage sludge through land application is widely practiced, but there are great variations in the percentage of land applied sludge in different countries and also over time in a single country (European Commission, 2002), The European Commission (DG Environment) has for some years been working on the revision of the Council Directive 86/278/EEC on the protection of the environment, and in the particular of the soil, when sewage sludge is used in agriculture (European Commission, 2000). This working document

© 2004 IWA Publishing. *2nd IWA Leading-Edge Conference on Water and Wastewater Treatment Technologies* edited by Mark Van Loosdrecht & Jonathan Clement. ISBN: 1 84339 508 8.

states that sludge should be used on land whenever possible and according to relevant Community or national legislation. Positive and negative aspects of land application are discussed, and the use of sludge should be carried out in such a way as to minimize the risk of negative effects to for instance human, animal and plant health. To achieve this, the concept of advanced treatments (hygienisation) and conventional treatments combined with land use restrictions has been introduced. This is further elaborated in a recent working document on sludge and biowaste intended as a basis for discussions in a multi-stakeholder process accompanying the development of a Soil Thematic Strategy (European Commission, 2003). Stakeholder discussions and a general public perception of the use of organic resources on land are important matters (Evans & Lowe, 2003), and there are increased focus on standards for pathogen elimination in sewage sludge (biosolids) and their adequacy for protecting public health (National Research Council, 2002).

Hygienisation criteria

Hygienisation of sewage sludge is not a well defined term, but in the most cases it is synonymous with the control of pathogens in sludge. Several references (notably U.S. EPA, 1992 and Strauch, 1998) have collated the pathogens that could be presented in sewage sludge, and these can be grouped in bacteria, viruses, protozoa, helminths, yeast and fungi. However, national sewage sludge regulations define different types of pathogens to be controlled, and in addition various indicator organisms and limit values or reduction efficiencies are included in the regulations.

It is not practicable to monitor treated sludge for the presence of all potential pathogens, and surrogates (indicator organisms) should be used for routine evaluation of sludge treatment plant performance and treated sludge quality. Surrogates should be organisms found in sludges that have similar resistance to treatment as pathogens. In a study for the European Commission on sludge treatments for pathogen reduction, Carrington (2001) suggests that *E.coli* (or enterococci) and *Clostridium perfringens* are used as indicator organisms. Numbers of *E.coli* should not exceed 1000 per gram dry solids, and it is tentatively recommended that spores of *Clostridium perfringens* should not exceed 3000 per gram dry solids.

The first Norwegian sewage sludge regulation with a general requirement for stabilisation and hygienisation of all sludges to be land applied, was launched in 1995 (Ministry of Health and Social Welfare and Ministry of the Environment, 1995). The following criteria were set for a hygienised sludge:

- No *Salmonella spp.* in 50 grams of sludge
- No viable helminth ova
- Less than 2 500 faecal coliforms (FC) per gram dry solids

In addition there are restrictions on land use (application rate, type of crops, etc.) (Walker, 2003).

It is recognized that the elimination of viable helminth eggs is the crucial requirement when designing and operating treatment processes for sludge hygienisation. However, the problem is how to document whether or not the plants are complying with this requirement since most plants have very low densities of such eggs in the raw sludge, and it might be difficult to find any eggs in the treated sludge to verify that the hygienisation process actually inactivate them. In addition, the test procedure for viable helminth ova takes several weeks and there are only a few laboratories in Norway capable of performing this analysis.

To overcome the problems, the Norwegian Water and Wastewater Works Association (NORVAR) has proposed a new approach, in which the plants shall document by their internal quality assurance systems that hygienisation processes are operated within conditions (mainly time-temperature combinations) that are pre-determined to comply with the hygienisation criteria of the regulation. However, the criteria for the inactivation of helminth ova varies between different countries (see Table 1). This situation is also discussed by Carrington (2001), although the third draft of the working paper on sludge (European Commission, 2000) does not include requirements for helminth ova. Some of the pathogen reduction processes employed in Norway have recently been tested and validated in full scale in order to determine their critical operating conditions regarding helminth ova inactivation (Paulsrud et al., 2003).

Table 1. Comparison of critical conditions for sludge treatment processes designed for helminth ova inactivation as one of the pathogen reduction criteria.

Processes	Critical process conditions (abbreviated)	
	Germany (ATV/VKS, 1988a)	USA (U.S. EPA, 1992, 1993)
Composting (windrow)	Dry solids content of sludge/bulking agent mixture: 40-60% ≥ 55°C for minimum 3 weeks	≥ 55°C for minimum 15 days Minimum 5 turnings of the windrow during the period with ≥ 55°C
Composting (within vessel or aerated static pile)	Dry solids content of sludge/bulking agent mixture: >30% ≥ 55°C for minimum 10 days, and within this period ≥ 65°C for minimum 48 hours Maturation in windrows for minimum 2 weeks with at least 1 turning	≥ 55°C for minimum 3 days
Thermophilic Aerobic Digestion	Minimum 2 reactors in series Hydraulic retention time: minimum 5 days 50°C for 23 hours or 55°C for 10 hours or 60°C for 4 hours	Solids retention time =10 days at 55°C - 60°C
Thermophilic Aerobic Pretreatment + Anaerobic Digestion	≥ 60°C for minimum 4 hours or similar conditions as pre-pasteurisation ≥ 30°C in succeeding digester(s)	No criteria for this process
Pre-pasteurisation + Anaerobic Digestion	All sludge particles less than 5mm 65°C for 30 min. or 70°C for 25 min. or 75°C for 20 min. or 80°C for 10 min.	≥ 70°C for minimum 30 min.
Thermal drying	No criteria for this process	Moisture content of sludge ≤ 10% Temperature of sludge particles > 80 °C or temperature of drying gas leaving the dryer >80 °C
Lime treatment (addition of qicklime to dewatered sludge)	≥ 55°C for 2 hours after lime/sludge mixing pH =12,5 ± 0,3 just after lime/sludge mixing	No criteria for this process

Techniques and experiences

General

In Norway the discussions of a new sewage sludge regulation, including requirements of hygenisation and stabilisation of all sludges to be applied on land, were started in 1989. During the early 90's many sewage treatment plants were upgraded with processes for combined hygienisation and stabilisation of the sludge. Design and operation of the sludge treatment plants were mainly based upon the German "ATV-Regelwerk" (ATV/VKS 1986, 1988a, 1988b) which included guidelines for processes which had been employed for several years in Central Europe (Strauch, 1989).

The following processes or process combinations have been in operation in Norway for more than ten years:

- Thermophilic aerobic digestion
- Thermophilic aerobic pre-treatment + mesophilic anaerobic digestion (dual digestion)
- Pre-pasteurisation + mesophilic anaerobic digestion
- Mesophilic anaerobic digestion + thermal drying
- Composting (windrow or in-vessel)
- Lime treatment
- Long-term storage of dewatered sludge

During the last 5-10 years we have also got experiences with:

- Thermal hydrolysis + mesophilic anaerobic digestion
- Thermophilic anaerobic digestion

NORVAR has initiated and partly funded several studies with the aim of collecting and summarising operational experiences with these sludge handling processes in order to recover information for future design and operation of such plants (Paulsrud & Nedland, 1996, 1997; Paulsrud et al., 2001 and Nybruket et al., 2003). The presentation below is a concentrate of all the information summarised in these studies. A common remark should be that processing sludge at thermophilic temperatures or above, create malodorous exhaust gases that have to be treated properly whenever possible.

Thermophilic aerobic digestion (liquid composting)

So far there is only one municipal wastewater treatment plant in operation in Norway (Vårnes WWTP) that employs this process (see Figure 1), but a few other liquid composting plants are handling food industry sludge and combinations of septic tank sludge and animal manure. The Vårnes plant was built as a full scale demonstration facility for handling primary-chemical sludge, and the design capacity is 10 000 p.e. The plant, which is a standard two-stage Fuchs system (German), was put into regular operation during April 1993, and a comprehensive sampling and monitoring programme was carried out for one year. The results from this operation period have been published previously (Paulsrud & Nedland, 1994) and can be summarised as follows together with the experiences from the last 10 years:

- In order to achieve and sustain hygienisation temperatures in an autothermal process, the volatile solids content in feed sludge should be at least 25 kg/m^3 (2,5% VS). For primary-chemical sludge this means a minimum dry solids concentration of 3.5-4% in the raw sludge.

- An average volatile solids reduction of 38% was achieved by the process in the one-year demonstration period. This has increased to an average of 43% VS-reduction in the succeeding years.
- Until year 2000 monthly samples for bacteriological analyses have revealed that faecal coliforms are reduced to levels below the Norwegian requirements (<2 500/gDS)m and no *Salmonella* have been detected in the digested sludge. In one period during autumn 2000 with heavy rainfall and very diluted sewage, the sludge did not contain enough organic solids to achieve temperatures above 50°C in the TAD-reactors. This resulted in high densities of faecal coliforms, and *Salmonella spp.* were detected in two samples.

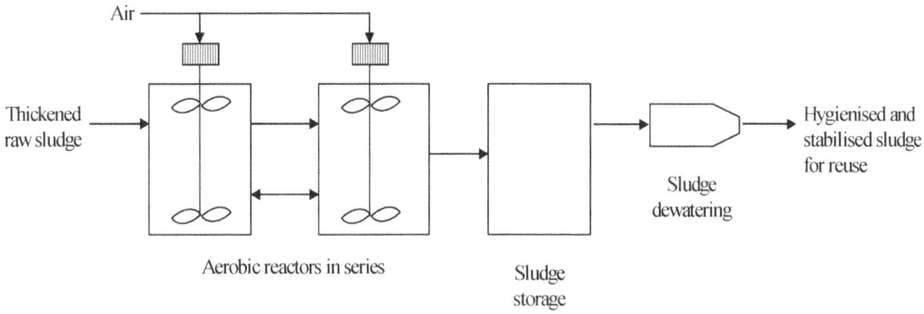

Figure 1 Schematic process diagram for Norwegian plants with thermophilic aerobic digestion

Thermophilic aerobic pre-treatment + mesophilic anaerobic digestion (dual digestion)

There are seven plants with this process combination which have provided information to the first NORVAR study (Paulsrud & Nedland, 1996). The pre-treatment stage of all these plants were contracted by UTB (former Swiss company), which was the market leader for such systems in Switzerland and Germany (Figure 2). All plants are designed for a semi-continuous operation, treating 16 batches of sludge per day, which means a detention time of 1,5 hours per batch at temperatures about 60 - 64°C. In the full scale validation test of this process, Paulsrud et al. (2003) proved that a batch detention time of 45 minutes at 61 – 62,5°C was sufficient for the inactivation of *Ascaris* eggs.

The general opinion among the plant managers is that this process combination is a reliable and effective method for complying with the Norwegian sewage sludge regulation. The microbiological data available support this statement (Paulsrud et al., 2001). However, the experience from one of the plants (Nordre Follo WWTP) gives a good example of possible recontamination of already hygienised sludge. After thermophilic aerobic pre-treatment + anaerobic digestion the sludge is stored in a large concrete tank below ground level before it is pumped to centrifuges for dewatering. In certain periods after plant start up, high densities of faecal coliforms were detected both in the sludge storage tank and in the sludge cake just after dewatering, even if they could not be detected in the sludge discharge from the digesters. The explanation was clear when they found a pipeline discharging washwater from septage trucks into the storage tank, introducing untreated septage to the hygienised sludge.

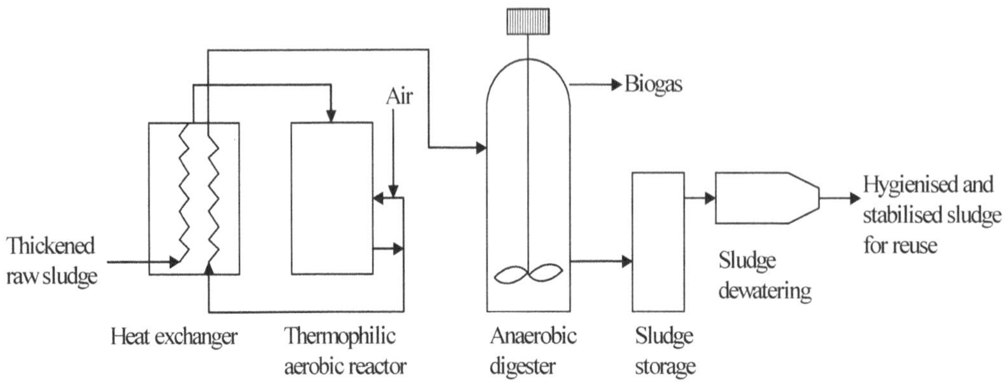

Figure 2. Schematic process diagram for Norwegian plants with dual digestion

Pre-pasteurisation + mesophilic anaerobic digestion

In Norway there are three plants with this process configuration, and they have been in operation for 10-15 years (Figure 3). They are all designed for a batch detention time of minimum 0,5 hours at 70°C, according to the criteria presented in Table 1. The experiences are very much the same as for the dual digestion concept, and again the hygienisation performance is very good. The pre-pasteurisation process was also included in the full scale validation test programme (Paulsrud et al., 2003), which documented that only 15 minutes at 65 – 66,5°C caused the inactivation of *Ascaris* eggs inserted into the process.

Paulsrud et al. (2001) presents the faecal coliform densities in weekly samples of hygienised and stabilised sludge from one of the plants during 1999 – sept. 2001. Five samples had values higher than 2 500 FC/gDS, and these samples correspond with periods where the temperature in the pasteurisation unit was reduced to below 65°C, and still operating at 0.5 hours detention time.

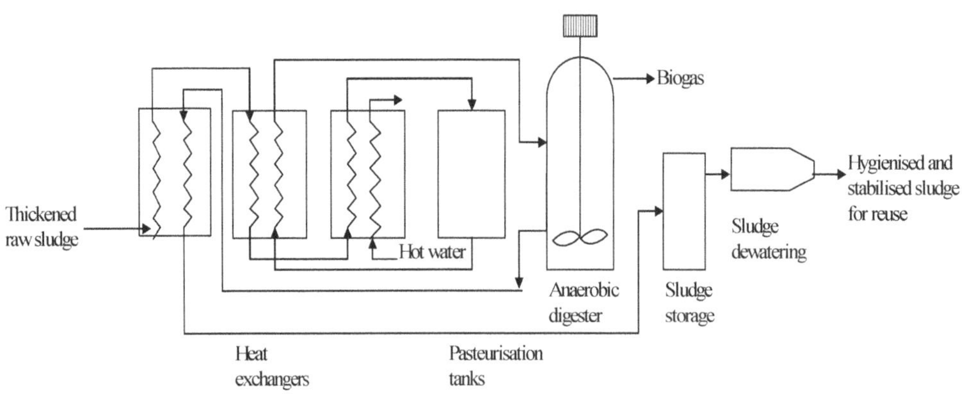

Figure 3. Schematic process diagram for Norwegian plants with pre-pasteurisation

Mesophilic anaerobic digestion + thermal drying

This process combination has become quite common in Norway in recent years and eight plants are now in operation (Figure 4). The plants comprise the largest sewage treatment plants in Norway, which contribute to about 50% of the total sludge production. Thermal drying means of course much more than sludge hygienisation, and major arguments for selecting this process have been reduced volumes for disposal, increased flexibility regarding disposal options and possibilities for choosing incineration as the last step if land application should be stopped.

So far there exist very little data about hygienisation performance at the sludge drying plants. The reason is that only one plant have been in regular operation for a longer period of time (> 5 years) due to improper design, unexperienced contractors, and a lot of mechanical problems with auxiliary equipment to the dryers. However, when producing pellets or granules with about 90% DS, it is expected that the operating conditions comply with the critical conditions presented in Table 1.

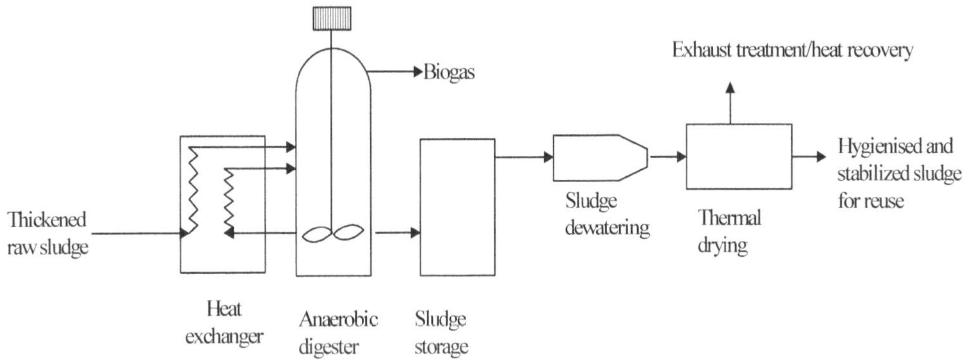

Figure 4. Schematic process diagram for Norwegian plants employing thermal drying of sludge.

Lime treatment

Addition of quicklime to dewatered sludge has become a popular process for hygienisation and temporary stabilisation at small and medium-sized plants (see Figure 5). Major design and performance data for 8 plants in operation in Norway are presented by Paulsrud et al. (2001). Only some plants have performed regular bacteria analyses, but continuous or frequent temperature readings in the limed sludge give an adequate process control. However, one plant has taken monthly samples since 1992 for determination of faecal coliforms and *Salmonella*, and these bacteria have not been detected in any of the samples. Another plant performed a test period of three months during the summer of 2000, taking weekly samples of limed sludge. Both faecal coliforms, *Salmonella* and helminth eggs were analysed, and all parameters were below detection limits. Both full scale and laboratory scale validation tests of the process (Paulsrud et al., 2003) have proved that at least 55°C for 2 hours is required for *Ascaris* inactivation.

The high lime dosage necessary for complying with the hygienisation criteria is the major problem for this process, involving high operation cost and limitations on land application rate, due to the lime content of the sludge. Some plants have upgraded their sludge dewatering system to produce a higher cake solids content and thereby reduce the necessary lime dosage to achieve temperatures above 55°C. Another plant has installed an electrical heating system to reduce necessary lime dosage.

The lime treated sludge has, in most cases, a granular consistency, which makes the sludge easy to spread on farmland. All the plants have reported that their total sludge production has been disposed of in agriculture after lime treatment.

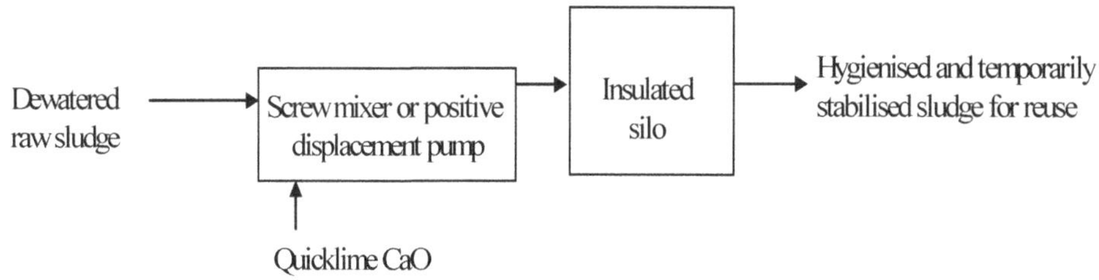

Figure 5. Schematic process diagram for Norwegian lime treatment plants

Windrow composting

Fourteen windrow composting plants were registered during the survey in 1995 (Paulsrud & Nedland, 1996), but today only ten plants are still in operation (Figure 6). Most of the plants are small, handling sludge from one or some wastewater treatment plants or from septic tanks, with a total load corresponding to 5 000 – 20 000 p.e. These composting plants are mostly located at the coastline in Southern and Western Norway with minor agricultural areas for sludge application, but where compost for non-agricultural use and land reclamation projects are quite popular.

Windrow composting is a process which may vary greatly in design and operation. Local conditions (type of bulking agent, dry solids content of sludge and available turning equipment) and operational strategy differ from plant to plant, and the process control is often insufficient to achieve a quality compost that meets the hygienisation criteria. Low bacteriological densities presented by Paulsrud & Nedland (1996) are for some of the plants based on compost produced under favourable conditions (dry and warm summer periods), while the situation is quite different during periods with heavy rain or in cold winters. Under such circumstances plants turning windrows with a standard front end loader and with a monthly or bi-monthly turning frequency, can hardly produce a hygienised compost (De Bertoldi et al., 1984).

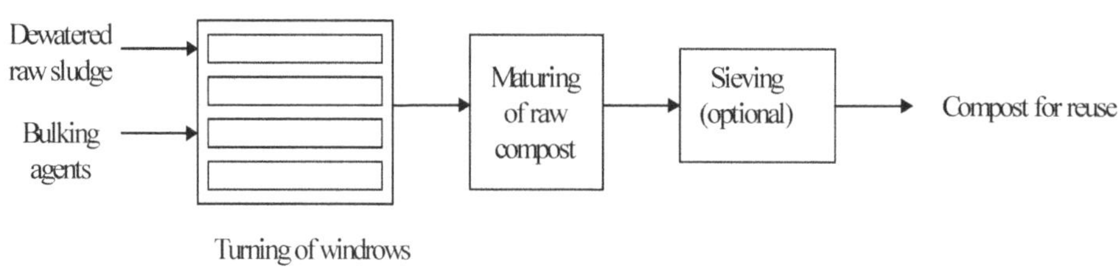

Figure 6. Schematic process design for Norwegian windrow composting plant.

Long-term storage of dewatered sludge

This is the most widespread method employed in Norway for septic tank sludge and sludge from small wastewater treatment plants. There are probably more than 100 such plants in the rural parts of Norway, and most of them are situated at sanitary landfill. Several of these plants are operated with turning the heaps once or twice a year to improve the sludge structure before final disposal, but with no bulking agents added, the temperature in the sludge seldom raise above 30°C.

These small plants are not regularly sampled and analysed for hygienic parameters, but available data clearly indicate that a storage period of minimum 6 months, of which minimum 2 months during summer time, leaves a sludge where *Salmonella* and faecal coliforms can not be detected, but helminth eggs can survive this treatment for several years. In 1999 a comprehensive survey of long-term storage plants and some composting plants were performed (Nedland, 1999). Eleven plants were visited and treated sludge was sampled and analysed for faecal coliforms, *Salmonella* and helminth eggs. The storage period varied among the plants (5 months – 6 years storage time), and at two of the plants with a sludge storage time of 5 months and 3 years, respectively, infective *Toxocara* eggs were detected, but no *Ascaris* eggs. At all plants investigated no *Salmonella* and <10 FC/g DS were detected.

Thermal hydrolysis + mesophilic anaerobic digestion

There are two such plants in operation in Norway: one treating sewage sludge (HIAS WWTP) and one for source separated organic household waste. Both plants are contracted by the Norwegian company Cambi, which have installed similar plants in Denmark, U.K., Ireland and Poland.

The HIAS plant has a capacity of 3500 tonnes DS/year and in the thermal hydrolysis process (see Figure 7) the sludge is exposed to temperatures of 160–170°C and a pressure of 7-8 bar for a holding time of 20 minutes. The treated sludge is sterilised, and due to the very good hygienic quality Norwegian health authorities have given permission for land application without any site restrictions, except those tied to the heavy metal content of the sludge. Another benefits of the process are reduced volume of the digester(s) due to pre-dewatering of raw sludge (~15%DS), increased volatile solids reduction, increased biogas production and improved dewaterability (Kepp et al., 1999; Fjærgaard & Sander, 2001). The main disadvantages are the high content of COD, NH_4 and PO_4 in the sludge liquor from digested sludge dewatering and the high investments costs of the process.

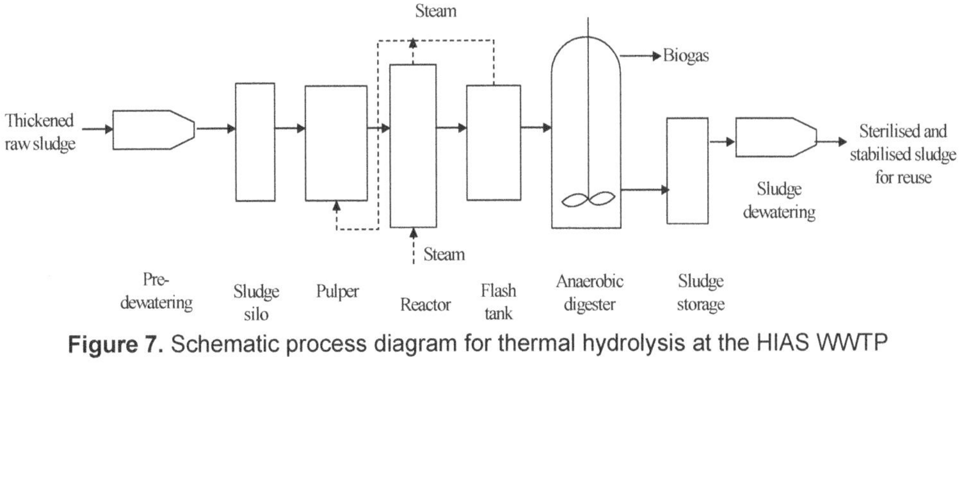

Figure 7. Schematic process diagram for thermal hydrolysis at the HIAS WWTP

Thermophilic anaerobic digestion

Today there are two plants in Norway operating their anaerobic digesters at thermophilic temperatures (50 - 55°C), but several existing plants having thermal dryers for sludge hygienisation are planning to convert their mesophilic digesters into thermophilic operation in order to achieve a hygienised sludge, also when the dryers are out of function. One of the plants (Bekkelaget WWTP in Oslo) was originally designed for thermophilic operation of their two digesters in series (with 12 days total detention time) as a redundant system for the thermal dryer which is the main hygienisation process. Based on three years of experience the thermophilic process is performing well, providing close control of the process parameters. Digested sludge samples have been analysed regularly for *Salmonella* and faecal coliforms, and they are complying with the hygienisation criteria as long as the temperature is above ~53°C in the digesters. No tests for viable helminth ova have been performed so far, but a validation programme with *Ascaris* eggs has been discussed.

Thermophilic anaerobic digestion has been practiced for years in the U.S.A. (Garber, 1954; Garber et al., 1975), but at that time the main purpose was not to hygienise the sludge. U.S. EPA (1979) stated that the process was not able to eliminate enteric viruses and helminth eggs, and they suggested that this may be due to the fact that plant-scale digesters normally are operated on a fill-and-draw basis. When most digesters are mixed continuously, severe short circuiting may occur, resulting in the removal of sludge that has been exposed to thermophilic temperatures for only a short time. Consequently, thermophilic anaerob digestion is not on the list of Processes to Further Reduce Pathogens (PFRPs), which is one of the alternatives (# 5) to meet the Class A pathogen requirements of U.S. EPA's Part 503 Rule (U.S. EPA, 1992; 1993). However, in recent years there have been a considerable activity in the U.S.A on designing thermophilic digestion processes that can meet the Class A pathogen requirements under Alternative 6: Use of a Process Equivalent to PFRP (WEF, 2002; WERF, 2003).

In the German guidelines (ATV/VKS, 1988a) thermophilic anaerob digestion was not recognized as a sludge hygienisation process, but again it is possible to achieve hygienisation conditions by changing the modes of sludge feeding and withdrawal to ensure that every sludge particle has a minimum residence time at thermophilic temperatures (Mitsdörffer et al., 1990).

Conclusions

- Design and operation of sludge hygienisation processes should be based on the specific criteria given in each sewage sludge regulation or directive. Requirements for elimination of viable helminth eggs and/or enteric viruses will impose stricter process conditions than criteria only for *Salmonella* and some indicator organisms (*E.coli*, *Clostridium perfringens*, etc.)
- More than ten years of full scale experience has documented that several treatment processes (i.e. thermophilic aerobic digestion, pre-pasteurisation, dual digestion and lime treatment) have proven to be reliable and cost-effective in complying with the Norwegian hygienisation criteria, which include elimination of viable helminth eggs.
- Pathogen control by low-tech methods (long-term storage of dewatered sludge and small scale windrow composting) is badly documented, while high-tech processes like thermal drying has experienced long periods out of operation (with no hygienisation) due to equipment failures.
- Converting mesophilic anaerobic digesters to thermophilic operation is an interesting option for sludge hygienisation, but there is still lacking information about the critical

process conditions (mainly time-temperature combinations) for different hygienisation criteria.

- Re-contamination of hygienised sludge should be avoided by close control of all internal streams and leakages originating from handling of raw sludges.
- Routine control of hygienisation processes should be established through quality assurance systems based upon HACCP principles (Hazard Analysis and Critical Control Point), including validated time-temperature combinations for each process.

References

ATV/VKS-Arbeitsgruppe 3.2.2 (1986). Entseuchung von Klärschlamm. Erster Arbeitsbericht, *Korrespondenz Abwasser* 11/86.

ATV/VKS-Arbeitsgruppe 3.2.2 (1988). Entseuchung von Klärschlamm. Zweiter Arbeitsbericht, *Korrespondenz Abwasser* 1/88.

ATV/VKS-Arbeitsgruppe 3.2.2 (1988). Entseuchung von Klärschlamm. Dritter Arbeitsbericht, *Korrespondenz Abwasser* 12/88.

Carrington, E.G. (2001). *Evaluation of sludge treatments for pathogen reduction – final report.* Report No. CO 5026/1, WRc Medmenham, UK.

DeBertoldi, M., Frasinetti, S., Bianchin, L. and Pera, A. (1984). Sludge hygienisation with different compost systems. In: *Inactivation of Microorganisms in Sewage Sludge by Stabilisation Processes*. D. Strauch, A.H. Havelaar and P.L'Hermelite (eds.). Elsevier, London.

European Commission (2000). *Working document on sludge, 3rd draft*, DG Environment E.3. Brussels, 27 April.

European Commission (2002). *Disposal and Recycling Routes for Sewage Sludge*. Economic sub-component report, DG Environment-B/2, Brussels, 29 January.

European Commission (2003). *Draft discussion document for the ad-hoc meeting on biowastes and sludges 15 – 16 January 2004.* DG Environment A.2, Brussels, 18 December.

Evans, T.D. and Lowe, N.R. (2003). Sustainable organic resources partnership goes live. In *Proceedings of the 8th European Biosolids and Organic Residuals Conference*, Wakefield, U.K., 23 – 26 November.

Fjærgaard, T. and Sander, O. (2001). Five years' experience with the Cambi process at HIAS. In *Proceedings of the Nordic Conference on Wastewater Treatment.* Copenhagen, Denmark, 17 – 19 January.

Garber, W.F. (1954). Plant scale studies of thermophilic digestion at Los Angeles. *Sewage and Industrial Wastes*, Vol 26, No. 10.

Garber, W.F.; Ohara, G.T.; Colebaugh, J.E. and Raksit, S.K. (1975). Thermophilic digestion at the Hyperion treatment plant, *JWPCF*, 47, No. 5, 950-961.

Kepp, U., Machenbach, I., Weisz, N. and Solheim, D.E. (1999). Enhanced stabilisation of sewage sludge through thermal hydrolysis – Three years of experience with full scale plant. Paper presented at IAWQ Specialised Conference on *Disposal and Utilisation of Sewage Sludge: Treatment Methods and Application Modalities*, Athens, Greece, 13 – 15 October.

Ministry of Health and Social Welfare and Ministry of the Environment (1995). *Regulation for the treatment, use and disposal of sewage sludge.* Oslo, Norway (in Norwegian).

Mitsdörffer, R.; Demharter, W. and Bischofsberger, W. (1990). Stabilisation and disinfection of sewage sludge by two-stage anaerobic thermophilic/mesophilic digestion. *Wat. Sci. Tech.*, Vol. 22, No. 7/8, pp. 289-290.

National Research Council (2002). *Biosolids Applied to Land: Advancing Standards and Practices.* National Academy Press, Washington, D.C. 20055, U.S.A.

Nedland, K.T. (1999). *Experiences with long-term storage and windrow composting of sewage sludge in Norway.* Aquateam-report no. 99-038, Aquateam, Oslo, September (in Norwegian).

Nybruket, S., Paulsrud, B. and Nedland, K.T. (2003). *Experiences with sewage sludge hygienisation methods in Norway.* VA-forsk report 2003-32, Svensk Vatten AB, Stockholm, Sweden (in Norwegian).

Paulsrud, B. and Nedland, K.T. (1994). Full scale experiences with thermophilic aerobic digestion. In: *Chemical Water and Wastewater Treatment III*, R. Klute and H.H. Hahn (eds.) Springer-Verlag, Berlin-Heidelberg.

Paulsrud, B. and Nedland, K.T. (1996). Full scale experiences with processes for stabilisation and disinfection of primary-chemical sewage sludge. In *Chemical Water and Wastewater Treatment IV*. H.H. Hahn, E. Hoffman and H. Ødegaard (eds.). Springer- Verlag, Berlin-Heidelberg-New York.

Paulsrud, B. and Nedland, K.T. (1997). Strategy for land application of sewage sludge in Norway. *Wat. Sci. Tech.*, Vol 36, No. 11, pp. 283-290.

Paulsrud, B., Gjerde, B. and Lundar, A. (2003). Full scale validation of helminth ova (*Ascaris suum*) inactivation by different sludge treatment processes. In Proceedings *of the IWA Specialist Conference. Biosolids 2003 Wastewater Sludge as a Resource*. Trondheim, 23-25 June.

Paulsrud, B., Nedland, K.T. and Nybruket, S. (2001). Control of pathogens in sewage sludge in Norway – full scale experience with different sludge treatment technologies. In *Proceedings of the 6th European Biosolids and Organic Residuals Conference*, Wakefield, U.K, 11 – 14 November.

Strauch, D. (1989). Improvement of the quality of sewage sludge – Microbiological aspects. In *Sewage Sludge Treatment and Use. New Developments, Technological Aspects and Environmental Effects*, A.H. Dirkzwager and P.L'Hermite (eds.). Elsevier Applied-Science, London – New York.

Strauch, D. (1998). Pathogenic micro-organisms in sludge. Anaerobic digestion and disinfection methods to make sludge usable as a fertiliser. *European Water Management*, 1(2) 12-26.

U.S. Environmental Protection Agency (1979). *Process Design Manual. Sludge Treatment and Disposal*, Technology Transfer, Municipal Environmental Research Laboratory, Cincinnati OH 45268, U.S.A.

U.S. Environmental Protection Agency (1992). *Environmental Regulations and Technology. Control of Pathogens and Vector Attraction in Sewage Sludge. EPA-625/R-91-013.* Office of Research and Development – U.S. EPA, Washington DC, U.S.A.

U.S. Environmental Protection Agency (1993). Part 503 – Standards for the Use and Disposal of Sewage Sludge. Rules and Regulations. *Federal Register*, 58, (32), 9248-9415.

Walker, J. (2003). Norway – Biosolids heaven? In *Proceedings of the 8th European Biosolids and Organic Residuals Conference*, Wakefield, U.K., 23-26 November.

Water Environment Federation, WEF (2002). How to Get Class A with Thermophilic Digestion: A Q & A Session with EPA Biosolids Scientists. *Biosolids Technical Bulletin*, Vol. 7, No. 2, March/April.

Water Environment Research Foundation, WERF (2003). Peer Review Project to Help Achieve Class A Standards, *Progress*, Vol. 14, Issue 2.

Supercritical Water Oxidation of Sewage Sludge in Combination with Phosphate Recovery

Magdalena Svanström*, Kristin Johansson* and Kjell Stendahl**

*Chemical Engineering and Environmental Science, Chalmers University of Technology, SE-412 96 Göteborg, Sweden, magdalena.svanström@chemeng.chalmers.se
**Feralco AB, Industrigatan 126, SE-252 32 Helsingborg, Sweden, kjell.stendahl@feralco.com

Abstract Supercritical Water Oxidation (SCWO) is an innovative and effective destruction method for organics in sewage sludge. The SCWO process leaves a slurry of inorganic ash in a pure water phase free from organic contaminants, which opens possibilities for a simple process to recover components like phosphates and/or coagulants from the sewage sludge, a process marketed as Aqua Reci. In a continuous pilot plant, digested sludge has been treated. The ash has been extracted in lab- and pilot scale with acids in order to recover phosphates and coagulants. The particle size of the inorganic contaminants in the water after the SCWO process is between 1–10 µm, which means that the material is very easy to extract. After an acid extraction, first iron phosphate can be separated and thereafter in a second step aluminium and finally heavy metals in a third step. By the acid dissolving process it is possible to recover phosphate, iron, aluminium, and heavy metals from the inorganic and the Aqua Reci process only leaves a silica residue representing less than 5% of the original volume. A cost calculation for a plant of a capacity of 10 000 tons DS is indicating 248 €/ton DS. The value of the process in the form of sludge volume reduction and more than 90% recovery of energy, phosphate and coagulants represents a value that will compensate for the cost of running an Aqua Reci process. In a life cycle assessment, the method is shown to have a positive environmental effect. Some emissions of greenhouse gases and acidifying substances can be avoided because production of district heat and artificial phosphorus fertilizer can be replaced by recovered oxidation heat and phosphate. Critical issues that need to be looked into if the environmental performance is to be improved are: formation of nitrous oxide in the SCWO reactor, and the use of MgO in the extraction process.

Keywords Phosphorus recovery, sewage sludge, supercritical water oxidation (SCWO), aluminium recovery, iron recovery, coagulant recovery, Life Cycle Assessment (LCA).

Introduction

Stockholm Water (the local water and sewage company in Stockholm) has shown interest in the Aqua Reci process, in which supercritical water oxidation (SCWO) is used to decompose organic material. The oxidation is followed by a chemical process in order to recover components in the inorganic residual ash, e.g. phosphates and coagulants. In this paper, a technical description of the Aqua Reci method is given together with an economic evaluation and an environmental assessment using life cycle assessment (LCA) methodology. In earlier LCA studies on sewage sludge SCWO, it has been shown that the energy system in which the process operates is of great importance (Svanström et al, 2004), and therefore, using LCA methodology is appropriate.

© 2004 IWA Publishing. *2nd IWA Leading-Edge Conference on Water and Wastewater Treatment Technologies* edited by Mark Van Loosdrecht & Jonathan Clement. ISBN: 1 84339 508 8.

Supercritical Water Oxidation (SCWO)

SCWO leads to complete destruction of organics: organic carbon is converted into carbon dioxide, organic and inorganic nitrogen into nitrogen gas (N_2), halogenated organics and inorganics into corresponding acids, sulphonated organics and inorganics into sulphuric acid. Metals are oxidised to their highest valency and phosphorus into P_2O_5. The oxidation reaction is exothermic, and consequently when the feed organic is concentrated enough, the reaction can produce sufficient energy to maintain the reactor temperature and even yield excess heat.

Inorganic ash, Figure 1, is very easy to extract, which means that metal oxides and phosphorus will easily dissolve in acid. Phosphorus can also be extracted in caustic soda and separated from metal oxides due to their insolubility in alkaline conditions.

Figure 1 Sewage sludge before and after SCWO

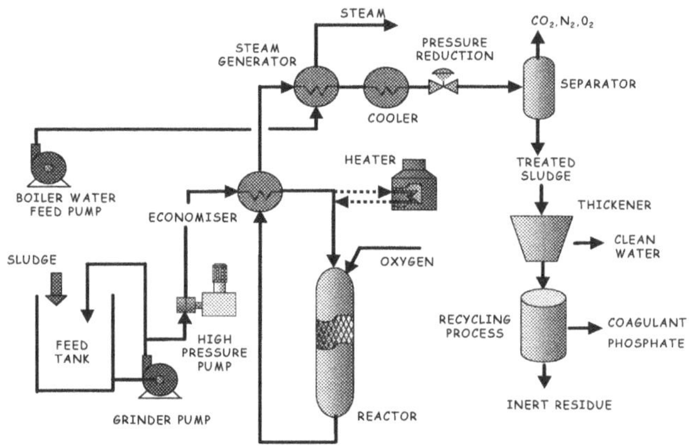

Figure 2 Flowsheet of the Aqua Reci process

In a pilot plant located in Karlskoga, Sweden, several tests have been conducted on digested sludge from sewage treatment plants in Stockholm and Karlskoga, all using iron as coagulant for phosphorus removal. Sludge at 15% DS (dry solids) was homogenised in a macerator before it was continuously pumped into the SCWO process. The SCWO process, Figure 2, consists of a long tube where the first part is an economiser in the form of a tube heat exchanger utilising heat from the reactor. The last part after the reactor is another tube heat exchanger, which can be used either to produce steam or hot water by utilising excess heat from the exothermic reaction. At the end of the process, gases (mainly carbon dioxide and nitrogen) are separated in an open separator tank. The total retention time is about 5 minutes, where 60 seconds represent the reactor part.

Aqua Reci experiments and results

Sludge used for the tests was dewatered digested sludge diluted from 25–30% to about 15% DS and homogenised using a macerator to make it pumpable. The COD content was on average 125 000 mg/l and the ammonia content 1 400 mg/kg in the 15% DS sludge. After the SCWO process, the COD level was, in all tests, reduced by more than 99.9%. If the temperature in the reactor exceeded 550°C, the ammonia nitrogen could be reduced by more than 99.9% to a value less than 10 mg/l. If the temperature was around 550°C, the ammonia nitrogen content reached 150 mg/l. Outgoing slurry from the process containing the inorganic components of the sludge had a DS content of 6% and was very easy to thicken to about 20–30% DS. Table 1 shows average concentrations of metals in water and solid phase, respectively, from SCWO of four different Stockholm sludges, and distribution of the inorganic components in the water phase and the solid phase leaving the SCWO reactor. The relatively low oxidation temperature and the fine particle size indicate high leachability of the inorganic residue and it is therefore an excellent raw material for production of phosphates and coagulants. The reason for the high aluminium content in the sludge is the use of zeolites in detergents.

Table 1 Distribution of inorganic components after SCWO in water (94% of inorganic mass) and solid (6% of inorganic mass) phase, respectively.

Water phase		Solid phase		Distribution between phases, %		
	mg/l		%		Water	Solid
Ca	360	Fe_2O_3	27.9	K	34	66
S	320	P_2O_5	19.1	Na	30	70
K	110	CaO	8.0	S	26	74
Na	64	Al_2O_3	8.2	Ca	13	87
P	24	SO_4	9.5	Mg	3	97
Mg	15	MgO	1.3	Cu	2	98
Al	2.0	K_2O	0.4	P	1	99
Cu	0.8	Na_2O	0.3	Fe	0	100
Mn	0.5	Insoluble	25	Heavy metals	0	100
Fe	0.4					
Heavy metals	ca 0.1					

Acid leaching and recovery of phosphate and metals

By leaching with sulphuric acid, the solution will drop to a pH of about 1.3 and will contain all the phosphorus and metals. One way to separate the components in the liquor is a 3-step neutralisation process with magnesium oxide. In the first step the pH is adjusted to around 2 where 3-valent iron phosphate ($FePO_4$, ferric phosphate) is practically insoluble, while all other metals will remain in the solution. Pure ferric phosphate can be separated. In a second neutralisation step at a pH of around 4.5, aluminium and residual iron will precipitate as alunite respectively jarosite ($NaAl_3(SO_4)_2(OH)_6$ respectively $NaFe_3(SO_4)_2(OH)_6$). Alunite and jarosite are very easy to filter with 60–70% DS in the filtercake. The final step is to further elevate the pH to 7–8 in order to precipitate all the heavy metals.

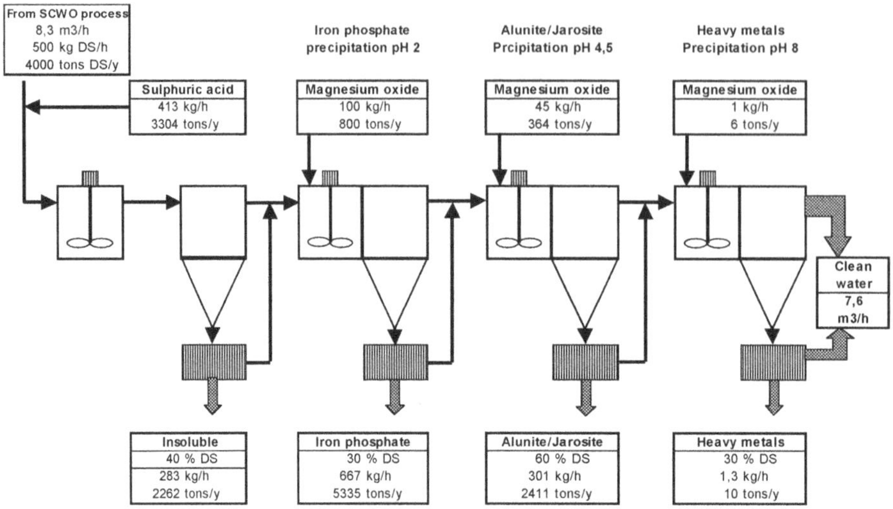

Figure 3 Process design and material flow for phosphate recovery from acid extraction. Numbers given are per year resp. day at 8 000 working hrs per year. The calculation is based on 10 000 tons of sludge DS, i.e. 67 000 tons as 15% DS coming into the SCWO process.

Even though iron phosphate can be used as a fertiliser, the economy is likely to be better if the iron phosphate can be separated into iron respectively phosphate salts. Two different ways of separating iron from phosphate have been tested – the alkaline respectively the acid way. In the alkaline case, phosphate can be dissolved with the help of caustic soda, leaving a sludge consisting of iron hydroxide. In this process (Figure 4), caustic soda can be recovered after precipitating the phosphorus with lime. The remaining iron hydroxide can be used as a raw material. After dissolving the ferric phosphate in hydrochloric acid, ferric chloride can be separated from phosphoric acid with the help of liquid extraction (Figure 5). The liquid extraction process will separate the iron from the phosphorus to practically 100%. As the phosphoric acid will be very pure it is possible to use it as a raw material for speciality phosphate production.

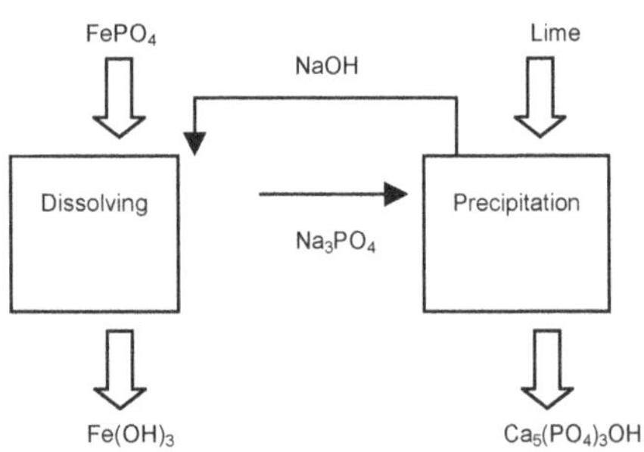

Figure 4 Alkaline separation process

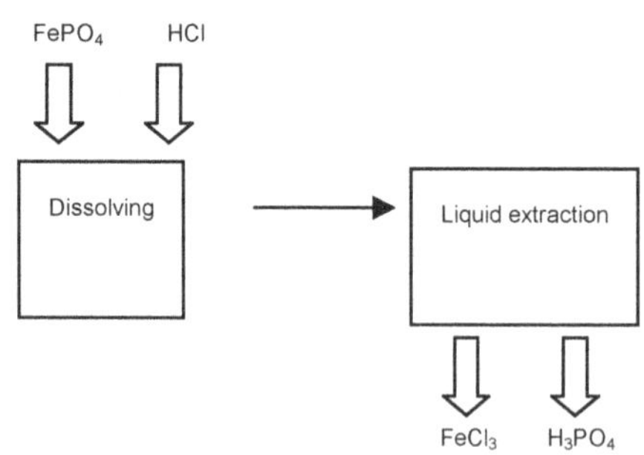

Figure 5 Acid separation process

Aqua Reci process for 10 000 tons of DS sludge per year

Based on the findings of the tests and using average sludge data from Table 1, an energy and material balance for the Aqua Reci process on digested sludge was made (Figure 6).

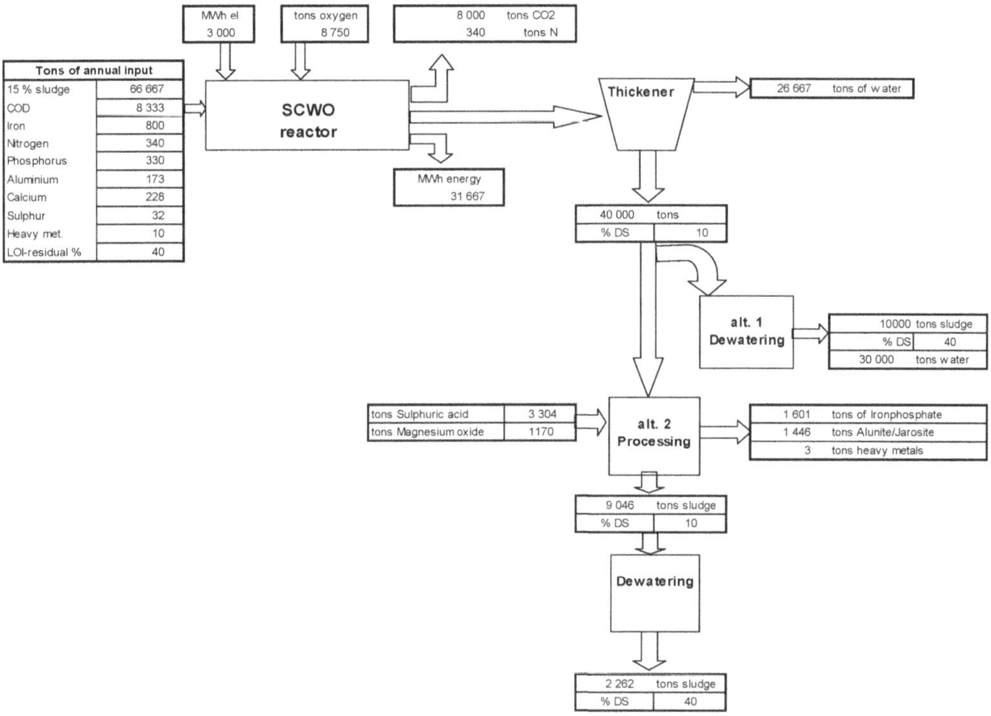

Figure 6 Material and energy balance for the Aqua Reci process with acid leaching

Cost calculation

A cost estimate for full-scale operation of a process with a capacity of 10 000 tons of DS per year was made. The sludge data is average data from the Stockholm sewage works. The total investment for the SCWO process has been estimated to 8.5 MEUR and for the recovery process to 600 000 EUR. The building, with a surface area of 1 000 m^2, is calculated to a cost at 1 MEUR. With a depreciation of 10 years and 5% real interest rate for the machinery and 20 years for the building, the total annual capital cost will be 1.2 MEUR. The operative cost has been calculated to 1.3 MEUR. This means a total cost of 248 EUR/ton DS. The positive side of the process is that it will produce 31 667 MWh energy, 1 782 tons of calcium phosphate, 6 667 tons of ferric chloride and 2 035 tons of aluminium sulphate. The value of the process in the form of sludge volume reduction and more than 90% recovery of energy, phosphate and coagulants represents a value that will compensate for the cost of running an Aqua Reci process.

Environmental Assessment of the Aqua Reci method

Life cycle assessment (LCA) is a structured way to quantitatively evaluate the total environmental impact from products or services. In LCA methodology, the life cycle of the product/service is studied from "cradle" to "grave". Environmental impacts in the form of natural resource consumption, emissions to air, water and land, and waste production are collected for all activities from raw material extraction, through transports, refining, manufacturing and use to waste management. Depending on the purpose of the study and on the availability of data, it may be wise to study only a part of the total life cycle or only certain environmental impacts.

The study was performed for sludge from a wastewater treatment plant (WWTP) in Stockholm, Sweden. The wastewater treatment was not considered in the LCA. The plant produces about 14 000 tons DS of digested sludge every year. The sludge contains 56% organic material. In the LCA study, the Aqua Reci plant is considered to be located at the WWTP. A system description of the LCA activities is shown in Figure 7.

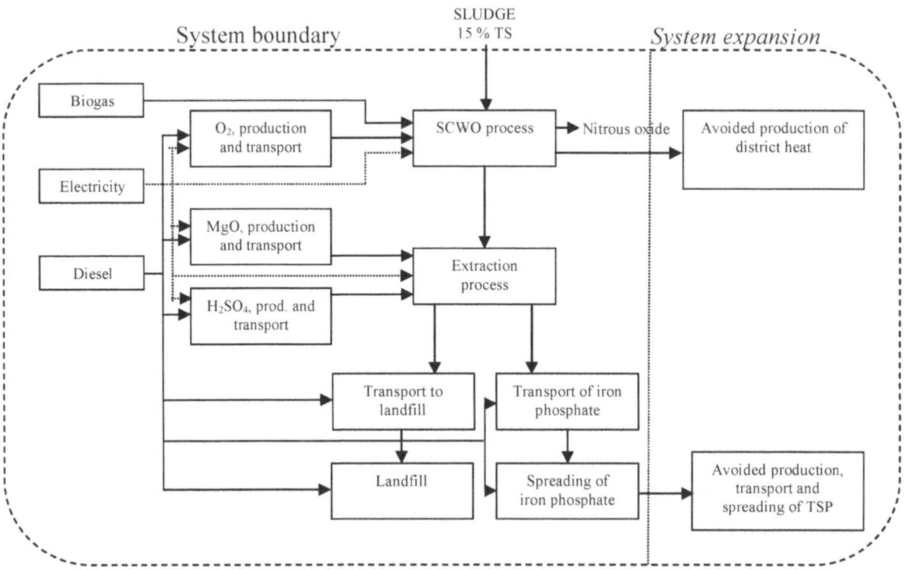

Figure 7 Life Cycle Assessment of the Aqua Reci method. System description.

SCWO process

Digested sludge of 15% DS is considered to enter the system. Locally produced biogas is used for start-up of the SCWO reactor once a month (55 kg). The fact that biogas is withdrawn from other uses was neglected. Electricity is consumed in the pumps (300 kWh/ton DS). Average electricity production in Sweden is considered (hydro 47%, nuclear 47%, etc). Oxygen (1000 kg/ton DS) is assumed to be produced in a large cryogenic production unit off site. Electricity consumption for oxygen production is assumed to be 250 MJ/ton oxygen. The oxygen is transported to the WWTP (120 km) with a heavy truck. Nitrous oxide (laughing gas) is produced in the reactor. A catalytic destruction unit is considered to treat the exhaust gases so that at least 99% of the nitrous oxide is destroyed (remaining: 0.22 kg/ton DS). Excess heat, after heat exchange for internal needs, is used for district heating (3200 kWh/ton DS) and is assumed to replace average district heat production in Sweden (renewables 24%, electricity 13%, oil 12%, peat 8%, coal 8%, natural gas 7%, other heat production 28%). Carbon dioxide is not collected.

Phosphorus extraction

To get maximum yield of phosphorus, acid extraction was considered. The inorganic rest (6100 ton DS/year) is dissolved in sulphuric acid (96-98%, 390 kg/ton DS sludge). The sulphuric acid is considered to be produced through incineration of liquid sulphur in Helsingborg, Sweden. The reaction is highly exothermic and gives an excess of energy (17 kWh average Swedish electricity and 780 kWh average Swedish district heat is replaced per ton of sulphuric acid). Sulphur is produced in Poland (60%, 1.9 MJ coal power/kg S) and is also extracted at a refinery in Lysekil, Sweden (40%, 0.0053 MJ LPG/kg S). The transport from Poland is assumed to be made by train (500 km, average European electricity) and cargo vessel (450 km) and from Lysekil (290 km) with cargo vessel. The sulphuric acid is transported to the WWTP by train (570 km) and truck (22 km). Magnesium oxide, used to raise the pH level (97 kg/ton DS sludge), is considered to be manufactured from dolomite (5 kg/kg MgO). Dolomite is extracted in Greece and transported to the MgO production facility by train (5 km). For processing of the dolomite to MgOH and further to MgO, electricity (European average, 6.6 MJ/kg MgO), heavy fuel oil (1.5 MJ/kg MgO) and natural gas (18 MJ/kg MgO) is expected to be consumed. MgO is transported from Greece to the WWTP by cargo vessel (3000 km) and truck (480 km). In the extraction process, some electricity is used for stirring and pumping (1.4 kWh/ton DS sludge, Swedish average electricity).

In this LCA, the extraction process was not taken further than to precipitation of ferric phosphate (8500 ton/year). The ferric phosphate is considered to be used directly in agriculture and will replace production, transport and spreading of triple super phosphate (TSP). Plant availability of the phosphorus in the ferric phosphate is expected to be 70% compared to the phosphorus in the TSP. Transport of ferric phosphate is considered to be made by truck (110 km). Spreading is assumed to be made by tractor with spreading equipment (23 MJ diesel/ton ferric phosphate). Data used for avoided production of TSP (130 kg/ton DS sludge) is for west European production. Avoided means of transportation for the TSP is assumed to be truck (50 km) and avoided TSP spreading is by tractor with spreading equipment (23 MJ diesel/ton TSP). The extraction process could be taken further to extract also coagulant etc., but in this case everything but the ferric phosphate was considered to be sent to landfill. Transport is assumed to be made by truck (200 km). Data for landfill management included in the study is for working vehicles and leaching water treatment.

Environmental impact assessment

To aid in the interpretation of the results, the parameters in the inventory list were classified according to contribution to global warming and to acidification. All emissions contributing to the two environmental themes were recalculated into one common unit and added up into one contribution from each activity in the LCA system. These characterizations are shown in Figures 8 and 9.

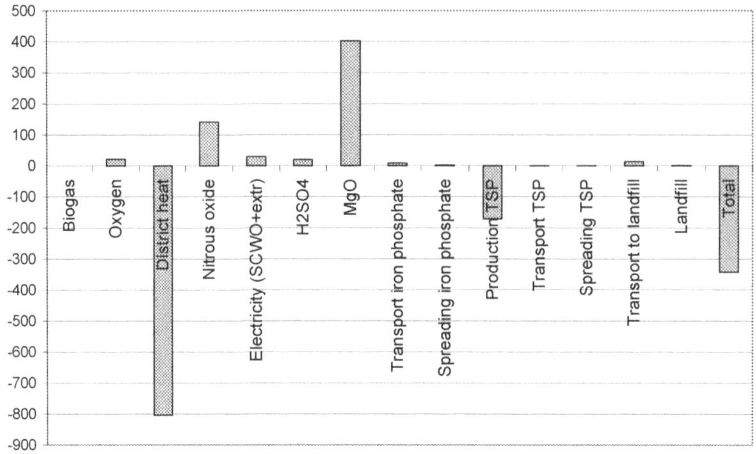

Figure 8 Global warming potential (GWP) characterization of the LCA results. Unit: kg carbon dioxide equivalents per ton DS sludge.

Figure 8 shows that for global warming, the major impacts are caused by MgO needed in the extraction process and by nitrous oxide formed in the SCWO reactor, even with a destruction efficiency as high as 99%. For MgO, the greenhouse gas emissions are created mainly during processing of dolomite ($CaMg(CO_3)_2$) to MgO. A considerable reduction in environmental impact could be obtained if another alkali was chosen for the extraction process, for example $MgCO_3$. For nitrous oxide, a combination of internal process measures and end-of-pipe technologies could decrease the emissions to a lower level. The total GWP impact for the system is below zero, meaning that the presented system gives savings in greenhouse gas emissions. This is due to the utilisation of heat and phosphorus, replacing some production of district heat and TSP. In fact, the opportunity to use excess heat in the district heating system is crucial for the beneficial turnout.

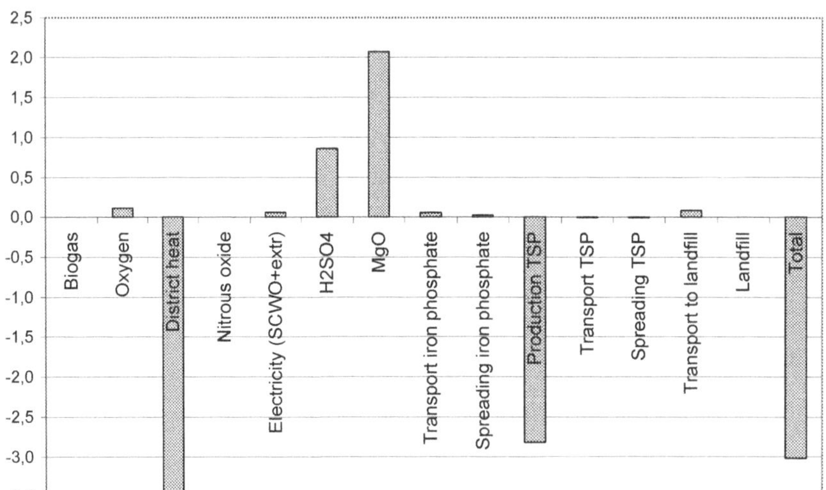

Figure 9 Acidification potential (AP) characterization of the LCA results. Unit: kg sulphur dioxide equivalents per ton DS sludge.

Figure 9 shows that for acidification, MgO production is still making up an important environmental impact. Sulphuric acid also presents an important environmental impact due to SO_2 emissions from the acid production in Helsingborg, coal power use in Poland and LPG combustion in Lysekil. Avoided district heat production is an important advantage also in the characterization. Avoided production of TSP gives a large decrease in acidifying emissions due to savings in SO_2 emissions when production of average European electricity is avoided. From Figures 8 and 9 it is clear that the landfilling has a small effect on global warming and acidification, even if a much larger amount is considered to go to landfill than if the complete extraction was carried out. The omission of the final steps of the extraction process in the LCA is not expected to have a large impact on the results.

Conclusions

In pilot scale it has been shown that the supercritical water oxidation process is feasible, not only to decompose organics to more than 99,9%, but also to oxidise ammonium into nitrogen. The total process is a closed process where the main reaction takes place in a tube reactor without any other emissions than carbon dioxide and nitrogen and small amounts of nitrous oxide. The excess heat from the oxidation process can be utilised as hot water or for steam production.

The residue from the oxidation process is an inorganic ash in pure water. The ash consists mainly of SiO_2, metal oxides and P_2O_5. Due to the relatively low temperature in the oxidation process, the inorganic ash is relatively easy to extract.

Acid leaching allows for separation of the phosphate and the metals from the inorganic ash. In a three-step precipitation process, it is possible to recover P, Fe, Al and heavy metals separately, leaving an ash consisting mainly of silicates representing less than 5% of the original volume, and pure water.

The cost calculation for a plant of a capacity of 10 000 tons DS is 248 €/ton DS (or 62 €/ton 25% sludge). The value from the process in the form of sludge volume reduction, more than 90%, recovery of energy, phosphate and coagulants represents a value that will compensate for the cost of running an Aqua Reci process.

In a life cycle assessment, the method is shown to have a positive environmental effect. Some emissions of greenhouse gases and acidifying substances can be avoided because production of district heat and artificial phosphorus fertilizer can be replaced by recovered oxidation heat and phosphate. Critical issues that need to be looked into if the environmental performance is to be improved are: formation of nitrous oxide in the SCWO reactor, and the use of MgO in the extraction process.

Literature

Anders Gidner, Mats Almemark, Lars Stenmark and Östen Ekengren. Treatment of sewage sludge by Supercritical Water Oxidation. 6th Annual Conference on Sewage Sludge, 16–17 February 2000, United Kingdom.

Feralco (2002). Pilotförsök att slutligt omhänderta slam från Stockholm Vattens vatten- respektive avloppsverk genom Aqua Reci-processen. Stockholm Vatten Rnr 03, februari 2002.

Patterson D.A., Stenmark, L. and Hogan F. Pilot-scale supercritical water oxidation of sewage sludge. 6th European Biosolids and Organic Residues Conference 12–14 November 2001. United Kingdom.

Stark K. (2002). Phosphorous recovery from sewage sludge by thermal treatment and use of acids and bases. Conference proceeding at 10th Gothenburg Symposium on chemical water and wastewater treatment 17–19 June, 2002.

Svanström, M., Fröling, M., Modell, M., Peters, W.A. and Tester, J. (2004). Environmental assessment of supercritical water oxidation of sewage sludge, *Resources, Conservation and Recycling*, in press.

Full Scale Application of high rate Digestion to improve Stabilisation

W. Trösch, B. Kempter-Regel

Fraunhofer-Institute for Interfacial Engineering and Biotechnology
Nobelstrasse 12. 70569 Stuttgart, Germany. E-mail: troesch@igb.fraunhofer.de

Abstract Organic loading rates for anaerobic sludge digestion can either be increased by a reduction of the retention time, or an increase of input volatile solid concentration or even better both of them. It is known that anaerobic bacterial consortia are able to convert organic matter with high efficiency at loading rates ten times higher than planned along the technical rules commonly used in Germany (ATV - rules). It is also known that anaerobic digestion follows first order reaction kinetics. If therefore a two stage process is used to convert organic matter the over all efficiency in the form of volume-time-yield can be increased tremendously in the first stage. But in the second stage in consequence the volume-time-yield falls to one third of the first stage. By an integration of a microfiltration unit, i.e. a dynamic ceramic disc filtration to the second stage reactor solid retention time can be separated from liquid retention time at least by the factor of three. Under these conditions second stage reactor can be operated under comparable conditions regarding the volume-time-yield as the first stage. Gas productivity runs at maximum supported by the concentration increase of residual substances and stabilisation is completed at highest degree. About seventy percent of volatile solids of sewage sludge are degraded under high load conditions, thus the equivalent reduction of disposal costs are welcome.

Keywords anaerobic digestion, high load system, dynamic microfiltration, ammonia recycling

1. Introduction

The activated sludge process to clean urban waste water produces two kinds of sludges, raw and excess sludge. Both of these can easily be digested by anaerobic bacterial consortia. Full scale processes have been successfully achieved at mesophilic temperatures, hydraulic retention times between 20 and 30 days and loading rates of about 1 kg volatile solids (vs)/m^3 d (Kapp, 1984; Loll, 1981). But under these conditions volume-time-yields are rather low and the sludge reduction is incomplete (Schäfer, 1998; Kempter-Regel et al., 2003). In recent years new full scale digestion processes have been installed. Regarding the first order reaction kinetics of the microbial disproportionation of organic matter to biogas two stage processes with short retention times well adapted to this kinetic could be realised (Kempter et al., 2000). Biogas productivity was superior to the standard operation and the process costs could be reduced by the reduction of disposal costs of residual sludge amounts. Laboratory scale digestion experiments for mixed sludges of aerobic activated sludge processes in real comparative studies have shown that for different domestic plants the digestion efficiency of mixed sludges is a linear function of organic loading rate when the retention time falls not below 5 days. When input sludge concentration after a standard sedimentation process

© 2004 IWA Publishing. *2nd IWA Leading-Edge Conference on Water and Wastewater Treatment Technologies* edited by Mark Van Loosdrecht & Jonathan Clement. ISBN: 1 84339 508 8.

exceeds not 3,5 % vs loading rates of up to 7 kg/m^3 d should be handled without problems. The full-scale high-load Heidelberg plant as first digestion stage shows successfully the new operation regime (Kempter-Regel et al., 2003). Digestion of 50 % of the volatile solids in the first stage and the total efficiency with a standard digestion tower in-line of about 60-65 % vs destruction are the results of the first yearly balance. Biogas yields are self-evident increased too.

Residual sludge components of waste water treatment plants contain a lot of organic components which are harmful for the environment (Ternes, 1998; Halling-Sorensen et al.,1998). Therefore the administration in Germany has published a recommendation to incinerate residual sludges from waste water treatment plants. But incineration of wet sludges is expensive and is not sustainable at all because the energy demand is higher than the energy content of the sludge. Thus highly efficient digestion processes at high volume-time-yields can contribute to reduce economically residual components to its maximum. Presented data of the optimisation of the digestion process are a prerequisite to increase the sustainability of the overall water cleaning operation.

2. Results

2.1 Influence of residence time reduction and first order reaction kinetics to the digestion process

Along with Monod-kinetics all bioprocesses where the main issue is cleaning of water streams from certain pollutions first order reaction comes into play. The appropriate reactor to guarantee constantly high cleaning grades is the pipe reactor where back mixing condition for the fluid does not occur and the residence time for a defined efficiency is given by the pipe length. But scale up of pipe reactors is complicated and the realisation too expensive. In Fig 1. inverse reaction constant of anaerobic sludge digestion is plotted against pollution concentration. The area under the curve for a defined degradation represents the volume needed for a defined efficiency. A totally mixed reactor used for the same efficiency has a volume demand which is represented by the big rectangle (V3) shown in Fig.1. It is easy to understand that a two stage fully mixed reactor concept achieves the same efficiency with half of the volume demand (V1 plus V2) and the main advantage is that such a concept has technical scale up potential furthermore.

Anaerobic degradation must be characterised by at least two reactions with different growth velocities. Acidification is the faster reaction compared to the methane generation. Therefore the least residence time of the first anaerobic digestion stage is always defined by growth velocities of methane bacteria. Investigated with different sludge qualities the minimum residence time for sewage sludge is about 5 days to prevent securely wash out for methanogens. Fig.2 shows digestion results gained with different municipal sewage sludges in a two stage plant and identic conditions. There is a linear dependency of digestion rate and the input concentration of vs at constant residence time of 5 days for each stage. According to these input concentration the organic loading rate was between 4 and 10 kg vs/m^3 d (Fig. 3) in stage 1. And there is no sign in the curve shape to be not elongated to even higher figures.

Surprisingly degradation level of volatile solids is also increased with increasing organic loading rate upto 10 kg vs/m^3 d. This is documented in Fig. 3.

But such an excellent digestion efficiency in the first stage is followed consequently by a low efficiency in the second stage which is also operated with a retention time of 5 days.

This follows exactly the theory. Compared to the volume-time-yield of the first stage the operation mode of the second stage seems to be a reactor volume wasting one.

2.2 Integration of dynamic microfiltration with ceramic modules to digestion

To enhance the efficiency in stage 2 a microfiltration unit was integrated. Thus the VS input of stage 2 was increased. Normal cross flow filtration with ceramic tube filters have been examined and showed bad performance especially in energy demand per m^3 filtrate. This high demand is caused by the high flow velocity inside the tubes to prevent fouling. Therefore a dynamic filtration unit with rotating microfiltration discs (RSF) which has been developed by Fraunhofer IGB was used to separate solid and liquid retention time for pre-digested sludges. High constant filtration rates could be achieved in long time experiments and the energy demand could be reduced by the factor of 8 compared to cross flow filtration in ceramic tube systems. Fig. 4 shows a scheme of RSF-filter unit.

Fig. 5 shows the results of the application of the separation of solid and liquid retention time for the second stage reactor. Increasing solid retention time shown as increasing vs input concentration in stage 2 at constant total liquid retention time results in a linear increase of digestion level. This is not expected because first order reaction kinetics predict a nonlinear dependence for heavily degradable substances.

2.3 Challenges of solid free filtrate from digestion process

The separation of liquid and solid retention time generates a filtrate stream free of solids due to a pore diameter of 0,2 micrometer and containing besides some volatile fatty acids ammonia up to 2 g/l. This amount of ammonia should not be discharged to the activated sludge process because of high cost for nitrification and denitrification. Under these condition it should be more economic to recycle ammonia by well known stripping technologies.

Fig.6 shows some air stripping experiments and the dependence of stripping efficiency from temperature. The second stream of the separation of solid and liquid retention time in the second stage of a digestion process is a concentrate stream of residual solids of up to 12 % total solids. This preconcentration reduces the liquid volume to further compactation by press. In addition running times of existing press capacity is reduced. Savings in personal cost are the consequence.

The calculation of mass balance for the digestion process at Heidelberg water works predicts the option to recycle about 400 kg/d ammonia. A pilot plant is now installed (Fig. 7) to test the integration of this new technique under the running conditions of a big waste water treatment plant.

3. Discussion

Standard organic loading rates of 1-2 kg vs/m^3 d in eye-shaped digestors may be sufficient for a stabilisation of sewage sludges regarding odour production along with disposal in agriculturally used fields or landfills. Under these conditions digestion grade is low and the biogas production rate is extremely low in relation to reactor volume needed: about 200-400l biogas/kg vs input (Zlokarnik, 1982). If maximum degradation of volatile solids at high volume-time-yield is the central issue, the knowledge of first order reaction kinetics of the digestion process predict the application of a two stage process. For practical reasons under this digestion mode two main possibilities for technical solution are possible. If a standard digestor with long retention time is existent, the in-line integration of a high load reactor with short retention time as first stage is recommended. There the fastly degradable portion of vs is completely converted to biogas (Kempter and Trösch 2000). In the second stage, where the acidification reaction is the velocity determining step, the hardly degradable fraction of sludge will be catalysed. Dependent of the length of retention time in that reactor stage the degradation of the asymptotic part up to the theoretical maximum along with first order reaction kinetic is performed.

When a technical plant for the digestion of sewage sludge is not available an alternative solution is to apply a two stage high load digestion.

Zehnder und Schraa (1988) assumed for hardly degradable substances the existence of a certain concentration for each of them which can not be degraded because the concentrations are too small to induce an uptake. In consequence a recommendation for a modification of monod equation has been formulated presented by Schulze and Lippe (1964) and Uden (1967).

Digestion residues of the first stage high load digestion solely contain such hardly degradable components when the history of sewage sludge is taken into account. With the enhancement of the concentration of these components by microfiltration this minimum concentration can be exceed and the concentration dependent (first order) part of normal Monod-kinetics can be achieved. Thus it should be possible to increase the level of digestion at short retention time compared to long retention time results. Linear dependency of the degradation level as function of the vs-loading rate in the second digestion stage with integrated microfiltration as shown in the results of this paper fits the theory completely.

The very low concentration of single components of sewage sludge in standard digestors at low loading rates is obviously the main reason for malfunction of technical stabilisation processes. Low energy efficiency is the consequence. By two stage digestion with integrated microfiltration at liquid retention times of 2x5 days highest degrees of digestion of organic matter up to 75% are possible (Kempter-Regel et al., 2003). The energy production is certainly dependent of the quality of sludges but the general tendency for all tested sewage sludges indicates that 600 l/kg vs_{in} can be reached. A volumetric productivity of 4 m^3/m^3 d was achieved for the first stage and 2 m^3/m^3 d for the second stage. It is remarkable that the second stage reactor with an integrated microfiltration can be operated with similar gas productivities as the first stage. There is no more a volume wasting mode remarkable. Under these condition residual sludge amount is reduced at its maximum and the dewaterability should be excellent because of the small amount of residual water binding organic matter.

Integrated microfiltration allows the sustainable recovery of nitrogen compounds by air stripping of the sludge water. Thus ammonia can be recycled as ammonia water or ammoniumsulfate if sulfuric acid is used as precipitation agent. Depending on the acid used different ammonia salt products on the demand of the market are possible. The advantage of such an operation mode is that the total input nitrogen load to the waste water treatment plant is reduced for 20-30%. Thus not only the energy demand of the plant can be reduced by this recycling technique but also upgrading of waste water treatment plants can be saved when they have to comply with the EC waste water directives.

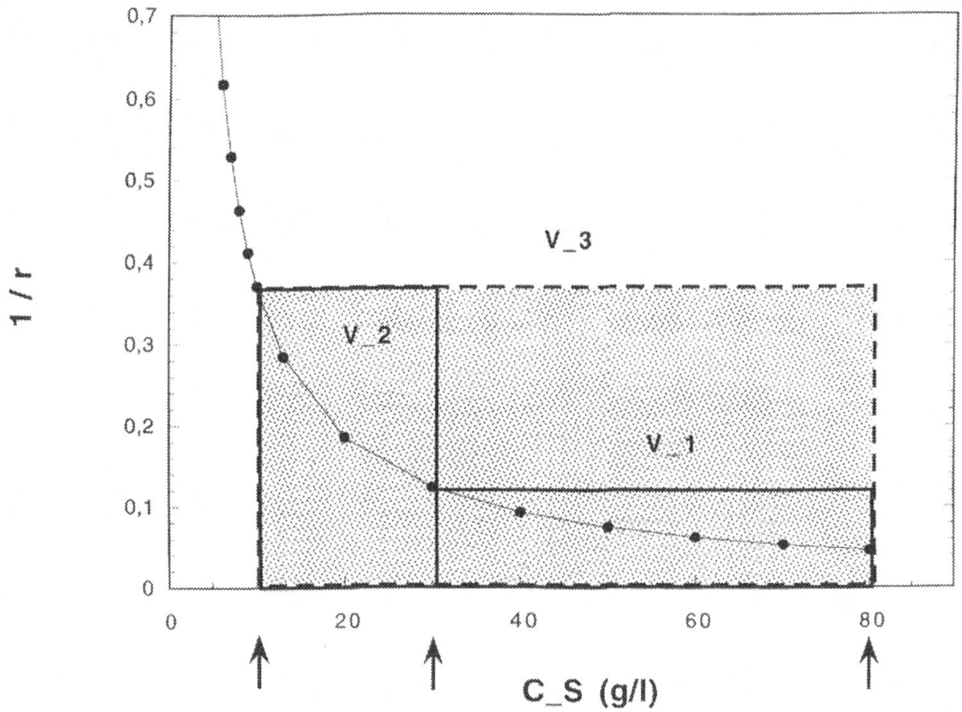

Figure 1 Reciprocal of the reaction rate r as a function of the substrate concentration C_S.

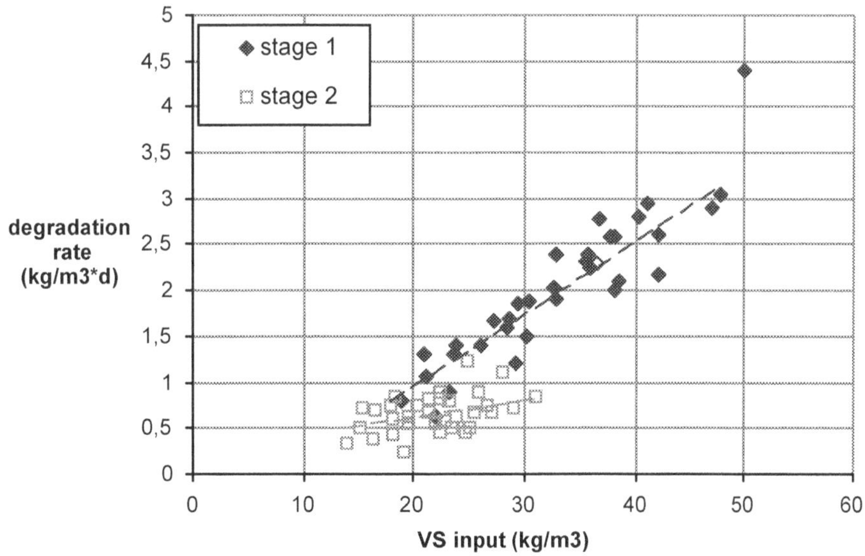

Figure 2 Degradation rate for stage 1 and stage 2 as a function of the concentration of the volatile solids in the input of stage 1 respectively stage 2 (with several municipal sewage sludges)

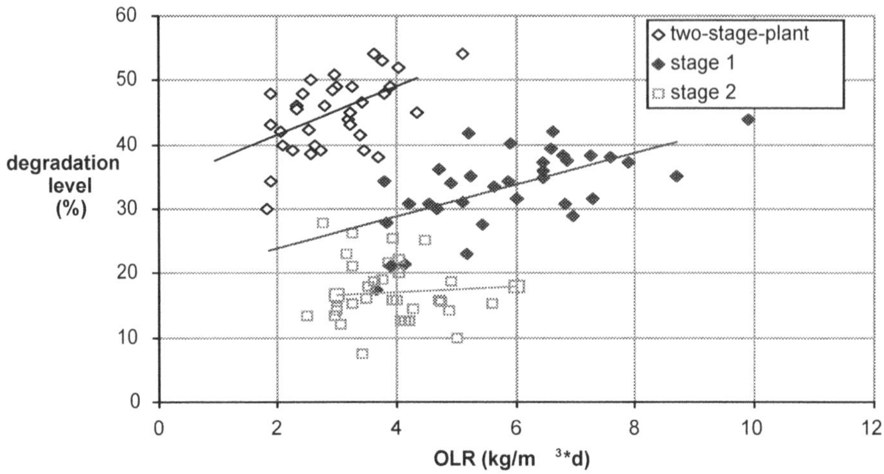

Figure 3 Degradation level for stage 1, stage 2 and total as a function of the organic loading rate in stage 1, stage 2 and total plant (with several municipal sewage sludges)

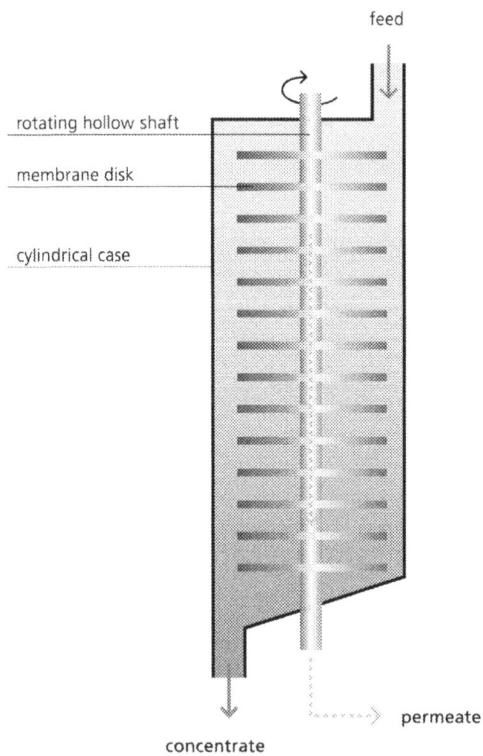

Figure 4 Scheme of the rotating-disc-filter unit (RSF)

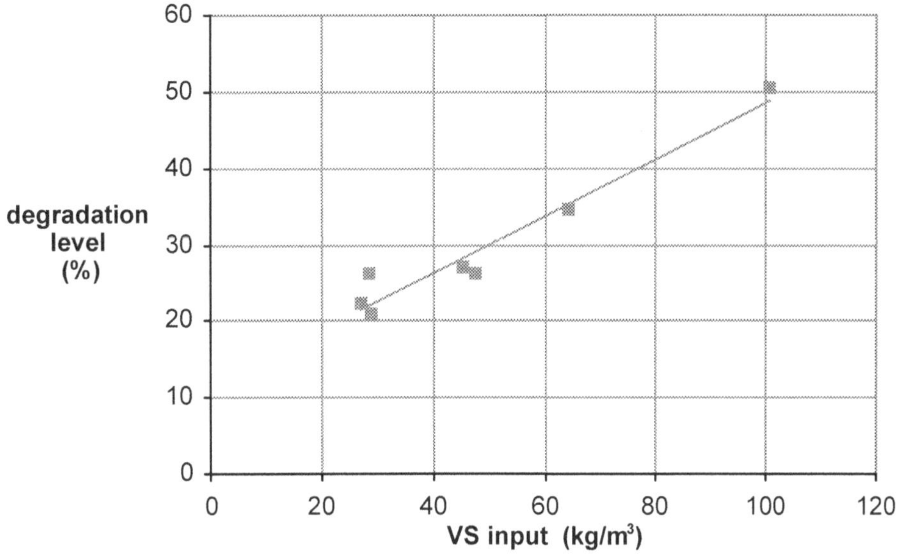

Figure 5 Degradation level in stage 2 as a function of VS input concentration in stage 2 which was increased with removing sludge water by use of microfiltration.

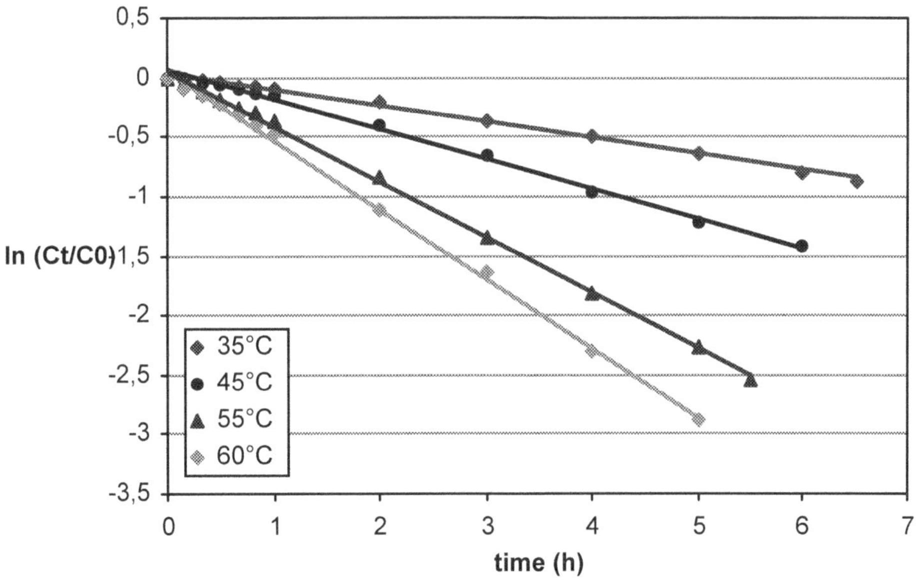

Figure 6 Decrease of ammonium-concentration in sludge water with air stripping at different temperatures.

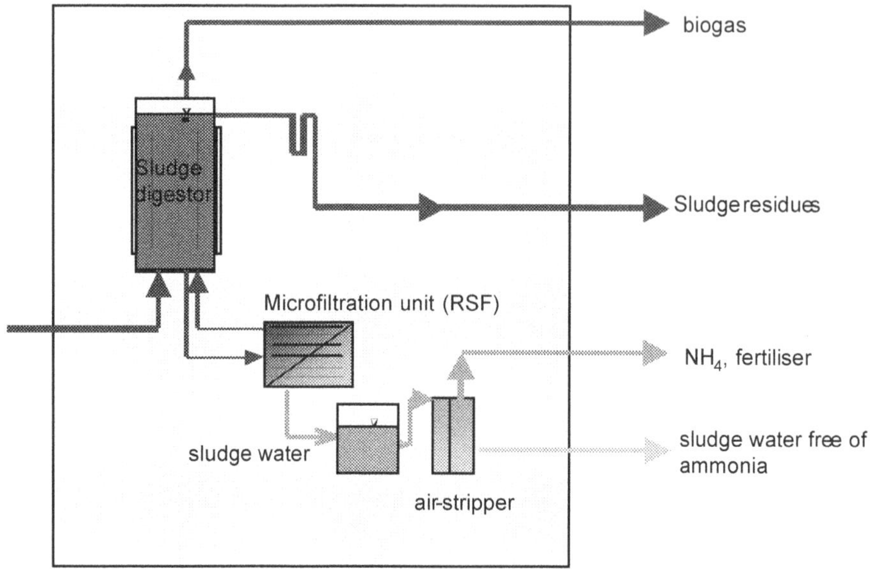

Figure 7 Scheme of the pilot plant in Heidelberg

4. References

Halling-Sorensen, B., Nielsen, S. N., Lansky, P. F., Ingerslev, F., Holten-Lützhoft, H. C., Jorgensen, S. E. (1998): Occurence, Fate and Effect of Pharmaceutical Substances in the Environment – a Review. Chemosphere 36

Kapp H. (1984): Schlammfaulung mit hohem Feststoffgehalt. Stuttgarter Berichte zur Siedlungswasserwirtschaft; 86

Kempter, B., Trösch, W. (2000): Optimiertes Vergärungsverfahren steigert Umsatz von organischen Abfällen. Wasser, Luft Boden, Zeitschrift für Umwelttechnik, 5/2000.

Kempter, B., Schmid-Staiger, U., Trösch, W. (2000): Verbesserter Abbau von kommunalen Klärschlämmen in einer zweistufigen Hochlast-Vergärungsanlage. KA Wasserwirtschaft Abwasser Abfall, 9/2000. Translation into Polish: Udoskonalony rozkład osadów ściekowych w dwustopniowych komorach fermentacyjnych GAZ, WODA I TECHNIKA SANITARNA, 1/2001 Translation into Czech: Dokonalejši rozklad kalů z městských čistíren odpadních vod dvoustupňovým vyhníváním při vysokém zatížení. Vodní hospodářství, 7/2001

Kempter-Regel, B., Oehlke, M., Weber, J., Trösch, W. (2003): Integration einer Hochlastfaulung in die herkömmliche Technik: Erste Bilanzierungsergebnisse der Schlammfaulung in Heidelberg. KA Wasserwirtschaft Abwasser Abfall, 11/2003.

Loll, U. (1981): Dimensionierungs- und Betriebswerte von Abwasser-Schlamm-Faulanlagen in der Bundesrepublik Deutschland. Dokumentation der Oswald-Schulze Stiftung Gladbeck 1981.

Schäfer, J. (1998): Verfahrenstechnische Untersuchungen zur Vergärung von Biomüll und Klärschlamm. Berichte aus Forschung und Entwicklung Nr. 12. Fraunhofer IRB Verlag.

Schulze, K. I. and Lipe R. S. (1964): Relationship between substrate concentration, growth rate and respiration rate of *Echerichia coli* in continuous culture. Arch. Mikrobiol. 48. 1 – 20.

Ternes, T. (1998): Occurance of Drugs in German Sewage Treatment Plants and Rivers. Water Res. 32

Uden U. van (1967): Transport-limited growth in the chemostat and its competitive inhibition; a theoretical treatment. Arch. Mikrobiol. 58. 145 – 154.

Zehnder A.J.B., Schraa G. (1988): Biologischer Abbau xenobiotischer Verbindungen bei niedrigen Konzentrationen und das Problem der Restkonzentration. Gwf-Wasser-Abwasser,129 (1988) 369-373

Zlokarnik M. (1982): Verfahrenstechnik der aeroben Abwasserreinigung. Chem.-Ing-Tech.54 939-952

The Use of Treated Wastewater for Nature: The Waterharmonica, a Sustainable Solution as an Alternative for Separate Drainage and Treatment

R. Kampf*, T. H. L. Claassen**

* Waterboard Hollands Noorderkwartier, P.O.Box 850, 1440 AW Purmerend, and Vrije Universiteit, Amsterdam, The Netherlands
** Friesland Water Authority, P.O. Box 36, 8900 AA, Leeuwarden, The Netherlands

ABSTRACT During the last decennium separation of wastes has matured into practical use. Separation not only reduces the amount of drinking water spilled in toilets, but makes reuse of nutrients attractive. The main drawback of separation is that it is rather expensive, especially in areas with an existing sewage system. Large centralised sewage treatment plants discharge waste water from large areas on one spot, with a considerable effect on surface water quality. It results in an ambivalent situation. Drinking water is becoming more and more expensive. After use in households and industry this water, collected together with rain water, is considered as a waste. It is transported in sewer systems and treated at huge costs in sewage treatment plants. But after discharging in surface water ecologists consider this cleaned water, originating from drinking water, as "dead water" compared with healthy surface water, not complying with the ideas about natural water quality, as described in the latest directives on ecological water quality of the European Union.

To bridge the gap between these two "worlds" (producing drinking water, transporting waste water on one hand and on the other hand the caretakers of surface waters) the *Waterharmonica* have been postulated. It is a low-tech concept based on ecological engineering, using food-chain approach to transfer well treated waste water into "healthy and useable" surface water. The main function is converting treated waste water into a usable natural surface water, but aspects like water buffering, recreation, buffer zones, nature development, water buffering during floods, creating fish spawning areas, growing fish for stocking wild populations are getting more and more important. In this paper the development of the concept is described. The *Waterharmonica* concept has resulted in a research and implementation programme (2003- 2004) of two of the main Waterboards in The Netherlands, supported by the Dutch Foundation for Applied Water Research.

This paper describes the development of the concept within integral water management.

KEYWORDS Integrated water management; ecological engineering; effluent; nature development; constructed wetland; food chains; *Daphnia*, Spoonbills

Introduction

In 1970 the Dutch Pollution of Surface Waters Act became into force. That was the start of a large scale effort to improve environmental quality in The Netherlands. From the very beginning the emission reduction approach (from point- as well as from diffuse sources) formed the main strategy for upgrading surface water quality. The water system approach was added later on as a second strategy focused on reaching surface water quality standards.

© 2004 IWA Publishing. *2nd IWA Leading-Edge Conference on Water and Wastewater Treatment Technologies* edited by Mark Van Loosdrecht & Jonathan Clement. ISBN: 1 84339 508 8.

Although tremendous efforts have been realised since then, surface water quality standards are still exceeded, especially for nutrients, transparency and chlorophyll. This discrepancy needs a reconciliation to decrease the gap between aquatic-ecological objectives and actual surface water quality, or -as a consequence- between emissions and receiving water system. The discharges from sewage treatment plants (STPs) still form, as point sources, an important percentage of the loads to Dutch surface waters. So, a further upgrading of wastewater before discharging into surface water seems to be needed.

Another development is coming up, that is the interest in reuse of effluents from STPs. The reasons for water reuse are quite various, as e.g. replenishing of ground water sources related to combating desiccation, as a source of fresh water in situations of water shortage or salty water, for irrigation in agriculture, and for recreational and nature development. However, actual effluent quality doesn't fit the specific standards for direct reuse; it does not matter for which intended use. So, in general additional treatment or polishing of the effluent seems to be necessary.

The concept of the Waterharmonica has been postulated to bridge the gap between both approaches, for the effluents of STPs, by upgrading the effluents to reanimated and useable water, feasible for many purposes. Creating short-circuits is possible in the lope of emissions of wastewater (within the water chain) and surface waters (within the water system). Environmental profits and contributions to sustainable management can be reached by incorporating this water cycle into regional plans and measures of integrated water management.

Texel

The island Texel is an attractive island for testing new policy for integrated water management. This island in the north western part of the Netherlands, is facing many problems and opportunities in water management. It is quite different from the mainland, it has very high natural values, and it is a well-known tourist resort and is still an agricultural stronghold. The island is enclosed by the North Sea and the Wadden Sea. Apart from a drinking water line from the mainland, there is no external fresh water supply. Basically it forms its own watershed; it is a small version of the water system on the main land. Until recently many measures in the water system on the island were taken without taking all these aspects into consideration. The agriculture requires lower ground water tables, leading to intrusion of brackish water and diminishing of the fresh groundwater water lens below the surface of the island. Contrary, nature conservation prefers higher groundwater levels and restoration of saline ground water at several natural areas. The high dikes of the island, a safeguard against seawater, also form a huge barrier for migrating fish. The De Cocksdorp siphon fish ladder has been an important step towards a sustainable water system with more opportunities for fish to migrate from the sea to the island water system (Wintermans, 1998).

Integrated water management learned that the way the water system on the island of Texel is managed is not wise. The idea appeared that "all water problems" should be tackled jointly with an integrated approach of "all the aspects of water affairs". In the Water for Texel Master Plan all parties on the island who have interest in water management (agriculture, nature, recreation) are taking their share. It directly appeared that such a plan makes sense and is worth the effort. The Water for Texel Master Plan aims on gaining knowledge on the very complex Texel water system and on enhancement of the natural values of the surface waters on Texel by separating the different flows and qualities. The total estimated costs of the master plan are about € 25 million.

The Everstekoog constructed wetland

The effluent from Everstekoog, located in the centre of the island, flowed to the north in the direction of a brackish area, with high natural values, before being pumped into the Wadden Sea. It was pointed out that it is much more attractive to use the effluent in an area with high agricultural values south of the Everstekoog STP. For this purpose a diversion channel has been constructed. To improve the effluent quality a full scale constructed wetland was added to the STP in 1994. A joint 4-year research project has been started in 1995 to monitor the efficiency (Schreijer et al, 2000).

Figure 1 Everstekoog: Aerial View of the constructed wetland, on the background the STP , (photo Simon Smit, Texel)

The STP Everstekoog is an oxidation ditch with a load of 45.000 P.E. (Population Equivalents) in summer due to the tourism and only 10.000 P.E. in winter (Figure 1). Dry weather flow in summer is 3000-4000 m^3/day; the maximum flow is 10.000 m^3/day. Phosphorus removal takes place simultaneously by dosing iron-chloride to the aeration basin. The quality of the effluent of the STP Everstekoog is typical for a well functioning oxidation ditch (very low loaded activated sludge plant): NH_4-N ca. 1 mg/l, NO_3-N ca 2.5 mg/l, COD ca. 30 mg/l. Due to simultaneous P-removal with FeSO4 total-P is ca. 1 mg/l. The constructed wetland consists of a presettling basin, nine parallel ditches with a length of 150 m and a discharge ditch (Figure 2). Total water surface is 13,000 $m^{2.}$ The mean total hydraulic retention time (HRT) in the constructed wetland was just over 2 days at dry weather flow in summer. In the first research period (1995-1996) all ditches received the same flow, in 1997/1998 different flow regimes through the ditches resulted in HRT of 1.8 up to 11 days

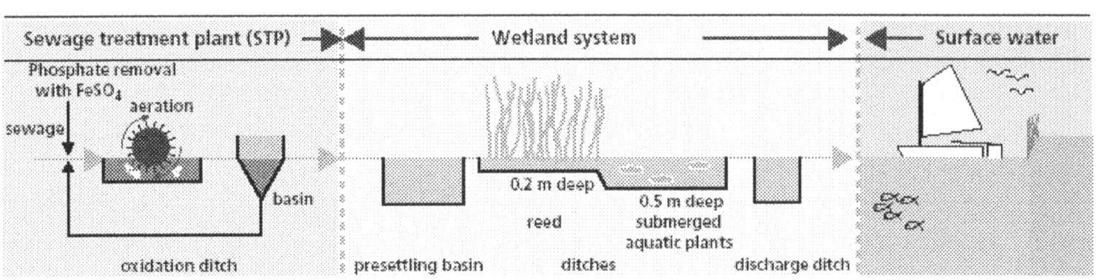

Figure 2 The Everstekoog constructed wetland in relation with the STP and surface water

The hydraulic retention time had a profound effect on nitrogen removal (mostly due to denitrification) in the constructed wetland. Ammonia levels in the effluent of the constructed wetland varied, but were mostly well under 1 mg/l, nitrate concentrations went down to < 0.5 mg/l at the longest hydraulic retention times, even in winter. The load of N and P with the STP effluent into the Everstekoog constructed wetland is around 5,000 kg N/ha/year and 700 kg P/ha/year. Nitrogen removal was calculated as 1,250 kg N/ha/year. This means that a surface area of at least 5,000/1,250 = 4 ha (instead of 1.3 ha) is needed for a complete nitrogen removal in the system. At the above mentioned N-load a HRT of 5.5 days will be needed for a low ammonium concentration in the effluent throughout the year (Toet, 2003), 10 days hydraulic retention time will lead to more natural values in the constructed wetland. Phosphorus removal in the Everstekoog constructed wetland was rather low, leading to the conclusion that when low P-concentrations in the effluent are preferred, biological or chemical removal in the activated sludge plant is more attractive.

To our opinion the "classic" water quality parameters do not describe the changes in the water in the constructed wetland aptly. Effluent from a STP is "dead water" from a surface water quality point of view. But, already in the presettling basin of the Everstekoog constructed wetland the water started to "live", started to resemble eutrophic, but clear surface water. Regularly the water turned red through high numbers of zooplankton, mainly *Daphnia magna* and also *Daphnia pulex*.

The oxygen level in the effluent of the STP was low, in the presettling basin it was only 2-3 mg/l. In the part of the ditches with submerged aquatic plants the daily oxygen pattern started to resemble the pattern of normal surface water. During daytime, the submerged aquatic plants and algae produced such an amount of oxygen that the level rose well above the saturation value (up to 20 – 30 mg/l). The high oxygen levels in the afternoon helped oxygen to penetrate deeper into the sediment. At the end of the day, the oxygen levels dropped sharply. This "solar energy process" for production of oxygen was also stable during longer periods, and functioned even under ice in wintertime.

The effluent of the constructed wetland was more turbid than the effluent of the STP, but it was a different kind of suspended solids. Instead of activated sludge flocks, the water contained algae, *Daphnia* and other small wildlife. An interesting observation was that the presettling basin and the ditches with a short retention time did not contain any fish, despite the high numbers of *Daphnia*. During daytime when the load of nitrogen is higher than the nitrification capacity of the STP ammonia levels will rise. This leads to high free NH_3, especially at higher temperatures and higher pH-values, which can be toxic for fish.

Only after a HRT of over 2 days in the ditches the water was suitable for fish such as Stickleback. On Texel both Threespine Stickleback (Gastrosteus acculeatus) and Ninespine Stickleback (*Pungitius pungitius*) occur, it is the main food of Spoonbills (*Platalea leucordia*), breeding in good numbers on the island. In ditches with more then 3 days HRT the number of Stickleback could be high, up to 25 per m^2. Constructed wetlands can be rich in wildlife (Kadlec and Knight, 1995 and Knight, 1996). Although the Everstekoog constructed wetland is situated in the agricultural part of the island natural values are attractive. The numerous fishes attract high numbers of birds. Especially Spoonbills (Platalea leucorodia) come to feed on small fish. In 1997 eleven bird species bred in the constructed wetland (Kampf et al., 1999; Kampf et al., 2002).

The high numbers of *Daphnia* in the constructed wetland effected also the disinfection capacity of the system. For disinfection from 500 – 1000 faecal coliforms per ml in the effluent of the STP to a level of 10, a HRT of 2 days will be sufficient. Possibly, due to wildlife in the system, the coli numbers were rarely below 1 per ml. For coli values of less then 10 per ml throughout the year, the HRT must be at least 4 days. To minimise the influence of storm water flows it is important to buffer as much water in the system as possible. For a surface flow system this can be done by means of an appropriate design of the weirs.

A surface flow constructed wetland, like the Everstekoog system, is a simple and attractive system. It is also cheap as long as land costs are not too high. The lay-out is like a Dutch polder landscape; the maintenance of the system resembles the maintenance of ditches and canals the waterboard is accustomed to already for centuries.

Spin-off: The "Kwekelbaarsjes system"; a food-chained based constructed wetland

The massive development of *Daphnia* and other zooplankton in the presettling basin of the Everstekoog constructed wetland in summertime had puzzled us first, but also led to some innovative ideas. The first question that arose was: how can all these *Daphnia* (only *Daphnia magna*) survive? As stated above fish is absent in this basin, but also the numbers of algae in this basin (<10 μg chlorophyll-a /l) were very low, not enough to maintain this population. Most aquaculture systems have an algae module involved, see for instance Borowitzka & Borowitzka (1988); Proulx & La (1985). This led to the hypothesis that the zooplankton lived mainly on bacteria, the so called "pin-point flocks" in the effluent. The number of *Daphnia* in the pond was determined by the availability of activated sludge flocks and loose bacteria, *Daphnia* indeed consumed activated sludge flocks, as the guts contents proofed (Kampf et al, 1999). The effect of *Daphnia* in a pond with well-treated wastewater can also be described by the filtration capacity. McMahon & Rigler (1965) and Lampert (1987) give values of up to 4 ml per *Daphnia* per hour. Thus, a population of 100 *Daphnia* / litre can filtrate 400 ml/l/h or 10 l/l/d, or "every drop of water" in the presettling basin will pass the body of a *Daphnia* 10 times per day.

It was concluded that growing *Daphnia* in a surface flow constructed wetland could be an interesting process. For the sewage treatment plant De Cocksdorp on the northern tip of Texel there could be another benefit. The improved effluent could be used as a lure flow for a fish trap to siphon fish from the sea across the dike. One of the problems on the island is that high dikes have been built for defence against the sea. This makes the island much more difficult to reach for fish migrating from the sea to land. Three Spined Stickleback grow up at sea and migrate back to inland waters to spawn, like Salmon. It is hardly a problem to migrate back to the sea, fish is easily pumped or flushed out with superfluous water. The new high dikes resulted thus in lower number of fish on the island. One of the most striking birds of Texel, the Eurasian Spoonbill, feeds mainly on sticklebacks. The Spoonbill is a highly valued species on Texel, a favourite bird of many people, inhabitants of the island as well of tourists. It can be considered as an indicator for the success of nature conservation and for water quality measures (Kampf et al., 2002).

All these aspects are brought together in the development of a step-wise food-chained water system, in a Dutch catchword (difficult to translate): the "kwekelbaarsjes system", described in Figure 3 (Kampf et al, 1999 and www.rekel.nl/kwekelbaarsjes).

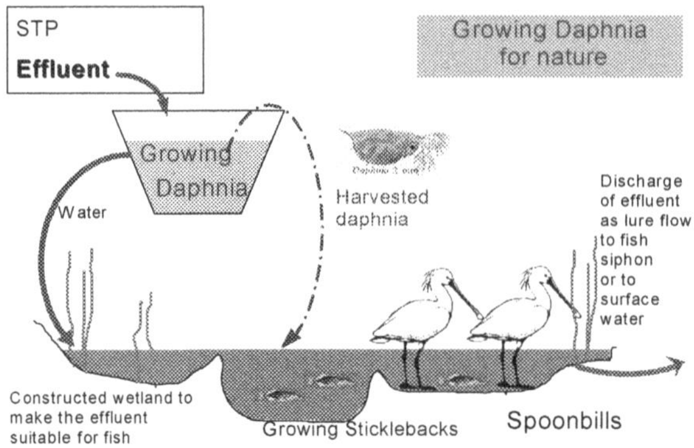

Figure 3 Sketch of the "kwekelbaarsjes system" to grow biomass on the "energy available" in treated wastewater

Basically, the system comprises a step-wise 'food-chain type' system to increase the ecological value of effluents from oxidation ditches. In the *Daphnia*-basin the sludge particles are used to culture *Daphnia*. The effluent flows to the constructed wetland to make the water suitable for fish. The harvested *Daphnia* can be transported to a deeper part of the "kwekelbaarsjes system" to be used as food for fish, such as Sticklebacks. This part is too deep for Spoonbills to forage. The last, shallow, part of this specially constructed wetland could be constructed as foraging area for Spoonbills. The Waterboard accepted this idea and gave permission for further development of the process. Interesting in this decision was that the Waterboard choose for a more or less uncertain and innovative ecological engineering process, instead of the original choice of demolition of the De Cocksdorp plant and the construction of a pipeline to the Everstekoog STP.

One of the uncertainties in the "kwekelbaarsjes process" is the "Growing *Daphnia*-module" (Figure 3). Despite the amount of knowledge about *Daphnia* available we could not find systematic knowledge about the process of growing *Daphnia* on treated wastewater. In the laboratory we focused on the possibility of production of *Daphnia* with activated sludge as the main food source. We have continued the research project with experiments aimed at the cultivation of *Daphnia* on effluent at a pilot scale. The experimental work is continued on the Everstekoog STP in four 20 m^3 ponds and four 2 m^3 mesocosms (see Foekema & Kampf, 2002 and www.rekel.nl/kwekelbaarsjes for more information. Results of the literature study and biological tests (Blankendaal et al., 2003) with effluents of nine STP's (including Everstekoog and De Cocksdorp) showed little eco-toxicological effects on the growth of *Daphnia* in pure effluents. More effects on the growth of algae have been found, as was expected based on earlier experiments. Special attention should be paid to avoiding overloading of STP's with a "food-chained constructed wetland" if this leads to lower removal efficiencies and thus increasing the risk of negative effects of the effluents, due to a lesser degree of removal of toxic compounds.

The research confirmed that the "kwekelbaarsjes system" near the village of De Cocksdorp, in connection with the siphon fish ladder, the extension of the STP of De Cocksdorp would be feasible. It is expected to lead to an innovative co-operation between engineering and nature (ecological engineering).

Prospects for food-chain based water systems

Demonstration of the values of the surface-flow constructed wetland on Texel lead to a slow, but steady increase of interests in The Netherlands of upgrading effluents with surface flow constructed wetlands. The first large scale Everstekoog type constructed wetland with reed beds and ponds was constructed at the STP Land van Cuijk, where an intensive monitoring programme is being carried out (Eijer-de Jong et al., 2002).

Friesland Water Authority is planning a similar project on the island of Ameland, an island comparable to Texel. The idea that could be worked out is not to discharge the effluent anymore into the Wadden Sea, but to keep it on the island itself. Another plan has been made for the STP Grou on the mainland of Friesland. After an Everstekoog-type surface-flow wetland, including some *Daphnia* ponds, a system of ponds could function as a fish spawning area (especially for Pike) for the surrounding canals, which are lacking natural values. The recently built constructed wetland Sint Maartensdijk is a combination of a root-zone constructed wetland for effluent polishing and a nature and recreation area (Ton, 2000).

The principles of ecological engineering are further applied in the Waterpark Groote Beerze, a combination of a constructed wetland and its surroundings. The extension of the STP Hapert (Waterboard De Dommel) is combined with a river restoration project. The effluent is polished by passage through root-zone reed-beds, open-ponds and wetland forest before it is discharged to the river Dommel (NN, 2001). Good examples elsewhere are the Ekeby constructed wetland in Sweden, the famous Arcata Marsh in northern California and the 7 ha Empuriabrava constructed wetland for polishing and reuse for nature and agriculture in the Costa Brava in Spain (Sala & Mujeriego, 2000). These and other examples will come available trough the project website www.waterharmonica.nl, see also Kampf et al (2003).

The results in a wider perspective: "the Waterharmonica"

Basically, a constructed wetland, including food-chain based water systems, like the "kwekelbaarsjes system" is a medium between "conventional" engineering (sanitary or process engineering in a sewage treatment plant, conventional polder and ditch management and waterway management in canals and lakes. For constructed wetlands it already has been tackled already on a theoretical way by Claassen (1996) of Friesland Water Authority. The basic idea is depicted in Figure 4.

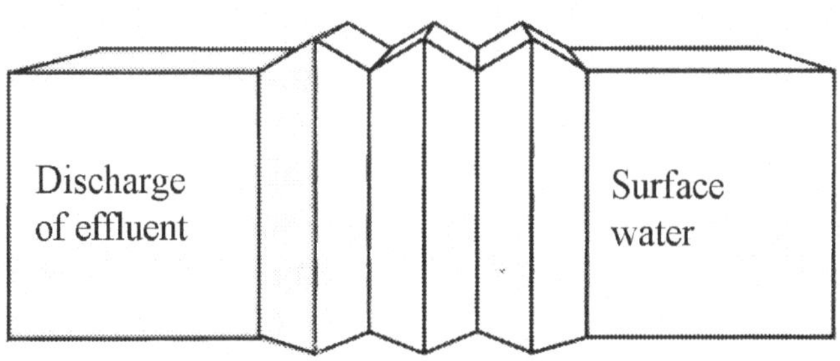

Figure 4 The "Waterharmonica" as a link between the sewage plant and surface water, based on Claassen, 1996

This model uses a "link" between the sewage treatment plant and the surface water, on which the effluent of the STP is discharged. In this link the different fields of engineering and ecology meet each other to improve the quality of the STP effluent into natural "living" surface water. The examples, described before (not only the Everstekoog constructed wetland and the "kwekelbaarsjes system", but also the other examples) fit very well with the concept of the "Waterharmonica". It seems to be an useful way of describing ecological engineering principles in water management. Multifunctional constructed wetlands seem to be good tools in water management focused on improving water quality, natural values, buffering water, recreation and using nutrient for agricultural production may all go hand in hand.

Much attention has been paid recently in sanitation concepts to separation of water flows, like separate discharge of rain water, but also to a further separation of wastes, to preventing discharge in sewer systems is emerging actually. In many cases it is not (yet) possible however to separate wastes at the source. In those cases it is obvious to choose for a clever solution, bearing in mind that wastewater was often "the best water we had". Before using it was rather expensive drinking water, plus rainwater that has been "mis-used" to discharge relative small volumes of wastes About 700 l faeces, urine and kitchen wastes are diluted into a total stream of 30.000 l waste water per person, annually. As is demonstrated in the Everstekoog research project it is possible to convert wastewater, after treatment in a well functioning STP, followed by a constructed wetland, into usable and valuable surface water. Even the nutrients and sludge particles in the effluent can be used beneficial, as demonstrated in the research carried out for the "kwekelbaarsjes system". This opens the door for natural constructed wetlands with recreational, natural and possibly even agricultural functions to convert treated wastewater in usable and biological healthy surface waters.

This concept has resulted in a research and implementation programme (2003- 2004), called the Waterharmonica, and financed by the Dutch Foundation for Applied Water Research.

The project is carried out by consultant Royal Haskoning (www.royalhaskoning.com). Main objectives of this project are to elaborate the concept of the "Waterharmonica", to promote demonstration projects in The Netherlands as well as abroad and to provide a basis for knowledge exchange in projects fitting in the "Waterharmonica": constructed wetlands to convert treated wastewater in usable and biological healthy surface waters, based on principles of ecological engineering. The Lettinga Associates Foundation (LeaF, www.ftns.wau.nl) of the Wageningen University is involved in assessing the use of the "Waterharmonica" for conversion of treated waste water into a natural resource in the developing world (Martijn et al, 2003).

REFERENCES

Blankendaal, V.G., Foekema, E.M. , P.C. Goedhart and Hoornsman, G. (2003) Ecotoxicologische aspecten bij de nabehandeling van rwzi-effluenten met behulp van biomassa kweek. Stowa report. 2003-12, Utrecht *

Borowitzka, M.A. and Borowitzka, L.J. (1988) Micro-algal biotechnology, Cambridge University Press.

Claassen, T.H.L. (1996) Het 3D-schakelsysteem: van tweesporenbeleid naar driesporenbeleid; ecotechnologisch van randverschijnsel naar centrumpositie. 25 jaar toegepast onderzoek waterbeheer, jubileumsymposium STOWA, 13 september 1996. [8], 141-153. *

Eijer-de Jong, J., H. Willers, B. Palsma, and Claessen, V. (2002) Monitoring moerassysteem rwzi Land van Cuijk. H2O 35: 26-29.

Foekema, E.M. and Kampf, R. (2002) Gebruik van effluent van de rwzi De Cocksdorp voor natuurdoeleinden: onderzoek naar de kweek van watervlooien in het kwekelbaarsjes systeem . DH02/007. Den Helder, MEP-TNO.*

Kadlec R.H. and Knight, R.L. (1995) Treatment Wetlands 415. Florida: CRC Press Inc, Boca Raton.

Kampf, R., Jak, R., and Groot, M. (1999) Growing Daphnia on effluent to improve the food situation of spoonbills on the island of Texel, do Daphnia rally eat sludge. 4th International Conference on Ecological Engineering for Wastewater Treatment, IEES, As, Norway, 7-11 June.*.

Kampf, R., Eenkhoorn, B.J., Foekema, E.M., and Dokkum, H.P. (2002) Can Spoonbills play a role in "integral water management" on Texel? Some questions, answers and practical solutions from the perspective of a Dutch waterboard". Eurosite-workshop on Management of Coastal Wetlands for Spoonbills and associated waders, Texel, The Netherlands, 19-22 April.*

Kampf, R., Graansma, J,, Dokkum, H.P. van, Foekma, E.M. and Claassen, T.H.L. (2003) Increasing the natural values of treated wastewater, the Waterharmonica: the missing link to transfer treated waste water into a usable surface water IWA-conference Efficient use and management of water for urban supply, Tenerife, 2 - 4 april 2003 and at the International Conference Constructed and Riverine Wetlands for Optimal Control of Wastewater at Catchment Scale Tartu, Estonia, 29 September - 2 October, 2003 *

Knight, R.L. (1996) Wildlife habitats and public use benefits of treatment wetlands. Water Science & Technology 35: 35-43.

Lampert, W. (1987) Feeding and nutrition in Daphnia. In Peters & Bernardi, ed. Memorie dell'Instituto Italiano di Idrobiologia 45.

Martijn, E., Kampf, R., Claassen, T.H.L. and Mels, A. (2003) Use of the "Waterharmonica" for conversion of treated waste water into a natural resource in the developing world, Ecological Engineering for Integrated Water Management, Harvard Graduate School of Design, Cambridge Mass., USA, 30 October – 2 November 2003 . *

McMahon, J.W. and Rigler, F.H. (1965) Feeding rate of Daphnia magna Straus in different foods labelled with radioactive phosphorus. Limnol. Oceanogr. 10: 105-113.

NN. Waterpark Groote Beerze (2001) speciaal magazine ter opening van het Waterpark Groote Beerze bij rwzi Hapert. Boxtel, Waterschap De Dommel.

Proulx, D. and La, J.D. (1985) Harvesting Daphnia Magna grown on urban tertiarily-treated effluents. Water Research 10 : 1319-1324.

Sala, L. and Mujeriego, R., (2000) Ecological water quality polishing and reuse: Costa Brava Spain, Scope Newsletter, 40, November 2000

Schreijer, M., Kampf, R., Verhoeven, J.T.A., and Toet, S. (2000) Nabehandeling van effluent tot bruikbaar oppervlaktewater in een moerassysteem met helofyten en waterplanten, Resultaten van een 4-jarig demonstratieproject op rwzi Everstekoog, Texel. Hoogheemraadschap Uitwaterende Sluizen en Universiteit Utrecht. *

Toet, S. (2003) A constructed wetland used for polishing effluent from a sewage treatment plant: processes and performance, dissertation, 8-9-2003. Utrecht University.

Ton, M. (2002) Aanleg helofytenfilter, Neerslag, 2000, 2

Wintermans, G.J.M. (1998) De hevel-vispassage op Texel. W.E.B. rapport 98-01 ed 443. Texel: Wintermans Ecologen Bureau.

** can be downloaded from www.waterharmonica.nl*

Demonstration Project for Separate Discharge and Treatment of Urine, Faeces and Greywater - Cost Comparison with the Conventional Wastewater System

A. Peter-Fröhlich[*], I. Kraume[*], F. Luck[**] , A. Lesouëf[***] and M. Oldenburg[****]

[*]Berliner Wasserbetriebe, Wastewater Works, P.O. Box 021098, D 10122 Berlin;
PH (+49)030-8644 1626; FAX (+49)030-8644 5077;
e-mail: anton.peter-froehlich@bwb.de
[**]KompetenzZentrum Wasser Berlin gGmbH, Berlin, Germany;
e-mail: francis.luck@kompetenz-wasser.de
[***]Vivendi Water, Wastewater Treatment Department, Paris, France;
e-mail: andre.lesouef@generale-des-eaux.net
[****]Otterwasser GmbH, Lübeck, Germany;
e-mail: oldenburg@otterwasser.de

Abstract The Berlin Centre of Competence for Water (Kompetenzzentrum Wasser Berlin) together with its partners Berliner Wasserbetriebe and Veolia Water has started a pilot project about new sanitation concepts. In order to define the experiments for testing new, sustainable sanitation concepts a pre-study has been performed. This study included a cost comparison between two new sanitation concepts with gravity and vacuum separation toilets and the conventional system. It could be demonstrated that the new sanitation concepts may have cost advantages depending on the situation. This was a further motivation starting a pilot project near Berlin testing the above mentioned toilet systems under realistic conditions. The operation of the gravity separation toilets concept started in October 2003.

Keywords Decentralised wastewater concepts; urine; gravity separation toilet; vacuum separation toilet; cost comparison

Introduction

The conventional centralised concepts for water supply and wastewater, developed for ages in industrialized countries, imply high costs and high water consumption, which make them not suitable as a sustainable solution especially for developing countries. Further development, testing and dissemination of alternatives to conventional wastewater systems are therefore becoming more and more indispensable for ecological, economic and societal reasons. More sustainable approaches should consider the reuse of treated water as well as the recycling of the nutrients if possible. Furthermore the energy consumption for wastewater discharge and treatment should be minimised. Such techniques and concepts are already available and in use, but further developments and validations are necessary.

These are the reasons why the Berlin Centre of Competence for Water (KWB) initiated with Berliner Wasserbetriebe (BWB) and Veolia Water (VW) a demonstration project for new sanitation concepts which should represent a relevant solution for:

- remote areas, where the connection to a central system (e.g. large sewer networks) would not be technically or economically interesting,

© 2004 IWA Publishing. *2nd IWA Leading-Edge Conference on Water and Wastewater Treatment Technologies* edited by Mark Van Loosdrecht & Jonathan Clement. ISBN: 1 84339 508 8.

- rapidly growing conurbations in developing countries,
- countries with scarce water resources and
- a contribution to the sustainable development with the recycling of nutrients and water.

Methods

The project is divided into two phases (***Phase I***, pre-study and ***Phase II***, pilot project). ***Phase I***, a theoretical approach, has been completed at the end of 2001.

- **A literature-based project review**, patent reviews and a collection of information about the various projects were made. Furthermore existing projects with separate treatment in Germany, Denmark and Sweden have been visited. These information and detailed economic investigations have been the prerequisite for continuing the project.
- **Cost comparisons between a conventional and two new sanitation concepts** for an intended new housing estate in the federal state Brandenburg near Berlin have been made. The housing estate should be realised stepwise from 672 up to 5,000 inhabitants within the next 10 years. For the economic calculation three different sanitation concepts have been compared for different cost levels:
  - **Conventional sanitation concept:** Conventional flush-water toilets with stop bottom, one sewer system, normal gravity sewer system for the area, pumping station with transport sewer to the existing sewer network, system operated by the public supplier.
  - **Separation sanitation concept (gravity, composting of faeces):** Gravity separation toilets with separate outlets for urine and faeces, collection and storage of the urine, transport to the farmer nearby and utilisation in agriculture, faeces transported by a gravity sewer system, aerobic treatment in a compost separator, utilisation of the compost in the horticulture on the area, transport of greywater in gravity sewer system, treatment in a constructed wetland, transport to the receiving water.
  - **Separation sanitation concept (vacuum, digestion of faeces):** Vacuum separation toilets, gravity urine transport, storage of the urine, transport to the farmer nearby and utilisation in agriculture, faeces transported by a vacuum sewer system, common treatment with ground bio waste in a biogas plant, biogas utilisation by the equipment of the energy concept, transport of the digested sludge to the farmer nearby and utilisation in the agriculture, transport of greywater in gravity sewer system, treatment in a constructed wetland, transport to the receiving water.

The comparison of these three sanitation concepts has been considered for four scenarios which are summarised in Table 1.

Table 1 Scenarios for the cost comparison of the 3 different sanitation concepts

	Inhabitants	Water Operator *
Scenario 1	672	Local company
Scenario 2	5,000	Local company
Scenario 3	672	Berliner Wasserbetriebe
Scenario 4	5,000	Berliner Wasserbetriebe

*This determines the costs of the "conventional" system and the costs for drinking water in all scenarios

A decision based on an economic point of view should consider three aspects:

- Costs of the investment
- Costs of the reinvestment
- Operation costs

These costs have been considered and the following assumptions were made for the cost calculations:

- Lifetime of the project: 50 years
- Duration of the components depending on their lifetime. Reinvestment after the end of the lifetime
- Real interest rate: 3.5 % per year
- Maintenance costs were calculated as a percentage rate of investment. Personal costs were taken in consideration separately.
- Operation costs divided into costs for- personal equipment- maintenance- water and wastewater- electricity- others equipment

The specific costs for water, wastewater, connection fees, energy and other costs were based on the information of the local company or of the Berliner Wasserbetriebe. With these assumptions and information the costs for the whole project period were calculated as dynamic prime costs (DPC). The DPC are the sum of money which is necessary for financing the whole project (investment, operation, reinvestment; see Figure 1) for the assumed lifetime based on today's cost level.

The calculations have been realised with the German guideline "Dynamische Kostenvergleichsrechnung" (dynamic co st comparison calculation) published by the "Länderarbeitskreis Wasser LAWA" (a working group of all German federal states concerning water management) (LAWA, 1998). This method can also be described as a life cycle cost analysis.

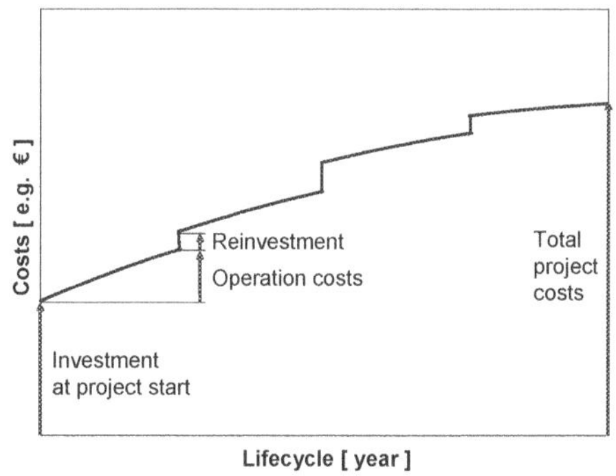

Figure 1 Demonstration of the total project costs (DPC)

Results

a) Literature survey and visits

The main results from the pre-study are:

- 17 new sanitation projects already exist in Western Europe (Table 2).
- The activities in relation to new sanitation concepts are increasing all over the world.
- The separation of urine, faeces and greywater based on the use of new toilet bowls, has proven to be feasible and accepted by the users (Johansson, 2001; Hellström and Thurdin, 1998; Swedenviro, 2001).

- Once urine is separated from the faeces, several configurations exist, differing between them by the collection and transport system chosen and the treatment of the three effluents (faeces, urine, greywater) (Otterpohl *et al.*, 1999).

Table 2 Existing new sanitation projects in Western Europe in 2001

No.	Project Name	Country	City	Project Start (year)	Toilet Installation	URL Project	Responsible Organisation
1	Hamburg-Allermöhe	Germany	Hamburg	1990	Composting toilet		
2	Hamburg-Braamwisch	Germany	Hamburg	1992	Composting toilet	www.oekologische-siedlung-braamwisch.de	Ökologische Siedlung Braamwisch e.V.
3	Kiel-Hassee	Germany	Kiel	1992	Composting toilet		Ökologische Siedlung Hassee
4	Öko-Technik-Park Hägewiesen	Germany	Hannover	1992	Solitar vacuum toilet	www.oeko-technik-park.de	BauBeCon AG mit Stadtwerke Hannover AG
5	As	Norway	As	1992	Drying toilet		
6	Ecological Village Björnsbyn	Sweden	Björnsby near Lulea	1994	Separation toilet		NLH (Norrbottens Läns Hushallningssällskap - the Agricultural Society of Norrbotten County)
7	Bielefeld Waldquelle	Germany	Bielefeld	1995	Composting toilet		
8	Palsternackan	Sweden	Stockholm	1995	Separation toilet		
9	Understenshöjden	Sweden	Stockholm	1995	Separation toilet		
10	Freiburg Vauban	Germany	Freiburg	1998	Vacuum toilet	www.vauban.de	
11	Gebers	Sweden	Skarpnäck	1998	Separation toilet (Drying toilet)	www.iees.ch/cs/cs_4.html	Fastighetsägare, BRF Konditorn, Gebersvägen 24, 128 65 Sköndal
12	Kiel-Vieburg	Germany	Kiel	1998	Composting toilet		
13	Hyldespäldet	Denmark	Kopenhagen	1999	Separation toilet		
14	Mön Museum	Denmark	Mön	1999	Separation toilet		
15	Wohnsiedlung Flintenbreite	Germany	Lübeck	1999	Vacuum toilet	www.flintenbreite.de	infranova GmbH & Co. KG, Flintenbreite 4, 23554 Lübeck
16	Lambertsmühle	Germany	Burscheid	2000	Separation toilet		Wupperverband
17	SolarCity Linz-Pichling	Austria	Linz	2001	Separation toilet		SBL Stadtbetriebe Linz

b) Cost comparison

The cost comparison between the conventional and new sanitation concepts shows that the new sanitation concepts have not only ecological advantages but can also have economical advantages (see below). The cost advantage is very depending on the specific conditions of the housing estate. For the chosen example (see methods) cost advantages occur in the most cases of the new sanitation concepts which are shown in Figure 2 to Figure 5.

In the *Scenario 1* (Figure 2) only the sanitation concept with gravity separation toilets is significantly less costly than the conventional system. With increasing inhabitants, shown in *Scenario 2* (Figure 3), both new sanitation concepts lead to cost savings compared to the conventional system after 3 and 9 years, respectively.

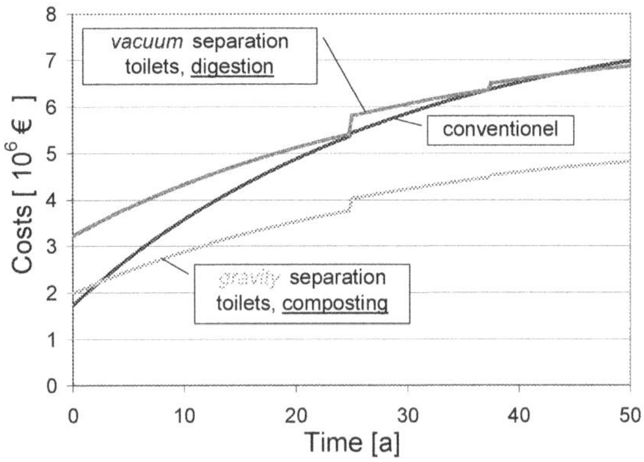

Figure 2 Total project costs for the conventional and the new sanitation concepts (*672* inhabitants); cost basis: *local company*

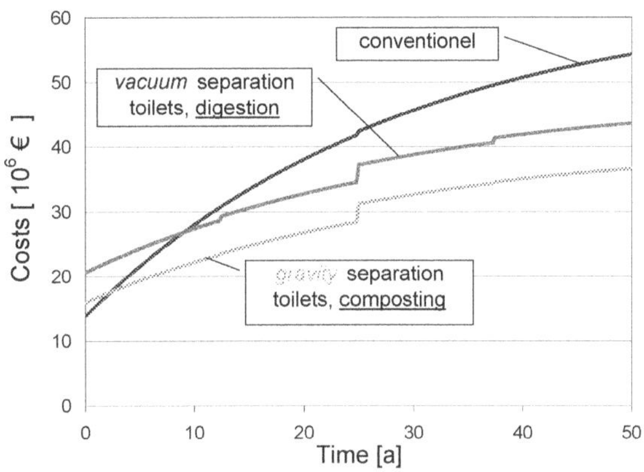

Figure 3 Total project costs for the conventional and the new sanitation concepts (*5,000* inhabitants); cost basis: *local company*

The results from *Scenario 3* (Figure 4) demonstrate only a benefit of the new sanitation concept with gravity separation toilets compared to the conventional system right from the project start. In the *Scenario 4* (Figure 5), which represents the enlarged settlement, both new sanitation concepts are cheaper than the conventional system during the whole operation time.

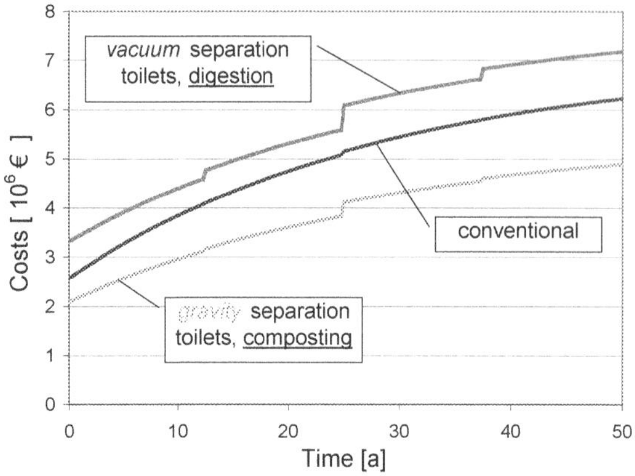

Figure 4 Total project costs for the conventional and the new sanitation concepts (*672* inhabitants); cost basis: *Berliner Wasserbetriebe*

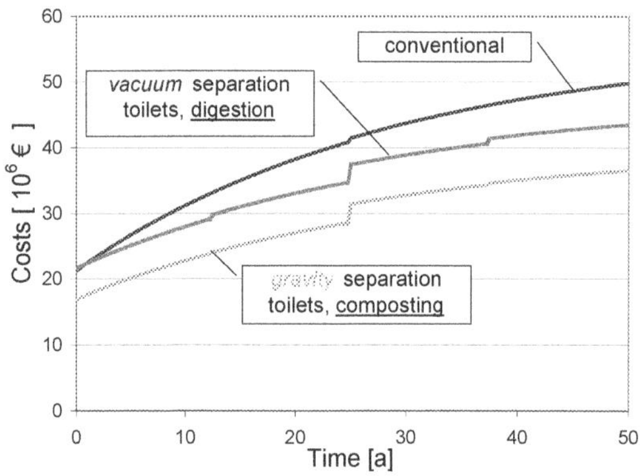

Figure 5 Total project costs for the conventional and the new sanitation concepts (*5,000* inhabitants); cost basis: *Berliner Wasserbetriebe*

The cost comparison between the two new sanitation concepts and the conventional system for the four chosen scenarios show in general that the advantages for the new sanitation concepts are increasing with the size of the settlement.

The results from this cost comparison were an additional motivation for the start of a pilot project testing:

- gravity separation toilets and
- vacuum separation toilets

in conjunction with different treatment configurations. The pilot project started in the year 2002 (***Phase II***). The new sanitation concepts will be tested in existing buildings (office building and apartment house) of the Stahnsdorf WWTP owned and operated by the Berliner Wasserbetriebe. The realisation of the new sanitation concept in the office building took place in the frame of a general restoration of this building. For the apartment house a stepwise realisation of a new sanitation concept is intended.

The general process scheme for the new sanitation concepts in the office building and in the apartment house can be seen in Figure 6.

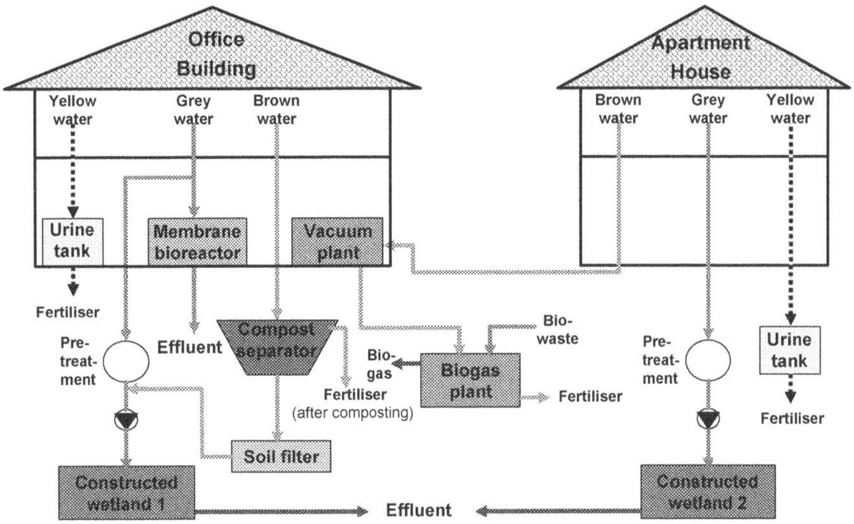

Figure 6 New sanitation concepts with *gravity* separation toilets in the office building and with *vacuum* separation toilets in the apartment house of the WWTP Stahnsdorf

In the new sanitation concept for the office building gravity separation toilets (10 toilets) are used. In the men toilets five waterless urinals installed additionally. The type of the toilet is shown in Figure 7.

With this type of toilet dilution of urine (yellow water), which reinforces the formation of "urine stones"(precipitation of Ca^{+}, Mg^{+}, PO_4^{3-}, mainly into struvite, hydroxyapatite and calcite; Udert *et al.*, 2002), can be prevented due to a moveable plug (see description in Figure 7). Hence formation of "urine stone" may be reduced which should prevent blockages in the urine pipes. Furthermore the volume of urine will not be increased by flush water. This is a further advantage for the urine treatment and handling.

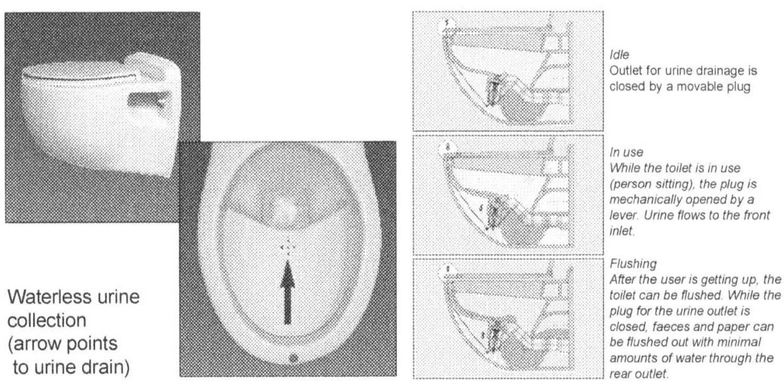

Figure 7 Gravity separation toilet (Roediger-No Mix Toilet; Roediger, 2001)

The faeces (brown water) will be drained and composted. The filtrate from the compost separator will be treated with a soil filter before mixing it with greywater. The greywater passes a septic tank before treatment in a constructed wetland. In parallel to the constructed wetland a membrane bioreactor will also be tested for greywater treatment. The urine flows into storage tanks. Different methods will be tested for handling and treatment of urine before using it as fertiliser. The methods may be adjusting different pH values during urine storage, extraction of the nutrients etc.

For the new sanitation concept for the apartment house (15 flats) vacuum separation toilets are taken into consideration. This type of toilet is under development. In this concept urine and greywater are discharged and transported by gravity, while faeces by transported by a vacuum system. Each flow is also treated separately. Urine will be treated as mentioned above. The faeces will be digested together with ground bio-waste. Digested sludge is also a fertiliser, e.g. for farmlands. Biogas can be used either in gas cookers or in a combined heat and power unit (CHPU). This topic will not be tested in this project. Greywater passes like in the case of the office building through a septic tank before treatment in a constructed wetland.

Since dish washing powders have a high content of phosphate (often more than 30 %) and dishwashing machines are more and more common, for both concepts a phosphate precipitation could also be necessary during greywater treatment.

The treated greywater can be used e.g. for irrigation in general. In this project the effluent of the membrane bioreactor will be investigated with respect to the different options of reuse as water with a lower quality than drinking water.

These two sanitation concepts are technical options belonging to the new approach, others are possible, e.g. composting of the faeces together with bio-waste should a production of biogas not be wished.

The type of greywater treatment for both new sanitation concepts will differ depending on the local situation. For large settlements an activated sludge tank etc. could be a more appropriate solution than a constructed wetland. The size of an activated sludge tank for greywater treatment could be however much smaller than for municipal wastewater treatment due to the far lower load of COD, nitrogen etc. (Otterpohl, 2001).

Conclusion

The results from the pre-study of this project enhanced the motivation for realising a pilot project testing gravity and vacuum separation toilets. Ten gravity separation toilets and a prototype of a vacuum separation toilet are in operation since the end of 2003. The necessary measures for realising the new sanitation concept for the apartment house are prefaced.

Although the cost comparison showed higher costs for the new vacuum sanitation concept compared to the gravity sanitation concept and in spite of a more complex operation, this system may be an appropriate solution especially in case of local water shortages. The flush water consumption should be about 6 l/(pd) compared to about 15 l/(pd) for the gravity separation toilet or about 25 – 40 l/(pd) for the conventional toilet with stop bottom.

Important objects of this pilot project (***Phase II***) are:

- Increasing of the knowledge of design, installation and costs of new sanitation techniques based on separation technologies
- Experience of the operation of new sanitation concepts by investigation of the various modules of the separation concept in different conditions

The interest for these new sanitation concepts is increasing in general and further new projects are realised [e.g. Christ 2003]

Acknowledgements

This project is supported by the LIFE financial instrument of the European Commission (Duration: 1 January 2003 – 30 June 2006, LIFE03 ENV/D/000025).

References

Christ, O. (2003) Hochtechnologie für Bürohaus – Stoffsztomorientierte Abwasserbehandlung. wwt 10-11, 26-32.

Hellström, D. and Thurdin, J. (1998) Drying as a urine dewatering method (Torkning som metod för att avvattna urin). Vatten, Vol. 54, 109-114.

Johansson, M. (2001) Urine Separation – Closing the Nutrient Cycle. Final report on the R&D project Source-Separated Human Urine – a future source of fertiliser for agriculture in the Stockholm region? Stockholm Vatten, AB Stockholmshem and HSB National Federation.

LAWA (1998) Leitlinien zur Durchführung dynamischer Kostenvergleichsrechnungen. Länderarbeitsgemeinschaft Wasser, Kulturbuchverlag Berlin GmbH.

Otterpohl, R., Abold, A. and Oldenburg, M. (1999) Source Control in Urban Sanitation and Waste Management: Ten Systems with Reuse of Resources. Wat. Sci. Tech., Vol. 39, No 5, 153-160.

Otterpohl, R. (2001) Black, brown, yellow, grey – the new colours of sanitation. Magazine of the International Water Association, WATER 21, October.

Roediger (2001) Roediger-No Mix Toilet. www.roevac.com.

Swedenviro (2001) Market survey – extremely low flush toilets. Swedenviro report, No. 2001:1, February.

Udert, K., Larsen, T. and Gujer, W. (2002) Biologically Induced Precipitation in Urine-collecting Systems and Urinal Traps. IWA conference in Melbourne, Preprints

Wastewater Treatment Plant of the Future - Decision Analysis Approach for Increased Sustainability

Glen T. Daigger* and George V. Crawford**

*CH2M HILL, 9191 South Jamaica Street, Englewood, CO 80112 USA
(E-Mail: gdaigger@ch2m.com)
**CH2M HILL, 255 Consumers Road, Toronto, Ontario M2J 5B6 Canada
(E-Mail: gcrawfor@ch2m.com)

Abstract A decision analysis approach to quantify the relative sustainability of a wide variety of wastewater management alternatives is presented. Decision analysis combines the values of the community to be served by the wastewater management system with technical evaluation of the alternatives relative to the identified criteria that define sustainability for the subject community. An iterative process is used so that a wide range of alternatives can be considered, with detailed evaluation focused on the most sustainable. Differing values by various stakeholders can be accommodated by considering different relative weights for the decision criteria and identifying those alternatives that are acceptable to each stakeholder.

Keywords Centralised, Decentralised, Decision Analysis, Reuse, Resource Recovery, Sustainable Development, Wastewater Treatment

Introduction

Technological developments provide a wide range of choices to environmental engineers seeking to select the "best" wastewater treatment plant configuration to serve the public. This is further complicated by the myriad of ways that wastewater collection and effluent management can be approached. This paper will briefly outline some of the available wastewater management systems and the wastewater treatment facilities that can support them. Evaluation and eventual selection for a particular application will then be addressed using the principles of sustainable development as the philosophical basis for defining the "best" approach for a particular application, with decision analysis as a supporting analytical tool. The overall approach can be summarised as follows:

1. **Identification of Stakeholders.** The various individuals and groups within the affected community that are interested and/or can affect implementation of wastewater management alternatives are identified and enlisted in the decision process. Consideration is also given to future members of the community, and to stakeholders outside the community who will be affected by the project.
2. **Frame the Issues.** Guiding principals, community values, and system goals and objectives are identified. These provide guidance in steps 3 and 4, below.
3. **Establish Decision Criteria and Value System.** Specific criteria are developed for each of the three categories of sustainable development considerations - economic, social and environmental – and for key technical issues as discussed below and displayed as the value hierarchy. General utility scales are established for each criteria, and relative weights are established for the various criteria for each stakeholder group. Each

© 2004 IWA Publishing. *2nd IWA Leading-Edge Conference on Water and Wastewater Treatment Technologies* edited by Mark Van Loosdrecht & Jonathan Clement. ISBN: 1 84339 508 8.

stakeholder group may have different values, which can be captured by different relative weights for the decision criteria. Alternatives can be evaluated using each set of values (weights) to identify the preferences of each stakeholder group.

4. **Identification of Project Alternatives.** A wide range of alternatives is identified.
5. **Evaluation of Project Alternatives.** The alternatives are evaluated relative to the established criteria. A first analysis may consider some potential "fatal flaws" – that is, levels of utility for some criteria that are just not acceptable. Alternatives with these fatal flaws can be screened out immediately. The remaining alternatives are evaluated relative to the established utility scales and value weightings to identify the most desirable alternatives consistent with the values of each stakeholder group. Several evaluations are conducted, each one using the values (weights) for each stakeholder group.
6. **Alternative Selection.** If sufficient information is available and consensus is reached by the various stakeholder groups, then a sustainable wastewater management option is selected. Alternatively, a refined set of alternatives may be developed and/or more detailed technical evaluations may be needed before a selection can be made. In this case the decision process returns to step 3 or 4 and a more refined and focused analysis is conducted. The key point is that the various stakeholder groups need not agree on values, only on the alternative to be implemented.
7. **Develop Implementation Plan.** An implementation plan is developed and ratified.

This general approach forms the basis for the decision process, with sustainable development principals implemented using decision analysis providing the necessary tools.

Range of Choices for Systems Alternatives

Historical approaches to wastewater management collected wastewater, removed it from the locality, and conveyed it to some location where it could be remotely discharged, thereby protecting public health. Wastewater was later treated to mitigate localized adverse environmental affects caused by the discharge. Today important alternatives exist to this collection/conveyance/discharge approach which offer advantages while also protecting public health. Some of these include:

1. **Reuse Rather Than Discharge.** The level of wastewater treatment required to protect the environment results in many instances in an effluent that can be reused, rather than discharged. In other instances water resources are in sufficiently short supply so that treatment required for reuse is cost-effective considering the value of the produced water.
2. **Centralised versus Decentralised versus Hybrid Systems.** Improvements in wastewater treatment technology allow smaller treatment facilities to produce high quality effluents that protect public health and allow for localized environmental discharge and thus eliminate the necessity to physically remove the wastewater from proximity with the public. Such systems can also facilitate reclamation and reuse by reducing reclaimed water conveyance needs. Hybrid systems consisting of a centralized "backbone" supplemented by decentralized reclamation facilities (the so-called "sewer mining" approach) are also possible.
3. **Source Separation.** Interesting experiments are on-going to consider the possibility of separating various waste streams at the household level and, consequently, producing a variety of waste streams that can be treated and reused in a more optimal fashion.

A broader range of environmental goals is also being expected of wastewater management systems (Daigger, 2003). Conventional wastewater collection and treatment is relatively energy intensive and contributes to the discharge of greenhouse gases. It also concentrates nutrients in the vicinity of the discharge and results in the production of residuals (biosolids) that must be managed. Interest exists in more "environmentally friendly" wastewater treatment systems that use less energy (or produce it), allow for biosolids recovery and reuse, and where nutrients are recovered for more beneficial introduction into natural nutrient cycles (Driver, *et al.*, 1999).

A wide range of wastewater treatment plant configurations can be conceived of, consistent with these goals and objectives. Three examples are presented below to illustrate the available range of choices: (1) High Quality Effluent (Reuse) Plant, (2) Low Energy Consumption Plant, and (3) Natural Treatment System Plant (Daigger, 2003). Flow diagrams showing major unit processes but excluding common elements (such as pump stations, utilities, thickening and dewatering facilities) are presented for each, along with a brief description.

High Quality Effluent (Reuse) Plant

This facility (Figure 1) includes preliminary treatment (screening and grit removal), primary treatment with flow equalization, membrane bioreactor biological treatment, and UV disinfection. Flow equalisation is needed with membrane bioreactor systems due to limitations on the peak flow rates that can be passed through such systems – primary effluent rather than raw sewage is equalized to minimize nuisance conditions within the flow equalization system. Biological and/or chemical nutrient removal may be incorporated into the membrane bioreactor, depending on effluent discharge requirements. Further treatment of the membrane bioreactor effluent by, for example, granular activated carbon, reverse osmosis, or ozone may be provided, depending on effluent quality requirements. Primary clarification may be deferred for smaller facilities.

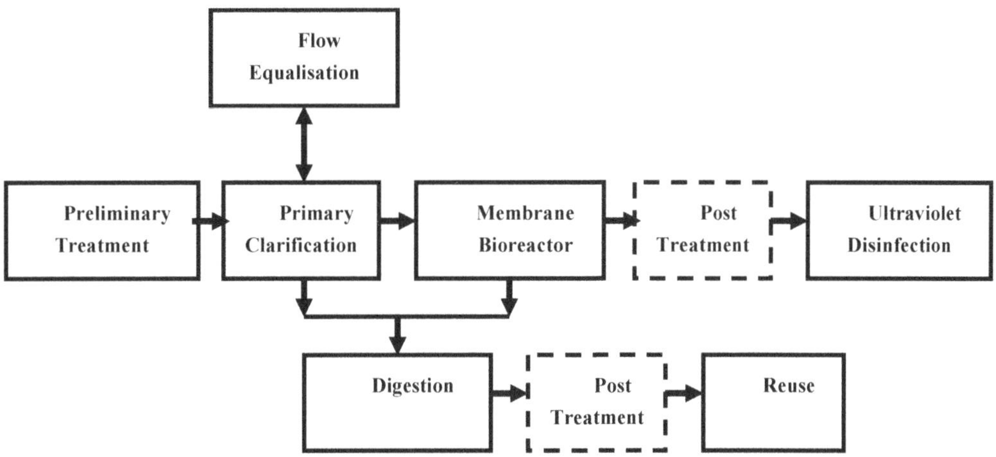

Figure 1. Process Schematic for High Quality Effluent (Reuse) Plant.

Produced sludges would be stabilized by digestion (either anaerobic or aerobic) and reused. Further processing of the digested biosolids by drying, composting, or alkaline stabilization, for example, may be provided to produce a more desirable product for reuse. Other resource recovery opportunities include energy (from the digestion process) and nutrients (from recycle streams). A facility such as this could be compact in nature. Depending on site-specific requirements, extensive odor and air emission control could be provided, and an aesthetically appealing facility could be designed.

Low Energy Consumption Plant

This facility (Figure 2) includes preliminary treatment (screening and grit removal), anaerobic treatment of the influent wastewater using, for example, an upflow anaerobic sludge blanket reactor, aerobic treatment using, for example, a fixed film process such as a trickling filter, followed by disinfection. Produced sludges are stabilized by anaerobic digestion, and the digested sludge is reused. Alternatively the aerobic sludge can be recycled to the upstream anaerobic treatment unit. Further treatment of the anaerobically digested biosolids to produce a product that is more attractive for reuse is not shown but can be provided. Extensive capture of produced biogas from the anaerobic treatment and anaerobic digestion processes would be practiced, and the digester gas would be used for electrical energy production either by a turbine generator or a fuel cell. A compact facility that is aesthetically pleasing and with extensive odor and air emission control can be provided.

Natural Treatment System Plant

This facility (Figure 3) includes preliminary treatment (screening and grit removal), a lagoon-based treatment systems that is shown for illustrative purposes here as an anaerobic/aerobic lagoon train, disinfection, followed by a wetland treatment system. Biosolids are periodically removed from the lagoon treatment system and reused. Several variations of this basic process train are possible. First, preliminary treatment can be provided in a less formal fashion than for the two previous plant configurations, for example with manually cleaned screens and a pond- or lagoon-based approach to grit removal. Secondly, the degree of pre-treatment prior to the wetlands treatment system can be varied significantly, depending on site constraints and effluent quality objectives. Raw wastewater can be applied directly to the wetlands if desired, but degree of pre-treatment is often cost-effective because of the significant decrease in wetlands area. Disinfection of the wastewater prior to introduction to the wetlands is desirable if habitat and recreational features are to be incorporated.

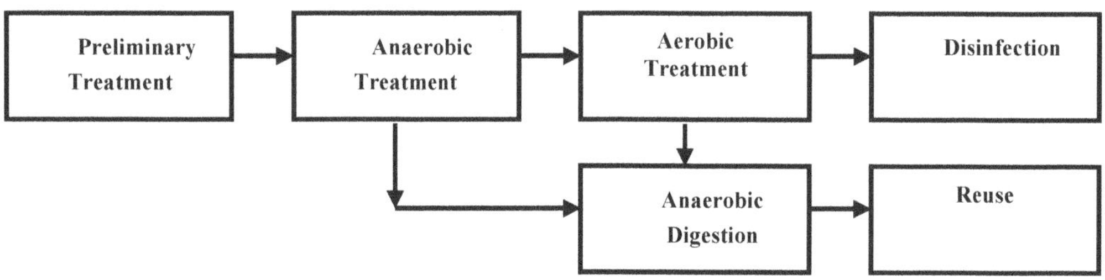

Figure 2. Process Schematic for Low Energy Consumption Plant.

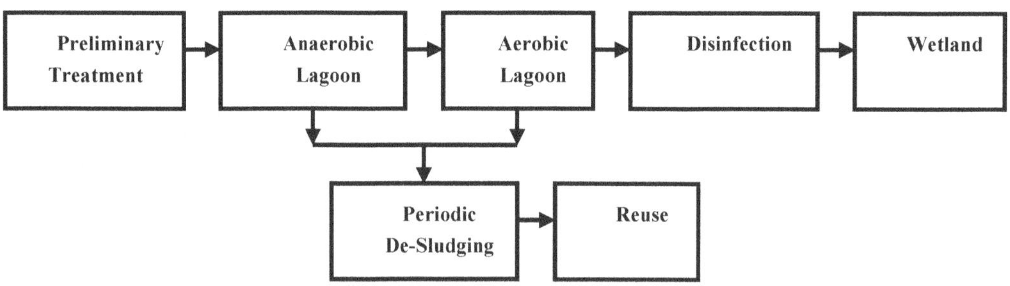

Figure 3. Process Schematic for Natural Treatment System Plant.

Table 1 summarises the relative advantages and disadvantages of these systems. While each provides significant advantages, none is without important disadvantages, at least from some perspectives. Importantly, Table 1 does not differentiate, or rank, the relative importance of any of the advantages or disadvantages. Involvement of the stakeholders in the assignment of relative importance, or value, to the various criteria is a fundamental principle of sustainable development. Thus, it is not clear from consideration of the technology, alone, which is the most sustainable.

Using Sustainable Development Principles as a Basis for Comparison and Selection of Alternatives

The above discussion illustrates that universally ideal treatment systems do not exist. It should also be obvious that the perception of "ideal" is not an absolute but is dependent on the values of the local community. Many options exist, well beyond those presented above, but defining the "best" depends on the context in which it will be used, including both the overall wastewater management system (with variations as discussed above), as well as the values of those who are stakeholders of (use, own, finance, or are affected by) the system. Some may implicitly assume that the selection of the "best" wastewater management system is a technical question - one only needs to define the proper objective function and perform the necessary technical analyses. However, selection of a wastewater management system, and the supporting wastewater treatment facility, is really a value decision, and decisions regarding relative sustainability are value based (Bradley and Daigger, 2000, 1999; Daigger, 2003). The proper (or best) decision depends on the values of those responsible for the decision, and on the inclusiveness of all stakeholders in that process. In the case of a wastewater management system it seems logical that this decision (the selection of the system) should be based on the values of the community that it serves, including both current and future generations who will be affected. It is certainly desirable that accurate technical analysis support the decision process, provided the analysis is in conjunction with stakeholder values. The community should make an informed decision concerning the ability of various options to truly best meet their goals and objectives. But, the decision basis, itself, is value based.

Table 1 Advantages and Disadvantages of Three Example Plant Configurations

High Quality Effluent (Reuse) Plant	Low Energy Consumption Plant	Natural Treatment System Plant
Advantages		
- Effluent Suitable for Wide Range of Uses, Including Indirect Potable. - Compact Facility – Easy to Address Aesthetics and Odors. - Automatic Control and Unattended Operation Quite Possible. - Moderate Operating Costs. - Modest Opportunities for Resource Recovery.	- Good Quality Effluent. - Low Energy Costs – Could be Net Energy Producer. - Automatic Control and Unattended Operation Quite Possible. - Modest Operating Costs. - Aesthetic and Odor Concerns Can be Addressed.	- High Quality Effluent Can be Produced. - Low Operation and Maintenance Costs. - Modest Capital Costs. - Significant Opportunities to Create Recreational, Wildlife Habitat, and Open Space Benefits. - Mechanically Simple.
Disadvantages		
- Relatively Capital	- Nutrient Removal	- Relatively Large Land

Table 1 Advantages and Disadvantages of Three Example Plant Configurations

High Quality Effluent (Reuse) Plant	Low Energy Consumption Plant	Natural Treatment System Plant
Intensive. - Relatively Equipment Intensive. - Energy Intensive.	Capability Reduced Due to Carbon Removal in Anaerobic Treatment. - Relatively Equipment Intensive.	Requirements, Especially in Comparison to Other Options. - Potential for Odor, Vectors, and Other Nuisances.

No value system is uniquely applicable to all decision-making. However, a common sustainable development framework can be established that is especially relevant to environmental projects concerning water use and quality. The principles of sustainable development, first developed by the business community to reflect that maximizing profits may not be the best long-term course (Nattrass and Altomare, 1999), has since been embraced for evaluation of public issues (ASCE, 1998). Instead of the single profit-oriented "bottom line" (or lowest cost for public projects), a "triple bottom line", considering economic, social, and environmental concerns, is embodied in the sustainable development concept. In this context, the most sustainable of the available options is the one that best balances economic, social, and environmental concerns (sometimes called assets). Thus, the essence of sustainability-based decision making is to assess the relative environmental, economic, and social performance of the various options and determining the one that best balances performance in these three areas, with "performance" being a combination of technical evaluations with value-based weighting to reflect relative importance (Bradley and Daigger, 2002, 1999; Daigger, 2003). This is known as relative (not absolute) sustainability. The optimum option (that best balances these three concerns) is the relatively most sustainable of all the options considered.

Decision Analysis as a Tool to Evaluate Sustainability

Evaluating the relative sustainability of various options is a daunting task. It requires a tool that can combine technical evaluations with community values in a decision-making framework. It also requires the evaluation and comparison of elements that cannot be measured (or are very difficult to measure) on the same scale. One approach, used by economists, is to monetise (that is, convert into monetary equivalents) all elements relevant to the decision. This can and has been done, but it can be a frustrating experience in some instances and is not well received by some stakeholders (how can one put a monetary value on….!!). Such an approach can also unfairly understate the financial impact on future stakeholders when certain resources will be in scarce supply, and/or at higher cost. Fortunately an alternate approach exists which was developed initially by scholars in the business schools but which has been adapted to environmental decision making. It is called decision analysis (Watson, 1987).

Decision analysis begins by identifying the major considerations that will drive a specific decision. These are already defined (environmental, economic, social) for a sustainable development approach to decision-making. One can also separate some specific technical considerations from the environmental considerations, creating a fourth category to recognize that technically flawed alternatives cannot be sustainable. The criteria that define each overall consideration are then identified so that specific, quantifiable definitions of each consideration are developed. The results can be displayed as a value hierarchy, as illustrated in Daigger (2003). Table 2 illustrates a sustainable development value hierarchy along with supporting criteria and provides definitions of those criteria. Taken together they define sustainability. Two important points. Each criterion must be measurable, even if by methods that cannot simply be added up, as long as the metric is relevant to that particular criterion. Second, each criterion should be independent. If they are not independent, the interdependent criteria will end up receiving a higher consideration, and value-weighting, than intended. As noted above, values ultimate drive decision-making when very different sets of considerations (such as environmental versus economic versus social concerns) are being considered.

Table 2 Sustainability Value Hierarchy and Criteria

Criteria	Description
Environmental Performance	
Surface water quality and quantity	Water quality and quality maintained (or improved) in source and receiving waters
Ground water quality and quantity	Water quality and quantity maintained or improved in source and receiving groundwater
Aquatic eco-systems	Water withdrawals and wastewater discharges do not threaten aquatic eco-systems, biodiversity, or individual species
Land-based ecosystems	Water withdrawals and wastewater discharges, or affects of urban sprawl, do not threaten land-based eco-systems, biodiversity, or individual species
Soil quality	Soil quality not deteriorated by discharges of wastewater through salt accumulation, pH alteration, or plugging with bioslimes
Air quality	Odors and emissions of biosols and toxic compounds controlled or reduced
Energy use	Energy use avoided or minimized: energy sources selected for reductions in pollutants, carbon emissions, and ecosystems impacts
Economic Performance	
Affordability	Ability of most members of the community to fund system implementation
Ability to finance.	Capacity of the community to finance the necessary capital improvements, considering initial and final population served
Ability to operate and maintain.	Capacity of the community to finance the necessary system O&M, considering initial and final population and time varying demands.
Ability to sustain.	Capacity of community to finance long-term system repair and replacement
Economic development.	System supports explicitly stated community economic development objectives
Social Performance	
Public health	The treatment system protects public health.
Values and beliefs	Promotes societal virtues such as the public trust.
Cultural resources	Preserves cultural traditions, ways of life, and physical heritage.
Public involvement	Community makes informed decisions. Actions reflect local values through public process where public has sense of ownership over decision making
Aesthetics	Preserves aesthetically-valued environments (e.g., beauty, open space, recreation, and wildlife viewing); does not degrade olfactory or audible aesthetics
Community development	Ability of community and all its members to develop and prosper to their highest potential through appropriate natural resource-based development
Technical Performance	
Knowledge Base.	The knowledge base for the technologies being considered is well

Table 2 Sustainability Value Hierarchy and Criteria

Criteria	Description
	established so that the alternative can be successfully implemented.
Site Impacts.	Sites to construct necessary facilities can be acquired.
Implementation.	All barriers to implementation can be successfully dealt with.

The relative values of an individual, or a group of individuals (a stakeholder group), can be surveyed, quantified and transformed into relative weights for the individual criteria within the value hierarchy. This set of weights can be used to evaluate a set of alternatives based on the values of that individual or stakeholder group. Different sets of weights, representing the values of various individuals or stakeholders can be used in a subsequent analysis to assess their preferences with regard to the identified alternatives.

The next step is to develop measures for the individual criteria within the value hierarchy. These are called utility scales and can initially be semi-quantitative in nature. Generally a 1 to 10 scale is used for each, with the rating associated with a quantitative or semi-quantitative measure. Developing the quantitative measures is a technical activity, while associating that measure with a utility (score) requires input from the stakeholders. It is often helpful to begin this exercise in a general way to capture the concept and then to add more quantitative rigor as the decision process proceeds. Table 3 provides examples of general utility scales to illustrate both the starting point and what is meant by a utility scale. For example, the utility scale referred to as "Standards" could apply to most of the Environmental Performance and public health criteria; that referred to as "Institutional Needs" could apply to many Economic Performance criteria; while that referred to as "Community Issues" could apply to many Social Performance criteria. Alternatives are then developed and evaluated relative to the utility scales. As mentioned above, the utility scales can also be used to establish minimum levels of utility for each criterion which provides the basis for a "fatal flaw" analysis. A composite score is calculated for each by multiplying the utility score for each criteria times the value-based weighting and summing the result to obtain a total score that represents the relative sustainability for that alternative.

Representatives of the community to be served by the wastewater system (the stakeholders) and the technical team conducting the analysis have separate and distinct roles. The technical team assembles the value hierarchy and examines it for completeness and independence (but with review from the stakeholders to ensure that it captures the range of relevant considerations). The stakeholders reflect their values by establishing the relative weights for the various criteria and co-operate with the technical team to identify wastewater management alternatives. The technical team evaluates each alternative relative to each criterion using the utility scales and computes the total score. The results are then reviewed with the stakeholders for action. As discussed above, when individual stakeholder groups have differing values, scores can be computed using each set of weights representing their individual values. The result is a series of analyses, each representing the preferences for each stakeholder group. Comparison of the higher-ranked alternatives for each stakeholder group provides the opportunity to identify common elements to formulate a single alternative that meets the essential concerns of each stakeholder group and allow selection of an alternative that all stakeholders can support. This approach has been used successfully in practice (BECC, 1996; Bradley, *et al.*, 2002).

Table 3 Example Utility Scales for Various Types of Criteria

Value	Standards	Institutional Needs	Community Issues
0	Not capable of meeting current or future goals or standards.	Not possible to meet institutional need.	Significant divergence or decrease in community issue.
4/3[1]	Meets current but not future goals or standards.	Introduction of new institution required.	Modest divergence or decrease in community issue.
7/5[1]	Meets current and can be modified to meet future goals and standards.	Requires use of available but not commonly used institutional approaches.	No impact on community issue
9/7[1]	Meets current and future goals and standards.		Modest divergence or increase in community issue
10	Exceeds current and future goals and standards.	Only available and routinely used institutions needed.	Significant divergence or increase in community issue.

[1] First value for Standards and Institutional Needs. Second value for Community Issues.

CONCLUSIONS

The relative sustainability of a wide variety of wastewater management options can be established using a decision analysis approach. Sustainability is ultimately based on the values of the community to be served by the wastewater management option, as reflected by the value hierarchy used in the decision analysis and the relative weights selected for the individual criteria. Technical assistance to the decision process includes structuring the value hierarchy and performing the technical analysis of alternatives relative to the established criteria. The decision analysis approach has been used successfully to assess the relative sustainability of alternatives and select the most sustainable.

REFERENCES

American Society of Civil Engineers and UNESCO/IHP (1998) *Sustainability Criteria for Water Resource Systems*, ASCE, in Reston VA.

BECC (1996). *Project Certification Criteria by the Border Environment Cooperation Commission*, November 9, 1996.

Bradley, B. R., Daigger, G. T., Rubin, R, and Tchobanaglous, G. (2002) Evaluation of onsite wastewater treatment technologies using sustainable development criteria. *Clean Techn. Environ. Policy* **4**, 87-99.

Bradley, B. R. and G. T. Daigger (1999) Decision making for sustainable development of water infrastructure. Paper Presented at the U. S. Environmental Protection Agency 6th National Drinking Water and Wastewater Treatment Technology Workshop, Kansas City, MO, August 2-4, 1999.

Bradley, B. R. and G. T. Daigger (2000) A Sustainable development approach to wastewater Infrastructure. *Proceedings of the 73rd Annual Conference and Exposition of the Water Environment Federation*, Water Environment Federation CD-ROM, Alexandria, VA.

Daigger, G.T. (2003) Tools for future success. *Wat. Envir. & Tech.* **15**(12), 38-45

Driver, J., D. Lijmbach, and I. Steen (1999) Why recover phosphorus for recycling, and how?. *Environmental Technology*, **20**, 651-662

Nattrass, B and M. Altomare (1999) *The Natural Step for Business: Wealth, Ecology, and the Evolutionary Corporation*, New Society Publishers, Gabriola Island, British Columbia, Canada.

Watson, S (1987) *Decision Synthesis*, Cambridge University Press, Cambridge.

Data-driven Approach to Sewage Inflow Prediction

Toshiaki Oka*, Hiroaki Tsutsui*, Junya Nishiguchi*

*Yamatake Corporation, 1-12-2 Kawana, Fujisawa, Kanagawa, 251-8522, Japan

Abstract Advanced operations in wastewater treatment, that are cost-minimizing operation, optimum maintenance planning and water quality control, can be realized by employing a data-driven approach. This paper describes the procedure and methodology of the proposed data mining. As an example of the practicable use of the data-driven approach, it introduces the prediction of sewage inflow volume to a treatment plant.

Keywords Data mining, Sewage inflow prediction, Infiltration/Inflow

1.Introduction

In the field of process automation, the accumulation of process historical data has become much easier due to advances in computer technology. While its original purpose was to record the data to ensure that processes were managed properly, the fact is that this vast amount of historical data, if used effectively, could make further improvements in product quality and facility maintenance. A data-driven approach offers the possibility of transforming such data into valuable know-how. However, due to many complex, interrelated factors involved in the process automation field, to employ conventional physical-based modeling would be impractical in terms of cost and accuracy.

For many years we have been investigating a kind of data-driven approach called quantitative data mining for making improvements in process automation (Tsutsui, 1996, Tsutsui et al., 1996, Matsuda et al, 1999). Unlike qualitative data mining, which extracts new knowledge from data, quantitative data mining relies on the basic knowledge of the process to find quantitative knowledge, such as prediction/estimation. This method of quantitative data mining has achieved improvements in petroleum processing and chemical manufacture.

Wastewater treatment is also affected by complex factors, such as natural phenomena and geographical features of the treatment area, and faces issues similar to petroleum processing and chemical manufacture. This paper introduces the use of quantitative data mining in predicting the volume of sewage inflow to a wastewater treatment plant (WWTP). The procedure, methodology and applications of quantitative data mining are explained, along with verification results. Future activities of quantitative data mining in locating the sources of infiltration/inflow (I/I) and predicting pollutant loads in real-time are also introduced.

2. Advanced operations employing quantitative data mining

By generating a prediction/estimation model of the WWTP, the following improvements in wastewater treatment operations can be expected:

© 2004 IWA Publishing. *2nd IWA Leading-Edge Conference on Water and Wastewater Treatment Technologies* edited by Mark Van Loosdrecht & Jonathan Clement. ISBN: 1 84339 508 8.

1) Cost-minimizing operation

Predictions of sewage inflow volume several hours in advance enable the WWTP to manage excess inflow without expanding its present capacity. By adjusting sewage inflow volume to the regulating reservoir according to predicted inflow, the WWTP can avoid the discharge of untreated sewage in wet weather. The plant can also make optimum biological conditions for wastewater treatment by preventing the inflow of diluted wastewater that can weaken or destroy sewage microorganisms (Oka et al., 2003). Further explanation is given in Section 5 Applications.

2) Optimum maintenance planning

I/I is caused by rainwater seeping into sanitary sewer pipes through cracks and pipe connections. Using data analysis to estimate the areas with high occurrences of I/I within the treatment area, municipalities can locate cracks in sewer pipes without a major investigation, and reduce the time and cost of sewer system rehabilitation. Details are given in Section 6 Future activities.

3) Water quality control

The WWTP measures Biological Oxygen Demand (BOD) to monitor the water quality of untreated wastewater that is discharged in wet weather; however, obtaining BOD measurements usually requires several days of lab analysis. By using accumulated data, including BOD measurements from lab analysis, to estimate current pollutant loads in real-time, it is possible for the WWTP to implement optimum operations. Details are given in Section 6.

3. Procedures for quantitative data mining

While there are various methods for data mining, this paper introduces a form of quantitative data mining using time series data as a modeling method for the prediction/estimation of behaviors and characteristics (Figure 3.1).

This procedure was created as a result of many years of trial and error in the process automation field, and its efficiency has already been verified in over seventy applications of petroleum processing and chemical manufacture. By checking and confirming the results of each step against the basic knowledge of the process while proceeding with the analyses, we were able to easily construct a model, and understand any problems in modeling and the problems' causes. Furthermore, the analysis results of new information given at each step become valuable know-how.

1) Defining the output

The output of prediction/estimation is decided according to the objective. For example, if the objective is efficient use of the regulating reservoir, the output is future inflow.

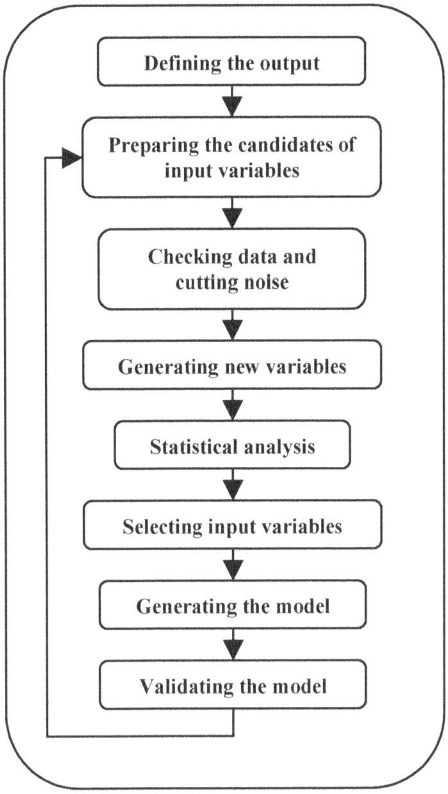

Figure 3.1 Procedure of data mining

2) Preparing the candidates of input variables

The candidates of input variables that affect the output are decided. Instead of focusing on candidates that affect the output, by analyzing all the measured data using the following steps, we can obtain new knowledge of the relationship between variables that was previously not assumed.

3) Checking data and cutting noise

By plotting all the data on a graph, we can check for noise and periodicity of the data. To eliminate noise and irregularities found in the measured data, an examination by statistical method or filtering, such as moving average, is used. Analyzing periodicity using frequency analysis, such as Fast Fourier Transfer (FFT) and Wavelet Transfer, is also effective.

4) Generating new variables

If the characteristics of the input candidates' dynamics have considerable influence on the output variables, new input variables that take into consideration the effects of time lag, differential and integration are added.

5) Statistical analysis

To analyze statistical features of the data, basic statistics, such as average, covariance, standard deviation and correlation coefficient, are used.

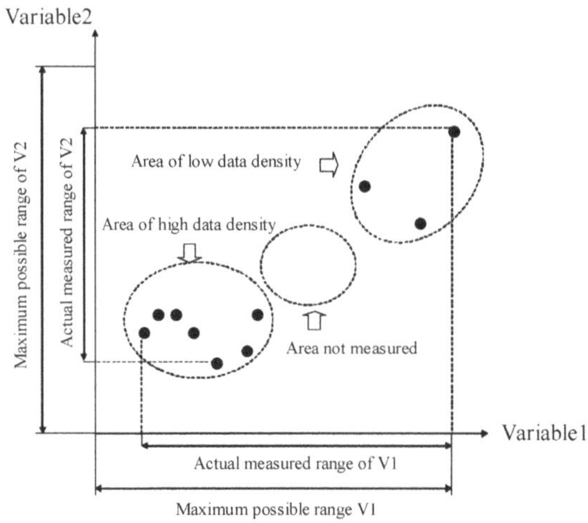

Figure 4.1 Features of actual data

6) Selecting input variables

In selecting input variables, the stepwise method (Flury et al., 1988) and hierarchical clustering (Flury et al., 1988) analysis are used. The stepwise method is used for selecting the input variables that affect the output variable by linear theory. It finds a set of variables that can describe the output variable in a multiple regression model without directly generating a model. Then, to analyze non-linear factors, hierarchical clustering analysis is used, a method with minimal bias between clusters. It adds variables that are independent from the linear variables to the input variables already selected by the stepwise method.

7) Generating the model

After the input variables are prepared, a model that shows the relationship between the input variables and output variables is generated by the set of data. The method for generating the model is discussed in Section 4.

8) Validating the model

In verifying the practicality of the generated model, it is necessary to prepare another set of data for validating the model. The practicality of the model is confirmed by checking the validation results of the predicted values and actual values on a graph, or calculating error. If the validation results are unsatisfactory, the modeling method is used to analyze the reasons for the unsatisfactory results and the analyses of the previous steps are repeated. Dealing with problems that occur when generating or using the model with the measured data is explained in Section 4.

4. Modeling via data mining

4.1 Data used for data mining

The data used in data mining is not specially gathered for modeling purposes. Therefore, the amount of information from the data will not necessarily be sufficient for modeling. What this means is that the data may have a bias – for example, having an area of high data density as well as another area of low data density. Also, the range in which each variable has been recorded does not necessarily coincide with the range that was supposed to be measured (Figure 4.1). When performing data mining it is important to be aware of the fact that modeling has to use imperfect data, with its possible biases and gaps.

4.2 Proposed modeling method

Conventionally, modeling with data has involved parameter identification using such approaches as the least squares method or neural networks, thus establishing the input-output function for the entire input-output space. As a result, using an incomplete population can cause problems when generating or using the model.

The modeling method proposed here does not identify the input-output function from learning data. Instead, it identifies the input space topology (distance space) that can estimate the margin of error in the model output (Figure 4.2). The input topological space is identified through quantization of the input space, and the quantum size is specified appropriately from learning data. The learning data is then converted into a case base model. Individual cases are generated for sets of input quanta and output values, which are derived by averaging the output values in learning data with an identical quantum in the input space. The calculation of output values – that is, the prediction/estimation – is performed by searching the case base for cases with the same topological distance that are nearest to the new input values, and then averaging their outputs. At this point, the distances measured in the input topology are used to calculate the margin of error included in the prediction/estimation, or output values. We have named this method Topological Case-Based Modeling (TCBM)(Tsutsui et al., 2001). By using TCBM, problems related to modeling with data of incomplete population can be solved:

(1) Generating a model using TCBM

In conventional functional modeling, parameters are specified by the entire functional range, which make them susceptible to any bias in the learning data. Since the parameter's values prioritize areas of high density data, this causes errors in prediction/estimation.

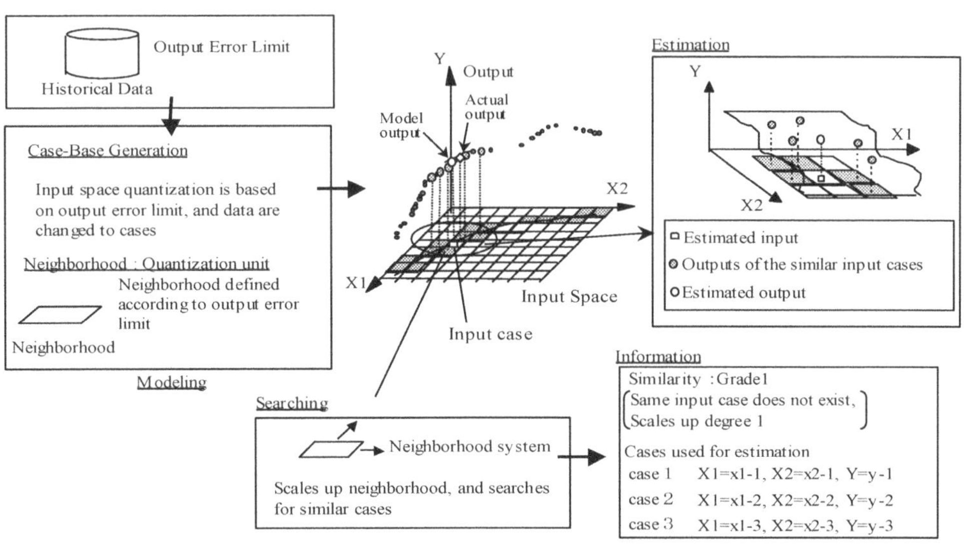

Figure 4.2 Overview of TCBM

TCBM has criteria (Output Distribution Limit, Continuity) for deciding whether or not it is possible to perform modeling with the specified accuracy from the learning data. These criteria are not easily affected by biases or gaps in the learning data since evaluation is performed locally and individually for each quantum in which an input exists. Also, since these criteria allow the evaluation of both noise processing insufficiency and input variable selection error, it is fairly easy to discover the cause of any problem that might occur during the model generation process.

(2) Employing a model using TCBM

With conventional functional models, even though their performance was satisfactory when the models were generated, actual performance will differ unless the input data is recorded in an area of high density area. Therefore, depending on the input data, the model's accuracy will differ.

TCBM has greater reliability because it calculates not only the output values from the input values in order to perform prediction/estimation, but also the margin of error from the identified topology. Furthermore, with TCBM the need for model revision can be recognized by frequent occurrences of a difference between the model output and measured output which exceeds the margin of error estimated from the topological distances. In such cases, the model can be revised by adding new data to the learning data and recalculating quantum size, or reselecting the input variables.

5. Verification of sewage inflow prediction

5.1 Outline

The Net-TCBM system (Oka et al., 2003) was developed for the purpose of predicting sewage inflow volume to a WWTP several hours in advance and supplying inflow predictions to the plant in real-time via the Internet so that the WWTP can make optimum use of its existing facilities in preventing sewage overflow in wet weather, etc. It enables the WWTP to access inflow predictions via simple equipment, which can be installed in a short time. Presently the Net-TCBM is installed in eight WWTPs in Japan. It is the first software product to receive a construction technology certification from the Japan Institute of Wastewater Engineering Technology, which certifies the use of this system in municipal WWTPs in Japan. To demonstrate its effectiveness in predicting sewage inflow volume, the certification test procedure and results are summarized.

5.2 Model generation for sewage inflow prediction

To perform quantitative data mining for the certification test, three years of historical data gathered at two WWTPs was used. The candidates of input variables for the prediction model include rainfall volume and temperature of the treatment area, inflow volume measured at the WWTP as well as the time and days of the week. A rain gauge with tipping bucket was used to measure rainfall. If the WWTP is not equipped with flow meters, such as an electromagnetic flow meter, the inflow volume to a WWTP can be calculated from historical data, such as pumped sewage volume and the water level of the inflow ditch, grit chamber and pump well (herein referred to as "inflow ditch, etc.").

Since sewage inflow to a treatment plant is temporarily stored in the inflow ditch, etc., inflow volume is considered to be roughly equal to pumped volume if the capacity of the inflow ditch, etc. is not large, and the water level of the inflow ditch, etc. is stable.

However, if the capacity of the inflow ditch, etc. is large, and the water level of the inflow ditch, etc. is not constant, inflow volume would not be considered equal to pumped volume (Figure 5.1). In such cases, inflow volume should be calculated from related data, as shown:

Sewage inflow volume = Pumped volume + Difference of water level x Cross section area

The cross section area will differ according to changes in water level; therefore it is necessary to check the effect of water level on the cross section area from blueprints of the WWTP. An immersion-type level meter was used to measure water level.

After acquiring all the candidates of input variables, pre-processing was performed by following the procedure for quantitative data mining in Section 3. To take into account the effect of rainfall volume on inflow volume, integrated variables and delayed variables were added as functions of new variable generation. Next, in selecting variables that have a strong effect on output, the stepwise method was used. These selected variables show the features of the treatment area of the WWTP. After that, TCBM was employed in constructing a prediction model using the first two years of historical data accumulated over a three-year period. The historical data of the third year was used to validate the prediction model by comparing the actual values and predicted values of inflow volume.

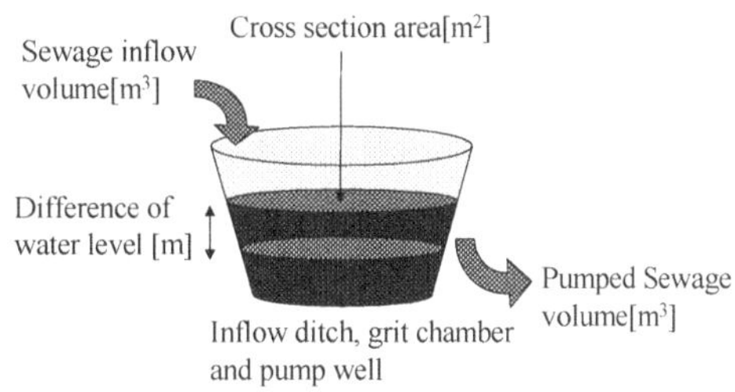

Figure 5.1 Calculating inflow volume

5.3 Results of verification

1) short-term prediction

Figure 5.2 shows the results for offline performance of inflow predictions of three hours ahead in wet weather together with actual inflow volume of three hours later. The same results are plotted on a scatter graph, as shown in Figure 5.3. The straight line through zero indicates the match between predicted inflow volume and actual inflow volume. The broken lines indicate the upper and lower sides of error in the measurement range of ±7%, which means that the points plotted within the broken lines are those with error in the measurement range of ±7% or less. The results of predictions of one, two and three hours ahead are shown in Table 1. This table indicates 1) less than 7% sum of Root Mean Squares(RMS*), 2) over 95% of all prediction items within ±17% error, and 3) less than ±17% error in 95% of all prediction items.

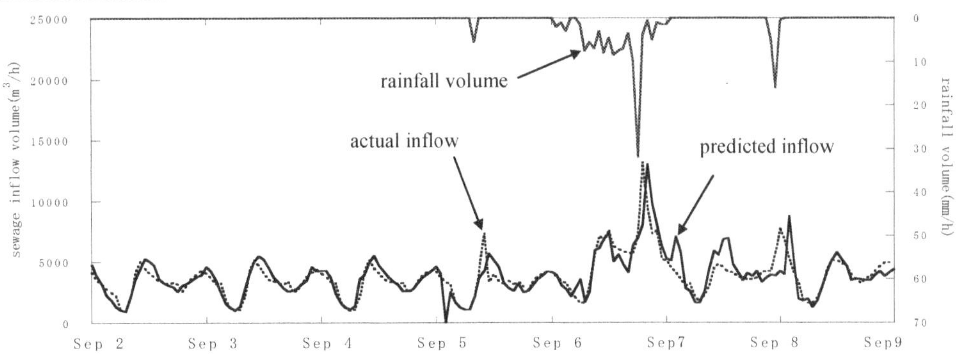

Figure 5.2 Results of inflow volume prediction

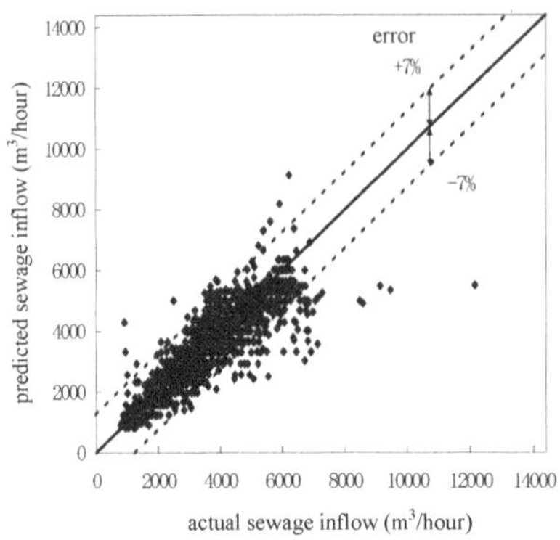

Figure 5.3 Comparison of actual and predicted inflows

From the results using real-time data, it was verified that online performance had a similar degree of accuracy to offline performance.
*The Root Mean Square is expressed as:

$$RMS(\%) = \frac{1}{Range}\sqrt{\frac{\sum e_i^2}{n}} \times 100$$

e_i: difference of prediction value and actual measured value, i: each data, n: data volume, *Range*: measurement range of sensor

2) long-term prediction

To verify predictions of inflow volume of five and sixteen hours ahead for operational guidance, meteorological data (rainfall volume and temperature) provided every hour by the Japan Meteorological Business Support Center and historical data accumulated at the WWTP and the Net-TCBM Server were used.

The results of predictions of five and sixteen hours ahead show 1) % sum of root-mean squares are 6.40% and 6.67%, respectively, 2) % of all prediction items within ±17% error are 98.4% and 98.3%, respectively, and 3) % error in 95% of all prediction items are 8.7% and 9.5%, respectively.

These results verify that the Net-TCBM system for predicting inflow volume to a WWTP has the accuracy required for practicable use.

Table 5.1 Offline performances

verification index	hours ahead	WWTP-A	WWTP-B
% of error in the sum of Root Mean Squares	one	2.29	4.22
	two	2.31	4.62
	three	2.31	5.03
% of all prediction items within ±17% of max. error	one	99.9	99.3
	two	99.9	99.3
	three	99.9	99.0
% of max. error in 95% of all prediction items	one	4.8	7.7
	two	4.8	7.9
	three	4.8	7.9
data volume	-	7300	1700

6.Future activities using a data-driven approach

6.1 Identifying I/I

Current measures for I/I investigation are costly: dividing the treatment area into blocks, smoke-testing and inflow measurements of sewer pipes are conducted block by block. We developed a method for estimating locations of I/I occurrences based on historical data. Using TCBM technology to estimate the volume of I/I in wet weather and comparing it to the rainfall pattern of treatment area, it is possible to indicate locations and degree of I/I occurrences on a map (Figure 6.1). These results can be utilized in reducing the cost of I/I investigation and enabling more efficient maintenance of aging sewer pipes. We are presently involved in joint development with municipalities in improving this function for practicable use.

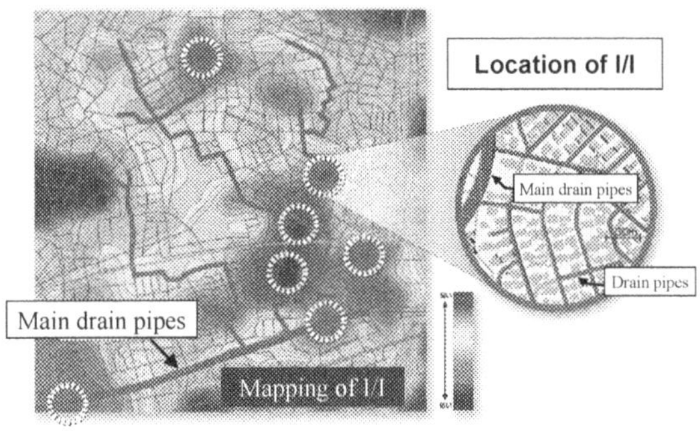

Figure 6.1 Identifying I/I

6.2 Estimating pollutant loads

In 2004 the Japanese government imposed a new environmental regulation that requires the BOD level of the total discharged wastewater in the treatment area of a combined sewer system be less than 40mg/liter by 2014. This regulation is aimed at raising the water quality of untreated wastewater that is discharged in wet weather and reducing environmental damage.

By using the Net-TCBM system for predicting sewage inflow volume, it is possible to receive estimations of current BOD levels online. These estimations are useful in adjusting the amount of sewage stored in the regulating reservoir and promoting optimum control of water quality.

7.Conclusion

This paper describes a novel approach to the operation and planning of wastewater treatment based on historical data. Our system for predicting the volume of sewage inflow has been verified to have the accuracy for practicable use, and is presently in use at eight WWTPs in Japan. In order to clarify the efficiency of our proposed approaches, it is necessary to further verify the merits of our activities in terms of cost effectiveness, such as the cost difference of the proposed measures versus expanding current facilities or building another regulating reservoir, smoke testing for I/I identification and actual measurement of BOD.

References

Flury, B. and Riedwgl, H. (1988) Multivariate Statistics - A Practical Approach. Chapman and Hall.

Matsuba, T., Tsutsui, H. and Kamimura, K. (1999) Development of Load Profile Prediction using TCBM and ARIMA Hybrid-Modeling. *Building Simulation '99 Conference.*

Oka, T., Tsutsui, H., Konda, T. and Sato, F. (2003) Sewage Inflow Prediction System Employing Topological Case-Based Modeling. *2003 IWA 9th Specialized Conference on the Design, Operation and Economics of Large Wastewater Treatment Plants, Conference Proceeding Poster Papers*, 179-182.

Tsutsui, H. (1996) Topological Case Based Modeling. *Building Optimization and Fault Diagnosis Source Book, IEA*, 187-197.

Tsutsui, H. and Kamimura, K. (1996) Chiller Condition Monitoring using Topological Case-Based Modeling. *ASHRAE*, 82-89.

Tsutsui, H. and Konda, T. (2001) The modeling and application of non-linear systems using data mining. *1st SIAM International Conference on DATAMINING.*

Anaerobic Treatment of Domestic Wastewater in Brazil: Overview

F.H.Passig*, J.R.Campos**

Departamento de Hidráulica e Saneamento, Escola de Engenharia de São Carlos (EESC), Universidade de São Paulo (USP) Avenida Trabalhador São-carlense,400 CEP – 13566-490. São Carlos (SP) Brasil. e-mail:* fhpassig@matrix.com.br ;** jrcampos@sc.usp.br

Abstract The anaerobic treatment process is recognized increasingly as the core method of an advanced technology for environmental protection and resource preservation. When combined with proper methane collection and usage as well as chemical P-precipitation in aerobic post treatment it provides a sustainable and appropriate wastewater treatment system for development countries in warmer climate region. In this paper, the main characteristics of anaerobic wastewater treatment are summarized.

Keywords Anaerobic Treatment; UASB reactor; sewage, Wastewater Treatment.

Introduction

The term "domestic wastewater" refers to the wastewater produced by a community, which may be originated from three different sources: domestic wastewater, generated from bathrooms and toilets, and activities such as cooking, washing, etc.; industrial wastewater, from industries using the same collection system for their effluents (treated or not), and rain-water (Van Haandel & Lettinga, 1994).

According to IBGE (2000), Brazil has 169.799.170 inhabitantes and 5507 municipal districts with 47,8% of municipal districts that not have collection of domestic wastewater, 32% of municipal districts that have collection and only 20.2% of municipal districts that have domestic wastewater collection and treatment (PNSB, 2002).

Brazil treats almost 16% of the volume of the produced domestic wastewater. Added with this number are the epidemiological profile, the social-economic characteristics of the communities and the lack of resources in this area that indicate the necessity of simplified wastewater treatment systems.

Anaerobic treatment of domestic wastewater in Brazil

The main types of reactors applied in the anaerobic treatment of domestic wastewater in Brazil are anaerobic ponds; sludge digesters, anaerobic filters, anaerobic sludge blanket reactors and recently the expanded bed or fluidized reactors.

Recent developments in anaerobic processes for wastewater treatment have to be credited to the success of the UASB reactor concept (Lettinga *et al.*, 1980).

The anaerobic reactors usage in the treatment of domestic wastewater had a significant increasing after the 80's with the anaerobic sludge blanket reactor development, mainly in the state of Parana, with the construction of about 200 unities. Nowadays, we can estimate that in Brazil there are more than three hundred UASB reactors treating domestic wastewater in different sizes and shapes, is found in all of Brazilian states. An example of this is

© 2004 IWA Publishing. *2nd IWA Leading-Edge Conference on Water and Wastewater Treatment Technologies* edited by Mark Van Loosdrecht & Jonathan Clement. ISBN: 1 84339 508 8.

Piracicamirim wastewater treatment system (Piracicaba/SP) that was built for 100,000 inhabitants (3 UASB reactors with 7500m^3 of useful volume). Another one is Atuba Sul wastewater treatment system (Curitiba/PR) that was built for 600,000 inhabitants with 16 UASB reactors with 2000m^3 of useful volume (Andrade Neto e Campos, 1999).

In Brazil, the PROSAB (Program on Basic Sanitation) is a specific program, supported by the Brazilian government, which involves seven research institutions. It aimed to investigate appropriate anaerobic treatment of domestic wastewater and soil disposal, considering the diversity of environmental and socio-economic conditions in distinct regions of the country. Details of the PROSAB results are presented in Campos *et al.* (1999).

The anaerobic systems advantages are: Low costs of installation, operation and maintenance; Good removal efficiency of organic matter and suspended solids; Simplicity of construction and operation; Flexibility of operation; Low energy consumption; Low space requirement; Low sludge production and Low nutrients and chemical requirement.

These characteristics, in addition with the Brazilian environmental conditions (temperature higher than 20^{0}C), have contributed to highlight the anaerobic systems for domestic wastewater treatment in that country.

The UASB reactors have been successfully used as the first unit of the anaerobic systems designed for domestic wastewater treatment at ambient temperatures higher than 20^{0}C. The COD removal efficiencies in the range of 65% to 80% have been obtained usually at an applied organic loading rate lower than 3 kg COD $m^{-3}.d^{-1}$ and hydraulic detention time ranging from 6 to 10 hours (Passig et al., 2000; Florencio et al., 2001; Torres and Foresti, 2001; Pontes et al., 2003; Kato et al., 2003).

The performance of UASB reactors in the treatment of domestic sewage permits to consider the anaerobic biotechnology to be established in tropical regions where the ambient temperature is higher than 20°C. The main design parameters are already set up, and the reactor performance after the start-up period can be predicted with reasonable security (Foresti, 2002). Nowadays, almost all alternatives analyses for wastewater treatment in Brazil include these reactors as one of the main options.

Despite of these advantages, UASB reactors have difficulties in producing effluents complying with the Brazilian environmental standards, mainly in reaching efficient organic matter removal (higher than 80% BOD_5 or less than 60 mg/L), and low nutrients and pathogens removal. Another consideration is the biogas production and a potential loss of the greenhouse gas methane dissolved in the effluent and due to unprofessional collection and usage. It is interesting to express that the UASB reactor substitutes: preliminary sedimentation, thinkers and digester in a unique unit.

Post-treatment of anaerobic reactor

To remove the remaining fraction of the organic matter and improve P-removal, it is necessary to apply a post-treatment with different biological reactors configurations and chemical P-precipitation, allowing the production of final effluents with compatible quality to the needs imposed by the national and local legal patterns of effluents emission and of rivers quality.

In Brazil, a specific program (PROSAB II – Program on Basic Sanitation) on the post-treatment of effluents from an anaerobic reactor is being carried out, involving twelve research institutions supported by the government. It aims to investigate appropriate post-treatment technologies considering the diversity of environmental and socio-economic conditions in distinct country regions. The alternatives include: soil/plant systems; ponds; biofilm reactors; activated sludge systems; dissolved-air flotation systems; filtration unities. The details of the main PROSAB results are presented in Chernicharo *et al.* (2001).

The use of domestic wastewater post-treatment (anaerobic-aerobic system) presents advantages with the reduction of energy costs and sludge production decreasing that permits the reduction of sludge dewatering disposal costs. An example of this application can be Piracicamirim wastewater treatment plant (Piracicaba/SP). The conception of this system was developed with anaerobic-aerobic reactors (UASB-activated sludge).

The global system achieved after 30 months of operation 87% of organic matter removal in terms of BOD and 75% of suspended solids concentration removal (Passig et al., 2000).

This plant was the first full scale one in Brazil to employ a UASB reactor followed by an activated sludge system and nowadays there are many others wastewater treatment plants in operation design which conception follows the same series (preliminary treatment, UASB reactor, activated sludge, and excess sludge recirculation to the UASB).

For the pathogens removal is necessary to use others units of disinfection that can be: polishing pond; chlorination- dechlorination; ozone, ultraviolet light or others. The application of disinfection is recent in Brazil, with few designed and constructed plants due to the lack of resources.

Biogas

Biogas produced by anaerobic process is a mixture of methane (CH_4), carbon dioxide (CO_2), nitrogen (N_2) and some trace gases of which hydrogen sulphide (H_2S) is the most evident. Among these gases, only CH_4 and H_2S are a real subject of concern. The first one is because of its elevated calorific value, high volumetric fraction in the gas stream and significant global warming potential (about 20 times more than of CO_2, when compared molecule by molecule). The H_2S is a concern because of its odorous, corrosive and hazardous properties (Pagliuso et al., 2002).

Methane is a fuel that when burned produces very low emissions of pollutants. Its flue gases will contain two moles of H_2O for each mol of CO_2 that is the best ratio among the industrial fuels from the global warming point of view. This trade off is much better than discharging methane directly into the atmosphere. Although its net radioactive forcing is null when it comes from a sewage treatment plant because in this case it is naturally recycled, CH_4 takes a long time to be oxidized to CO_2 in the atmosphere. Consequently it is better to do it at once through its combustion. More important however is that if energy is recovered in these processes, an equivalent amount of fossil fuel will not be used and greenhouse gas emissions will again be prevented (WEF, 1995).

There are several options for the H_2S odour treatment, among them chemical solution and oxidation, biological degradation, the use of iron salts, absorption by activated carbon, odour masking and neutralization and thermal oxidation (incineration). Each of these options has its advantages and disadvantages and a variable degree of success. Additionally it should be considered that although H_2S is the principal agent of odour it is not the single one and mercaptans might also be present in the gas. In this case two or more chemical treatment devices might be necessary if this technique is chosen. Chemical and thermal options are perhaps the most extensively used today. The first is usually expensive and has an efficiency range between 60 and 90%, sometimes more (WEF, 1995).

Thermal oxidation or destruction is considered one of the most effective techniques for odour control because it removes not only H_2S but also methyl mercaptan and practically any other objectionable odorous gas that could be present in the biogas. The week point of thermal destruction is the cost of the fuel when an external source is needed. However this is not the case for the sewage treatment plants because the biogas has a high concentration of methane (WEF, 1995).

Conclusion

The UASB reactors have been successfully used as the first unit of the anaerobic systems designed for domestic wastewater treatment at room temperatures higher than 20^{0}C.

The conception: preliminary treatment, UASB reactor, activated sludge for nitrification (if dilution in receiving water is low)? and chemical P-precipitation, and the excess sludge recirculation to the UASB produces effluents with good quality (organic matter and suspended solids removal).

Methane is a fuel that when burned produces very low emissions of pollutants. But a proper collection and usage system is essential to reduce the imission of methane to the atmosphere as much as possible.

Biogas must be collected and treated to minimize the odor impact.

Our main goal is to increase the quantity of the municipal districts attended by domestic treatment, first of all organic matter and after that nutrient removal (mainly phosphate) and disinfection is needed in our country to give better sanitation conditions to the citizens.

Acknowledgements

The authors acknowledge the support obtained from the Brazilian Scientific and Technological Development Council – CNPq ; FINEP and FAPESP

References

Andrade Neto, C.; Campos, J.R. (1999) In: Campos, J.R. "Tratamento de esgoto sanitário por processo anaeróbio e disposição controlada no solo". Capitulo 1–Introdução. ABES, 435p. (in portuguese).

Chernicharo, C.A.L., Van Haandel, A., Cybis, L.F. and Foresti, E. (2001). Post-treatment of anaerobic effluents in Brazil: state of the art. In Preprints of "9th Word congress of Anaerobic Digestion, Vol I, 747-752.

Foresti, E. (2002). Anaerobic treatment of domestic sewage: established technologies and perspectives Wat. Sci. Tech. 45(10), 181-186.

Florerêncio,L., Kato,M.T. and Moraes, J.C. (2001). Domestic sewage treatment in full-scale UASB plant at Mangueira Recife,Pernambuco. Wat. Sci. Tech. 44(4), 71-77.

IBGE (2000). Instituto Brasileiro de Geografia e Estatistica, - Censo 2000 (in portuguese)

Kato, M.T., Florencio, L. and Arantes, R.F.M. (2003). Post-treatment of UASB effluent in expanded granular sludge bed reactor type using flocculent sludge. Wat. Sci. Tech. 48(6), 279-284.

Lettinga, G., van Velsen, A.F.M., Hobma, S.W., de Zeew, W. and Klapwijk, A. (1980). Use of the Up-flow Sludge Blanket (USB) concept for biological waste water treatment, specially for anaerobic treatment. Biotechn. Bioeng., 22, 699–734.

Pagliuso, J.D., Passig, F.H. and Villela, L.C.H. (2002). Odour treatment and energy recovery in anaerobic sewage treatment plants In: VII Taller y Simpósio Latino Americano sobre Digestão Anaeróbia. México.

Passig, F.H., Villela, L.H. and Ferreira, O.P. (2000). Piracicamirim sewage treatment plant – Conception utilizing anaerobic process – Evaluation of operational conditions and compatibility of the process. In: "Preprints of the VI Latin American Workshop and seminar an anaerobic digestion". Vol. I.53-59

PNSB, 2002 – Programa Nacional de Saneamento Básico - Instituto Brasileiro de Geografia e Estatistica, - Censo 2000 (in portuguese).

Pontes, P.P., Chernicharo, C.A.L., Frade, E.C. and Porto, M.T.R. (2003). Performance evaluation of UASB reactor used for combined treatment for domestic sewage and excess aerobic sludge from trickling filter. Wat. Sci. Tech. 48(6), 227-234.

Van Haandel, A.C. and Lettinga, G. (1994). Tratamento anaeróbio de esgotos. Um manual para países de clima quente (in portuguese).

Torres, P. and Foresti, E. (2001). Domestic sewage treatment in a pilot system composed of UASB and SBR reactors. Wat. Sci. Tech. 44(4), 247–25.

WEF, 1995 Odour Control in Wastewater Treatment Plants, Manual of Practice N22.

Keyword Index

IWA Publishing's authorised EU representative for General Product Safety Regulations is Diane D'Arras, 15 rue Duret, 75116 Paris, France, e-mail: safety@iwap.co.uk.

Printed and bound by CPI Group (UK) Ltd, Croydon, CR0 4YY
12/05/2026
02108863-0011